Environmental Contamination Remediation and Management

There are many global environmental issues that are directly related to varying levels of contamination from both inorganic and organic contaminants. These affect the quality of drinking water, food, soil, aquatic ecosystems, urban systems, agricultural systems and natural habitats. This has led to the development of assessment methods and remediation strategies to identify, reduce, remove or contain contaminant loadings from these systems using various natural or engineered technologies. In most cases, these strategies utilize interdisciplinary approaches that rely on chemistry, ecology, toxicology, hydrology, modeling and engineering.

This book series provides an outlet to summarize environmental contamination related topics that provide a path forward in understanding the current state and mitigation, both regionally and globally.

Topic areas may include, but are not limited to, Environmental Fate and Effects, Environmental Effects Monitoring, Water Re-use, Waste Management, Food Safety, Ecological Restoration, Remediation of Contaminated Sites, Analytical Methodology, and Climate Change.

Amir Parnian • Majeti Narasimha Vara Prasad
James N. Furze
Nematollah Jaafarazdeh Haghighi Fard
Editors

Bioremediation of Petroleum Contaminated Soil and Sediment

Springer

Editors
Amir Parnian
National Salinity Research Center (NSRC)
Agricultural Research, Education and Extension Organization (AREEO)
Yazd, Iran

James N. Furze
Royal Geographical Society (with the Institute of British Geographers)
London, UK

Majeti Narasimha Vara Prasad
Department of Plant Sciences, School of Life Sciences
University of Hyderabad, (An Institution of Eminence)
Hyderabad, Telangana, India

Nematollah Jaafarazdeh Haghighi Fard
Environmental Technologies Research Center, Medical Basic Sciences Research Institute
Ahvaz Jundishapur University of Medical Sciences
Ahvaz, Iran

ISSN 2522-5847 ISSN 2522-5855 (electronic)
Environmental Contamination Remediation and Management
ISBN 978-3-031-95523-5 ISBN 978-3-031-95524-2 (eBook)
https://doi.org/10.1007/978-3-031-95524-2

This Springer imprint is published by the registered company Springer Nature Switzerland AG
The registered company address is: Gewerbestrasse 11, 6330 Cham, Switzerland

This volume is dedicated to family, friends, students, and all who remain concerned for the future of Earth's ecosystems.

–Amir Parnian
–Majeti Narasimha Vara Prasad
–James N. Furze
–Nematollah Jaafarazdeh Haghighi Fard

Foreword

Alignment of the unrelenting march of industrialization with sustenance of the Earth and environmental concerns requires the use of positive, or at least "less negative" technologies to provide potential compensation for our destructive habits of extraction. The technique of bioremediation stands as a beacon of future hope and creativity. We deliver the collective wisdom of researchers, practitioners, and professionals dedicated to restoration and regeneration of areas scarred by residues of petroleum exploitation.

This volume's fundamental theme engages with natural resource flow within the precarious balance of Earth's ecosystems. Nature-based solutions recognize the potential of biological methods to restore what has been damaged and pave a sustainable course for the future. With relation to biological agents and contaminated matrices, an environmentally sustainable path and restoration plans for our ecosystems may be envisaged.

As we stand at the intersection of environmental crises and technological advancement, Bioremediation of Petroleum Contaminated Soil and Sediment provides a prism through which we may envisage the coexistence of economic prosperity and a green environment. We hope that this book will motivate and prepare the next generation of environmental stewards to continue ecological redemption of the planet.

Yazd, Iran	Amir Parnian
Hyderabad, Telangana, India	Majeti Narasimha Vara Prasad
London, UK	James N. Furze
Ahvaz, Iran	Nematollah Jaafarazdeh Haghighi Fard
13 November 2025	

Preface

The editors are honored to write a preface for *Bioremediation of Petroleum Contaminated Soil and Sediment*. Dr. Amir Parnian and the editorial team contribute to the development of sustainable and functional environmental systems, leading education of soil and sediment bioremediation in the harshness of cascaded impact on our terrestrial environment.

Gaining a practical understanding of bioremediation of soil and sediment contaminated with petroleum is required to master major nodes of environmental speciality that guide us in maintaining our local and global ecosystems. The well-written chapters in this book feature a variety of topics, emphasizing the significance of petroleum pollution in ecosystems based on soil and sediment or remediation of petroleum-polluted materials, which when addressed collectively permit balanced advancement.

From the fundamental viewpoints of science, policy relations and imperatives, community implementations, and soil/sediment bioremediation systems learning and intuition, this book represents new ideas in science and technology. Recalling essential and emerging topics provides insight for ecosystem bioremediation and development into the future.

Expanding our ability to reduce risks to Earth's ecosystems calls for rigorous, disciplined thought, creative analysis, and sympathetic consideration of community needs and issues. These elements provide keys to the main themes of protecting structural integrity and renewal of soil and sediment-based systems, crucial aspects of Earth systems and resource dynamics in variable scenarios.

We endeavor to offer readers who are interested in learning, utilizing, or simply enjoying the key components that enable soil bioremediation to sustain life for an unlimited amount of time, an exciting and engaging experience.

Special features of the volume include:

- Bioremediation of petroleum and related contaminants in soil and sediment
- Management and guidance for resource use and collaboration in human and naturalized ecosystems
- A balance of qualitative and quantitative chapters written by leading authors

- Problems and solutions that are community and subjectively driven, rewarding sustainable ownership directly to the people the volume is intended to support

We would like to give recognition to the editorial collective composed of four members, representing disciplines and subjective fields of study included in the volume, in addition to teams of contributing authors from around the world, reviewers situated in between the authors, communities, governmental representatives, and representatives of organizational units (from many of our "United" core bodies). We share and focus our efforts on the writers, publishers, and community members who kindly donated their time and assistance to the creation of this book. We sincerely thank you for your support.

Bioremediation of Petroleum Contaminated Soil and Sediment is written for a global audience, encompassing all sectors. The volume is composed of both descriptive and illustrative chapters, enabling management of different resource types and uses, maintaining the dynamic integrity of bioremediation systems over time and space.

The target audience for this book includes academics, professionals, students, legislators, and community activists. It provides direction for every industry, promoting prudent resource management and preservation of bioremediation systems.

Yazd, Iran — Amir Parnian
Hyderabad, Telangana, India — Majeti Narasimha Vara Prasad
London, UK — James N. Furze
Ahvaz, Iran — Nematollah Jaafarazdeh Haghighi Fard
13 November 2025

Introduction

In order to ease understanding of soil and sediment contamination associated with petroleum pollution in our Earth's environment, this book acknowledges suitable and currently available bioremediation strategies. The volume has a rigorous educational framework for soil and sediment bioremediation, resulting in the development of long-lasting and functional ecological infrastructures.

Environmental restoration can be achieved by the tried-and-true method of bioremediation. Human civilizations are concerned with petroleum contamination, which bioremediation technology may be able to address. Nonetheless, there are a number of approaches accessible in this field, which raise questions among users regarding the efficacy of the outcomes, timeliness, and cost-effectiveness.

Results from co-composting, mycoremediation, phytoremediation, and numerous other bio-based techniques are found to be effective. Such bioremediation techniques, however, are confusing to consumers and contain gray areas. The editors have endeavored to compile and offer vital yet often disregarded pointers and advice to effectively provide bioremediation of petroleum residue in soil and sediment in this publication.

Using bioremediation in large-scale projects necessitates a higher level of knowledge concentration than has been presented in previously published books, essays, and keynote speeches. *Bioremediation of Petroleum Contaminated Soil and Sediment* offers promise to this often neglected field on actual, workable techniques toward remediation and restoration of our environment for future generations.

Yazd, Iran
Hyderabad, Telangana, India
London, UK
Ahvaz, Iran
13 November 2025

Amir Parnian
Majeti Narasimha Vara Prasad
James N. Furze
Nematollah Jaafarazdeh Haghighi Fard

Contents

Contributors

Affan Ahmed Environmental and Evolutionary Biology, Department of Biology, University of Louisiana at Lafayette, Lafayette, LA, USA

Aftab Ahmed School of Biological Sciences, University of the Punjab, Lahore, Pakistan

Subhamoy Banerjee Department of Basic Science and Humanities, Institute of Engineering & Management, University of Engineering and Management, Kolkata, West Bengal, India

Pedro C. Barrulas Chemistry and Biochemistry Department, School of Sciences and Technology, Évora University, Évora, Portugal

University of Évora, HERCULES Laboratory, Évora, Portugal

Tina Basu Department Basic Science and Humanities, Institute of Engineering & Management, University of Engineering and Management, Kolkata, West Bengal, India

Kyanoush Behrahi University of Applied Science and Technology, Tehran, Iran

Hossein Beyrami National Salinity Research Center, Agricultural Research, Education and Extension Organization (AREEO), Yazd, Iran

Marco Aurélio Carbone Carneiro Federal University of Lavras (UFLA), Lavras, MG, Brazil

Elisabete P. Carreiro LAQV-REQUIMTE, School of Sciences and Technology, University of Évora, Évora, Portugal

Bipasha Datta Department of Genetics and Plant Breeding, School of Crop Improvement, College of Post Graduate Studies in Agricultural Sciences, Central Agricultural University, Umiam, Meghalaya, India

Juliano de Carvalho Cury Federal University of São João del-Rei, São João del-Rei, MG, Brazil

Daniela de Fátima Pedroso Federal University of Lavras (UFLA), Lavras, MG, Brazil

Lander de Jesus Alves Postgraduate Program in Biology and Biotechnology of Microorganisms, State University of Santa Cruz (UESC), Ilhéus, Bahia, Brazil

Suzana Rodrigues de Moura Postgraduate Program in Biology and Biotechnology of Microorganisms, State University of Santa Cruz (UESC), Ilhéus, Bahia, Brazil

João Carlos Teixeira Dias Postgraduate Program in Biology and Biotechnology of Microorganisms, State University of Santa Cruz (UESC), Ilhéus, Bahia, Brazil

Jessé Valentim dos Santos Federal University of Lavras (UFLA), Lavras, MG, Brazil

Nematollah Jaafarazdeh Haghighi Fard Environmental Technologies Research Center, Medical Basic Sciences Research Institute, Ahvaz Jundishapur University of Medical Sciences, Ahvaz, Iran

Ghulam Fatima Institute of Biochemistry and Biotechnology, University of Veterinary & Animal Sciences, Lahore, Pakistan

Bakhtawar Fayyaz Atta Ur Rahman School of Applied Biosciences (ASAB), National University of Sciences and Technology (NUST), Islamabad, Pakistan

James N. Furze Royal Geographical Society (with the Institute of British Geographers), London, UK

Laboratory of Biotechnology and Valorization of Natural Resources, Faculty of Sciences-Agadir, Department of Biology, Ibn Zohr University, Agadir, Morocco

Control and Systems Engineering Department, University of Technology-Iraq, Baghdad, Iraq

Earth Systems Protection Foundation, Somerset, UK

Eduardo Gross Postgraduate Program in Biology and Biotechnology of Microorganisms, State University of Santa Cruz (UESC), Ilhéus, Bahia, Brazil

R. K. Gupta Department of Zoology, Chaudhary Charan Singh Haryana Agricultural University, Hisar, Haryana, India

Kaynat Iftekhar Atta Ur Rahman School of Applied Biosciences (ASAB), National University of Sciences and Technology (NUST), Islamabad, Pakistan

Biswajit Jena Faculty of Agriculture and Allied Sciences, C V. Raman Global University, Bhubaneswar, Odisha, India

Rinnu Joseph Department of Botany, St. George's College, Aruvithura, India

Farhad Keykhosravi National Iranian Oil Company (NIOC), Tehran, Iran

Mudassir Khan Atta Ur Rahman School of Applied Biosciences (ASAB), National University of Sciences and Technology (NUST), Islamabad, Pakistan

Mahdieh Khorshidi Department of Plant Protection, Faculty of Agriculture, University of Tabriz, Tabriz, Iran

Ghulam Kubra Atta Ur Rahman School of Applied Biosciences (ASAB), National University of Sciences and Technology (NUST), Islamabad, Pakistan

Anupam Kumari Department of Plant Pathology, BPSAC, Purnea, Bihar Agricultural University, Sabour, Bihar, India

Rahul Kumar Department of Zoology, Chaudhary Charan Singh Haryana Agricultural University, College of Agriculture, Bawal, Rewari, Haryana, India

Mehdi Mahbod Water Sciences and Engineering Department, Faculty of Agriculture, Jahrom University, Jahrom, Iran

Vandana K. Manoj Department of Botany, St. George's College, Aruvithura, India

Carolina S. Marques LAQV-REQUIMTE, School of Sciences and Technology, University of Évora, Évora, Portugal

Eilin Maria Martin Department of Botany, St. George's College, Aruvithura, India

Ruchira Mukherjee Department of Basic Science and Humanities, Institute of Engineering & Management, University of Engineering and Management, Kolkata, West Bengal, India

Aleeza Munawar Department of Horticulture, Pir Mehr Ali Shah Arid Agriculture University, Rawalpindi, Pakistan

Zain Mushtaq Department of Soil Science, Faculty of Agricultural Sciences, University of the Punjab, Lahore, Pakistan

Malavika M. Nair Department of Botany, St. George's College, Aruvithura, India

Fábio Carvalho Nunes Academic Department, Federal Institute Baiano (IF BAIANO), Alagoinhas and Catu Bahia, Brazil

Amin Parnian Young Researchers Club, Masjed Soleyman Branch, Islamic Azad University, Masjed Soleyman, Iran

Parnian Sazeh Chehelsotoun (PSC) Company, Isfahan, Iran

Amir Parnian National Salinity Research Center (NSRC), Agricultural Research, Education and Extension Organization (AREEO), Yazd, Iran

Ana Paula Honrado Pinto Mediterranean Institute for Agriculture, Environment and Development, MED & Global Change and Sustainability Institute, CHANGE, Institute for Advanced Studies and Research, Évora, Portugal

Chemistry and Biochemistry Department, School of Sciences and Technology, Évora, Portugal

Manoj Kumar Prajapati United University Rawatpur Jhalwa, Prayagraj, Uttar Pradesh, India

Majeti Narasimha Vara Prasad Department of Plant Sciences, School of Life Sciences, University of Hyderabad, (An Institution of Eminence), Hyderabad, Telangana, India

Flávio Henrique Silveira Rabêlo Luiz de Queiroz College of Agriculture, University of São Paulo, Piracicaba, Brazil

Suganthi Rajendran Department of Environmental Sciences, JSA College of Agriculture and Technology, Cuddalore, India

Poornima Ramesh Department of Environmental Sciences, Vanavarayar Institute of Agriculture, Coimbatore, India

Rachel Passos Rezende Postgraduate Program in Biology and Biotechnology of Microorganisms, State University of Santa Cruz (UESC), Ilhéus, Bahia, Brazil

Raquel Milagros Rodríguez-Rodríguez Federal University of Lavras (UFLA), Lavras, MG, Brazil

Smithsonian Tropical Research Institute, Panama, Panamá

Muhammad Abu Bakar Saddique Institute of Urban Agriculture, Chinese Academy of Agricultural Sciences, Chengdu National Agricultural Science and Technology Center, Chengdu, China

Sudip Sarkar ICAR-Research Complex for Eastern Region, Patna, Bihar, India

Suranajana Sarkar Department of Microbiology, School of Life Sciences, Swami Vivekananda University, Bara Kanthalia, West Bengal, India

Abin Sebastian Department of Botany, St. George's College, Aruvithura, India

Salman Selmy Department of Soils and Water, Faculty of Agriculture, Assiut University, Assiut, Egypt

Anamita Sen School of Agriculture Sciences and Forestry, Sardar Bhagwan Singh University, Balawala, Dehradun, Uttarakhand, India

Shweta Sharma Department of Physiology, AIIMS, New Delhi, India

Aline Oliveira Silva Federal University of Lavras (UFLA), Lavras, MG, Brazil

Dilbag Singh School of Agriculture Sciences, Galgotias University, Greater, Noida, India

Nibha Singh Department of Biochemistry, Jamia Hamdard University, Delhi, India

Love Singla Department of Microbiology, Maharaja Agrasen University, Baddi, India

Yousef Soleimani Gachsaran Oil and Gas Production Company (GOGPC), National Iranian Oil Company (NIOC), Ghachsaran, Iran

Maria Clara Bessa Souza Postgraduate Program in Biology and Biotechnology of Microorganisms, State University of Santa Cruz (UESC), Ilhéus, Bahia, Brazil

Rakesh Subramanian Srirangarayan Department of Crop Physiology, Tamil Nadu Agricultural University, Coimbatore, India

P. M. Sumayya Department of Botany, St. George's College, Aruvithura, India

Davi Santos Tavares Federal University of Lavras (UFLA), Lavras, MG, Brazil

Mehdi Tayebi Aghajari Oil and Gas Production Company (AGOPC), Omidiyeh, Iran

Mohamad Zakipour Arvandan Oil and Gas Company, Khorramshahr, Iran

Elaheh Zallaghi National Iranian South Oil Company (NISOC), Ahvaz, Iran

Gholamreza Zareian Soil and Water Research Department, Fars Agricultural and Natural Resources Research Center, AREEO, Shiraz, Iran

Chapter 1
Pollution, Ecosystem Degradation, Human Vulnerability and Projection

Mehdi Mahbod

Abstract The Earth and all life are harmonically linked. Ecosystems are shaped by the interrelationship of living and non-living. The Earth has many habitats, which has led to a variety and variability of life through natural selection and evolution. Additionally, biodiversity in an ecosystem directly refers to its strengths and resilience to changes and stressors. Pollution is a type of human-made stress that causes a decrease in biodiversity and ecosystem productivity. Pollution changes life's playground. Moreover, by affecting ecosystem productivity, it damages or destroys the life of dependent organisms. Humans depend on surrounding ecosystems and biodiversity, any pollution endangers society. This chapter reviews the story of environmental pollution and its relationship to the ecosystem, biodiversity, and humans.

Keywords Contamination · Harmful · Environmental-resistance · Biodiversity · Impact

1.1 Introduction

Environmental pollution refers to introducing dangerous or excessive pollutants into Earth's environment. Pollutants can manifest in several forms, including chemical compounds, energy like noise or heat, and even light. Pollution is typically categorized as either point-source or nonpoint-source pollution. Point source pollution emits dangerous compounds into a water body, such as an oil spill. Nonpoint source pollution is diffuse contamination that spreads over a vast region and does not come from a specific source, like pesticides carried by the wind.

Pollution significantly affects the natural components of our planet: air, water, and soil. Air pollution is the emission of chemicals and particulates into the

M. Mahbod (✉)
Water Sciences and Engineering Department, Faculty of Agriculture, Jahrom University, Jahrom, Iran

A. Parnian et al. (eds.), *Bioremediation of Petroleum Contaminated Soil and Sediment*, Environmental Contamination Remediation and Management,
https://doi.org/10.1007/978-3-031-95524-2_1

atmosphere. Carbon monoxide, sulfur dioxide, chlorofluorocarbons (CFCs), and nitrogen oxides are common gaseous pollutants from industry and motor vehicles. When nitrogen oxides and hydrocarbon vapors react to sunlight, photochemical ozone and smog are formed (Glencross et al. 2020). Water pollution involves releasing waste products and toxins into surface runoff, river drainage systems, groundwater through leaching, liquid spills, wastewater discharges, eutrophication, and littering. Soil contamination, leading to water pollution, arises from elevated nutrients like nitrogen and phosphorus in the soil, resulting in harmful impacts on plant life and, consequently, on animals and humans (Schwarzenbach et al. 2010).

Pollution has diverse and intricate consequences. It presents a substantial risk to human health, leading to various ailments such as respiratory issues and cancer. It impacts ecosystems, causing a reduction in biodiversity and the disturbance of food chains. Pollution can result in physical harm to the ecosystem, such as the whitening of coral reefs due to rising ocean temperatures from global warming (Vrijheid et al. 2016). Global societies are implementing efforts to address pollution concerns. The activities involve establishing and utilizing sustainable energy sources, reducing waste, promoting recycling programs, and establishing protected conservation zones. Legislation and regulations have been implemented to reduce pollution, including emission requirements for automobiles and industry and the prohibition of specific pesticides and plastics (Shinwari et al. 2024). Environmental pollution is a worldwide issue that impacts all aspects of life on our planet. A diverse approach is necessary to properly handle this problematic issue (Fuller et al. 2022). Progress is being achieved despite hurdles, and with ongoing effort and innovation, creating a less polluted world is a feasible objective (Sharif et al. 2023). The future well-being of our planet relies on our comprehension of the issue and our determination to act.

1.2 Environmental Pollution and Ecosystems

1.2.1 Impact of Various Forms of Pollution on Ecosystems

Ecosystems are intricate relationships between living and non-living elements within a defined geographical area. They offer a range of functions and advantages to humans and other organisms, including food, water, climate regulation, and biodiversity. Ecosystems can be harmed by several forms of pollution, leading to adverse effects on their organization, operation, and well-being (Wang et al. 2024). Human activities, such as the discharge of pollutants from factories or farms or the accumulation of trash in cities, can contribute to pollution just as much as natural disasters like wildfires or volcanic eruptions (Connell and Miller 2022). Pollution can impact ecosystems in various ways, contingent on the pollutant's nature, origin, quantity, and duration. Air, water, and soil pollution are the primary contaminants that impact ecosystems and have consequences for human and environmental health (Pepper et al. 2011).

Air pollution refers to the existence of gaseous substances, particles, or chemical compounds in the Earth's atmosphere that can cause harm to living creatures or impact the climate. Air pollution impacts ecosystems by diminishing the availability and quality of air necessary for the survival of plants and animals, altering weather and climate patterns that influence species distribution and diversity, and causing acid rain that harms vegetation and soil (Pepper et al. 2011). Fossil fuel combustion releases pollutants into the air, increasing atmospheric concentrations of greenhouse gases such as carbon dioxide and methane. This, in turn, can cause the atmosphere to retain heat and eventually cause global warming (Jacobson 2012). Global warming can change the temperature and precipitation patterns of ecosystems, impacting the development and survival of plants and animals (Harvey 2018). Air pollution can lead to ozone depletion, a decrease in the stratosphere's ozone layer that protects the Earth from damaging ultraviolet radiation. Ozone depletion can heighten ecosystems' vulnerability to UV rays, leading to DNA and cell damage in living species, resulting in mutations, malignancies, and illnesses (Zhang et al. 2019).

Water pollution is the introduction of harmful substances or energy into water bodies, such as rivers, lakes, oceans, and groundwater, which can damage aquatic life or render the water unsuitable for human use (Babuji et al. 2023). Water pollution impacts ecosystems by diminishing the availability and standard of water necessary for aquatic and terrestrial organisms, changing the characteristics of water bodies, influencing the structure and operation of marine ecosystems, and leading to eutrophication, characterized by the overgrowth of algae and plants from nutrients in sewage and fertilizers (Chaudhry and Malik 2017). Eutrophication can lower water oxygen levels and lead to dead zones unsuitable for life (Nazari-Sharabian et al. 2018). Water pollution from oil spills can damage marine life, including fish, birds, and mammals, by covering their bodies and feathers with oil. This can hinder their ability to move, breathe, and regulate body temperature (Malik et al. 2020). Contaminated water can lead to diseases like cholera, typhoid, and dysentery, impacting human health due to water pollution (Lin et al. 2022).

Soil pollution occurs when soil is contaminated by substances or energy that can harm plants, animals, or humans. Soil pollution can impact ecosystems by diminishing soil quality and fertility essential for plant and animal growth, disrupting the nutrient cycle and organic matter decomposition, influencing the productivity and stability of terrestrial ecosystems, and inducing salinization, the buildup of salt in soil from irrigation and evaporation (Steffan et al. 2018). Salinization can diminish soil moisture and fertility, rendering it unsuitable for plant cultivation. Soil pollution caused by pesticides can harm plants, animals, and humans by killing helpful insects and microorganisms, accumulating in the food chain and leading to bioaccumulation and biomagnification, and entering the human body to cause poisoning, cancers, and congenital disabilities (Singh 2021).

Therefore, various forms of pollution can negatively impact ecosystems and the diversity of species they sustain. Pollution can diminish the accessibility and standard of natural resources, disturb ecological processes and relationships, and jeopardize the survival and adjustment of species. Hence, avoiding and managing pollution is crucial to safeguard the ecosystem. Implementing legislation, creating solutions, and adopting sustainable practices can reduce the development and emission of pollutants. Restoring and maintaining ecosystems affected by pollution is also crucial. By doing this, we can guarantee the health and welfare of humans and other living creatures on Earth.

1.2.2 Impacted Ecosystems Case Studies

The initial case study examines the Kissimmee River in Florida, USA, impacted by water pollution from flood control and agricultural practices. The Kissimmee River is a 214-kilometer tributary of Lake Okeechobee that feeds into the Everglades, a significant wetland habitat with a diverse range of plant and animal species. The US Army Corps of Engineers altered the river's course to minimize flooding and supply water for cultivation in the 1960s. Consequently, 70% of the floodplain wetlands along the river were lost, leading to decreased habitat and food availability for various species like wading birds, fish, and alligators. The increased fertilizer and sediment load in the river led to algae blooms and decreased water quality. A comprehensive restoration project began in 1992 to return the river to its original flow and function. The project included removing levees, filling canals, and reconnecting oxbows and marshes. The project has demonstrated favorable outcomes, including the restoration of wetland vegetation, the growth of wildlife populations, and the enhancement of water quality (Rand and Gardinali 2004; Koebel Jr 1995).

The second case study examined the Great Basin shrublands in the western USA, impacted by soil contamination from invasive plants and changed fire patterns. The Great Basin shrublands are dry ecosystems that span around 600,000 square kilometers of land, primarily in Nevada, Utah, and Idaho. The area is covered mainly by sagebrush and other indigenous shrubs, grasses, and forbs, serving as a habitat and food source for various animals, including the greater sage-grouse, a bird species at risk. The shrublands have been overtaken by cheatgrass, an exotic annual grass from Eurasia in the late 1800s. Cheatgrass surpasses natural plants, diminishes soil moisture and nutrients, and escalates the occurrence and severity of wildfires. Cheatgrass flourishes post-fire, outpacing the slow recovery of native plants, leading to a feedback cycle that promotes cheatgrass dominance and diminishes biodiversity. Various management strategies, including cheatgrass removal, native plant reseeding, herbicide and biological control agent use, and manipulation of fire regimes, have been implemented to restore the shrublands (Snyder et al. 2019).

The third case study examines the Chesapeake Bay in the eastern USA, which is impacted by water pollution from nutrient and sediment loading originating from upstream agricultural and urban activities. The Chesapeake Bay is the biggest estuary in the United States, covering a watershed area of approximately 166,000 km2, extending over six states and the District of Columbia. The ecosystem sustains various marine organisms, including blue crabs, oysters, and striped bass, and offers numerous economic and recreational advantages to multiple individuals. The bay has deteriorated due to high levels of nitrogen and phosphorus from fertilizers, animal manure, sewage, and rainwater runoff, leading to eutrophication and hypoxia. Eutrophication is the excessive development of algae and plants caused by an abundance of nutrients, leading to less sunlight penetration and decreased oxygen levels in water. Hypoxia is the state of reduced oxygen levels in water, leading to asphyxiation and mortality in fish and other aquatic organisms. In 1983, a multi-stakeholder collaboration was established to restore the bay, comprising federal, state, and local governments, non-governmental organizations, academic institutions, and commercial sectors. The partnership has taken steps including reducing fertilizer and sediment inputs, restoring wetlands and oyster reefs, and enforcing laws and incentives (Hobbs III 2004; Kemp et al. 2005).

The fourth case study examines the impact of air pollution on the longleaf pine forest in the southeastern USA caused by logging and fire suppression. The longleaf pine trees are fire-dependent forest ecosystem spanning around 37 million hectares of land, extending from Virginia to Texas. Numerous unique and threatened species inhabit the area, including the red-cockaded woodpecker, the gopher tortoise, and the pitcher plant. The forest has shrunk to less than 3% of its original size primarily because of harvesting for lumber and naval products and fire prevention for land use and farming. Logging has eliminated the majority of mature trees. At the same time, fire prevention has enabled the encroachment of hardwoods and shrubs, which hinder the growth of pine seedlings and change the forest's structure and makeup. Various conservation measures have been implemented to restore the forest, including protecting existing stands, restoring native trees, reintroducing prescribed fires, and establishing wildlife corridors (Moser and Wade 2005; Bowman and Chen 2022).

These four case studies demonstrate how various forms of pollution can harm ecosystems and the biodiversity they sustain. Pollution can diminish the accessibility and excellence of natural resources, disturb ecological processes and relationships, and jeopardize the survival and adjustment of species. Hence, avoiding and managing pollution is crucial to safeguard the ecosystem. Implementing legislation, creating solutions, and adopting sustainable practices can reduce the development and release of pollutants. Restoring and protecting ecosystems can help repair damage caused by pollution.

1.3 Environmental Pollution and Biodiversity

1.3.1 Explanation of Biodiversity

The term "biodiversity" describes the variety and breadth of life on Earth. The concept encompasses the variety of genes, species, and ecosystems, along with their interactions (Cardinale et al. 2012). Biodiversity is crucial for preserving the well-being and operation of natural ecosystems while also offering a range of advantages and services to humans, including food, medicine, recreation, and cultural value. Biodiversity might be harmed due to habitat loss, overexploitation, pollution, invasive species, and climate change (Tilman et al. 2017). Hence, it is crucial to preserve and safeguard biodiversity for the benefit of our world and humanity.

1.3.2 The Relationship Between Pollution and Biodiversity Decline

Pollution can impact biodiversity in several ways, contingent on the characteristics of the pollutant, such as type, source, quantity, and duration (Sigmund et al. 2023). This section will explore the impact of three primary sources of pollution—air, water, and soil—on biodiversity and the resulting repercussions of this decline.

Air pollution can impact biodiversity by diminishing the availability and quality of air necessary for the survival of plants and animals, altering weather and climate patterns that influence species distribution and diversity, and inducing acid rain that harms vegetation and soil (Pawankar and Wang 2021). Burning fossil fuels can lead to increased levels of greenhouse gases like methane and carbon dioxide in the atmosphere, resulting in trapped heat and causing global warming. Global warming can change the temperature and precipitation patterns in ecosystems, impacting the development and survival of plants and animals (Kinney 2018). Air pollution can lead to ozone depletion, the decrease in the ozone gas layer in the stratosphere protecting the Earth from damaging ultraviolet radiation. Ozone depletion can heighten ecosystems' exposure to UV rays, leading to DNA and cell damage in living species, resulting in mutations, malignancies, and diseases (Umar and Tasduq 2022).

Water pollution can impact biodiversity by diminishing the availability and quality of water necessary for aquatic and terrestrial organisms, changing the physical, chemical, and biological characteristics of water bodies, influencing the structure and operation of marine ecosystems, and leading to eutrophication, characterized by the overgrowth of algae and plants from nutrients in sewage and fertilizers (Singh et al. 2021). Eutrophication can deplete oxygen levels in water, leading to the formation of dead zones devoid of life. Water pollution from oil spills can harm marine life, like fish, birds, and mammals, by covering their bodies and feathers with oil, which can hinder their movement, breathing, and insulation (Wang et al. 2021). Contaminated water can lead to diseases including cholera, typhoid, and dysentery, impacting human health.

Soil pollution can impact biodiversity by diminishing soil quality and fertility necessary for the growth and well-being of plants and animals, disrupting the nutrient cycle and organic matter decomposition, influencing the productivity and stability of terrestrial ecosystems, and inducing salinization, the buildup of salt in the soil from irrigation and evaporation (Tibbett et al. 2020). Salinization can diminish soil moisture and fertility, rendering it inappropriate for plant growth (Haj-Amor et al. 2022). Soil pollution from pesticides can harm plants, animals, and humans by killing beneficial insects and microbes, accumulating in the food chain, producing bioaccumulation and biomagnification, and entering the human body, leading to poisoning, malignancies, and congenital disabilities. Pollution has a detrimental effect on biodiversity and ecosystems by disturbing ecological processes and threatening species' survival. Regulations, sustainable practices, and ecosystem restoration are essential for preventing and regulating pollution.

1.3.3 Species Impacted by Pollution

Pollution can impact various species differently based on the pollutant's kind, source, quantity, and duration. This section contains four examples of species affected by multiple types of outlined pollution. The causes, impacts, and solutions for each case are examined.

The initial example is the California condor, a severely endangered avian species impacted by air pollution resulting in lead poisoning. The California condor is North America's most giant airborne bird, boasting a wingspan that may reach up to 3 meters. The creature consumes carrion, including deer, cattle, and sheep, and has a lifespan of up to 60 years. The population of the California condor has significantly decreased because of habitat loss, poaching, and poisoning. Lead is a primary source of toxicity for condors when they consume corpses contaminated with lead bullets. Lead can result in neurological impairment, kidney failure, and mortality in condors. In 1987, a captive breeding and reintroduction program was started to prevent species extinction. The program included capturing the surviving wild condors, breeding them in zoos, and releasing the young back into the wild. The program has effectively boosted the California condor population from 22 in 1987 to over 400 in 2019. Lead poisoning continues to pose a significant risk, prompting initiatives to decrease the use of lead ammunition and to oversee and address the impacted condors (Romanov et al. 2022).

The coral reef, a rich and productive marine ecosystem, has been impacted by water pollution due to climate change and ocean acidification. Coral reefs consist of colonies of tiny creatures known as corals, which produce calcium carbonate skeletons that create the reef's framework. Coral reefs support diverse marine life and offer human benefits like sustenance, tourism, and protection for coastlines. Coral reefs are facing stress due to factors like overfishing, sedimentation, and pollution. Carbon dioxide, generated by burning fossil fuels, is a primary source of pollution that dissolves in seawater, leading to ocean acidification. Ocean acidification decreases the pH of seawater, hindering the ability of corals and calcifying organisms to form their skeletons. Heat pollution, stemming from global warming, elevates the sea surface temperature. Coral bleaching can occur because of elevated temperatures, resulting in the expulsion of symbiotic algae that supply corals with nutrients and color. Coral bleaching can damage and kill corals, ultimately causing the deterioration and collapse of coral reefs. Efforts to safeguard and revive coral reefs include decreasing greenhouse gas emissions, establishing protected areas, and transferring healthy corals to degraded reefs (Gil-Agudelo et al. 2020).

The third example is the honey bee, an essential pollinator and agricultural agent impacted by soil pollution from pesticides. Honey bees reside in colonies, including thousands of people, queens, workers, and drones. Honey bees gather pollen and nectar from flowers to create honey and wax. They also pollinate various crops like fruits, vegetables, and nuts, improving their productivity and quality. The honey bee population has decreased dramatically in recent years due to issues like parasites, illnesses, and habitat destruction (Iwasaki and Hogendoorn 2021). Pesticides, chemicals utilized in agriculture to eliminate pests and weeds, are a significant source of pollution (Bird et al. 2021). Pesticides have the potential to pollute the soil, water, and plants that honey bees depend on, impacting their well-being and behavior. Pesticides can result in acute poisoning, leading to the direct death of honey bees, or chronic poisoning, affecting their learning, memory, navigation, and reproduction. Pesticides can compound the susceptibility of honey bees by interacting with other stressors like diseases and climate change (Tome et al. 2020). Efforts to protect and improve the honey bee population include strategies like minimizing pesticide usage, advocating for organic agriculture and bee-friendly flora, and backing beekeeping and research initiatives (Panziera et al. 2022).

The fourth example is the polar bear, a captivating and threatened mammal species impacted by air pollution by persistent organic pollutants. The polar bear is the largest land carnivore and a dominant predator in the Arctic, preying on seals and other animals on the sea ice. The polar bear has adaptations for the cold and harsh habitat, including thick hair, fat, and giant paws. The survival of polar bears is jeopardized by hunting, oil spills, and climate change. Persistent organic pollutants (POPs) are manufactured compounds that defy decomposition and can spread over long distances in the atmosphere, making them a significant source of pollution. Persistent organic pollutants (POPs) encompass pesticides like DDT and endosulfan and industrial chemicals such as PCBs and dioxins. POPs can build up in the fatty tissues of polar bears, impacting their well-being and ability to reproduce. POPs can lead to hormonal imbalance, immunological suppression, and cancer in polar bears. POPs can be passed from the mother to the children via the placenta and milk, impacting the cubs' growth and survival. Efforts to decrease polar bear's exposure to Persistent Organic Pollutants (POPs) include measures like the Stockholm Convention, a worldwide agreement designed to prohibit or limit the manufacture and utilization of POPs (Dominique et al. 2020; Vorkamp et al. 2022). These four instances demonstrate how various types of pollution can negatively impact different species and the ecosystems that sustain them. Pollution can diminish the accessibility and caliber of natural resources, disturb ecological processes and interactions, and jeopardize the survival and adaption of species.

1.4 Environmental Pollution and Humans

1.4.1 Effects of Pollution on Society and the Economy

Pollution can have substantial socio-economic effects by impacting the livelihoods, incomes, and well-being of many individuals, particularly those who are poor and vulnerable. According to the World Health Organisation, air pollution results in over seven million preventable deaths annually, primarily in low- and middle-income countries (Shaddick et al. 2020). Air pollution can result in economic consequences, including decreased productivity from illness and absenteeism, higher spending on healthcare and medicine, and diminished quality of life and well-being (Lelieveld et al. 2020). Air pollution can impact climate and weather patterns, causing severe occurrences like droughts, floods, and heat waves. This can harm infrastructure, agriculture, and natural resources and raise the likelihood of disasters and conflicts (Manisalidis et al. 2020).

Water pollution can impact human health by spreading waterborne illnesses, including cholera, typhoid, and dysentery, and exposing individuals to harmful substances like arsenic, mercury, and lead (Madhav et al. 2020). The World Health Organization states that water pollution causes around 1.8 million fatalities yearly, with a higher incidence among children under 5 years old (Ajibade et al. 2021). Water pollution can result in economic repercussions, including decreased income and livelihoods from damaged fisheries, tourism, and recreational activities, higher costs for water treatment and purification, and diminished access to and quality of freshwater resources crucial for drinking, sanitation, irrigation, and hydropower (Bastos et al. 2021).

Soil contamination can impact human health by transmitting diseases like hookworm, roundworm, and whipworm and exposing individuals to harmful substances, including pesticides, heavy metals, and radioactive elements (Brevik et al. 2020). Soil contamination contributes to around 24% of the worldwide disease burden, particularly in low- and middle-income nations, as reported by the Food and Agriculture Organization (Rodríguez-Eugenio et al. 2018). Soil pollution can result in economic consequences, including decreased productivity and income from degraded agricultural land, higher soil remediation and fertilization costs, and compromised food security and nutrition from contaminated crops and livestock (Martinho 2020). Pollution can harm human communities and economies by diminishing natural resources, raising health care and environmental cleanup expenses, and impacting worker and consumer productivity and well-being. Hence, mitigating and managing pollution while safeguarding the ecosystem is crucial. Implementing rules, creating solutions, and adopting sustainable practices can reduce the development and release of pollutants. Monitoring and treating impacted people is also vital. It is also beneficial to examine the situations of communities affected by pollution.

1.4.2 Case Studies on Communities Impacted by Pollution

The initial case study involves the Ogoni people in Nigeria, who are impacted by oil pollution caused by international oil corporations. The Ogoni people are an ethnic minority residing in the Niger Delta, an area abundant in oil and gas reserves. Since the 1950s, the oil sector has extensively damaged the delta by activities like oil spills, gas flaring, pipeline vandalism, and land degradation. Oil contamination has poisoned the soil, water, and air, impacting the health and livelihoods of the Ogoni people, who rely on farming and fishing for sustenance. The oil contamination has led to social and political disputes, with the Ogoni community protesting oil firms and the government for their entitlements and reparations. In 1990, the Ogoni people established the movement for the Survival of the Ogoni People (MOSOP) to promote environmental justice, human rights, and self-determination. In 1995, the Nigerian government executed nine Ogoni leaders, which included the MOSOP president Ken Saro-Wiwa, leading to global condemnation and sanctions. In 2011, the United Nations Environment Program (UNEP) completed an environmental evaluation of Ogoni land and proposed a thorough cleanup and restoration plan. The plan was initiated in 2016 (Elum et al. 2016; Jaja and Obuah 2019; Abayomi et al. 2021).

The second case study examines the impact of water contamination on the Flint community in Michigan, USA, resulting from a change in water supply and the absence of corrosion control measures. Flint is a city with a primarily African American population and low-income residents, which has been impacted by economic downturn, poverty, and high crime rates. In 2014, the Flint River replaced the Detroit River as the city's primary water source, opting for a more cost-effective option despite its higher pollution levels and corrosiveness. The city neglected to implement corrosion control treatment on the water, resulting in lead pipes releasing lead into the water. The lead-contaminated water exposed residents, particularly children, to elevated levels of lead, resulting in potential brain damage, developmental delays, and behavioral issues. The water also harbored other impurities like bacteria, chlorine, and trihalomethanes, which can potentially induce infections, skin irritations, and cancer. The citizens' complaints regarding the water's color, smell, and taste were dismissed by city and state officials until independent testing and investigations confirmed the issue. In response, the city resumed using Detroit water in 2015, provided citizens with bottled water and filters, and replaced the lead pipes in 2019 (Roy and Edwards 2019; Pauli 2020).

The third case study examines the Shipibo-Konibo people in Peru, impacted by mercury contamination from illicit gold mining. The Shipibo-Konibo people are an Indigenous community residing in the Amazon rainforest, where they engage in their traditional culture and spirituality. Illegal gold mining has threatened their area and lifestyle, escalating after the 2008 global financial crisis. Mercury is used by illegal gold miners to extract gold from the Earth, contaminating the streams and the food chain, and affecting the fish, plants, and animals crucial to the Shipibo-Konibo people. Mercury contamination has led to health issues like kidney failure, neurological illnesses, and birth deformities among the Shipibo-Konibo community, who have challenges in accessing healthcare and education. Mercury contamination has led to environmental harm, including deforestation, soil erosion, and biodiversity loss, impacting the rainforest's ecosystem services and resilience. The Shipibo-Konibo people have united with other Indigenous groups, NGOs, and researchers to request acknowledgment, safeguarding, and reparation from the government and the global community in response to the issue (Markham and Sangermano 2018; Torres and De-la-Torre 2022).

The fourth case study examines the impact of chemical contamination on the Bhopal community in India caused by a gas leak from a pesticide facility. Bhopal is a heavily populated and impoverished city in central India where a pesticide plant owned by the Union Carbide Corporation (UCC) was situated. In 1984, a gas leak from the facility spewed a poisonous cloud of methyl isocyanate (MIC) and other chemicals, causing the deaths of thousands of people in the city shortly after. Hundreds of thousands of individuals were damaged and crippled by the gas leak, experiencing respiratory, ocular, and skin issues, along with chronic ailments like cancer, tuberculosis, and diabetes. The gas leak resulted in contamination of the soil, water, and air, impacting the health and livelihoods of the survivors and their descendants, who are now dealing with shame and discrimination. The gas leak resulted in social and economic consequences, including financial loss, educational setbacks, missed opportunities, and the deterioration of family and community bonds. In 1989, the UCC and the Indian government agreed to a settlement to compensate the disaster victims. However, the deal was widely criticized for being insufficient and unjust. Through judicial, political, and social initiatives, the survivors and their allies have pursued justice, accountability, and rehabilitation (Sriramachari 2004; Pandya 2023).

These four case studies demonstrate the severe impact of various types of pollution on human societies, particularly those marginalized, vulnerable, or located near pollution sources. Pollution can lead to health issues, environmental harm, social disputes, and economic setbacks, impacting the dignity and welfare of numerous individuals. Hence, avoiding and managing pollution while safeguarding the ecosystem is crucial. This can be achieved by enforcing legislation, creating solutions, and embracing sustainable practices to reduce the production and release of pollutants. It also involves acknowledging, empowering, and assisting communities affected by environmental issues in their fight for fairness.

1.5 Future Projections

1.5.1 Anticipated Patterns in Environmental Contamination

Environmental pollution is projected to rise because of causes like population expansion, urbanization, industrialization, energy consumption, and changes in land use. The OECD Environmental Outlook to 2050 predicts a 50% rise in greenhouse gas emissions by 2050, resulting in a 3–6 °C rise in world average temperature by the end of the century. By 2050, global water consumption is expected to rise by 55%, beyond the sustainable availability of freshwater resources. By 2050, global soil erosion will likely increase by 10%, impacting the productivity and quality of agricultural land. The trends will substantially affect the environment, human well-being, and achieving the Sustainable Development Goals (OECD 2012).

Environmental pollution leads to climate change, which is the modification of the Earth's climate system caused by the rise of greenhouse gases in the atmosphere, like methane, carbon dioxide, and nitrous oxide. Climate change can impact weather and climate patterns, including temperature, precipitation, and wind, leading to extreme events, including floods, droughts, heat waves, and storms. The Intergovernmental Panel on Climate Change (IPCC) reports that the global average temperature has increased by around 1.1 °C from the pre-industrial era and is expected to climb by 1.5 °C between 2030 and 2052 if present emissions persist. The IPCC cautions that increasing the temperature to 1.5 °C would necessitate swift and unparalleled changes in energy, land, urban, and industrial systems. Going above 1.5 °C will heighten the dangers and consequences of climate change (Masson-Delmotte et al. 2021). Climate change can impact sea level and ice by melting glaciers and ice sheets, causing saltwater to expand as it warms, and changing ocean currents and circulation. Based on the greenhouse gas emission scenario, the following sea level and ice changes are anticipated by 2100 (Horton et al. 2020).

Under a low-emission scenario limiting global temperature rise to 1.5 °C above pre-industrial levels, the global mean sea level is to increase by approximately 0.4 meters (1.3 feet), Arctic sea ice is forecasted to decrease by around 43%, and the Antarctic ice sheet is anticipated to lose about 0.16 trillion tones of ice annually (Frederikse et al. 2020). Under a high-emission scenario with a global temperature increase of 4 °C above pre-industrial levels, the global mean sea level is to increase by approximately 0.8 meters (2.6 feet), the Arctic sea ice is forecasted to vanish entirely during the summer, and the Antarctic ice sheet is anticipated to lose around 1.2 trillion tones of ice annually. These can cause harmful effects on the environment and human health, including biodiversity loss, ecological degradation, disease spread, population displacement, and conflict escalation (Dayan et al. 2021).

Environmental pollution leads to acid rain, the deposition of acidic chemicals like sulfur dioxide and nitrogen oxides from the atmosphere. Acid rain can impact the quality and equilibrium of natural elements like water, soil, and vegetation, leading to issues such as lake and river acidification, nutrient, and metal leaching from soil, crop and forest damage, and corrosion of buildings and monuments. Acid rain can impact human health by inducing respiratory and skin issues and polluting drinking water and food supplies (Grennfelt et al. 2020; Shi et al. 2021).

One result of environmental pollution is eutrophication, characterized by the overgrowth of algae and plants in water bodies caused by the influx of nutrients like phosphorus and nitrogen from agricultural and urban runoff. Eutrophication can deplete oxygen levels in water, leading to the formation of dead zones devoid of life. Eutrophication can generate harmful compounds like cyanobacteria and microcystins, negatively impacting aquatic ecosystems and human well-being. Eutrophication can affect water bodies' visual and recreational appeal by forming algal blooms, unpleasant odors, and increased turbidity (Kakade et al. 2021). One further result of environmental pollution is soil degradation, which is the decline in soil's physical, chemical, and biological characteristics caused by the depletion of organic matter, topsoil erosion, soil compaction, soil salinization, and soil contamination. Soil degradation can impact the quality and production of agricultural land by diminishing soil fertility, moisture, and structure and heightening its susceptibility to pests and diseases. Soil deterioration can affect the ecosystem services and functions of soil, including water and nutrient cycle regulation, carbon and water storage, biodiversity support, and human health protection (Prăvălie et al. 2021).

Environmental pollution is projected to rise because of causes like population expansion, urbanization, industrialization, energy consumption, and changes in land use. Environmental pollution can cause harmful impacts on the environment and human health, including climate change, acid rain, eutrophication, and soil degradation. Hence, avoiding and managing environmental contamination is crucial to safeguard the environment. Implementing legislation, creating solutions, and adopting sustainable practices can reduce the development and release of pollutants. Monitoring and treating affected habitats and populations are also crucial steps.

1.5.2 Possible Future Effects on Ecosystems, Biodiversity, and Human Health

One potential future consequence is the depletion of ecosystem services, which are the advantages that humans derive from ecosystems, including supplying, regulating, sustaining, and cultural functions. Ecosystem services are crucial for human well-being by fulfilling necessities and improving the quality of life (Kosanic and Petzold 2020). Ecosystem services are deteriorating due to human activities such as pollution, climate change, habitat loss, overexploitation, invading species, and developing illnesses (Turner et al. 2016). Pollution can diminish the quality and accessibility of water, air, and soil, impacting the availability of freshwater, food, and energy (Javan et al. 2023). Climate change can modify ecosystems' temperature and precipitation patterns, impacting species' distribution and variety and the stability and ability to recover ecosystems. Habitat loss can lead to fragmentation and destruction of natural habitats, resulting in decreased population size and genetic diversity and heightened risk of extinction (Mantyka-pringle et al. 2012). Overexploitation can exhaust ecosystems' natural resources and biomass, surpassing their potential to regenerate and maintain sustainability. Invasive species can outcompete and replace native species, changing the composition and operation of ecosystems and introducing novel viruses and parasites. Emerging diseases can impact the health and survival of plants, animals, and humans, changing ecological processes and interactions and heightening the possibility of pandemics (Maja and

Ayano 2021). When these components come together, it may result in combined and increasing effects, causing a decline in ecosystem services and subsequent implications on human welfare.

Another possible future consequence is the reduction of biodiversity, which refers to the range and diversity of life on Earth, encompassing the variety of genes, species, and ecosystems, along with the relationships between them. Biodiversity is crucial for preserving the well-being and operation of natural ecosystems while offering a range of human advantages and services, including sustenance, medication, leisure, and cultural enrichment. Biodiversity is decreasing rapidly because of human activities such as pollution, climate change, habitat destruction, overexploitation, invading species, and new illnesses (Díaz and Malhi 2022). Pollution can impact species' genetic diversity and adaptation through mutations, malignancies, and diseases, decreasing individuals' fitness and reproductive success (Ukaogo et al. 2020). Climate change can impact the species diversity and distribution of ecosystems by changing temperature and precipitation patterns and causing discrepancies in species' timing and biological processes. Habitat loss can impact ecosystem diversity and function by diminishing the size, connectivity, and integrity of natural habitats and heightening species' isolation and susceptibility. Overexploitation can impact species' population variety and abundance by decreasing their biomass and genetic diversity, raising their mortality, and increasing the danger of extinction. Invasive species can impact ecological community diversity and composition by competing with and displacing native species and changing ecosystems' trophic structure and function. Emerging diseases can impact the health and survival of species by causing illness and death and interrupting ecological processes and relationships. These factors can interact and build up, leading to decreased biodiversity and subsequent implications on ecosystem services and human health.

Pollution, climate change, habitat loss, overexploitation, invading species, and new illnesses are issues that may harm ecosystems, biodiversity, and human health in the future. The effects can significantly influence policy decisions and actions, thereby hindering the attainment of the Sustainable Development Goals. These goals are a collection of worldwide objectives and benchmarks designed to eradicate poverty, safeguard the environment, and promote peace and prosperity for all by 2030. Hence, it is crucial to prevent and manage these factors to protect the environment. This can be achieved through enforcing regulations, creating solutions, and adopting sustainable practices that reduce the production and release of pollutants. Additionally, efforts should be made to restore and preserve ecosystems, biodiversity, and human health harmed or ruined by these factors. By taking this action, we can ensure the well-being and health of people and other living things on Earth.

1.6 Conclusion

Environmental pollution is a worldwide issue that harms the environment and human health. It can result from natural events like volcanic eruptions or wildfires and human activity like industrial emissions, agricultural runoff, or municipal garbage. Pollutants can significantly affect different environmental characteristics, such as air, water, and soil pollution. Air pollution can lead to respiratory issues, cardiovascular ailments, and brain abnormalities. Climate change is caused by increased greenhouse gas concentrations, which impact weather patterns, sea levels, ice cover, ocean circulation, and biogeochemical cycles. It can also lead to severe occurrences such as droughts, floods, heat waves, and storms. Water pollution is the introduction of harmful substances or energy into water bodies, which can adversely affect aquatic life and render the water unsuitable for human consumption. It can lead to waterborne illnesses, expose individuals to hazardous substances, and generate areas devoid of life known as dead zones.

Eutrophication is the overabundant growth of algae and plants in water sources caused by fertilizer influx from agricultural and urban runoff. Eutrophication leads to decreased oxygen levels and the formation of lifeless zones. Soil pollution is the presence of harmful substances or energy in the soil and can negatively impact plants, animals, and humans. It can lead to soil-transmitted diseases, expose individuals to toxic compounds, and degrade soil's physical, chemical, and biological characteristics. It is essential to tackle environmental pollution to safeguard the health and welfare of humans and other organisms on Earth. Soil pollution can negatively impact plants, animals, and humans. It can lead to soil-borne illnesses, expose individuals to hazardous substances, and deteriorate soil. Soil pollution can affect the quality and production of agricultural land by diminishing fertility, moisture, and structure. Environmental pollution presents a significant danger to the environment and human welfare, jeopardizing the existence of species, the stability of ecosystems, and human dignity. Enforcing legislation, establishing sustainable practices, and overseeing the treatment of affected areas and populations is crucial. It is essential to promote additional research and initiatives about environmental pollution. Increasing knowledge and educating individuals about environmental pollution's origins, effects, and remedies can inspire them to show concern and take action for the environment. Advocating for research and innovation in environmental pollution can create and share new knowledge and technology to prevent, diminish, and manage pollution. Developing and enforcing policies and procedures about environmental pollution involves creating rules, standards, and incentives to reduce the production and release of pollutants and monitoring and treating affected areas and populations. Environmental pollution is a significant issue that can be mitigated through various measures and policies focused on preventing, decreasing, and managing pollution.

1.7 Suggestions

The following suggestions are made to combat pollution and help to minimize pollution and vulnerability:

(i) The mitigation of environmental pollution necessitates a comprehensive strategy incorporating legislative measures, sustainable methodologies, and remedial endeavors. An essential element pertains to implementing and enforcing comprehensive environmental legislation that governs industrial emissions, agricultural methods, and waste disposal. Enhancing monitoring and enforcement systems is necessary to guarantee adherence to pollution control measures and establish mechanisms for holding polluters responsible. Furthermore, sustainable practices in various sectors, such as industries, agriculture, and urban areas, can mitigate pollution. Promoting the use of renewable energy sources, green infrastructure, and environmentally sustainable production practices has the potential to reduce environmental consequences substantially.

(ii) It is crucial to allocate resources towards the repair of contaminated ecosystems. Cleanup activities and restoration programs have the potential to effectively repair ecologically impacted areas and enhance the overall quality of the environment. The dissemination of knowledge regarding environmental pollution is of paramount importance. Through the implementation of educational initiatives and communication campaigns, individuals can acquire knowledge and comprehension, enabling them to make well-informed decisions and engage in collaborative efforts to mitigate pollution in their communities and everyday routines. Another crucial aspect is the allocation of resources toward research and innovation. The advancement of novel technologies, techniques, and solutions aimed at preventing, mitigating, and managing environmental pollution holds the potential to contribute significantly to establishing a more sustainable future. It is crucial to work with stakeholders to create thorough policies and plans. Effective and inclusive solutions can be achieved by actively including governments, corporations, non-governmental organizations (NGOs), and communities in policymaking.

(iii) It is imperative to cultivate international cooperation and collaboration to tackle transboundary contamination concerns effectively. By promoting collective accountability for environmental stewardship, nations can collaborate to address worldwide environmental issues. It is essential to incentivize businesses and individuals to embrace cleaner technology and practices. A potential strategy to promote environmental stewardship and foster innovation involves the implementation of incentives, subsidies, and awards. Acknowledging and incentivizing endeavors that decrease pollution and conserve the environment can foster beneficial transformation. Ultimately, strategizing for the future is crucial. Creating policies and action plans that prioritize preventive measures, resilience-building, and adaptive management approaches to protect ecosystems, biodiversity, and human well-being for future generations is critical.

References

Abayomi O, Olayemi TE, Ogungbade T (2021) Environmental pollution and its ecological consequences on The Niger Delta: a review of the literature. Afr J Environ Nat Sci Res 4:27–42

Ajibade FO, Adelodun B, Lasisi KH, Fadare OO, Ajibade TF, Nwogwu NA et al (2021) Environmental pollution and their socioeconomic impacts. In: Microbe mediated remediation of environmental contaminants. Woodhead Publishing, pp 321–354

Babuji P, Thirumalaisamy S, Duraisamy K, Periyasamy G (2023) Human health risks due to exposure to water pollution: a review. Water 15(14):2532

Bastos MI, Roebeling PC, Alves FL, Villasante S (2021) Water pollution threatening marine, coastal and estuarine systems: a review of environmental-economic approaches for the assessment of development strategies. In: Disaster management and human health risk VII: reducing risk, improving outcomes, vol 1, pp 135–147

Bird G, Wilson AE, Williams GR, Hardy NB (2021) Parasites and pesticides act antagonistically on honey bee health. J Appl Ecol 58(5):997–1005

Bowman KA, Chen X (2022) Current standing of longleaf pine trees under climate change. J Bot Res 4(1):28–39

Brevik EC, Slaughter L, Singh BR, Steffan JJ, Collier D, Barnhart P, Pereira P (2020) Soil and human health: current status and future needs. Air Soil Water Res 13:1178622120934441

Cardinale BJ, Duffy JE, Gonzalez A, Hooper DU, Perrings C, Venail P et al (2012) Biodiversity loss and its impact on humanity. Nature 486(7401):59–67

Chaudhry FN, Malik MF (2017) Factors affecting water pollution: a review. J Ecosyst Ecography 7(1):225–231

Connell DW, Miller GJ (2022) Chemistry and toxicology of pollution: ecological and human health. Wiley

Dayan H, Le Cozannet G, Speich S, Thiéblemont R (2021) High-end scenarios of sea-level rise for coastal risk-averse stakeholders. Front Mar Sci 8:569992

Díaz S, Malhi Y (2022) Biodiversity: concepts, patterns, trends, and perspectives. Annu Rev Environ Resour 47:31–63

Dominique M, Letcher RJ, Rutter A, Langlois VS (2020) Comparative review of the distribution and burden of contaminants in the body of polar bears. Environ Sci Pollut Res 27:32456–32466

Elum ZA, Mopipi K, Henri-Ukoha A (2016) Oil exploitation and its socioeconomic effects on The Niger Delta region of Nigeria. Environ Sci Pollut Res 23:12880–12889

Frederikse T, Buchanan MK, Lambert E, Kopp RE, Oppenheimer M, Rasmussen DJ, Wal RSVD (2020) Antarctic ice sheet and emission scenario controls on 21st-century extreme sea-level changes. Nat Commun 11(1):390

Fuller R, Landrigan PJ, Balakrishnan K, Bathan G, Bose-O'Reilly S, Brauer M et al (2022) Pollution and health: a progress update. Lancet Planet Health 6(6):e535–e547

Gil-Agudelo DL, Cintra-Buenrostro CE, Brenner J, González-Díaz P, Kiene W, Lustic C, Pérez-España H (2020) Coral reefs in the Gulf of Mexico large marine ecosystem: conservation status, challenges, and opportunities. Front Mar Sci 6:807

Glencross DA, Ho TR, Camina N, Hawrylowicz CM, Pfeffer PE (2020) Air pollution and its effects on the immune system. Free Radic Biol Med 151:56–68

Grennfelt P, Engleryd A, Forsius M, Hov Ø, Rodhe H, Cowling E (2020) Acid rain and air pollution: 50 years of progress in environmental science and policy. Ambio 49:849–864

Haj-Amor Z, Araya T, Kim DG, Bouri S, Lee J, Ghiloufi W et al (2022) Soil salinity and its associated effects on soil microorganisms, greenhouse gas emissions, crop yield, biodiversity and desertification: a review. Sci Total Environ 843:156946

Harvey LD (2018) Global warming. Routledge

Hobbs CH III (2004) Geological history of Chesapeake Bay, USA. Quat Sci Rev 23(5–6):641–661

Horton BP, Khan NS, Cahill N, Lee JS, Shaw TA, Garner AJ et al (2020) Estimating global mean sea-level rise and its uncertainties by 2100 and 2300 from an expert survey. NPJ Clim Atmos Sci 3(1):18

Iwasaki JM, Hogendoorn K (2021) How protection of honey bees can help and hinder bee conservation. Curr Opin Insect Sci 46:112–118
Jacobson MZ (2012) Air pollution and global warming: history, science, and solutions. Cambridge University Press
Jaja JM, Obuah E (2019) The politics of the Ogoni cleanup: challenges and prospects. Afr Res Rev 13(3):101–113
Javan K, Altaee A, BaniHashemi S, Darestani M, Zhou J, Pignatta G (2023) A review of interconnected challenges in the water–energy–food nexus: urban pollution perspective towards sustainable development. Sci Total Environ 912:169319
Kakade A, Salama ES, Han H, Zheng Y, Kulshrestha S, Jalalah M et al (2021) World eutrophic pollution of lake and river: biotreatment potential and future perspectives. Environ Technol Innov 23:101604
Kemp WM, Boynton WR, Adolf JE, Boesch DF, Boicourt WC, Brush G et al (2005) Eutrophication of Chesapeake Bay: historical trends and ecological interactions. Mar Ecol Prog Ser 303:1–29
Kinney PL (2018) Interactions of climate change, air pollution, and human health. Curr Environ Health Rep 5:179–186
Koebel JW Jr (1995) An historical perspective on the Kissimmee River restoration project. Restor Ecol 3(3):149–159
Kosanic A, Petzold J (2020) A systematic review of cultural ecosystem services and human wellbeing. Ecosyst Serv 45:101168
Lelieveld J, Pozzer A, Pöschl U, Fnais M, Haines A, Münzel T (2020) Loss of life expectancy from air pollution compared to other risk factors: a worldwide perspective. Cardiovasc Res 116(11):1910–1917
Lin L, Yang H, Xu X (2022) Effects of water pollution on human health and disease heterogeneity: a review. Front Environ Sci 10:880246
Madhav S, Ahamad A, Singh AK, Kushawaha J, Chauhan JS, Sharma S, Singh P (2020) Water pollutants: sources and impact on the environment and human health. In: Sensors in water pollutants monitoring: role of material. Springer, Berlin, pp 43–62
Maja MM, Ayano SF (2021) The impact of population growth on natural resources and farmers' capacity to adapt to climate change in low-income countries. Earth Syst Environ 5:271–283
Malik DS, Sharma AK, Sharma AK, Thakur R, Sharma M (2020) A review on impact of water pollution on freshwater fish species and their aquatic environment. In: Advances in environmental pollution management: wastewater impacts and treatment technologies, vol 1. Agro Environ Media—Agriculture and Ennvironmental Science Academy, Haridwar, pp 10–28
Manisalidis I, Stavropoulou E, Stavropoulos A, Bezirtzoglou E (2020) Environmental and health impacts of air pollution: a review. Front Public Health 8:14
Mantyka-pringle CS, Martin TG, Rhodes JR (2012) Interactions between climate and habitat loss effects on biodiversity: a systematic review and meta-analysis. Glob Chang Biol 18(4):1239–1252
Markham KE, Sangermano F (2018) Evaluating wildlife vulnerability to mercury pollution from artisanal and small-scale gold mining in Madre de Dios, Peru. Trop Conserv Sci 11:1940082918794320
Martinho VJPD (2020) Exploring the topics of soil pollution and agricultural economics: highlighting good practices. Agriculture 10(1):24
Masson-Delmotte VP, Zhai P, Pirani SL, Connors C, Péan S, Berger N et al (2021) Ipcc, 2021: summary for policymakers. In: Climate change 2021: the physical science basis. Contribution of working group i to the sixth assessment report of the intergovernmental panel on climate change. Ipcc
Moser WK, Wade DD (2005) Fire exclusion as a disturbance in the temperate forests of the USA: examples from longleaf pine forests. Scand J For Res 20(S6):17–26
Nazari-Sharabian M, Ahmad S, Karakouzian M (2018) Climate change and eutrophication: a short review. Eng Technol Appl Sci Res 8(6):3668–3672

OECD (2012) OECD environmental outlook to 2050: the consequences of inaction. OECD Publishing, Paris. https://doi.org/10.1787/9789264122246-en
Pandya JS (2023) Crip life amidst debilitation: medicalization, survival, and the Bhopal gas leak. Disabil Stud Q 43(1)
Panziera D, Requier F, Chantawannakul P, Pirk CW, Blacquière T (2022) The diversity decline in wild and managed honey bee populations urges for an integrated conservation approach. Front Ecol Evol 10:767950
Pauli BJ (2020) The Flint water crisis. Wiley Interdiscip Rev Water 7(3):e1420
Pawankar R, Wang JY (2021) Climate change, air pollution, and biodiversity in Asia Pacific and impact on respiratory allergies. Immunol Allergy Clin N Am 41(1):63–71
Pepper I, Gerba CP, Brusseau ML (2011) Environmental and pollution science. Elsevier
Prăvălie R, Patriche C, Borrelli P, Panagos P, Roşca B, Dumitraşcu M et al (2021) Arable lands under the pressure of multiple land degradation processes. A global perspective. Environ Res 194:110697
Rand GM, Gardinali PR (2004) South Florida ecosystems. Ecotoxicology 13:179–184
Rodríguez-Eugenio N, McLaughlin M, Pennock D (2018) Soil pollution: a hidden reality. FAO, Rome, p 142
Romanov MN, Da Y, Chemnick LG, Thomas SM, Dandekar SS, Papp JC, Ryder OA (2022) Towards a genetic linkage map of the California condor, an endangered New World vulture species. Animals 12(23):3266
Roy S, Edwards MA (2019) Preventing another lead (Pb) in drinking water crisis: lessons from the Washington DC and Flint MI contamination events. Curr Opin Environ Sci Health 7:34–44
Schwarzenbach RP, Egli T, Hofstetter TB, Von Gunten U, Wehrli B (2010) Global water pollution and human health. Annu Rev Environ Resour 35:109–136
Shaddick G, Thomas ML, Mudu P, Ruggeri G, Gumy S (2020) Half the world's population are exposed to increasing air pollution. NPJ Clim Atmos Sci 3(1):23
Sharif F, Hussain I, Qubtia M (2023) Energy consumption, carbon emission and economic growth at aggregate and disaggregate level: a panel analysis of the top polluted countries. Sustainability 15(4):2935
Shi Z, Zhang J, Xiao Z, Lu T, Ren X, Wei H (2021) Effects of acid rain on plant growth: a meta-analysis. J Environ Manag 297:113213
Shinwari R, Wang Y, Gozgor G, Mousavi M (2024) Does FDI affect energy consumption in the belt and road initiative economies? The role of green technologies. Energy Econ 132:107409
Sigmund G, Ågerstrand M, Antonelli A, Backhaus T, Brodin T, Diamond ML et al (2023) Addressing chemical pollution in biodiversity research. Glob Chang Biol 29(12):3240–3255
Singh A (2021) Soil salinization management for sustainable development: a review. J Environ Manag 277:111383
Singh V, Shukla S, Singh A (2021) The principal factors responsible for biodiversity loss. Open J Plant Sci 6(1):011–014
Snyder KA, Evers L, Chambers JC, Dunham J, Bradford JB, Loik ME (2019) Effects of changing climate on the hydrological cycle in cold desert ecosystems of the Great Basin and Columbia plateau. Rangel Ecol Manag 72(1):1–12
Sriramachari S (2004) The Bhopal gas tragedy: an environmental disaster. Curr Sci 86(7):905–920
Steffan JJ, Brevik EC, Burgess LC, Cerdà A (2018) The effect of soil on human health: an overview. Eur J Soil Sci 69(1):159–171
Tibbett M, Fraser TD, Duddigan S (2020) Identifying potential threats to soil biodiversity. PeerJ 8:e9271
Tilman D, Clark M, Williams DR, Kimmel K, Polasky S, Packer C (2017) Future threats to biodiversity and pathways to their prevention. Nature 546(7656):73–81
Tome HV, Schmehl DR, Wedde AE, Godoy RS, Ravaiano SV, Guedes RN et al (2020) Frequently encountered pesticides can cause multiple disorders in developing worker honey bees. Environ Pollut 256:113420

Torres FG, De-la-Torre GE (2022) Mercury pollution in Peru: geographic distribution, health hazards, and sustainable removal technologies. Environ Sci Pollut Res 29(36):54045–54059

Turner KG, Anderson S, Gonzales-Chang M, Costanza R, Courville S, Dalgaard T et al (2016) A review of methods, data, and models to assess changes in the value of ecosystem services from land degradation and restoration. Ecol Model 319:190–207

Ukaogo PO, Ewuzie U, Onwuka CV (2020) Environmental pollution: causes, effects, and the remedies. In: Microorganisms for sustainable environment and health. Elsevier, pp 419–429

Umar SA, Tasduq SA (2022) Ozone layer depletion and emerging public health concerns-an update on epidemiological perspective of the ambivalent effects of ultraviolet radiation exposure. Front Oncol 12:866733

Vorkamp K, Carlsson P, Corsolini S, de Wit CA, Dietz R, Gribble MO et al (2022) Influences of climate change on long-term time series of persistent organic pollutants (POPs) in Arctic and Antarctic biota. Environ Sci: Processes Impacts 24(10):1643–1660

Vrijheid M, Casas M, Gascon M, Valvi D, Nieuwenhuijsen M (2016) Environmental pollutants and child health—a review of recent concerns. Int J Hyg Environ Health 219(4–5):331–342

Wang H, Molinos JG, Heino J, Zhang H, Zhang P, Xu J (2021) Eutrophication causes invertebrate biodiversity loss and decreases cross-taxon congruence across anthropogenically-disturbed lakes. Environ Int 153:106494

Wang Y, Zhao J, Fu Z, Guan D, Zhang D, Zhang H et al (2024) Innovative overview of the occurrence, aging characteristics, and ecological toxicity of microplastics in environmental media. Environ Pollut 346:123623

Zhang J, Wei Y, Fang Z (2019) Ozone pollution: a major health hazard worldwide. Front Immunol 10:2518

Chapter 2
Petroleum Pollution in Soil and Sediment

Amir Parnian, Nematollah Jaafarazdeh Haghighi Fard, James N. Furze, Mehdi Tayebi, Zain Mushtaq, and Salman Selmy

Abstract Pollution occurs when contaminants are released into the Earth's environment and interfere with health, quality, and natural ecosystem function. Ecosystem sustainability is key in reducing population vulnerability; affecting economy, food and energy security, and the surrounding environment. Soil and sediment pollution is the critical environmental issue impacting ecosystem function, local water quality, and exposes human health to risk. Anthropogenic activity is the major source of petroleum hydrocarbons, which lead to environmental devastation. To combat petroleum hydrocarbon contamination in soil and sediment, this chapter presents available environmental remediation techniques and outlines major sources of environmental health hazards.

A. Parnian (✉)
National Salinity Research Center (NSRC), Agricultural Research, Education and Extension Organization (AREEO), Yazd, Iran

N. J. H. Fard
Environmental Technologies Research Center, Medical Basic Sciences Research Institute, Ahvaz Jundishapur University of Medical Sciences, Ahvaz, Iran

J. N. Furze
Royal Geographical Society (with the Institute of British Geographers), London, UK

Laboratory of Biotechnology and Valorization of Natural Resources, Faculty of Sciences-Agadir, Department of Biology, Ibn Zohr University, Agadir, Morocco

Control and Systems Engineering Department, University of Technology-Iraq, Baghdad, Iraq

Earth Systems Protection Foundation, Somerset, UK

M. Tayebi
Aghajari Oil and Gas Production Company (AGOPC), Omidiyeh, Iran

Z. Mushtaq
Department of Soil Science, Faculty of Agricultural Sciences, University of the Punjab, Lahore, Pakistan

S. Selmy
Department of Soils and Water, Faculty of Agriculture, Assiut University, Assiut, Egypt

A. Parnian et al. (eds.), *Bioremediation of Petroleum Contaminated Soil and Sediment*, Environmental Contamination Remediation and Management,
https://doi.org/10.1007/978-3-031-95524-2_2

Keywords Hydrocarbon · Ecosystem · Contaminant · Security · Risk

2.1 Introduction (What Is Pollution?)

The process of releasing pollutants or contaminants into the Earth's atmosphere that compromise the existence and well-being of humans, animals, the standard of living; and indeed regular, organic processes of ecosystems is known as pollution. Although pollution in our immediate surroundings can occur naturally, since the Industrial Revolution, human activity has been the main source of ecosystem contamination (Mushtaq et al. 2024). According to Ukaogo et al. (2020), the most prevalent pollutants found in the air, water, soil, or sediment matrix are waste products from the manufacturing, transportation, and energy production processes that characterize modern society (Table 2.1).

2.2 What Is Environmental Pollution?

Environmental pollution refers to the diffusion of hazardous substances or products into the surrounding environment, leading to adverse effects on ecosystems, plants, animals, and human health. Moreover, environmental pollution encompasses the contamination of each phase (air, water, soil, and sediment) by anthropogenic materials, including chemicals, organic and physical substances (Parnian et al. 2023).

The consequences of environmental pollution are far-reaching, affecting all forms of life and the planet's natural cycle. Biodiversity loss, habitat destruction, and increased health risk ensue for humans, including respiratory, mental, and/or physiological-based disease. Addressing environmental pollution requires concerted efforts to reduce gas emissions, manage waste and wastewater responsibly, and promote environmentally friendly practices (Ukaogo et al. 2020). Through

Table 2.1 Environmental pollution categories

Pollution categories	Description	References
Air pollution	Characterized by the presence of toxic substances in the Earth's atmosphere, primarily caused by car emissions, industrial discharges, and energy production	Sofia et al. (2020)
Water pollution	Occurs when chemicals or waste materials, often from industrial processes, agriculture, or sewage treatment, enter water bodies, leading to ecosystem degradation and health hazards	Parnian et al. (2021)
Soil/sediment pollution	Involves the contamination of soil with hazardous materials, which can result from industrial waste disposal, agricultural chemicals, and oil spills	Parnian et al. (2022a)

education, policy change, and technological innovation, we effect the impacts of pollution and safeguard our environment for future generations.

2.3 Soil/Sediment Pollution

Ecosystem resilience is key in decreasing population vulnerability, affecting the economy, food security, energy security, and ecological stability (Pirasteh-Anosheh et al. 2021). Environmental pollution is a serious problem in the modern era and challenges natural environmental sustainability (Parnian et al. 2020). Environmental contamination poses a threat to aquatic ecosystems, the terrestrial ecosystem (plants, soil, and sediments), and the atmosphere and is imperative for life on Earth (Kumar et al. 2022).

Soil and sediment degradation is critical affecting ecosystems, water quality, and human health. Pollution occurs when hazardous substances, including petroleum-based materials, heavy metals, pesticides, and industrial waste, accumulate in soil or sediment. Pollutants originate from various sources, such as agricultural runoff, industrial discharges, mining activities, and unsuitable waste disposal. The presence of contaminants in soil and sediments has detrimental effects on plant growth, reducing agricultural productivity and leading to food insecurity. Pollutants also leach into groundwater, making it unsafe for consumption and use. Sediment pollution in aquatic environments destroys habitats, harms aquatic life, and disrupts ecological balance in our surrounding environment. One of the most concerning aspects of soil and sediment contamination is its persistence and bioaccumulation. Many contaminants do not break down easily and accumulate in the tissues of living organisms, moving up the food chain and posing significant health risks to plants, animals, and humans (Ferreira et al. 2022).

2.4 Soil Pollution Has Health Impacts

Soil pollution poses significant risks to human health, primarily through exposure to contamination present in the soil. Health effects vary depending on the type and concentration of contaminants, the route of exposure (such as inhalation, ingestion, or dermal absorption), and individual vulnerability (Table 2.2).

The spread of soil pollution also leads contamination of food, as crops grown in polluted soil absorb and accumulate harmful substances. This not only affects food safety but can also lead to the ingestion of contaminants, further exacerbating health risks. In addition, contaminated soil also impacts the health of livestock and grazing wildlife. Pollutants accumulate in their bodies, posing a risk to those who consume their meat or products. The long-term effects of soil pollution on ecosystems and human health highlight the importance of proper waste management and ensuring the health of our soil for future generations (Ferreira et al. 2022).

Table 2.2 Some soil pollution has health impacts

Number	Soil Pollutant's health impact	Description	References
1	Acute	Short-term exposure to high levels of soil pollutants leads to immediate health issues such as intoxication or diarrhea.	Egendorf et al. (2020)
2	Chronic	Long-term exposure results in more severe health problems, including various forms of cancer.	Maddela et al. (2022)
3	Organ Damage	Certain soil contaminants, like heavy metals, cause damage to vital organs such as liver, kidneys, and the nervous system.	Priyadarshanee et al. (2022)
4	Developmental Disorders	Exposure to polluted soil during critical development periods leads to developmental disorders in children.	Kao et al. (2021)
5	Neurological Disorders	Neurotoxic contaminants in soil affect the brain and nervous system, leading to neurological disorders.	Iqubal et al. (2020)

2.5 Soil/Sediment Pollutant Types

Soil and sediment pollutants encompass a wide array of substances that can have detrimental effects on ecosystems and human health. These pollutants are typically categorized based on their origin and chemical characteristics (Table 2.3).

The impact of pollution is profound, affecting not only the immediate environment but also broader ecological systems. They alter the soil chemistry, reduce fertility, and contaminate food chains. Sediment pollution smother aquatic habitats and affect the health of coral reefs and coastal ecosystems. Addressing these pollutants requires comprehensive environmental policies, effective waste management, and public education on sustainable practices to prevent further contamination and promote remediation efforts. Implementing these strategies is crucial in ensuring the long-term health and sustainability of our natural ecosystems. By reducing the input of pollutants into our waterways, we can help to mitigate the negative effects of eutrophication and harmful algal blooms. Through collaborative efforts between government agencies, industries, and communities, we can work towards a cleaner and healthier environment for all. It is essential to take action now to protect our ecosystems for future generations (Namdari et al. 2017; Rastegari Mehr et al. 2016).

2.6 Petroleum Contamination in Soil/Sediment

Petroleum hydrocarbons, which severely contaminate the environment, are mostly produced by human industrial activities. Persistent pollution is despised by all living species across a wide range of environments and significantly diminishes

Table 2.3 The primary types of soil and sediment pollutants

Number	Pollutant types	Description	References
1	Organic	Pesticides, herbicides, and industrial chemicals like polychlorinated biphenyls (PCBs) and polycyclic aromatic hydrocarbons (PAHs). Origins include agricultural runoff, industrial waste, and improper disposal of household chemicals.	Bolan et al. (2023)
2	Inorganic	Heavy metals such as lead, mercury, cadmium, and arsenic, which can be released into the environment through mining, industrial processes, and the use of contaminated water for irrigation.	Kypritidou et al. (2024)
3	Pathogens	Soil and sediments become polluted with bacteria, viruses, and parasites from sewage spills, animal waste, and contaminated water sources. These pathogens pose direct health risks to humans and wildlife.	Samaddar et al. (2021)
4	Radioactive	Nuclear industry activities, medical facilities, and natural sources introduce radioactive materials into soil and sediment, leading to long-term environmental and health hazards.	Devi and Chauhan (2020)
5	Nutrients	Excessive amounts of nutrients, particularly nitrogen and phosphorus from fertilizers, lead to soil degradation and water quality issues, including eutrophication and harmful algal blooms.	Schulte-Uebbing et al. (2022) and Brownlie et al. (2021)

environmental variety. Additionally, the presence of petroleum hydrocarbons contaminates the ecosystem, making it unstable and dangerously susceptible to biotic and abiotic complexes' stresses and regular changes (Ambaye et al. 2022). The constant release of hydrocarbons into the environment poses a threat to biodiversity and ecosystem stability. Organisms struggle to adapt to or overcome the toxic effects of petroleum pollution, leading to a decline in population numbers and species diversity (Kuppusamy et al. 2020a). Industries must implement effective measures to reduce hydrocarbon emissions and prevent further adverse effects on the environment.

2.7 Risks of Petroleum Contamination

Petroleum hydrocarbons are organic compounds derived from crude oil and its byproducts. They impact ecology and life on Earth. Damage to structure and function of DNA is a consequence which results in mutations and cancer. Cellular membrane disruption, which impairs vital cell functions and processes and affects organism health, rate of reproduction, and sustainability during their life cycle, is an adverse result of petroleum hydrocarbon pollution. It is believed that environmental contamination with petroleum hydrocarbons poses a hazardous and carcinogenic risk to both people and animals (Gao et al. 2022).

Pollutants with petroleum hydrocarbons are dangerous compounds derived from crude oil and its byproducts. They have the potential to seriously harm human health and well-being, as well as the biodiversity and health of other living things in the ecosystem. Petroleum-hydrocarbon pollutants have an indirect effect on people by upsetting the stability and balance of the environment, and have a direct impact through ingestion, inhalation, or skin contact. It is essential to manage the risk of contamination in order to improve environmental remediation and avoid and reduce pollution from petroleum hydrocarbons (Kuppusamy et al. 2020b).

Both humans and animals develop cancer as a result of petroleum hydrocarbon exposure. Additionally, it reduces the species diversity regeneration. The enormity of the issue warrants mitigation efforts (Kuppusamy et al. 2020b; Ambaye et al. 2022). For endangered communities that struggle with hazardous environmental disorders caused by petroleum contamination, remediation is essential for ecosystem sustainability.

To clean up soil and sediment contaminated with petroleum hydrocarbons, certain techniques have been designed and put into action. These techniques include landfill, which isolates contaminated soil in a designated area or controlled burial site; fixation, which immobilizes the pollutants by adding chemical or physical agents; bioremediation, which uses microorganisms to degrade the pollutants; and thermal processing, which uses heat to evaporate, degrade, decontaminate, or incinerate the organic contaminants. Each technique has its advantages and drawbacks, and the most appropriate method often depends on the specific characteristics of the contamination site. Landfilling is a common but costly option, while fixation can be effective in immobilizing contaminants but may not fully eliminate them. Bioremediation is a more sustainable option that is effective over time, while thermal processing is often a quicker but more energy-intensive solution. Overall, a combination of these techniques may be necessary to effectively remediate soil and sediment contaminated with petroleum hydrocarbons (Parnian et al. 2022a, b).

2.8 The Potential Source of Petroleum Pollution in Soil/Sediment

The primary cause of petroleum hydrocarbon pollution in the environment (soil, sediment, water, and air) is human activity, including oil production, transportation, storage, and consumption (Parnian et al. 2020). These activities face the risk of causing spills, leakage in storage tanks and pipelines, and emissions of oil and its derivatives into the air, water, sediment, and soil. The well-being and sustainability of ecosystems, as well as the creatures who rely on them, are significantly jeopardized by these contaminants. In addition to decreasing environmental diversity and resilience, petroleum contaminants make ecosystems more vulnerable to other stresses, like climate change (Parnian et al. 2021). Human industrial and recreational activities are to blame for contamination caused by petroleum hydrocarbons. Extreme environmental contamination is a result of the exploration, production, storage, refinement, and transportation of oil and gas. There are several environmental hazards connected to oil and gas-related economic activities that generate waste. The soil, air, and aquatic ecosystems are vulnerable to dangerous threats (Gao et al. 2022). Oil spills, leakages, and improper disposal of waste products are common occurrences in the oil and gas industry, leading to severe pollution in the surrounding environment. These pollutants not only harm the flora and fauna in the affected areas but also pose a significant risk to human health through the contamination of food and water sources. The long-term consequences of this environmental degradation are devastating, impacting biodiversity and ecosystem functioning for generations to come. Policymakers and industry stakeholders must take immediate action to mitigate these risks and prevent further damage to our planet.

2.9 Petroleum Pollution Control

Petroleum hydrocarbon pollution is a byproduct of human activity in the industry, services, and agricultural sectors. This pollution has an impact on ecosystem productivity and stability. As they supply resources and services, stable ecosystems are crucial to human populations and economic health (Parnian et al. 2023). One of the primary causes of ecosystem disruption and degradation that results from human interference is petroleum hydrocarbon contamination. The oil and gas industry, which includes the extraction, processing, refining, storage, and transportation of petroleum products, is one of the primary causes of pollution in the environment. Workers, managers, designers, and/or managers frequently engage in these activities with ignorance and carelessness, which results in spills and accidents that pose significant environmental concerns. Contamination impacts the biodiversity and functional parts of the ecosystems that depend on the contaminated soil, air, and water, as well as their quality and health. Contamination of petroleum hydrocarbons is a serious environmental issue that requires management both before and after it occurs. Stringent laws, regulations, and guidelines must be put in place to guarantee the safe operation of petroleum-related operations and prevent contamination. This entails offering thorough instruction and training to everyone working in the field, underscoring the value of risk management and environmental preservation. Regular evaluations and inspections should be carried out to find any potential hazards or systemic flaws so that prompt action can be taken to avoid incidents. However, in the unfortunate occurrence of contamination, quick action and thoughtful management plans need to be taken to lessen the harm to the environment and rebuild ecosystems and surrounding areas. This could entail applying remediation procedures to restore/recover the polluted soil, along with professional equipment/acting groups to control/manage/remediate the petroleum spill (Alexandersen et al. 2014). Furthermore, communication with local authorities and communities is crucial to keep them informed and involved in the cleanup process. It is important to work together towards the common goal of restoring the affected area and minimizing the impact on the environment. By taking swift and decisive action, we can prevent further damage and ensure a more sustainable future for all.

2.10 Conclusion

The chapter has underscored the criticality of understanding the types of soil/sediment pollutants, including organic and inorganic compounds, pathogens, and radioactive materials, and their pathways through the environment. The health implications of soil pollution are significant, with soil contaminants posing risks of acute and chronic diseases, organ damage, and developmental and neurological disorders. In conclusion, this chapter has illuminated the pervasive issue of soil petroleum pollution, its origins, and its profound impact on the environment and human health. Soil petroleum pollution, a consequence of industrialization and modern human activities, manifests in various forms and affects soil and sediment. The latter contributes to the degradation of ecosystems and the quality of life for both humans and animals. The chapter advocates for a proactive stance on pollution management, emphasizing the necessity of remediation strategies, stringent regulations, and public education to mitigate adverse effects. Ultimately, the chapter calls for a collective commitment to sustainable practices and policies that prioritize environmental health, ensuring the preservation of natural resources and the well-being of all life forms. It is a call to action for responsible stewardship of the planet, safeguarding it against the relentless threat of soil petroleum pollution for generations to come.

2.11 Suggestions

The following suggestions are made:

I. **Endangered societies should stay informed**: Education about petroleum-based pollutants and their sources, in parallel with understanding the impact of soil petroleum pollution and its impact on the environment and health, can empower individuals/communities/policymakers, and governments to make informed decisions and advocate for change.

II. **Adopt Sustainable Practices**: Implement eco-friendly habits in society and also encourage personal actions that contribute to the collective effort to reduce soil petroleum pollution.

III. **Engage in Community Action**: Participate in community initiatives aimed at soil pollution control and ecosystem protection, which can include joining clean-up drives, supporting local environmental policies, and promoting awareness.

References

Alexandersen DK, Headley T, Prigent S, Mahmutoglu I (2014) Ex-situ bioremediation of hydrocarbon contaminated soil-an example from Oman. In: SPE middle east health, safety, security, and environment conference and exhibition. SPE, p SPE-170411

Ambaye TG, Chebbi A, Formicola F, Prasad S, Gomez FH, Franzetti A, Vaccari M (2022) Remediation of soil polluted with petroleum hydrocarbons and its reuse for agriculture: recent progress, challenges, and perspectives. Chemosphere 293:133572

Bolan S, Padhye LP, Mulligan CN, Alonso ER, Saint-Fort R, Jasemizad T, Wang C, Zhang T, Rinklebe J, Wang H, Siddique KH (2023) Surfactant-enhanced mobilization of persistent organic pollutants: potential for soil and sediment remediation and unintended consequences. J Hazard Mater 443:130189

Brownlie WJ, Sutton MA, Reay DS, Heal KV, Hermann L, Kabbe C, Spears BM (2021) Global actions for a sustainable phosphorus future. Nat Food 2(2):71–74

Devi V, Chauhan RP (2020) Estimation of natural radionuclide and exhalation rates of environmental radioactive pollutants from the soil of northern India. Nucl Eng Technol 52(6):1289–1296

Egendorf SP, Gailey AD, Schachter AE, Mielke HW (2020) Soil toxicants that potentially affect children's health. Curr Probl Pediatr Adolesc Health Care 50(1):100741

Ferreira SL, da Silva Junior JB, dos Santos IF, de Oliveira OM, Cerda V, Queiroz AF (2022) Use of pollution indices and ecological risk in the assessment of contamination from chemical elements in soils and sediments–practical aspects. Trends Environ Anal Chem 35:e00169

Gao H, Wu M, Liu H, Xu Y, Liu Z (2022) Effect of petroleum hydrocarbon pollution levels on the soil microecosystem and ecological function. Environ Pollut 293:118511

Iqubal A, Ahmed M, Ahmad S, Sahoo CR, Iqubal MK, Haque SE (2020) Environmental neurotoxic pollutants. Environ Sci Pollut Res 27:41175–41198

Kao CS, Wang YL, Chuang TW, Jiang CB, Hsi HC, Liao KW, Chien LC (2021) Effects of soil lead exposure and land use characteristics on neurodevelopment among children under 3 years of age in northern Taiwan. Environ Pollut 286:117288

Kumar R, Qureshi M, Vishwakarma DK, Al-Ansari N, Kuriqi A, Elbeltagi A, Saraswat A (2022) A review on emerging water contaminants and the application of sustainable removal technologies. Case Stud Chem Environ Eng 6:100219

Kuppusamy S, Maddela NR, Megharaj M, Venkateswarlu K, Kuppusamy S, Maddela NR, Megharaj M, Venkateswarlu K (2020a) Ecological impacts of total petroleum hydrocarbons. In: Total petroleum hydrocarbons: environmental fate, toxicity, and remediation. Springer, Cham, pp 95–138

Kuppusamy S, Maddela NR, Megharaj M, Venkateswarlu K, Kuppusamy S, Maddela NR, Megharaj M, Venkateswarlu K (2020b) Impact of total petroleum hydrocarbons on human health. In: Total petroleum hydrocarbons: environmental fate, toxicity, and remediation. Springer, Cham, pp 139–165

Kypritidou Z, Kelepertzis E, Kritikos I, Kapaj E, Skoulika I, Kostakis M, Vassilakis E, Karavoltsos S, Boeckx P, Matiatos I (2024) Geochemistry and origin of inorganic contaminants in soil, river sediment and surface water in a heavily urbanized river basin. Sci Total Environ 927:172250

Maddela NR, Ramakrishnan B, Kakarla D, Venkateswarlu K, Megharaj M (2022) Major contaminants of emerging concern in soils: a perspective on potential health risks. RSC Adv 12(20):12396–12415

Mushtaq Z, Muzammil A, Bellitürk K, Anwar W, Akhter A, Khan HAA, Parnian A, Rahman SU (2024) Role of rhizobacteria in phytoremediation of heavy metal. In: Heavy metal remediation: sustainable nexus approach. Springer Nature Switzerland, Cham, pp 183–211

Namdari A, Roayaei E, Jaafarzadeh N, Shariat M (2017) Measurement the significant heavy metals of petroleum desalination influent in an Iranian on-shore desalination plant. Pet Sci Technol 35(7):681–686

Parnian A, Parnian A, Pirasteh-Anosheh H, Parvizi H (2020) Potential of a pyrolyzed sugarcane industry waste for spillage oil absorption. J Nat Resour Conserv Manag 1(2):163–168

Parnian A, Furze JN, Parnian A, Mayad EH (2021) Water purification plantations for oil and gas industries in Iran. Environ Sci Pollut Res 28(45):64193–64198

Parnian A, Parnian A, Pirasteh-Anosheh H, Furze JN, Prasad MNV, Race M, Hulisz P, Ferraro A (2022a) Full-scale bioremediation of petroleum-contaminated soils via integration of co-composting. J Soils Sediments 22(8):2209–2218

Parnian A, Furze JN, Parnian A (2022b) Drilling waste management based on new methods of bioremediation and solar desalination. In: Earth systems protection and sustainability, vol 1. Springer International Publishing, Cham, pp 79–97

Parnian A, Mahbod M, Prasad MNV (2023) Microplastics remediation–possible perspectives for mitigating saline environments. In: Microplastics in the ecosphere: air, water, soil, and food. John Wiley & Sons, Inc, Hoboken, pp 465–476

Pirasteh-Anosheh H, Parnian A, Spasiano D, Race M, Ashraf M (2021) Haloculture: a system to mitigate the negative impacts of pandemics on the environment, society and economy, emphasizing COVID-19. Environ Res 198:111228

Priyadarshanee M, Mahto U, Das S (2022) Mechanism of toxicity and adverse health effects of environmental pollutants. In: Microbial biodegradation and bioremediation. Elsevier, Amsterdam, pp 33–53

Rastegari Mehr M, Keshavarzi B, Moore F, Sacchi E, Lahijanzadeh AR, Eydivand S, Jaafarzadeh N, Naserian S, Setti M, Rostami S (2016) Contamination level and human health hazard assessment of heavy metals and polycyclic aromatic hydrocarbons (PAHs) in street dust deposited in Mahshahr, southwest of Iran. Hum Ecol Risk Assess Int J 22(8):1726–1748

Samaddar S, Karp DS, Schmidt R, Devarajan N, McGarvey JA, Pires AF, Scow K (2021) Role of soil in the regulation of human and plant pathogens: soils' contributions to people. Philos Trans R Soc B 376(1834):20200179

Schulte-Uebbing LF, Beusen AH, Bouwman AF, De Vries W (2022) From planetary to regional boundaries for agricultural nitrogen pollution. Nature 610(7932):507–512

Sofia D, Gioiella F, Lotrecchiano N, Giuliano A (2020) Mitigation strategies for reducing air pollution. Environ Sci Pollut Res 27(16):19226–19235

Ukaogo PO, Ewuzie U, Onwuka CV (2020) Environmental pollution: causes, effects, and the remedies. In: Microorganisms for sustainable environment and health. Elsevier, Amsterdam, pp 419–429

Chapter 3
Contaminated Soil and Sediment Bioremediation Techniques

Lander de Jesus Alves, Fábio Carvalho Nunes, Maria Clara Bessa Souza, Suzana Rodrigues de Moura, João Carlos Teixeira Dias, Rachel Passos Rezende, Eduardo Gross, and Majeti Narasimha Vara Prasad

Abstract The contamination of sediments and soils by petroleum and its derivatives represents a significant global challenge. Conventional physical and chemical remediation methods often not enough to overcome the challenges posed by contamination, because the approaches are expensive, ineffective and environmental risks. As a result, many scientists focus on bioremediation techniques, which are cheaper, easier to handle and ecological. The bioremediation of soils and sediments mineralize (biodegradation) and transforming organic petroleum contaminants into inorganic compounds, carbon dioxide, and water. In addition, phytoremediation strategies can immobilize, extract, transform, accumulate, stabilize contaminants or to mediate a slower and safer release into the environment. This chapter discusses techniques for bioremediation of petroleum contaminated soils and sediments, emphasizing the advantages, disadvantages and indicating possibilities for future research.

Keywords Petroleum · Phytoremediation · Petroleum · Organic · Future

---

L. de Jesus Alves (✉) · M. C. Bessa Souza · S. Rodrigues de Moura · J. C. T. Dias
R. P. Rezende · E. Gross
Postgraduate Program in Biology and Biotechnology of Microorganisms, State University of Santa Cruz (UESC), Ilhéus, Bahia, Brazil
e-mail: ljalves@uesc.br; mcbsouza.ppgbbm@uesc.br; srmoura@uesc.br; jctdias@uesc.br; rachel@uesc.br; egross@uesc.br

F. C. Nunes
Academic Department, Federal Institute Baiano (IF BAIANO), Alagoinhas and Catu Bahia, Brazil
e-mail: fabio.nunes@ifbaiano.edu.br

M. N. V. Prasad
Department of Plant Sciences, School of Life Sciences, University of Hyderabad (An Institution of Eminence), Hyderabad, Telangana, India
e-mail: mnvsl@uohyd.ac.in

A. Parnian et al. (eds.), *Bioremediation of Petroleum Contaminated Soil and Sediment*, Environmental Contamination Remediation and Management,
https://doi.org/10.1007/978-3-031-95524-2_3

3.1 Introduction

Although international organizations and corporations are making considerable efforts to make the world's energy matrix free of fossil fuels (Prasad et al. 2023), where there has been a considerable increase in the global production of energy produced by green sources, in the last decade there has also been an increase in fossil fuel consumption (Nunes et al. 2024; Ritchie et al. 2022). This situation is possibly due to an increase in demand for energy in many countries and is linked to an increase in population and consumption of industrialized products, negatively affecting the energy transition process and keeping the demand for fossil fuels active (Nunes et al. 2024).

Soil and sediment contamination caused by polycyclic aromatic hydrocarbons (PAHs) is a global problem, as they are toxic, genotoxic, mutagenic, and potentially carcinogenic organic pollutants (Tsibart and Gennadiev 2013). Contamination typically occurs when there are spills in the extracting, producing, storing, transporting, and refining oil, but it can also be caused by aircraft, freighters, cars and diesel or gasoline engines (Ambaye et al. 2023) (Fig. 3.1).

Contamination by PAHs can have several negative impacts on the soil/sediment system, for example: (a) it affects the circulation of air and solutions; (b) alters ionic exchanges (cationic and anionic); (c) harms the dynamics of flora, especially because it hinders photosynthesis, the assimilation of nutrients and contaminates with heavy metals; (d) harms fauna, which can die due to intoxication and/or difficulty obtaining oxygen.

PAHs can harm soil fertility, as they affect the balance between the solid, liquid, and gaseous phases (Nunes et al. 2023) (Fig. 3.2), hindering several environmental functions performed by soils, some of them fundamental for the balance of different ecosystems (Nunes et al. 2020). Furthermore, it can compromise the quality of surface and groundwater, plant growth, yield, grain quality and even contaminate food, which could affect the web of life and human health (Ambaye et al. 2022a, b).

Conventional physical and chemical remediation approaches are not sufficient to overcome the challenges posed by PAHs contamination, as they are expensive, ineffective and/or harmful to the environment. Therefore, scientists have dedicated themselves to studying bioremediation techniques, which is a recovery process that normally uses bacteria, fungi and algae to degrade pollutants present in soils and sediments.

Bioremediation techniques for oil-contaminated soils/sediments are cheaper, easier to handle, environmentally friendly and aim to mineralize organic contaminants (biodegradation), transforming them into inorganic compounds, carbon dioxide and water. Additionally, phytoremediation strategies can be used to immobilize, extract, transform, accumulate, stabilize contaminants, or mediate a slower and safer release into the environment. This chapter discusses bioremediation techniques for petroil-contaminated soils and sediments, emphasizing the advantages, disadvantages and indicating possibilities for future research.

Fig. 3.1 Soils/sediments contaminated by petrol in environment disaster on the east coast of Brazil (Beach in the city of Ilhéus, Bahia in 2019. (**a**) Schematic representation of an environment contaminated by petroleum aromatic hydrocarbons; (**b**) oil on the beach; (**c**) getting study material; (**d**, **e**) detail of oil samples). (Source: the authors)

3.2 Soil/Sediment Contamination with PAHs

PAH contamination occurs when petroleum products such as crude oil, diesel or gasoline leak or spill onto soils and sediments. Contamination can occur due to disasters on oil platforms, pipeline ruptures, leaks on ships, gas stations, industrial and commercial facilities or even in homes. PAHs can penetrate soils/sediments, causing damage to the environment and posing a potential risk to human health.

The negative impacts of oil contamination include (a) compaction and sealing of the soil/sediment, hindering the development of biota and plant growth; (b) change in soil/sediment composition, limiting the capacity to retain water, solutions, and nutrients; (c) contamination of surface and underground waters, posing a risk to

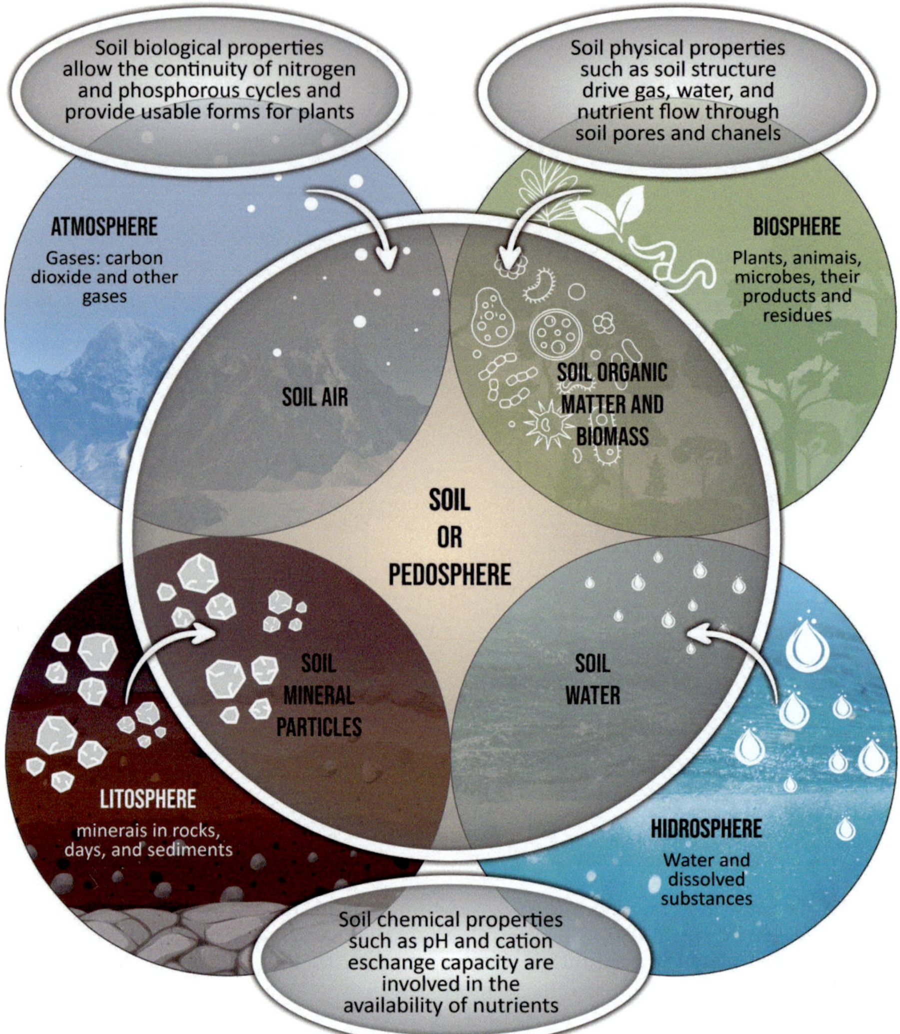

Fig. 3.2 Interactions between terrestrial systems and soils responsible for their fertility. (Source: Adapted from Weil and Brady (2017) by Nunes et al. (2023))

animals, plants, and humans; (d) air pollution, posing a risk to fauna, plants, and humans.

There are several methods to recover petroil-contaminated soils/sediments, including scraping, washing, or replacing soils/sediments, in addition to bioremediation and consortium of different techniques. Such techniques will be discussed in later items, but the focus of this chapter will be bioremediation.

3.2.1 PAHs: Properties and Sources

PAHs are a group of ubiquitous organic pollutants with at least two aromatic rings (Fig. 3.3), toxic and some of these substances are mutagenic, carcinogenic, and teratogenic (Tsibart and Gennadiev 2013). Due to their chemical structure, PAHs have hydrophobic properties (they are poorly soluble in water), which is why they accumulate on the surface of solid materials such as soil, sediments, and solid waste.

Various conventional techniques have been employed for soil/sediment remediation, including electrokinetic remediation, soil washing, stabilization, solidification, washing, chemical reduction, oxidation, and landfilling, but most of them cause problems of high secondary pollution, soil erosion and infertility (Ambaye et al.

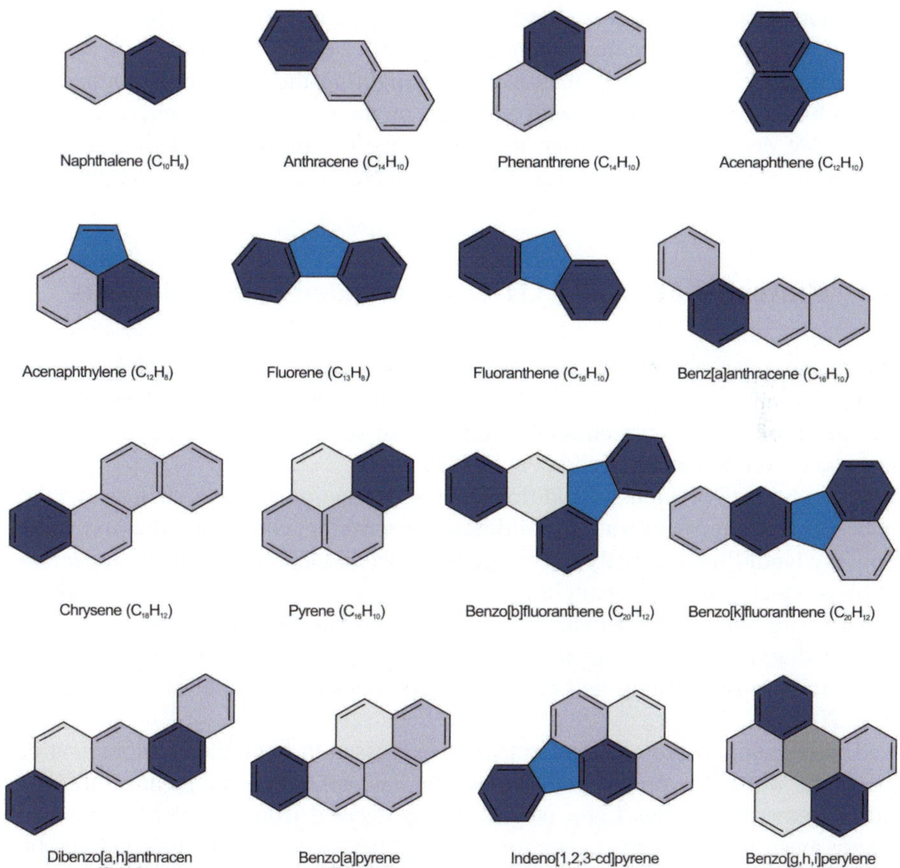

Fig. 3.3 Structure and chemical formula of polycyclic aromatic hydrocarbons (PAHs) considered priority pollutants by United States Environmental Protection Agency (USEPA). (Source: Adapted from Sayara and Sanchez (2020))

2023; Sayara and Sanchez 2020). Furthermore, they require large investments to be applied on a large scale (Xu et al. 2012).

Soil is a complex ecological system, representing a vital component of terrestrial ecosystems and fundamental for nutritional and food security (Nunes et al. 2020). However, soil is subject to degradation or decline in quality, due to different anthropogenic activities (Chen et al. 2015) and contaminants, including PAHs. Soil contamination concerns the world and PAHs have been registered as pollutants of priority importance, as they are hydrophobic, recalcitrant, genotoxic, mutagenic, potentially carcinogenic (Fu et al. 2012; Sayara and Sanchez 2020) and ubiquitous in atmospheric systems, hydrosphere, pedosphere, and biosphere.

Great efforts have been directed around the world to remediate PAHs in soils/sediments and other environmental systems. From 1995 onwards, the growth of research on the subject became evident and continued to expand around the world, with China, USA, India, and Canada dominating the number of publications.

Although there has been a significant increase in the number of research in the field of bioremediation in this century, there is still a need for expansion in developing countries. In this sense, it is worth highlighting the considerable number of research carried out in Brazil, with emphasis on the prospecting of microorganisms capable of degrading recalcitrant molecules, for use in bioremediation processes (Buzanello et al. 2014; Maciel et al. 2007; Matos Neto et al. 2023; Santos et al. 2014).

3.3 Bioremediation of PAH-Contaminated Soils/Sediments

Human progress, especially since the industrial revolution, has resulted in significant environmental damage in the form of pollution. Recalcitrant molecules with a high level of toxicity can currently be found in soils and sediments, in bodies of water and even in the air (Kalia et al. 2022).

As mentioned previously, traditional decontamination methods such as incineration, soil washing, stabilization, solidification, washing, chemical reduction, oxidation, and landfilling are costly and can generate secondary risks due to the appearance of by-products that can remain in the environment and cause problems of high secondary pollution (Ambaye et al. 2023; Sayara and Sanchez 2020). In this scenario, bioremediation emerges as a possible solution, helping to reduce costs and increase the efficiency of cleaning and remediation processes (Prasad et al. 2021; Kumar et al. 2018).

The meaning of bioremediation can be expressed through the etymological origin of the word. The prefix 'bio' comes from ancient Greek and means "life". The suffix 'remediation' has Latin origins and is derived from the word 'remediare' whose meaning is to correct, cure, or resolve. Therefore, it can be inferred that bioremediation is a process that uses living organisms with the aim of correcting the environment. Biological metabolism is the central point in this process, as it is capable of degrading complex organic contaminants into simple molecules such as CO_2,

H_2O and inorganic compounds, or less complex molecules with little or no potential to generate environmental damage (Shivalkar et al. 2021; Speight and Arjoon 2012).

In general, there is no standard bioremediation methodology to be used in all cases of contamination. Although there are different bioremediation techniques, impacted/polluted environments must be carefully evaluated to designate the best strategy to be employed. The choice of biological remediation technique depends on the assessment of the nature of the pollutant in question, pollution intensity, type of environment, possible recovery costs, time, and the environmental legislation in force at the location (Azubuike et al. 2016). Other criteria may also be relevant when choosing the bioremediation technique, including nutritional concentrations of the affected area, temperature, pH, and other biotic and abiotic factors (Azubuike et al. 2016).

It is worth noting that given the existing diversity of bioremediation techniques, there is an abundance of studies regarding their use in cases of contamination by hydrocarbons. This fact is related to: (1) frequent occurrence of spillage events/contamination of soil and water with this pollutant; (2) great adaptation of this technique compared to others used to treat other contaminants; (3) the great dependence on oil and derivatives as a source of energy and raw material (Azubuike et al. 2016).

3.3.1 Organisms in the Degradation of Environmental Contaminants

Numerous species are capable of degrading aromatic compounds, mainly low molecular weight PAH. However, some xenobiotic molecules of organic origin, such as PAH and other fractions, are resistant to degradation (Kalia et al. 2022). In this case, microorganisms stand out because they have a high capacity to remove highly recalcitrant xenobiotics (Hossain et al. 2022).

Achromobacter, *Acinetobacter*, *Alcaligenes*, *Arthrobacter*, *Aspergillus*, *Bacillus*, *Corynebacterium*, *Flavobacterium*, *Fusarium*, *Micrococcus*, *Mucor*, *Mycobacterium*, *Nitzschia*, *Nocardia*, *Penicillium*, *Pseudomonas*, *Rhodococcus*, *Rhodotorula*, *Scenedesmus*, *Sporobolomyces*, *Yarrowia*, *Alcanivorax*, *Enterobacter*, *Gordonia*, *Staphylococcus* and *Streptomyces*, are some of the genera of microorganisms most cited as agents in the degradation of hydrocarbons (Kalia et al. 2022).

Considering the specific characteristics of each microorganism, as well as their adaptations to the environments, they are often adapted to the environmental conditions of their places of origin. For example, in environments with high salinity, organisms that are more adapted to this condition are found. In turn, for greater success in a bioremediation strategy in soils with high salinity, it is necessary to search in saline environments for microorganisms that are resistant to salinity (Kalia et al. 2022).

Additionally, it is necessary to highlight that microbial consortia (microbial cultures containing several microorganisms), when compared to microbial isolates, present greater adaptability to different environmental conditions and, consequently, greater potential for contaminant degradation (Kalia et al. 2022; Santos et al. 2014).

As an example of efforts to isolate microorganisms capable of integrating microbial consortia with greater efficiency in oil degradation, we can highlight the effort of researchers from the State University of Santa Cruz (UESC) in Brazil, which has maintained consortiums for more than 20 years oil-degrading microbials under laboratory conditions (Maciel et al. 2007). Biological samples collected from a landfarm located in areas of the first oil refinery in Brazil, the Landulfo Alves refinery (currently called Mataripe Refinery, located in the municipality of São Francisco do Conde, in the Recôncavo Baiano region—State of Bahia), are maintained under constant agitation and periodic feeding with oil, in a minimal mineral medium, with oil as the only carbon source (Matos Neto et al. 2023) (Fig. 3.4). It is also worth highlighting efforts to prospect microorganisms from extreme locations, capable of degrading oil, where researchers from UESC collected samples from the ocean floor of the Antarctic continent and, since then, prospecting and identification of microorganisms efficient in degrading oil have been carried out (Fig. 3.5). The sample collection campaigns in Antarctica were developed within the scope of the Brazilian Antarctic Program (PROANTAR), which seeks to invest in scientific research on the Antarctic continent (MCTI 2024; PROANTAR 2024).

The selective cultivation of microorganisms carried out at UESC are important examples of efforts to select organisms that are efficient in the control and

Fig. 3.4 Maintenance and use of oil-degrading microorganisms in the laboratory. (**a**) Maintenance of microbial consortium in minimal mineral medium under constant agitation with periodic supplementation with petroleum; (**b**) example of microbial diversity from a microbial consortium created with soil samples from Landfarm of oily sludge; (**c**) Growth of microorganism isolated from microbial consortium, showing signs of oil degradation; (**d**) Production of biosurfactant in the laboratory; (**e**) Carrying out an emulsification test to screen biosurfactant production

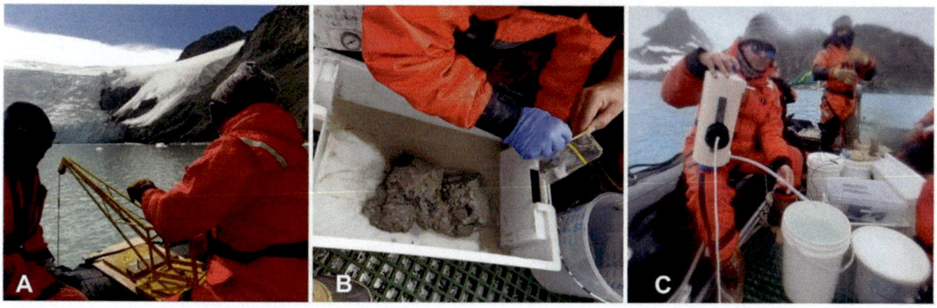

Fig. 3.5 Team from the State University of Santa Cruz (UESC) in a sample collection campaign in Martel Creek on the Antarctic continent. (**a**) Dredging substrate from the ocean floor; (**b**) Oceanic substrate; (**c**) Seawater sampling

bioremediation of harmful compounds. In this context, the identification or even the creation of selective pressure environments capable of selecting microorganisms are strategies that have shown success in identifying microorganisms, creating hydrocarbon-degrading enzymes and isolating biosurfactants (Maciel et al. 2007; Matos Neto et al. 2023; Santos et al. 2014).

In some cases, it is worth considering that the choice of microbial inoculum to be used in the soil may encounter some problems in terms of colonization, in cases where the pre-existing microbial community in the soil is well adapted and does not cover the growth of the inoculum applied. As previously mentioned, in such cases it is essential to consider the selection of microorganisms adapted from the microbiome isolated from contaminated soil that has been subjected to prolonged exposure to pollutants (Baldan et al. 2015), as well as the use of innovative tools for identifying microorganisms such as metagenomics (Sierra-García et al. 2014).

The degradation of hydrocarbons encompasses the action of several enzymes including mono-oxygenases (degrading alkanes, alkenes and cycloalkenes), dioxygenases (degrading alkanes), cytochrome P450 enzymes (eukaryotic and prokaryotic) (acting on C10 alkane substrates -C16, fatty acids and alkanes in cycles), dehydrogenases, laccases, aldehyde dehydrogenases and catechol 2,3-dioxygenases (act on polycyclic aromatic hydrocarbons), manganese peroxidases (act on benzo(a) pyrenes), and alkane monooxygenases (act in n-hexadecanes) (Kalia et al. 2022).

Finally, when the subject is biodegradation of contaminants present in the soil, it is also necessary to consider that the restriction on mass transfer and the bioavailability of these contaminants are barriers to the access of microorganisms for metabolism, directly causing a reduction in degradation. In this sense, biosurfactants can be used with the aim of increasing the bioavailability of these molecules, increasing access to microorganisms and their respective enzymes, accelerating biodegradation (Kalia et al. 2022).

3.3.2 Production of Biosurfactants

The degradation of hydrocarbons involves the action of several enzymes, including mono-oxygenases (degrading alkanes, alkenes and cycloalkenes), dioxygenases (degrading alkanes), cytochrome P450 enzymes (eukaryotic and prokaryotic, which act on C10-C16 alkane substrates, fatty acids and cycloalkenes), dehydrogenases, laccases, aldehyde dehydrogenases and catechol 2,3-dioxygenases (act on polycyclic aromatic hydrocarbons), manganese peroxidases (act on bezo(a)pyrenes), and alkane monooxygenases (act in n-hexadecanes) (Kalia et al. 2022).

When it comes to the biodegradation of contaminants present in the soil, the restriction on mass transfer and the bioavailability of these contaminants are barriers to the access of microorganisms for metabolism, directly causing a reduction in degradation. In this context, biosurfactants can be used with the aim of increasing the bioavailability of these molecules, increasing access to microorganisms and their respective enzymes, accelerating biodegradation (Kalia et al. 2022).

Surfactants of microbial origin are produced when bacteria and fungi are exposed to specific conditions. For example, when there are hydrophobic carbon sources in the substrate, the production of biosurfactants favors the growth of microorganisms, reducing the surface tension at the phase boundary and leaving the substrate available. With a polar head and non-polar tail, surfactants are surface-active agents capable of altering the surface and interface characteristics of the medium into which they are inserted (Gudiña et al. 2013).

Biosurfactants are classified according to their chemical structure and microbial origin, they are divided into: lipopeptides and lipoproteins, glycolipids, phospholipids, and polymeric substances. Lipopeptides are the most promising surface-active agents, they work efficiently in oil recovery and have potential therapeutic applications, such as antitumor and antiviral (Abdel-Mawgoud et al. 2008). And they also have other important properties, such as demulsifying and antimicrobial agents, they have significant anti-biofilm and anti-adhesive actions (Goswami and Deka 2019).

The use of biosurfactants to degrade petroleum and its derivatives is the most current and environmentally beneficial strategy (Patel et al. 2019). Due to their amphiphilic characteristic, surfactants act as emulsifying agents, reducing the surface tension of interfaces and forming micelles that will help remove hydrophobic contaminants. These compounds are classified by their chemical composition, molecular weight and considered as multifunctional molecules of the twenty-first century, as, in addition to having ecological importance, they have biotechnological importance in sectors such as the clinical and industrial sectors (Santos et al. 2016).

In industry, for example, biosurfactants can be efficiently used in the formulation of cosmetic products (Rodrigues 2015). Due to their chemical nature, these compounds allow the simultaneous use of fat-soluble and water-soluble substances, and can be used in the processes of wetting, emulsification, demulsification, detergency, foaming, dispersion, solubilization of hydrophobic substances (Banat et al. 2010). Furthermore, when compared to synthetic surfactants, biosurfactants are economically accessible substances and a sustainable alternative (Gudiña et al. 2013; Vecino et al. 2017).

In the agricultural sector, some studies relate the use of biosurfactants to the removal of heavy metals in polluted effluents (Hidayati et al. 2014; Sarubbo et al. 2015). The presence of heavy metals is also a problem in food production, causing serious damage to human health. In this context, the use of biosurfactants can help eliminate the absorption of these compounds in vegetables (Jimoh and Lin 2019). In agriculture, since oils and fats are routinely added to animal feed, biosurfactants can be effective in helping to digest nutrients and, thus, avoiding losses in feed conversion and growth rate (Naughton et al. 2019).

With the increase in antibiotic resistance in the last decade, biosurfactants also have therapeutic and biomedical relevance as they have bactericidal, bacteriostatic, inhibition and interruption of biofilm formation, synergistic effects, and adjuvants with antibiotics (Naughton et al. 2019). Biosurfactants can inhibit bacterial and fungal growth. *Staphylococcus saprophyticus* produces biosurfactant with antibacterial activity against *Klebsiella pneumoniae*, *Escherichia coli*, *Vibrio cholera*, *Bacillus subtilis* and *Staphylococcus aureus* (Kim et al. 1998; Lotfabad et al. 2010). Furthermore, biosurfactants may have immunomodulatory, antitumor, and pharmacological properties (Perfumo et al. 2018).

Due to their structure and biophysical properties, biosurfactants are also a broad-spectrum alternative for application in cancer biotherapy and chemotherapy. Some glycolipids are capable of selectively inhibiting the proliferation of cancer cells and causing cell lysis through apoptosis pathways. Although in vitro antitumor activity has already been demonstrated, in vivo efficiency still needs to be better studied (Gudiña et al. 2013). Another potential clinical use is in passive immunization. In these cases, biosurfactants are used as drug carriers when drug treatment options are limited. Liposomes have a wide application based on this drug delivery approach, including vaccination (Loew et al. 2011).

Biosurfactants also have anti-adherent activity, this property is important to prevent the formation of biofilm, considering that the first step towards the formation of biofilm is the attachment of bacteria to the surface. Biosurfactants play an important role in preventing biofilm formation on silicon, titanium, and polystyrene plate surfaces. The action of biosurfactants probably occurs through binding to cell wall components or the cell surface, modifying the hydrophobicity of the outer membrane and reducing bacterial adhesion properties (Ciandrini et al. 2016).

3.3.3 Bioremediation: In Situ and Ex Situ

The location of application of bioremediation strategies is used as a parameter to classify this approach. In this way, in situ bioremediation is applied at the native site of contamination and can encompass approaches such as bioaugmentation, bioattenuation, and biostimulation. Ex situ bioremediation involves the removal of contaminated material for treatment in places suitable for this purpose, such as bioreactors, biopiles and landfarming (Arjoon and Speight 2022; Shivalkar et al. 2021).

3.3.3.1 In Situ

Bioattenuation or Intrinsic Bioremediation

Among in situ bioremediation strategies, bioattenuation represents an approach where there are no external interventions. In this case, the microorganisms present at the contamination site will naturally complete the biodegradation process of recalcitrant or xenobiotic compounds through microbial metabolic activity native to the contamination site (Arjoon and Speight 2022). However, to apply this methodology, it is necessary to assess the risk of application by exploring the time required for remediation so that there is a guarantee that the pollutant does not reach the point of maximum exposure for the closest animal and human populations (Azubuike et al. 2016).

The autochthonous microbiome of the polluted site is generally the focus for resolving environmental disorders resulting from contamination. However, for this to occur, ideal conditions for microbial growth and metabolism are necessary (Azubuike et al. 2016). Bacteria, fungi, and algae capable of metabolizing polycyclic aromatic hydrocarbons and other contaminating compounds are abundantly found in the most diverse ecosystems (Hossain et al. 2022).

The metabolic activity of indigenous microorganisms is one of the factors that makes bioattenuation a long-term strategy. This fact results from the need for adaptation and dependence on adverse chemical, physical and environmental factors for biodegradation to occur effectively in this environment (Arjoon and Speight 2022). The combination of bioremediation strategies has shown great potential for increasing the efficiency of this process (García-Delgado et al. 2015). Because of this, it is necessary in many cases to combine other strategies including bioaugmentation, biostimulation and bioventilation (Arjoon and Speight 2022).

Bioventilation or Bioventation

Aerobiosis can favor the degradation of several contaminants in bioattenuation. The injection of oxygen into the soil, known as bioventilation, and the injection of hydrogen peroxide (H_2O_2) are relevant in acceleration of natural attenuation as they increase metabolic activity (Shivalkar et al. 2021).

In bioventilation, air wells are injected above the water table in order to supply oxygen with the pressure necessary for the gas to dissipate in the soil and meet the biological demand for microbial oxygen. In addition to oxygen, other important nutrients can be supplied including nitrogen and phosphorus. Bioventilation's success is linked to the number of air injection points at appropriate rates, capable of achieving uniform air distribution. Soils with low permeability can benefit from the addition of pure oxygen, such as ozonization, which can assist in the partial oxidation of recalcitrant compounds, and, consequently, accelerate biodegradation (Azubuike et al. 2016).

In addition to being used to aerate unsaturated areas, bioventilation can be used in cases of anaerobic bioremediation, including the treatment of sites polluted with chlorinated compounds that are recalcitrant in the presence of oxygen. Therefore, replacing the addition of oxygen, proportions of other gases are injected, including H_2 and CO_2 as electron donors to reduce chlorinated vapor (Azubuike et al. 2016).

Bioslurping

According to Bala et al. (2022) the bioslurping technique involves the direct addition of oxygen gas to stimulate microbiological degradation together with vacuum-mediated pumping, bioventilation and soil vapor extraction, with the aim of recovering contaminated soil and underground areas.

One of the main challenges related to this bioremediation methodology involves the formation of voids in soils with high porosity and changes in the water table capable of forming zones of waterlogged soil that can hinder the access of air added by the technique (Hussain et al. 2022).

Bioaspargment

In a similar way to bioslurping, the Bioaspargment technique is based on pumping air into the soil with the aim of increasing microbial activity in the soil and consequently increasing the removal of contaminants from polluted areas (Bala et al. 2022).

This technique is most common in removing diesel and kerosene from water supplies. The efficiency of this methodology is intrinsically dependent on soil characteristics, such as porosity. Another point that influences the effectiveness of this technique is the biodegradability of the contaminant (Hussain et al. 2022).

Phycoremediation

Algae are associated with the degradation of a wide range of organic contaminants and a promising production of biomass with a high capacity for accumulation, detoxification, and degradation of the most diverse types of xenobiotics and pollutants. This biomass has potential application as an input for energy production (Baghour 2019).

Microalgae capable of producing biosurfactants are also being explored as a possible application in environments contaminated with hydrocarbons (Baghour 2019).

Phytoremediation

Phytoremediation is characterized by the use of plants for the bioremediation of contaminated soils, sediments and groundwater, being considered economical and effective in long-term applications, and also having the advantage of providing an aesthetic improvement to the site (Ali et al. 2013; Gatliff et al. 2016; Kalia et al. 2022; Nunes et al. 2022; Prasad 2015).

3.3.3.2 Ex Situ

Biopiles and Compost Piles

Biopiles are reliable systems to mitigate the concentration of contaminants that are based on the transfer of contaminated soils to a site adapted to stimulate the aerobic microbial metabolism of soils through aeration, addition of nutrients, minerals, and moisture control (Baldan et al. 2015). In the meantime, the biopile allows bioremediation to occur in a reduced time, as heating systems can be incorporated to increase the temperature and consequently microbial activities and the availability of contaminants (Azubuike et al. 2016).

Another positive aspect of biopiles is the ability to operate large volumes of contaminated soil in a limited space and being possible to design pilot scales of biopiles for laboratory studies (Azubuike et al. 2016).

Windrows

Windrows are characterized by aerobic processes applied to both soils and sediments containing biodegradable organic pollutants capable of mitigating the toxicity of some materials and contamination by PAHs. To proceed with this aerobiosis, periodic rotation or mixing is essential to increase native or transient microbial degradation. However, the effectiveness of this type of treatment depends on the type of soil (Patel et al. 2022). However, when the pollutant contains toxic volatile compounds, windrowing is not the best choice.

Bioreactors

Bioreactors are containers capable of providing favorable parameters for the growth and maintenance of microorganisms, mediating the chemical reactions necessary for bioremediation, with the possibility of monitoring. To this end, several factors are controlled by biostimulation that can influence the process, such as temperature, pH, mixing rate, aeration, and nutritional supplementation (Shivalkar et al. 2021).

Sludge bioreactors have been developed to better biodegrade sediment/soil. In these containers, the soil is subjected to an aqueous solution of 10 to 30% w/v where it is mixed mechanically or pneumatically. With this mixture, there is an increase in mass transfer and the contact area between the microorganisms used in the bioaugmentation of the bioreactor, the pollutants present in the contaminated soil and the nutrients from the biostimulation. Consequently, degradation rates increase when compared to other treatments such as in situ and ad-situ bioremediation (Robles-González et al. 2008).

One of the main negative aspects of bioreactors for the degradation of petroleum contaminants is the high cost resulting from a high need for energy for maintenance (Arjoon and Speight 2022). Furthermore, the construction and operation process are also very costly (Robles-González et al. 2008).

The indication of bioremediation by bioreactors occurs mainly when it is not possible to use natural attenuation or in situ biostimulation due to the type of soil and type of contaminant. Another factor that may motivate the use of this higher cost approach is when bioremediation is essential to be carried out in a short period, due to the pressure exerted by regulatory and environmental monitoring bodies in the face of the associated contamination case (Robles-González et al. 2008).

On the other hand, bioreactors are also remembered for positive aspects, such as the control of numerous variables that can favor a more accelerated process. Another positive aspect is the fact that this methodology avoids interfering external factors such as changes in temperature and humidity due to climate change, characteristics of *in situ* bioremediation (Patel et al. 2022).

Landfarming

Landfarming is a method designed for ex-situ bioremediation aerobically in the surface portion of the soil (Maddela et al. 2023). In landfarming, contaminated sediment or soil is transferred from the contamination site to a specific location where the contaminated soil is excavated and spread in a layer 50–60 cm deep (Di Marcantonio et al. 2023; Iturbe-Espinoza et al. 2023).

The Landfarming area is generally sealed to collect leachate. Landfarming soil is generally treated with water, nutrients, additives (such as pH correctors) and even soil aeration techniques to provide aerobiosis in the contaminated region to increase microbial activity (biostimulation) (Di Marcantonio et al. 2023; Iturbe-Espinoza et al. 2023). Bioaugmentation in landfarming may be responsible for further improving the process (Kim et al. 2022).

Aerobic biodegradation mediated by the combination of biological and physicochemical strategies with the aim of degrading contaminants in the soil is the basis of the Landfarming technique (Iturbe-Espinoza et al. 2023). Since 1970, landfarming has been used to remediate sites contaminated by various contaminants, especially HPAS (polycyclic aromatic hydrocarbons) derived from petroleum (Maddela et al. 2023). Landfarming is a reference when it comes to ex-situ bioremediation due to its low incorporation and operation cost (Di Marcantonio et al. 2023).

3.4 Bioremediation Challenges and Possibilities

The choice of microbial inoculum to be used in the soil may encounter some problems in terms of colonization, in cases where the pre-existing microbial community in the soil is well adapted and does not cover the growth of the applied inoculum. In these cases, it is essential to consider the selection of microorganisms adapted from the microbiome isolated from contaminated soil that has been subjected to prolonged exposure to pollutants (Baldan et al. 2015).

The application of microbial inocula associated with bioremediation by bioaugmentation presents some important challenges to be observed, such as genetic pollution. Some microorganisms, no matter how effective they are in degrading hydrocarbons due to the production of biosurfactants and enzymes, can promote the dissemination of genes capable of polluting and modifying the native microbiome. In this case, we can mention *Pseudomonas aeruginosa*, considered an opportunistic pathogen capable of spreading virulence genes and resistance to multiple antibiotics such as penicillin, cephalosporins, and tetracyclines (Hu et al. 2023).

The concern associated with genetic pollution of microorganisms used in bioremediation is because these genes have the potential for dissemination through bodies of water, soil and contact with workers. To overcome this situation, enzymatic degradation, and whole cell biocatalysis are being studied (Hu et al. 2023).

As future perspectives for bioremediation, the study and execution of multiple techniques may be necessary to increase the effectiveness of this approach by merging different types of bioremediations to reduce the limitations of individual techniques, time and cost of operation (Azubuike et al. 2016), as well as investing in the prospecting of microorganisms and molecules capable of favoring the degradation of contaminating elements.

The use of genetically modified organisms, for example, appears to be a promising aspect in the development of more efficient metabolic routes capable of expanding the range of substrates and the stability of metabolic activities. On the other hand, horizontal gene transfer and the possibility of uncontrolled dissemination of GMOs are important factors limiting the application of this methodology *in situ* (Azubuike et al. 2016).

It is also important to highlight the importance of metagenomics as a strategy for investigating non-cultivable microorganisms existing in reservoirs and contaminated areas (Sierra-García et al. 2014). Still, within the countless possibilities and challenges in bioremediation, two lines of research and development will be highlighted here for their great relevance and potential for innovation. These are the use of nanotechnologies in bioremediation and the combined use of biological and chemical systems.

3.4.1 Nanotechnology in Bioremediation

The technique for removing environmental pollutants using nanomaterials and nanoparticles produced by bacteria through nanotechnology is known as nanobioremediation (Ramezani et al. 2021). This approach is promising in bioremediation by helping to reduce the toxicity of the pollutant, increasing the surface area, and reducing the activation energy, increasing microbial metabolic efficiency, which directly influences the time and course of the bioremediation process (Azubuike et al. 2016).

Nanoparticles of biogenic origin are recent and promising applications for nanobioremediation, representing sustainable options. Among its benefits, the possibility of use without associated adverse effects and the high efficiency attributed to the vast surface area when the particles are reduced to nanoscale stand out (Kapoor et al. 2021). The precise control of contaminants at the atomic and molecular levels is another relevant aspect of this type of approach. In complex conditions of contaminant composition and contamination site, nanomaterials can target specific contaminants in addition to presenting high stability and sensitivity (Patra Shahi et al. 2021).

Biogenic nanoparticles are more efficient and stable than chemically synthesized nanoparticles (Kapoor et al. 2021). The applications of nanoparticles in bioremediation are reported for various compounds including dyes, catalytic dehalogenation, and heavy metal removal, in addition to the bioremediation of petroleum hydrocarbons and pesticides (Kapoor et al. 2021; Patra Shahi et al. 2021).

In the remediation of contaminated soils and waters, biological nanoparticles can be used to accelerate the recovery process of the contaminated environment (Patra Shahi et al. 2021). As an illustration, we can mention the combination of nanomaterials with phytoremediation to accelerate the enzymatic bioremediation process. Furthermore, substances of a complex nature such as long-chain hydrocarbons, for example, can be resistant to microbial and plant degradation, and, because of this, the fusion of nanotechnology and biotechnology can accelerate the breakdown of these compounds into compounds of lower complexity to provide bioavailability for decomposition by microorganisms and plants (Ramezani et al. 2021).

When the subject of nano bioremediation refers to contaminating polycyclic aromatic hydrocarbons, the literature still does not present a consensus regarding the interaction mechanisms between nanoparticles and microorganisms to optimize the process. In the meantime, the possible mechanism presented is that microorganisms are responsible for secreting specific enzymes and are immobilized in nanoparticles in which PAHs are adsorbed and carry out degradation more efficiently. In addition, the structures of the bacterial outer layer, such as S layers and lipopolysaccharides (LPS), enable interaction with contaminants sorbed by nanoparticles (Rajput et al. 2023).

3.4.2 *Bioelectrochemical Systems (BES)*

Bioelectrochemical systems (BES) combine biological and electrochemical methods in the context of a redox reaction. This is a promising technology for the remediation of soils/sediments contaminated by PAHs, because the commonly used physicochemical and biological approaches are slow, ineffective and/or harmful to the environment (Ambaye et al. 2023). Furthermore, bioremediation may be ineffective when pollutant concentrations exceed certain limits and be affected by environmental factors (Sayara and Sanchez 2020).

Numerous researchers have dedicated themselves to studying the effectiveness of BES in remediating soils/sediments contaminated by PAHs (Table 3.1). The works have different characteristics and evaluated the effectiveness of BES remediation in different contaminants, using different bacteria and reactors in different working conditions. For Anthracene, for example, experiments under different conditions and using varied species can be observed in the literature, such as *Proteobacteria*, *Actinobacteria*, *Chloroflexi*, *Firmicute*, *Thermomonas*, *Proteiniphilum*, *Nocardioides*, *Pseudoxanthomonas*, and *Geobacter*. However, the

Table 3.1 Application of BES for contaminated soil remediation with petroleum and derivates

Type of pollutant	References
Anthracene	Yu et al. (2017, 2019)
Aliphatic hydrocarbon	Cheng et al. (2017)
Crude oil	Wang et al. (2020)
Diesel	Wang et al. (2019)
Metolachlor	Li et al. (2019)
Petroleum hydrocarbon	Chen et al. (2022), Zhang et al. (2020, 2021), Li et al. (2016, 2018, 2019), and Lu et al. (2014)
Phenanthrene	Liang et al. (2020), Sharma et al. (2020), Zhao et al. (2019), Liu et al. (2019), Yu et al. (2019), Xu et al. (2019), Adelaja et al. (2014), and Yan et al. (2012)
Phenol	Luo et al. (2009)
Pyrene	Liang et al. (2020), Yu et al. (2019), Xu et al. (2019), and Liu et al. (2019)
Total petroleum hydrocarbon	Ambaye et al. (2022a, b) and Mohanakrishna et al. (2019)

study carried out by Yu et al. (2021) achieved the highest remediation efficiency (61.7%), using *Proteobacteria*, *Actinobacteria*, *Chloroflexi*, and *Firmicute*, with a microbial fuel cell operating in conditions of 30 °C, pH 7.62, at 24.0 mW/m^2 for 115 days.

For Pyrene, the study carried out by Xu et al. (2019) achieved the highest remediation efficiency (59.8%), using *Proteobacteria*, *Firmicutes*, *Bacteroidetes*, *Actinobacteria*, *Chloroflexi*, *Planctomycetes* and *Acidobacteria*, with a microbial fuel cell operating for 60 days at a maximum of 56.7 mV. For Phenanthrene, the study carried out by Yan et al. (2012) was the one that obtained the highest efficiency (99.47%), using *Proteobacteria Nitrospira* and *Chloroflexi*, with a microbial fuel cell operating for 240 days at pH 7.8, temperature of 25 °C and 17.1 mV.

For Petroleum hydrocarbon, Li et al. (2016) had the best result, 100% remediation efficiency, using a microbial fuel cell for 144 days, at 30 °C, pH 8.3 and 203 mA/m^2. For Total petroleum hydrocarbon, Ambaye et al. (2022a, b) had the best result (72.5% efficiency), with a microbial fuel cell operating for 22 days, at 20 °C, pH 7.0 and 9.5 A/m^2. Wang et al. (2019) studied the removal of diesel, using a microbial fuel cell in sandy soil (at pH 6.9) and clayey soil (at pH 73) for 248 days, obtaining efficiency of 59% and 45%, respectively. In Crude oil, Wang et al. (2020) obtained 76% efficiency in a 140-day experiment.

For Aliphatic hydrocarbon, Cheng et al. (2017) had 83% efficiency, in a 21-day experiment, at 23 °C. For Phenol, Luo et al. (2009) achieved 90.1% efficiency, under conditions of pH 7.0, for 120 h in a closed circuit. For Metolachlor, Li et al. (2019) achieved an efficiency of 36% in a 23-day experiment, using *Mortierella*, *Pseudogymnoascus*, *Kernia*, and *Debaryomyces* bacteria.

In the BES used by the authors, single-chamber air cathode reactors based on biochar were preferred, due to their lower cost. It is important to point out that several factors can affect the performance of BES, as observed in the articles cited.

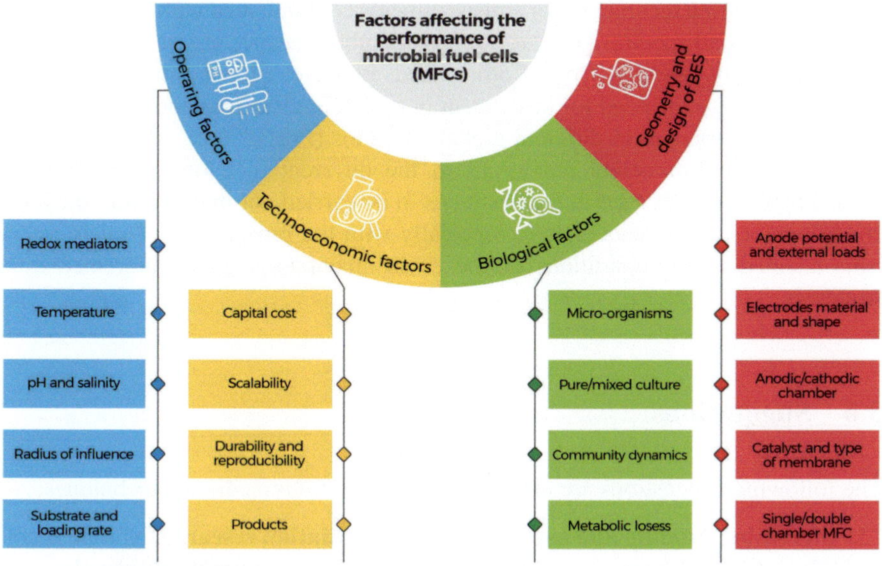

Fig. 3.6 The most important parameters affecting MFC performance during electricity generation and remediation efficiency. (Source: Adapted from Ambaye et al. (2023))

Such factors can be organized into four categories: (1) Operational; (2) Technical-economic; (3) Biologicals; (4) Geometric and Design (Fig. 3.6).

Additional research is essential to develop BES reactors that are more efficient in soil/sediment remediation, energy production, large-scale application, application in different terrain conditions, climates, and seasonal variations, as well as resolving other factors that affect the performance (Ambaye et al. 2023).

3.5 Considerations

The constant efforts to build a global energy matrix that is increasingly free from the use of fossil fuels must necessarily involve the use of renewable and efficient energies, as well as the preservation of biodiversity and restoration of ecosystems (Prasad et al. 2023). In this sense, the use of a microbial approach in the remediation and mitigation of toxic elements in natural environments proves to be an indispensable tool for the recovery of impacted areas and contaminated soils (Alves et al. 2020).

The use of bioremediation for the recovery of soils and sediments contaminated with hydrocarbons has proven to be an important methodology and remediation, especially when compared to chemical or physical treatment strategies, which are often more expensive, less efficient, and less sustainable (Azubuike et al. 2016).

Bioremediation techniques for soils/sediments contaminated with oil, in addition to having a broad spectrum of use, are considerably cheaper, easier to handle, and environmentally friendly. On the other hand, its development and application is directly related to the need for investment in research, dissemination and sharing of successful experiences in different global scenarios (Prasad et al. 2021). The great diversity of environmental conditions in the different locations where impacts caused by oil are observed requires studies to be carried out that consider the local conditions of each impacted area, especially with the prospecting of microorganisms adapted to the conditions of the environment targeted by bioremediation strategies.

3.6 Suggestions

The following suggestions are made to assist use of bioremediation techniques:

I. **Increasing funding and support for bioremediation research, as well as collaboration between research institutions and environmental agencies, can accelerate the discovery and application of effective bioremediation techniques.**
II. **In this sense, the importance of sharing scientific knowledge is highlighted, as well as the wide dissemination of research results and successful case studies of bioremediation, favoring the exchange of knowledge and experiences between scientists, policy makers and professionals across the world.**
III. **Multidisciplinary and interinstitutional actions, such as those developed within the scope of the Brazilian Antarctic Program (PROANTAR) to prospect for microorganisms from extreme environments capable of degrading oil, are examples of government efforts aimed at developing bioremediation. However, to guarantee the development of multidisciplinary and interinstitutional research, it is extremely important to maintain regular and adequate financial resources.**
IV. **Support for policies and regulation of research actions in bioremediation is necessary. Governments and international collaborations must create policies that encourage the use of bioremediation over traditional engineering, chemical or physical soil remediation methods.**

Acknowledgements The authors thank biologist Mr. Jackson M. Ministro and Mr. Leandro Fonseca for your help with figures. Thanks also to Federal Institute Baiano (IFBAIANO), Graduate Program in Biology and Biotechnology of Microorganisms (PPGBBM); State University of Santa Cruz (UESC) and Brazilian Federal Agency for Support and Evaluation of Graduate Education (CAPES).

References

Abdel-Mawgoud AM, Aboulwafa MM, Hassouna NAH (2008) Characterization of surfactin produced by *Bacillus subtilis* isolate BS5. Appl Biochem Biotechnol 150(3):289–303

Adelaja O, Keshavarz T, Kyazze G (2014) Enhanced biodegradation of phenanthrene using different inoculum types in a microbial fuel cell. Eng Life Sci 14:218–228. https://doi.org/10.1002/elsc.201300089

Ali H, Khan E, Sajad MA (2013) Phytoremediation of heavy metals—concepts and applications. Chemosphere 91:869–881. https://doi.org/10.1016/j.chemosphere.2013.01.075

Alves L d J, Nunes FC, Santos I, Loureiro DM, Casaes PA, Gross E, Majeti P (2020) Microbial approach for alleviation of potentially toxic elements in agricultural soils. In: Climate change and soil interactions. Elsevier, pp 271–303. https://doi.org/10.1016/B978-0-12-818032-7.00010-2

Ambaye TG, Chebbi A, Formicola F, Prasad S, Gomez FH, Franzetti A, Vaccari M (2022a) Remediation of soil polluted with petroleum hydrocarbons and its reuse for agriculture: recent progress, challenges, and perspectives. Chemosphere 293. https://doi.org/10.1016/j.chemosphere.2022.133572

Ambaye TG, Formicola F, Sbaffoni S, Franzetti A, Vaccari M (2022b) Insights into rhamnolipid amendment towards enhancing microbial electrochemical treatment of petroleum hydrocarbon contaminated soil. Chemosphere 307, Part 4:136126. https://doi.org/10.1016/j.chemosphere.2022.136126

Ambaye TG, Vaccari M, Franzetti A, Prasad S, Formicola F, Rosatelli A, Hassani A, Aminabhavi TM, Rtimi S (2023) Microbial electrochemical bioremediation of petroleum hydrocarbons (PHCs) pollution: recent advances and outlook. Chem Eng J 452, Part 3:139372. https://doi.org/10.1016/j.cej.2022.139372

Arjoon KK, Speight JG (2022) Petroleum biodegradation and oil spill bioremediation, 1st edn. CRC Press. https://doi.org/10.1201/9781003042631

Azubuike CC, Chikere CB, Okpokwasili GC (2016) Bioremediation techniques–classification based on site of application: principles, advantages, limitations and prospects. World J Microbiol Biotechnol 32(11):180. https://doi.org/10.1007/s11274-016-2137-x

Baghour M (2019) Algal degradation of organic pollutants. In: Martínez LMT, Kharissova OV, Kharisov BI (eds) Handbook of ecomaterials. Springer International Publishing, pp 565–586. https://doi.org/10.1007/978-3-319-68255-6_86

Bala S, Garg D, Thirumalesh BV, Sharma M, Sridhar K, Inbaraj BS, Tripathi M (2022) Recent strategies for bioremediation of emerging pollutants: a review for a green and sustainable environment. Toxics 10(8):484. https://doi.org/10.3390/toxics10080484

Baldan E, Basaglia M, Fontana F, Shapleigh JP, Casella S (2015) Development, assessment and evaluation of a biopile for hydrocarbons soil remediation. Int Biodeterior Biodegrad 98:66–72. https://doi.org/10.1016/j.ibiod.2014.12.002

Banat IM, Franzetti A, Gandolfi I et al (2010) Microbial biosurfactants production, applications and future potential. Appl Microbiol Biotechnol 87:427–444

Buzanello EB, Rezende RP, Sousa FMO, de Lima Silva Marques E, Loguercio LL (2014) A novel Bacillus pumilus-related strain from tropical landfarm soil is capable of rapid dibenzothiophene degradation and biodesulfurization. BMC Microbiol 14:257. https://doi.org/10.1186/s12866-014-0257-8

Chen M, Xu P, Zeng GM, Yang CP, Huang DL, Zhang JC (2015) Bioremediation of soils contaminated with polycyclic aromatic hydrocarbons, petroleum, pesticides, chlorophenols and heavy metals by composting: applications, microbes and future research needs. Biotechnol Adv 33:745–755. https://doi.org/10.1016/j.biotechadv.2015.05.003

Chen X, Han T, Miao X, Zhang X, Zhao L, Sun Y, Ye H, Li X, Li Y (2022) Ferrihydrite enhanced the electrogenic hydrocarbon degradation in soil microbial electrochemical remediation. Chem Eng J 446. https://doi.org/10.1016/j.cej.2022.136901

Cheng Y, Wang L, Faustorilla V, Megharaj M, Naidu R, Chen Z (2017) Integrated electrochemical treatment systems for facilitating the bioremediation of oil spill contaminated soil. Chemosphere 175:294–299. https://doi.org/10.1016/j.chemosphere.2017.02.079

Ciandrini E, Campana R, Casettari L et al (2016) Characterization of biosurfactants produced by Lactobacillus spp. and their activity against oral streptococci biofilm. Appl Microbiol Biotechnol 100:6767–6777. https://doi.org/10.1007/s00253-016-7531-7

Di Marcantonio C, Chiavola A, Noce A, Straccamore E, Giannuzzi A, Jirillo J, Gallo F, Boni MR (2023) A sustainable approach to enhance heavy hydrocarbons removal in landfarming treatment. Biodegradation 34(5):417–430. https://doi.org/10.1007/s10532-023-10025-6

Fu PP, Xia Q, Sun X, Yu H (2012) Phototoxicity and environmental transformation of polycyclic aromatic hydrocarbons (PAHs)-light-induced reactive oxygen species, lipid peroxidation, and DNA damage. J Environ Sci Health C 30:1–41. https://doi.org/10.1080/10590501.2012.653887

García-Delgado C, Alfaro-Barta I, Eymar E (2015) Combination of biochar amendment and mycoremediation for polycyclic aromatic hydrocarbons immobilization and biodegradation in creosote-contaminated soil. J Hazard Mater 285:259–266. https://doi.org/10.1016/j.jhazmat.2014.12.002

Gatliff E, Linton PJ, Riddle DJ, Thomas PR (2016) Phytoremediation of soil and groundwater. In: Bioremediation and bioeconomy. Elsevier, Amsterdam, pp 589–608

Goswami M, Deka S (2019) Biosurfactant production by a rhizosphere bacteria Bacillus altitudinis MS16 and its promising emulsification and antifungal activity. Colloids Surf B: Biointerfaces 178:285–296

Gudiña EJ, Rangarajan V, Sen R, Rodrigues LR (2013) Potential therapeutic applications of biosurfactants. Trends Pharmacol Sci 34(12):667–675

Gudiña EJ, Teixeira JA, Rodrigues LR (2016) Biosurfactants produced by marine microorganisms with therapeutic applications. Mar Drugs 14:1–15

Hidayati N, Surtiningsih T, Nimatuzahroh (2014) Removal of heavy metal Pb, Zn and Cu from sludge waste of paper industries using biossurfactant. J Bioremed Biodegr 5:255

Hossain MF, Akter MA, Sohan MSR, Sultana DN, Reza MA, Hoque KMF (2022) Bioremediation potential of hydrocarbon degrading bacteria: isolation, characterization, and assessment. Saudi J Biol Sci 29(1):211–216. https://doi.org/10.1016/j.sjbs.2021.08.069

Hu F, Wang P, Li Y, Ling J, Ruan Y, Yu J, Zhang L (2023) Bioremediation of environmental organic pollutants by Pseudomonas aeruginosa: mechanisms, methods and challenges. Environ Res 239:117211. https://doi.org/10.1016/j.envres.2023.117211

Hussain A, Rehman F, Rafeeq H, Waqas M, Asghar A, Afsheen N, Rahdar A, Bilal M, Iqbal HMN (2022) In-situ, ex-situ, and nano-remediation strategies to treat polluted soil, water, and air—a review. Chemosphere 289:133252. https://doi.org/10.1016/j.chemosphere.2021.133252

Iturbe-Espinoza P, Brown DM, Weedon JT, Braster M, Brandt BW, Bonte M, Van Spanning RJM (2023) Microbial communities associated with landfarming amendments during bioremediation of crude oil in Niger Delta soils. Appl Soil Ecol 191:105058. https://doi.org/10.1016/j.apsoil.2023.105058

Jimoh AA, Lin J (2019) Biosurfactant: a new frontier for greener technology and environmental sustainability. Ecotoxicol Environ Saf 184:109607

Kalia A, Sharma S, Semor N, Babele PK, Sagar S, Bhatia RK, Walia A (2022) Recent advancements in hydrocarbon bioremediation and future challenges: a review. 3 Biotech 12(6):135. https://doi.org/10.1007/s13205-022-03199-y

Kapoor RT, Salvadori MR, Rafatullah M, Siddiqui MR, Khan MA, Alshareef SA (2021) Exploration of microbial factories for synthesis of nanoparticles – a sustainable approach for bioremediation of environmental contaminants. Front Microbiol 12:658294. https://doi.org/10.3389/fmicb.2021.658294

Kim K, Jung SY et al (1998) Suppression of inflammatory responses by surfactin, a selective inhibitor of platelet cytosolic phospholipase A2. Biochem Pharmacol 55:975–985

Kim SH, Woo H, An S, Chung J, Lee S, Lee S (2022) What determines the efficacy of landfarming for petroleum-contaminated soils: significance of contaminant characteristics. Chemosphere 290:133392. https://doi.org/10.1016/j.chemosphere.2021.133392

Kumar V, Shahi SK, Singh S (2018) Bioremediation: an eco-sustainable approach for restoration of contaminated sites. In: Singh J, Sharma D, Kumar G, Sharma NR (eds) Microbial bioprospecting for sustainable development. Springer Singapore, pp 115–136. https://doi.org/10.1007/978-981-13-0053-0_6

Li X, Wang X, Zhang Y, Zhao Q, Yu B, Li Y, Zhou Q (2016) Salinity and conductivity amendment of soil enhanced the bioelectrochemical degradation of petroleum hydrocarbons. Sci Rep 6:1–11. https://doi.org/10.1038/srep32861

Li X, Zhao Q, Wang X, Li Y, Zhou Q (2018) Surfactants selectively reallocated the bacterial distribution in soil bioelectrochemical remediation of petroleum hydrocarbons. J Hazard Mater 344:23–32. https://doi.org/10.1016/j.jhazmat.2017.09.050

Li X, Li Y, Zhang X, Zhao X, Sun Y, Weng L, Li Y (2019) Long-term effect of biochar amendment on the biodegradation of petroleum hydrocarbons in soil microbial fuel cells. Sci Total Environ 651:796–806. https://doi.org/10.1016/j.scitotenv.2018.09.098

Liang Y, Zhai H, Liu B, Ji M, Li J (2020) Carbon nanomaterial-modified graphite felt as an anode enhanced the power production and polycyclic aromatic hydrocarbon removal in sediment microbial fuel cells. Sci Total Environ 713. https://doi.org/10.1016/j.scitotenv.2019.136483

Liu B, Zhai H, Liang Y, Ji M, Wang R (2019) Increased power production and removal efficiency of polycyclic aromatic hydrocarbons by plant pumps in sediment microbial electrochemical systems: a preliminary study. J Hazard Mater 380. https://doi.org/10.1016/j.jhazmat.2019.120896

Loew S, Fahr A, May S (2011) Modelling the release kinetics of poorly water-soluble drug molecules from liposomal nanocarriers. J Drug Deliv 2011:376548

Lotfabad TB, Abassi H et al (2010) Structural characterization of a rhamnolipid-type biosurfactant produced by Pseudomonas aeruginosa MR01: enhancement of dirhamnolipid proportion using gamma irradiation. Colloids Surf B Biointerfaces 81:397–405

Lu L, Huggins T, Jin S, Zuo Y, Ren Z (2014) Microbial metabolism and community structure in response to bioelectrochemically enhanced remediation of petroleum hydrocarbon-contaminated soil. Environ Sci Technol 48(7):4021–4029. https://doi.org/10.1021/es4057906

Luo H, Liu G, Zhang R, Jin S (2009) Phenol degradation in microbial fuel cells. Chem Eng J 147:259–264. https://doi.org/10.1016/j.cej.2008.07.011

Maciel BM, Dias JCT, Freire dos Santos AC, Argôlo Filho RC, Fontana R, Loguercio LL, Rezende RP (2007) Microbial surfactant activities from a petrochemical landfarm in a humid tropical region of Brazil. Can J Microbiol 53:937–943. https://doi.org/10.1139/W07-052

Maddela NR, Ramakrishnan B, Venkateswarlu K, Megharaj M (2023) Landfarming and its microbiology. In: Encyclopedia of soils in the environment. Elsevier, pp 533–546. https://doi.org/10.1016/B978-0-12-822974-3.00217-2

Matos Neto G, Marques EDLS, Oliveira LKS, Rezende RP, Dias JCT (2023) Searching for bacteria able to metabolize polycyclic aromatic sulfur compounds in 12-years periodically fed bioreactor. Arch Microbiol 205:336. https://doi.org/10.1007/s00203-023-03674-x

MCTI (2024) Ciência Antártica. http://cienciaantartica.mcti.gov.br/. Accessed 31 May 2024

Mohanakrishna G, Al-Raoush RI, Abu-Reesh IM, Pant D (2019) A microbial fuel cell configured for the remediation of recalcitrant pollutants in soil environment. RSC Adv 9:41409–41418. https://doi.org/10.1039/C9RA06957G

Naughton PJ, Marchant R et al (2019) Microbial biosurfactants: current trends and applications in agricultural and biomedical industries. J Appl Microbiol 127(1):12–28

Nunes FC, Lander JA, Carvalho CCN, Gross E, Soares TM, Prasad MNV (2020) Soil as a complex ecological system for meeting food and nutritional security. In: Prasad MNV, Pietrzykowski M (eds) Climate change and soil interaction, vol 1, 1st edn. Elsevier, Amsterdã, pp 229–269

Nunes FC, Alves LDJ, Prasad MNV, De Carvalho CCN, Nolasco EC (2022) Phytoremediation strategies for rehabilitation of soils affected by red mud: the Mariana tailing dam collapse (Minas Gerais, Brazil). Eurasian Soil Sci 55:673–685. https://doi.org/10.1134/S1064229322050052

Nunes FC, Carvalho CCN, Alves LJ, Prasad MNV (2023) Soil degradation: a major challenge in the twenty-first century. In: Prasad MNV, Kumar C (eds) Agroecological approaches for sustainable soil management, vol 1, 1st edn. John Wiley and Sons, Incorporated, London, pp 1–21

Nunes FC, de Jesus Alves L, Bitterncourt P, Prasad MNV (2024) Nonfossil energy targets for environmental sustainability. In: Vara Prasad MN, Erickson LE, Nunes FC, Ramadan BS (eds) Decarbonization strategies and drivers to achieve carbon neutrality for sustainability. Elsevier, pp 27–44. https://doi.org/10.1016/B978-0-443-13607-8.00031-6

Patel S, Homaei A, Patil S, Daverey A (2019) Microbial biosurfactants for oil spill remediation: pitfalls and potentials. Appl Microbiol Biotechnol 103(1):27–37

Patel AK, Singhania RR, Albarico FPJB, Pandey A, Chen C-W, Dong C-D (2022) Organic wastes bioremediation and its changing prospects. Sci Total Environ 824:153889. https://doi.org/10.1016/j.scitotenv.2022.153889

Patra Shahi M, Kumari P, Mahobiya D, Kumar Shahi S (2021) Nano-bioremediation of environmental contaminants: applications, challenges, and future prospects. In: Bioremediation for environmental sustainability. Elsevier, pp 83–98. https://doi.org/10.1016/B978-0-12-820318-7.00004-6

Perfumo A, Banat IM, Marchant R (2018) Going green and cold: biossurfactants from low-temperature environments to biotechnology applications. Trends Biotechnol 36:277289

Prasad MNV (2015) Phytoremediation crops and biofuels. In: Lichtfouse E (ed) Sustainable agriculture reviews. Springer International Publishing, Cham, pp 159–261

Prasad MNV, de Jesus Alves L, Nunes FC (2021) Global remediation industry and trends. In: Handbook of assisted and amendment: enhanced sustainable remediation technology. John Wiley & Sons, Ltd, pp 1–28. https://doi.org/10.1002/9781119670391.ch1

Prasad MNV, Smol M, Freitas H (2023) Achieving sustainable development goals via green deal strategies. In: Vara Prasad MN, Smol M (eds) Sustainable and circular management of resources and waste towards a green deal. Elsevier, pp 3–23. https://doi.org/10.1016/B978-0-323-95278-1.00002-4

PROANTAR (2024) Programa Antártico Brasileiro [WWW Document]. CIRM. https://www.marinha.mil.br/secirm/pt-br/proantar/sobre. Accessed 31 May 2024

Rajput VD, Kumari S, Minkina T, Sushkova S, Mandzhieva S (2023) Nano-enhanced microbial remediation of PAHs contaminated soil. Air Soil Water Res 16:117862212311700. https://doi.org/10.1177/11786221231170099

Ramezani M, Rad FA, Ghahari S, Ghahari S, Ramezani M (2021) Nano-bioremediation application for environment contamination by microorganism. In: Panpatte DG, Jhala YK (eds) Microbial rejuvenation of polluted environment, vol 26. Springer Singapore, pp 349–378. https://doi.org/10.1007/978-981-15-7455-9_14

Ritchie H, Roser M, Rosado P (2022) Energy [WWW Document]. Our world in data. https://ourworldindata.org/energy. Accessed 27 Aug 2023

Robles-González IV, Fava F, Poggi-Varaldo HM (2008) A review on slurry bioreactors for bioremediation of soils and sediments. Microb Cell Factories 7(1):5. https://doi.org/10.1186/1475-2859-7-5

Rodrigues LR (2015) Microbial surfactants: fundamentals and applicability in the formulation of nano-sized drug delivery vectors. J Colloid Interface Sci 449:304–316

Santos ACF, Rezende RP, Brendel M, Souza SS, Gonçalves ACS, Dias JCT (2014) Evaluation of the biodegradability of petroleum in microcosm systems by using mangrove sediments from Camamu Bay, Bahia, Brazil. Genet Mol Res 13:2048–2059. https://doi.org/10.4238/2014.March.24.9

Santos DKF, Rufino RD et al (2016) Biosurfactants: multifunctional biomolecules of the 21st century. Int J Mol Sci 17:401

Sarubbo LA, Rocha JR et al (2015) Some aspects of heavy metals contamination remediation and role of biossurfactants. Chem Ecol 31:707–772

Sayara T, Sanchez A (2020) Bioremediation of PAH-contaminated soils: process enhancement through composting/compost. Appl Sci 10:3684. https://doi.org/10.3390/app10113684

Sharma M, Nandy A, Taylor N, Venkatesan S, Kollath V, Karan K, Thangadurai V, Tsesmetzis N, Gieg L (2020) Bioelectrochemical remediation of phenanthrene in a microbial fuel cell using an anaerobic consortium enriched from a hydrocarbon-contaminated site. J Hazard Mater 389. https://doi.org/10.1016/j.jhazmat.2019.121845

Shivalkar S, Singh V, Sahoo AK, Samanta SK, Gautam PK (2021) Bioremediation: a potential ecological tool for waste management. In: Bioremediation for environmental sustainability. Elsevier, pp 1–21. https://doi.org/10.1016/B978-0-12-820318-7.00001-0

Sierra-García IN, Alvarez JC, de Vasconcellos SP, de Souza AP, dos Santos Neto EV, de Oliveira VM (2014) New hydrocarbon degradation pathways in the microbial metagenome from Brazilian petroleum reservoirs. PLoS One 9:e90087. https://doi.org/10.1371/journal.pone.0090087

Speight JG, Arjoon KK (2012) Bioremediation of petroleum and petroleum products. Scrivener Publishing, Beverly

Tsibart A, Gennadiev A (2013) Hidrocarbonetos aromáticos policíclicos em solos: Fontes, comportamento e significado de indicação. Eurasian Soil Sci 46:728–741

Vecino X, Cruz JM, Moldes AB, Rodrigues LR (2017) Biosurfactants in cosmetic formulations: trends and challenges. Crit Rev Biotechnol 37(7):911–923

Wang F, Gu Y, O'Brien JP, Yi SM, Yalcin SE, Srikanth V, Shen C, Vu D, Ing NL, Hochbaum AI, Egelman EH, Malvankar NS (2019) Structure of microbial nanowires reveals stacked hemes that transport electrons over micrometers. Cell 177:361–369. https://doi.org/10.1016/j.cell.2019.03.029

Wang H, Cui Y, Lu L, Jin S, Zuo Y, Ge Z, Ren ZJ (2020) Moisture retention extended enhanced bioelectrochemical remediation of unsaturated soil. Sci Total Environ 724. https://doi.org/10.1016/j.scitotenv.2020.138169

Xu P, Zeng GM, Huang DL, Feng CL, Hu S, Zhao MH, Lai C, Wei Z, Huang C, Xie GX et al (2012) Use of iron oxide nanomaterials in wastewater treatment: a review. Sci Total Environ 424:1–10. https://doi.org/10.1016/j.scitotenv.2012.02.023

Xu P, Xiao E, Zeng L, He F, Wu Z (2019) Enhanced degradation of pyrene and phenanthrene in sediments through synergistic interactions between microbial fuel cells and submerged macrophyte Vallisneria spiralis. J Soils Sediments 19:2634–2649. https://doi.org/10.1007/s11368-019-02247-0

Yan Z, Song N, Cai H, Tay J, Jiang H (2012) Enhanced degradation of phenanthrene and pyrene in freshwater sediments by combined employment of sediment microbial fuel cell and amorphous ferric hydroxide. J Hazard Mater 199–200:217–225. https://doi.org/10.1016/j.jhazmat.2011.10.087

Yu B, Tian J, Feng L (2017) Remediation of PAH polluted soils using a soil microbial fuel cell: influence of electrode interval and role of microbial community. J Hazard Mater 336:110–118. https://doi.org/10.1016/j.jhazmat.2017.04.066

Yu B, Li Y, Feng L (2019) Enhancing the performance of soil microbial fuel cells by using a bentonite-Fe and Fe3O4 modified anode. J Hazard Mater 377:70–77. https://doi.org/10.1016/j.jhazmat.2019.05.052

Yu B, Feng L, He Y, Yang L, Xun Y (2021) Effects of anode materials on the performance and anode microbial community of soil microbial fuel cell J of Hazardous Materials 401123394-10.1016/j.jhazmat.2020.123394

Zhang X, Li X, Zhao X, Chen X, Zhou B, Weng L, Li Y (2020) Bioelectric field accelerates the conversion of carbon and nitrogen in soil bioelectrochemical systems. J Hazard Mater 388. https://doi.org/10.1016/j.jhazmat.2019.121790

Zhang X, Li R, Song J, Ren Y, Luo X, Li Y, Li X, Li T, Wang X, Zhou Q (2021) Combined phyto-microbial-electrochemical system enhanced the removal of petroleum hydrocarbons from soil: a profundity remediation strategy. J Hazard Mater 420. https://doi.org/10.1016/j.jhazmat.2021.126592

Zhao L, Deng J, Hou H, Li J, Yang Y (2019) Investigation of PAH and oil degradation along with electricity generation in soil using an enhanced plant-microbial fuel cell. J Clean Prod 221:678–683. https://doi.org/10.1016/j.jclepro.2019.02.212

Chapter 4
You May Need to Cure the Water Repellency of the Contaminated Materials!

Mehdi Mahbod, Amin Parnian, and Amir Parnian

Abstract Soil quality is related to ecosystem sustainability and can be negatively affected by anthropogenic hazards. Soil quality and ecosystem health are harmed by petroleum contaminations normally caused by oil operations, such as exploration, drilling, transfer, transport, refinery, and distribution. Soil water repellency (SWR) of petroleum-contaminated soils is a retarding factor for bioremediation operations. This study is a pilot-scale SWR reduction of saline-contaminated soil obtained from a polluted area in Iran. The initial total petroleum hydrocarbon (TPH) content of the soil was between 6.9 and 17.1 g kg^{-1} and was contaminated with heavy crude oil. Initial SWR was between 1500 and 12,500 S. The recovery procedure commenced with sugarcane bagasse and compost mixtures were added to soils. After irrigation and aeration of piles, soils recovered approximately in 7 days. The water repellency reduced from extremely water repellent (number 9 in the classification of SWR) to slightly water repellent (number 2–3 in the classification of SWR), which was about 6 S in the moderately, 8 S in highly and 12 S in very highly polluted soil piles.

Keywords Soil Recovery · Petroleum Contamination · Organic

M. Mahbod (✉)
Water Sciences and Engineering Department, Faculty of Agriculture, Jahrom University, Jahrom, Iran

A. Parnian
Young Researchers Club, Masjed Soleyman Branch, Islamic Azad University, Masjed Soleyman, Iran

Parnian Sazeh Chehelsotoun (PSC) Company, Isfahan, Iran

A. Parnian
National Salinity Research Center (NSRC), Agricultural Research, Education and Extension Organization (AREEO), Yazd, Iran

A. Parnian et al. (eds.), *Bioremediation of Petroleum Contaminated Soil and Sediment*, Environmental Contamination Remediation and Management,
https://doi.org/10.1007/978-3-031-95524-2_4

4.1 Introduction (Petroleum Hydrocarbon Contamination and Soil Water Repellency)

The property of soil water repellency (SWR) is a dynamic characteristic of soil that serves to diminish or impede the infiltration of water into the soil (Mao et al. 2019). The initial observation of peat soils in the Netherlands occurred during the eleventh century when dikes were employed to reclaim coastal lowlands for agricultural purposes (de Bakker 2013). Water molecules in natural soils experience a loss of cohesion due to the attraction forces between soil particles and water molecules. Consequently, these water molecules tend to persist as water droplets and migrate over the surface of the particles, resulting in the moistening of the soil. If a water-repellent coating, such as organic matter, resins, petroleum compounds, etc., is present on a soil particle, the attraction interactions between water and soil are significantly diminished or eradicated (Smettem et al. 2021; Parnian et al. 2022). Consequently, water will persist in the form of spherical droplets on the surface of the soil. The soil will exhibit aridity, preventing water infiltration, thereby earning the designation of being water-repellent (Sandor et al. 2021).

The phenomenon of water repellency results in an uneven distribution of water within the soil, hence impeding the ease of soil wetting and diminishing the soil's capacity for water infiltration (Olorunfemi et al. 2014). This ultimately restricts the existence of creatures, and in severe instances, the region becomes completely barren of any living species (Schnabel et al. 2013; Hervé-Fernández et al. 2023). The water repellency of soil is shown in several significant ways. First and foremost, it hinders the process of water infiltration and retention within the soil, hence impeding the vital absorption and storage of moisture that is crucial for the growth of plants (Gavrilescu 2021). As a result, this phenomenon gives rise to heightened runoff and erosion, since the soil's capacity to absorb and hold water is compromised, resulting in its removal through augmented surface flow (Lowe et al. 2021). The reduced availability of water and the damaged structure of soil lead to a decline in plant growth and productivity, presenting difficulties for both agricultural and ecological systems (Li et al. 2019). Furthermore, the process of soil compaction is heightened as a result of decreased water content, hence worsening the challenges associated with root penetration and nutrient absorption (Coelho et al. 2005). Moreover, the heightened vulnerability of the soil to waterlogging and flooding intensifies these problems, resulting in a recurring pattern of deteriorated soil quality and reduced land productivity (Movasat and Tomac 2021). Therefore, it is vital to comprehend and address the issue of soil water repellency in order to maintain the well-being of ecological systems and enhance agricultural output.

The phenomenon of soil water repellency is a multifaceted matter that can exert a substantial influence on the overall well-being and production of soil (Müller et al. 2014). Through comprehending the origins and consequences of soil water repellency, land managers can implement measures to mitigate its adverse effects (García-Carmona et al. 2020). Water repellency is a significant issue associated with soils contaminated with oil, leading to suboptimal water infiltration and retention

(Abou Najm et al. 2021). This phenomenon results in a decrease in the quantity of water accessible to living organisms, hence impeding the pace of bioremediation and constraining the viability of soil-dwelling organisms (Braun et al. 2010). The water repellency of soil particles is attributed to the existence of hydrophobic compounds, such as oil and grease. These chemicals form a protective layer on the soil particles, impeding the attachment of water to them. Consequently, water tends to flow off the soil's surface instead of into it (Doerr et al. 2000). A range of approaches have been suggested to tackle this difficulty, encompassing the utilization of lime, surfactants, organic matter, and zeolite supplements.

Lime, namely calcium carbonate or calcium hydroxide, is frequently employed as a means of enhancing soils characterized by low pH values (Ameyu 2019). In addition, lime has been demonstrated to decrease SWR via modifying soil characteristics, such as elevating soil pH and enhancing cation exchange capacity (Chen et al. 2022). Research has shown that the application of lime can enhance the process of water infiltration and decrease the amount of runoff in soils that are resistant to water. Nevertheless, the efficacy of lime amendment can be contingent upon factors such as soil composition, pace of application, and prevailing environmental circumstances (Khattab et al. 2007; Olsson et al. 2019; Babu and Poulose 2018). In addition, an overabundance of lime application can result in soil alkalization, a process that might potentially have adverse consequences on soil microbial populations and the availability of nutrients (Sakala et al. 2007). Surfactants, including non-ionic and anionic surfactants, are commonly employed to mitigate the effects of SWR by diminishing the surface tension of water and augmenting the processes of wetting and infiltration. These substances can infiltrate hydrophobic soil layers and enhance the dispersion of water throughout the soil profile (Ogunmokun et al. 2020). Studies have demonstrated that the application of surfactants can successfully reduce soil water SWR in different soil types and land uses (Kostka et al. 1997; Cisar et al. 2000; Oostindie et al. 2008; Lehrsch and Sojka 2011). Nevertheless, the efficacy of surfactants in the long run may be constrained due to their vulnerability to degradation and leaching (Lowe et al. 2017). In addition, the utilization of surfactants may give rise to apprehensions regarding the potential environmental consequences and toxicity towards non-target creatures (Badmus et al. 2021). Organic matter amendments, including compost, manure, and biochar, have been suggested as viable strategies for mitigating SWR and improving soil structure and fertility in a sustainable manner (Täumer et al. 2005). Organic matter has the ability to enhance soil aggregation, porosity, and moisture retention, resulting in a decrease in SWR and an increase in water infiltration rates. Numerous studies have documented favorable outcomes associated with the use of organic matter amendments in the mitigation of SWR in both agricultural and natural environments (e.g. Abel et al. 2013; Šařec and Novák 2016; Sun et al. 2018). The effectiveness of organic matter amendments can be influenced by various factors, including the rate of application, the specific type of organic material used, and the qualities of the soil (Müller and Deurer 2011). Moreover, the accessibility and expense of organic modifications can restrict their extensive implementation in specific areas (Sharma et al. 2019; Sun et al. 2021). Zeolites are minerals that have a small number of pores and have the ability to exchange cations and retain water (Belviso 2020). The potential of these amendments in alleviating SWR has been examined due to their ability to enhance water retention and facilitate ion exchange activities in hydrophobic soils (Farzadian et al. 2015). According to existing research, the use of zeolite has been found to enhance soil wettability, water infiltration, and nutrient retention in soils that exhibit water-repellent properties (Gholizadeh-Sarabi and Sepaskhah 2013; Nakhli et al. 2017; Adamczuk et al. 2022). Nevertheless, the efficacy of zeolite amendments can be influenced by various factors, including the specific type of zeolite, its particle size, and the manner of administration. Furthermore, there have been expressed concerns about the environmental ramifications associated with the extraction and refinement of zeolite materials (Jakkula and Wani 2018; Mondal et al. 2021; Asghari et al. 2022).

The examination of the efficacy of lime, surfactant, organic matter, and zeolite amendments in soil degradation caused by oil contamination is of utmost importance. The comprehension of their efficacy can provide valuable insights for the development of sustainable remediation approaches, the reduction of soil water repellency, the improvement of soil health, and the restoration of ecological function in regions impacted by oil spills, thereby ensuring the protection of both the environment and human welfare. The lack of water in soils contaminated with oil presents considerable difficulties for the adjacent ecological system, resulting in adverse consequences for many life forms (Ossai et al. 2020). The diminished accessibility of water primarily hampers the growth and productivity of plants, hence constraining their capacity to flourish and carry out their ecological functions (da Silva Correa et al. 2022). The reduction in vegetation cover intensifies the potential for soil erosion, as plants have a vital function in maintaining soil stability and mitigating erosion through their root systems (Kermani and Ebadi 2012; Mohammadi et al. 2018). Moreover, the convergence of arid, oil-contaminated soils with elevated temperatures might amplify the susceptibility to wildfires, hence presenting hazards to both terrestrial environments and human habitations. Furthermore, the adverse impact of water shortages and oil contamination on habitat quality has significant implications for wildlife populations, causing disruptions to ecological processes and potentially resulting in a decrease in biodiversity (Williams et al. 2014; Young et al. 2017). The imperative to tackle the difficulties presented by water shortages in soils contaminated with oil is crucial in order to alleviate these detrimental consequences and safeguard the well-being and adaptability of impacted ecological systems.

Situated in the southwestern part of Iran, the province of Khuzestan holds around 5.42% of the world's oil reserves. Furthermore, it is worth noting that crude oil constitutes 80% of the country's total oil production. As a result, the ecological consequences of oil spills have the potential to extend beyond the immediate vicinity of Khuzestan. This chapter presents a comprehensive investigation into the many approaches that can be employed to enhance the water repellency of petroleum-contaminated soil in the region of Khuzestan.

4.2 Materials and Methods

4.2.1 Study Area and Contaminated Soils

The Jofir region in Hoveyzeh County, Khuzestan Province, served as the study area. Figure 4.1 displays the geographical coordinates of Hoveyzeh County. A total of 1200 cubic meters of soil that had been contaminated was gathered through the utilization of heavy machinery, which encompassed a loader, excavator, and truck. The polluted soils were categorized into three groups based on the visual assessment of contamination: moderate, high, and extremely high contamination. To assess the extent of contamination, a total of five soil samples were obtained from each of the three piles after the process of soil separation. The investigation of the samples' physical and chemical features was conducted following the guidelines outlined in Table 4.1. According to the Iranian rules for soil testing, the electrical conductivity values of the contaminated soils exceed 16 dS m^{-1}, hence classifying them as highly saline soils.

1200 m^3 of contaminated dirt was gathered from a four-hectare area using heavy machinery such as loaders, excavators, and trucks. The contamination was visually

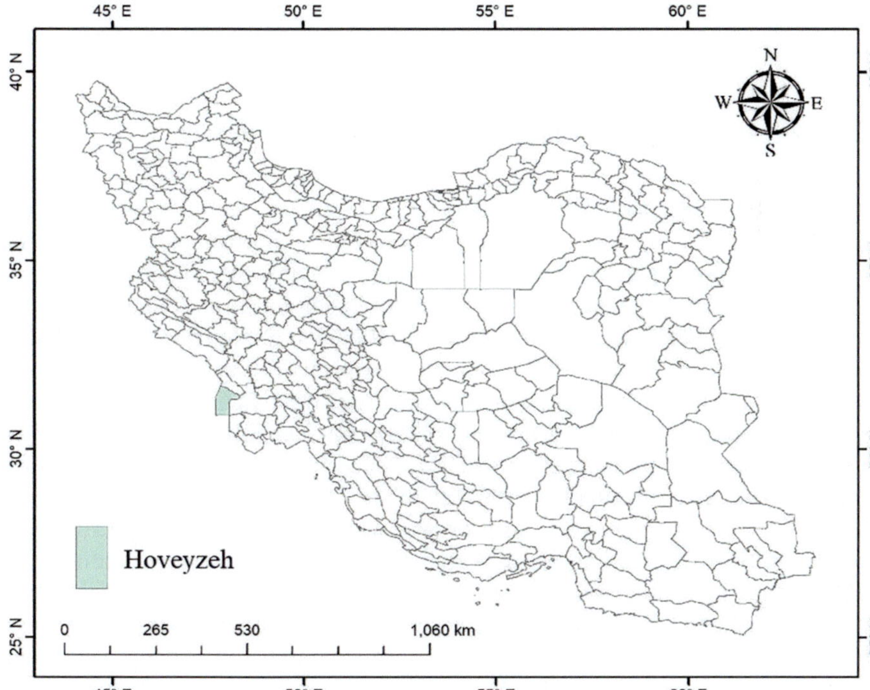

Fig. 4.1 The geographical location of the site is southwest of Iran, 48.187 E, 31.450 N, 11 m above sea level

Table 4.1 The physical-chemical properties of the polluted soils before the remediation process (Parnian et al. 2022)

	ECe (dS m^{-1})	pH	TPH (g kg^{-1})	Water repellency (S)	Texture (USDA)	Gravel (%)
Moderately polluted	18.6 ± 1.2	6.95 ± 0.23	6.9–7.4	1500–2500	SCL	<1
Highly polluted	17.3 ± 2.0	7.00 ± 0.20	9.6–12.8	8000–9500	SCL	<1
Very highly polluted	19.1 ± 1.5	7.05 ± 0.15	15.6–17.1	10,500–12,500	SCL	<1

The standard error (±SE) is used for means comparison ($n = 5$)
EC electrical conductivity, *TPH* total petroleum hydrocarbons

assessed, and the contaminated soils were categorized into heaps based on the levels of pollution, ranging from moderate to high and extremely high. This was done based on the study of the contaminants. After the dirt was separated, five soil samples were collected from each of the three heaps to assess the extent of contamination. Based on the information provided in Table 4.1, the samples were examined to determine their physical and chemical characteristics. The polluted soils exhibited electrical conductivity values exceeding 16 dS m^{-1} (Table 4.1). Following Iranian regulations for soil assessment (No. 466), all of the soils were classed as falling within the category of highly saline soils.

4.2.2 *Procedure for Reducing Soil Water Repellency*

Various amendments, such as surfactants, organic matter, zeolite, and lime, were employed to address the issue of soil water repellency. Surfactants play a crucial role in enhancing soil penetration and dissolving pollutants by reducing the surface tension of water. The presence of organic matter in soil has been found to improve soil structure and fertility, as well as to bind pollutants, minimizing their mobility and potential impact. Zeolite exhibits a notable capacity for absorbing diverse pollutants due to its substantial surface area, hence concurrently augmenting soil drainage and aeration. Lime plays a crucial role in soil rehabilitation and ecological restoration endeavors by increasing soil pH, hence fostering favorable conditions for plant growth and reducing the solubility of specific pollutants.

The soils were supplemented with organic matter amendments, specifically a combination of sugarcane bagasse and compost at a weight-to-weight ratio of 5% (with a volume-to-volume ratio of 1:1). Table 4.2 provides an overview of the characteristics exhibited by this blend. The mixing materials utilized in this study were obtained from the Amir Kabir Sugarcane Agro-industrial Co., a local sugarcane processing facility situated on the Ahvaz-Khoramshar old road in the Khuzestan province of Iran. Following this, the aforementioned materials underwent periodic irrigation and aeration using durable road construction equipment over a period of 2 weeks.

Table 4.2 The characteristics of compost and sugarcane bagasse that are utilized in the process of bioremediation for soils contaminated with heavy oil (Parnian et al. 2022)

Material	$pH_{(1:10)}$	$EC_{(1:10)}$ (dS m^{-1})	TOC (%)	C:N	Water content (%)
Compost	11.0 ± 0.80	18.2 ± 0.15	17.5 ± 0.07	2.2 ± 0.08	7.25 ± 0.13
Sugarcane bagasse	9.5 ± 1.40	130.9 ± 2.20	56.4 ± 0.56	1.8 ± 0.10	8.54 ± 0.22

± standard error ($n = 5$)
EC electrical conductivity, *TOC* total organic carbon, *C:N* carbon to nitrogen ratio

Table 4.3 Soil water repellency classification based on water drop infiltration time (Letey et al. 2000)

Water repellency	Infiltration time (s)	SWR class
Wettable	≤5	1
Slightly water repellent	6–10	2
	11–60	3
	61–180	4
Strongly water repellent	181–300	5
	301–600	6
	601–900	7
Severely water repellent	901–3600	8
Extremely water repellent	≥3600	9

4.2.3 Soil Water Repellency Classification

The method of water infiltration time is considered to be a straightforward and pragmatic approach for quantifying the water repellency of soil. The approach in question was devised by Van't Woudt in the year 1959. This experimental procedure involves the application of three droplets of water onto the surface of the soil, followed by the measurement of the duration required for the water to permeate the soil. The classification of soil water repellency is determined by the water drop infiltration time, as indicated in Table 4.3. When the contact angle between water and soil particles is below 90°, water is able to permeate the soil. Hence, the temporal duration obtained by this approach corresponds to the duration necessary for the contact angle to transition from a value of 90° to a value below 90°. According to this approach, soils are classified into five categories of water repellency: non-water repellent, mildly water repellent, partially water repellent, very water repellent, and highly water repellent. This method is very common and can classify soils based on their water repellency. This method has also been widely used in recent studies (Bagarello et al. 2020; Banet et al. 2021; Caltabellotta et al. 2022; Fu et al. 2021).

4.3 Results and Discussion

The effect of increasing surfactant concentration on SWR was investigated. Table 4.4 shows the water repellency of certain soil samples both prior to and after the introduction of surfactant, zeolite, and lime amendments. Accordingly, the efficiency of each amendment is calculated. Surfactant concentrations of 1, 2, and 8 g/L were added to soil samples with different pollution levels of moderately polluted, highly polluted, and very high polluted. The results showed that the amount of SWR reduction was 5%, 12%, and 19% in these soil samples, respectively. Therefore, it can be said that increasing surfactant at these concentrations did not have an effective impact on reducing soil hydrophobicity. Also, applying higher concentrations was not economically justified on a large scale. The increase in zeolite by 2%, 10%, and 30% (w/w) to soil samples with pollution levels of moderately polluted, highly polluted, and very high polluted showed that the amount of SWR in these soil samples decreased by 5%, 8%, and 50%, respectively. Based on these results, it can be concluded that adding 2% and 10% (w/w) zeolite to the soil did not have much effect on the amount of SWR, and on the other hand, adding 30% (w/w) zeolite to the soil was not economically justified. Similar results were obtained for lime as for surfactant. Increasing the lime concentration by 2%, 4%, and 8% (w/w) in the soil caused a decrease in SWR of 15%, 12%, and 11% in the samples that were moderately polluted, highly polluted, and very highly polluted, respectively. Therefore, it can be concluded that increasing lime at this level does not have a significant and noticeable effect on reducing SWR.

Due to the reasonable price of organic matter, more treatments were tested in the soil. Therefore, at this stage, only highly polluted soil samples with an initial SWR of 9000 s were studied. Accordingly, 1%, 5%, 10%, and 20% (w/w) of organic matter were added to the soil (Table 4.5). Then, considering that, the organic matter was mixed with the soil for 7 days, a control treatment was also considered in order to investigate the role of soil mixing on SWR, in which no organic matter was added.

Table 4.4 The water repellency of selected soil samples before and after the application of surfactant, zeolite, and lime amendments

Surfactant concentration (g kg^{-1})	1	2	8
Water repellency pre-application	2200	9000	11,000
Water repellency post-application	2090	7920	8910
Recovery efficiency (%)	5%	12%	19%
Zeolite (%)	2%	10%	30%
Water repellency pre-application	2200	9000	11,000
Water repellency post-application	2090	8280	5500
Recovery efficiency (%)	5.00%	8.00%	50.00%
Lime (%)	2.00%	4.00%	8.00%
Water repellency pre-application	2200	9000	11,000
Water repellency post-application	1870	7920	9790
Recovery efficiency (%)	15.00%	12.00%	11.00%

Table 4.5 The water repellency of selected soil samples before and after the application of organic matter

Organic matter (%)	0.00%	1.00%	5.00%	10.00%	20.00%
Water repellency pre-application	9000	9000	9000	9000	9000
Water repellency post-application	7650	8955	900	450	180
Recovery efficiency (%)	15.00%	50.00%	90.00%	95.00%	98.00%

Table 4.6 The water repellency of selected soil samples before and after the application of organic matter on different days

Time (mixing)	1	3	5	7	10
Water repellency pre-application	9000	9000	9000	9000	9000
Water repellency post-application	8550	7650	8955	900	90
Recovery efficiency (%)	5.00%	15.00%	50.00%	90.00%	99.00%

The results showed that mixing the soil for 7 days was able to reduce the SWR of the soil sample by 15%. While adding only 1% organic matter to the soil along with mixing for 7 days was able to reduce the SWR by 50%, which is a significant result. At higher concentrations of 5%, 10%, and 20%, it was observed that the soil hydrophobicity decreased by 90%, 95%, and 98%, respectively. Therefore, with these percentages added and mixing for 7 days, the hydrophobicity problem of highly polluted soil was eliminated.

In order to investigate the effect of soil mixing on the effect of organic matter on highly polluted soil samples, the amount of SWR reduction was measured on days 1, 3, 5, 7, and 10 with the addition of 5% organic matter. The decrease in hydrophobicity on days 1, 3, and 5 was 5%, 15%, and 50%, respectively. Therefore, up to 5 days of mixing, the amount of hydrophobicity reduction was not sufficient and appropriate. The results showed that from day 7, the amount of SWR reduction was significant so it reached 99% on day 10 (Table 4.6). This means that with lower amounts of organic matter but more mixing, the reduction in hydrophobicity increases.

4.4 Conclusion

In this study, the effect of various amendments on SWR was investigated. Surfactant, zeolite, and lime at different concentrations did not significantly reduce hydrophobicity, and higher concentrations were not economically feasible. Organic matter, however, showed promise. Mixing soil for 7 days with 1% organic matter resulted in a 50% reduction in hydrophobicity. Increasing organic matter content with continued mixing further enhanced the reduction, completely eliminating hydrophobicity in highly polluted soil samples. Additionally, the effect of organic matter became more pronounced with longer mixing times. These findings suggest that organic matter, particularly when combined with sufficient mixing time, can be an effective and economical strategy for remediating soil water repellency.

4.5 Suggestions

Here are some suggestions for future research:

I. Investigate the effects of the studied amendments (surfactant, zeolite, lime) at even higher concentrations, potentially beyond what was considered economically feasible in this study. This could be valuable for specific scenarios where cost is less of a concern than achieving maximum hydrophobicity reduction.
II. Explore the long-term effects of using organic matter on soil hydrophobicity. This study only investigated the effects over a 7-day period. It would be beneficial to understand how these amendments perform over longer timescales.
III. Test the effectiveness of organic matter on a wider range of soil types. The current study focused on highly polluted soil samples. Examining how organic matter affects hydrophobicity in various soil compositions would provide broader applicability of the findings.
IV. Conduct research on the potential negative consequences of using organic matter for soil remediation. While the study highlights its effectiveness, there might be unforeseen drawbacks, such as impacting soil fertility or introducing unwanted microorganisms.
V. Explore potential combinations of amendments. The study focused on individual amendments. Investigating if combining organic matter with other amendments (surfactant, zeolite, lime) at lower concentrations could achieve similar results while potentially offering economic or efficiency benefits.

Acknowledgments This publication is a detailed account of a technique for "bioremediation of oil-contaminated soil". Parnian Sazeh Chehelsotoun Co. (PSCCO) in Ahvaz, Iran provided financial support for the project. We express our sincere gratitude for the assistance provided by Mr. Reza Parnian, the Chief Executive Officer of PSCCO, and Mr. Mohamad Zaki-Pour, the Project Supervisor of AOGC.

References

Abel S, Peters A, Trinks S, Schonsky H, Facklam M, Wessolek G (2013) Impact of biochar and hydrochar addition on water retention and water repellency of sandy soil. Geoderma 202:183–191

Abou Najm MR, Stewart RD, Di Prima S, Lassabatere L (2021) A simple correction term to model infiltration in water-repellent soils. Water Resour Res 57(2):e2020WR028539

Adamczuk A, Gryta A, Skic K, Boguta P, Jozefaciuk G (2022) Effect of different minerals on water stability and wettability of soil silt aggregates. Materials 15(16):5569

Ameyu T (2019) A review on the potential effect of lime on soil properties and crop productivity improvements. J Environ Earth Sci 9(2):17–23

Asghari HR, Bochmann G, Tabari ZT (2022) Effectiveness of biochar and zeolite soil amendments in reducing pollution of municipal wastewater from nitrogen and coliforms. Sustainability 14(14):8880

Babu N, Poulose E (2018) Effect of lime on soil properties: a review. International Research Journal of Engineering and Technology 5(5)

Badmus SO, Amusa HK, Oyehan TA, Saleh TA (2021) Environmental risks and toxicity of surfactants: overview of analysis, assessment, and remediation techniques. Environ Sci Pollut Res 28:1–20

Bagarello V, Basile G, Caltabellotta G, Giordano G, Iovino M (2020) Testing soil water repellency in a Sicilian area two years after a fire. J Agric Eng 51(2):64–72. https://doi.org/10.4081/JAE.2019.988

Banet G, Turaani AK, Farber R, Armoza-Zvuloni R, Rotem N, Stavi I, Cahan R (2021) The effects of biostimulation and bioaugmentation on crude oil biodegradation in two adjacent terrestrial oil spills of different age, in a hyper-arid region. J Environ Manag 286:112248. https://doi.org/10.1016/J.JENVMAN.2021.112248

Belviso C (2020) Zeolite for potential toxic metal uptake from contaminated soil: a brief review. Processes 8(7):820

Braun B, Böckelmann U, Grohmann E, Szewzyk U (2010) Bacterial soil communities affected by water-repellency. Geoderma 158(3–4):343–351

Caltabellotta G, Iovino M, Bagarello V (2022) Intensity and persistence of water repellency at different soil moisture contents and depths after a forest wildfire. J Hydrol Hydromech 70(4):410–420. https://doi.org/10.2478/JOHH-2022-0031

Chen C, Jiang Y, Sun B, Zhou H, Hallett PD (2022) Organic manure and lime change water vapour sorption of a red soil by altering water repellency and specific surface area. Eur J Soil Sci 73(2):e13223

Cisar JL, Williams KE, Vivas HE, Haydu JJ (2000) The occurrence and alleviation by surfactants of soil-water repellency on sand-based turfgrass systems. J Hydrol 231:352–358

Coelho COA, Laouina A, Regaya K, Ferreira AJD, Carvalho TMM, Chaker M et al (2005) The impact of soil water repellency on soil hydrological and erosional processes under Eucalyptus and evergreen Quercus forests in the Western Mediterranean. Soil Res 43(3):309–318

da Silva Correa H, Blum CT, Galvão F, Maranho LT (2022) Effects of oil contamination on plant growth and development: a review. Environ Sci Pollut Res 29(29):43501–43515

de Bakker H (2013) Major soils and soil regions in the Netherlands. Springer Science & Business Media, New York

Doerr SH, Shakesby RA, Walsh R (2000) Soil water repellency: its causes, characteristics and hydro-geomorphological significance. Earth Sci Rev 51(1–4):33–65

Farzadian M, Hojati S, Sayyad GA, Enayatizamir N (2015) Decreasing water repellency in a petroleum-contaminated soil using zeolite. J Sci Technol Agric Nat Resour 19(72):57–67

Fu Z, Hu W, Beare MH, Müller K, Wallace D, Wai Chau H (2021) Contributions of soil organic carbon to soil water repellency persistence: characterization and modelling. Geoderma 401:115312. https://doi.org/10.1016/J.GEODERMA.2021.115312

García-Carmona M, Arcenegui V, García-Orenes F, Mataix-Solera J (2020) The role of mosses in soil stability, fertility and microbiology six years after a post-fire salvage logging management. J Environ Manag 262:110287

Gavrilescu M (2021) Water, soil, and plants interactions in a threatened environment. Water 13(19):2746

Gholizadeh-Sarabi S, Sepaskhah AR (2013) Effect of zeolite and saline water application on saturated hydraulic conductivity and infiltration in different soil textures. Arch Agron Soil Sci 59(5):753–764

Hervé-Fernández P, Muñoz-Arriagada R, Glucevic-Almonacid C, Bahamonde-Vidal L, Radic-Schilling S (2023) Influence of rangeland land cover on infiltration rates, field-saturated

hydraulic conductivity, and soil water repellency in southern Patagonia. Rangel Ecol Manag 90:92–100
Jakkula VS, Wani SP (2018) Zeolites: potential soil amendments for improving nutrient and water use efficiency and agriculture productivity. Sci Rev Chem Commun 8(1):1–15
Kermani M, Ebadi T (2012) The effect of oil contamination on the geotechnical properties of fine-grained soils. Soil Sediment Contam Int J 21(5):655–671
Khattab SA, Al-Mukhtar M, Fleureau JM (2007) Long-term stability characteristics of a lime-treated plastic soil. J Mater Civ Eng 19(4):358–366
Kostka SJ, Cisar JL, Short JR, Mane S (1997) Evaluation of soil surfactants for the management of soil water repellency in turfgrass. Int Turfgrass Soc Res J 8:485–494
Lehrsch GA, Sojka RE (2011) Water quality and surfactant effects on the water repellency of a sandy soil. J Hydrol 403(1–2):58–65
Li Y, Yao N, Tang D, Chau HW, Feng H (2019) Soil water repellency decreases summer maize growth. Agric For Meteorol 266:1–11
Lowe MA, McGrath G, Mathes F, Leopold M (2017) Evaluation of surfactant effectiveness on water repellent soils using electrical resistivity tomography. Agric Water Manag 181:56–65
Lowe MA, McGrath G, Leopold M (2021) The impact of soil water repellency and slope upon runoff and erosion. Soil Tillage Res 205:104756
Mao J, Nierop KG, Dekker SC, Dekker LW, Chen B (2019) Understanding the mechanisms of soil water repellency from nanoscale to ecosystem scale: a review. J Soils Sediments 19:171–185
Mohammadi S, Homaee M, Sadeghi SH (2018) Runoff and sediment behavior from soil plots contaminated with kerosene and gasoil. Soil Tillage Res 182:1–9
Mondal M, Biswas B, Garai S, Sarkar S, Banerjee H, Brahmachari K et al (2021) Zeolites enhance soil health, crop productivity and environmental safety. Agronomy 11(3):448
Movasat M, Tomac I (2021) Assessment of physical properties of water-repellent soils. J Geotech Geoenviron 147(9):06021010
Müller K, Deurer M (2011) Review of the remediation strategies for soil water repellency. Agric Ecosyst Environ 144(1):208–221
Müller K, Deurer M, Jeyakumar P, Mason K, van den Dijssel C, Green S, Clothier B (2014) Temporal dynamics of soil water repellency and its impact on pasture productivity. Agric Water Manag 143:82–92
Nakhli SAA, Delkash M, Bakhshayesh BE, Kazemian H (2017) Application of zeolites for sustainable agriculture: a review on water and nutrient retention. Water Air Soil Pollut 228:1–34
Ogunmokun FA, Liu Z, Wallach R (2020) The influence of surfactant-application method on the effectiveness of water-repellent soil remediation. Geoderma 362:114081
Olorunfemi IE, Ogunrinde TA, Fasinmirin JT (2014) Soil hydrophobicity: an overview. J Sci Res Rep 3(8):1003–1037
Olsson Å, Persson L, Olsson S (2019) Influence of soil characteristics on yield response to lime in sugar beet. Geoderma 337:1208–1217
Oostindie K, Dekker LW, Wesseling JG, Ritsema CJ (2008) Soil surfactant stops water repellency and preferential flow paths. Soil Use Manag 24(4):409–415
Ossai IC, Ahmed A, Hassan A, Hamid FS (2020) Remediation of soil and water contaminated with petroleum hydrocarbon: a review. Environ Technol Innov 17:100526
Parnian A, Parnian A, Pirasteh-Anosheh H, Furze JN, Prasad MNV, Race M, Hulisz P, Ferraro A (2022) Full-scale bioremediation of petroleum-contaminated soils via integration of co-composting. J Soils Sediments 22(8):2209–2218. https://doi.org/10.1007/s11368-022-03229-5
Sakala GM, Rowel DL, Pilbeam CJ (2007) Alkalinization and re-acidification patterns of a limed acid soil amended with plant residues. Afr Crop Sci Conf Proc 8:1579–1584
Sandor R, Iovino M, Lichner L, Alagna V, Forster D, Fraser M et al (2021) Impact of climate, soil properties and grassland cover on soil water repellency. Geoderma 383:114780
Šařec P, Novák P (2016) Influence of biological transformation of organic matter on improvement of water infiltration ability of modal luvisol. In: 6th international conference on trends in agricultural engineering, TAE, pp 627–632

Schnabel S, Pulido-Fernández M, Lavado-Contador JF (2013) Soil water repellency in rangelands of Extremadura (Spain) and its relationship with land management. Catena 103:53–61

Sharma B, Vaish B, Monika, Singh UK, Singh P, Singh RP (2019) Recycling of organic wastes in agriculture: an environmental perspective. Int J Environ Res 13:409–429

Smettem KRJ, Rye C, Henry DJ, Sochacki SJ, Harper RJ (2021) Soil water repellency and the five spheres of influence: a review of mechanisms, measurement and ecological implications. Sci Total Environ 787:147429

Sun D, Yang H, Guan D, Yang M, Wu J, Yuan F et al (2018) The effects of land use change on soil infiltration capacity in China: a meta-analysis. Sci Total Environ 626:1394–1401

Sun Y, Xiong X, He M, Xu Z, Hou D, Zhang W et al (2021) Roles of biochar-derived dissolved organic matter in soil amendment and environmental remediation: a critical review. Chem Eng J 424:130387

Täumer K, Stoffregen H, Wessolek G (2005) Determination of repellency distribution using soil organic matter and water content. Geoderma 125(1–2):107–115

Williams CJ, Pierson FB, Robichaud PR, Boll J (2014) Hydrologic and erosion responses to wildfire along the rangeland–xeric forest continuum in the western US: a review and model of hydrologic vulnerability. Int J Wildland Fire 23(2):155–172

Young AM, Higuera PE, Duffy PA, Hu FS (2017) Climatic thresholds shape northern high-latitude fire regimes and imply vulnerability to future climate change. Ecography 40(5):606–617

Chapter 5
Microbial Degradation of Petroleum Hydrocarbons in Soil and Sediment: Mechanisms, Challenges, Limitations, and Case Studies

Love Singla

Abstract Soil is the primary habitat for many living organisms. However, human activities and modernization have led to soil pollution, primarily from substances like xenobiotics and persistent pollutants, which have had detrimental effects on soil composition and characteristics. Oil spills pose significant risks to both human health and environmental integrity, impacting economic stability, public health, ecosystems, and societal well-being. Bioremediation techniques are eco-friendly and have a sustainable nature. Leveraging microorganisms for bioremediation shows particular promise, especially in the restoration of soil contaminated with petroleum hydrocarbons (PHCs), including total petroleum hydrocarbons (TPHs), semi-volatile organic compounds (SVOCs), and volatile organic compounds (VOCs). Microbial-based approaches offer advantages such as environmental compatibility, cost-effectiveness, minimal risk of secondary contamination, and ease of application and cultivation. It is imperative to prevent and degrade these pollutants to safeguard soil properties, microbial communities, and overall ecosystem health. This chapter delves into the mechanisms underlying the degradation capabilities of specific microbial taxa (such as bacteria, fungi, and algae) employed in the bioremediation of petroleum-contaminated soil, drawing insights from a case study conducted in the Borhola oil fields of Assam, India, where a remarkable 75% degradation of hydrocarbon contaminants was achieved within a year. Furthermore, the chapter addresses the challenges encountered during the bioremediation process and underscores the inherent limitations of microbiological methodologies.

Keywords Microbes · Bioremediation · Pollution · Sustainability · Community

---

L. Singla (✉)
Department of Microbiology, Maharaja Agrasen University, Baddi, India

A. Parnian et al. (eds.), *Bioremediation of Petroleum Contaminated Soil and Sediment*, Environmental Contamination Remediation and Management,
https://doi.org/10.1007/978-3-031-95524-2_5

5.1 Introduction

The genesis of environmental contamination caused by petroleum and its products is human activity. Petroleum incurs significant losses during extraction, transportation, and storage, resulting in millions of tons lost annually (Apulu et al. 2021). There are several sources (untreated industrial wastes, improper disposal practices, continental runoff, etc.) that lead to the addition of various contaminants of a chemical nature (heavy metals, metalloids, aliphatic and polycyclic aromatic hydrocarbons, and halogenated compounds) into the marine environment (Abbaspour et al. 2020). These contaminants pose a significant threat to the diversity of marine life and the supply of ecosystem benefits and services that contribute to human well-being. The adverse effects are particularly pronounced in coastal ecosystems that experience significant human activity and have limited water movement, resulting in a substantial build-up of chemical pollutants in the sediment (Ansari et al. 2018). In 2005, it was seen that around 1 million hectares area of Russian land was contaminated with poison with different levels of toxicity that required re-cultivation (Bulanova et al. 2005).

Various physicochemical methods have been devised to decrease the levels of pollutants in water, such as ion exchange chromatography, reverse osmosis, and ultrafiltration (Crini and Lichtfouse 2018). Similarly, for the treatment of contaminated sediments, ex-situ approaches involving chemical, electrochemical, and thermal strategies have been developed (Peng et al. 2018). Figure 5.1 illustrates many techniques for removing contaminants from soil.

Nevertheless, it is imperative to discover sustainable and environmentally friendly methods for addressing the pollution in sediments. The toxicity of petroleum to all living organisms has been widely recognized for a considerable period.

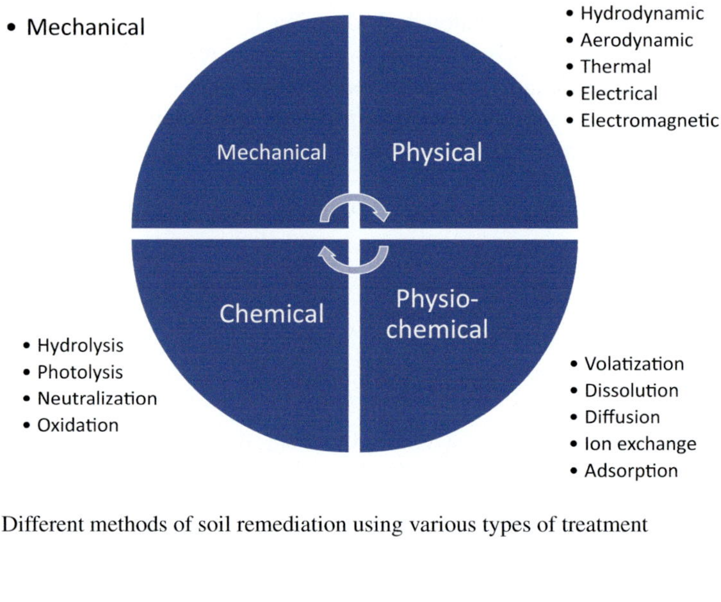

Fig. 5.1 Different methods of soil remediation using various types of treatment

The interaction of petroleum with living cells leads to the weakening of the lipid bilayer of the plasma membrane (Broniatowski et al. 2017). Petroleum has the ability to hinder the genetic processes within a cell, in addition to its effects on membranes (Hisamuddin et al. 2022). Considering the aforementioned, petroleum is considered a hazardous environmental contaminant due to its significant toxic nature and extensive distribution throughout the biosphere. Based on detrimental effects, petroleum, its byproducts, and waste materials are classified as the second most harmful substances, following radiation. This chapter will explore the application of microorganisms, such as bacteria, fungi, and algae, in bioremediation, which involves the decontamination of petroleum-polluted soils.

5.2 Petroleum Contamination

Petroleum spills significantly contribute to hydrocarbon pollution in soil. According to a survey, the annual worldwide marine contamination by petroleum and its products is estimated at 600,000 metric tons. According to estimates, petroleum contamination has affected around 3.5 million European locations (Adipah 2018). Approximately 4.7 million hectares of terrestrial area in China have petroleum-products content that exceeds the established safe threshold (Wang et al. 2017). Different nations and regions employ diverse methods for sampling and transporting petroleum, resulting in varying sources and levels of contamination. Figure 5.2 shows different types of petroleum pollutant origin present in soil. Moreover, pollutants are released into the nearby and underlying soil in both horizontal and vertical directions, as well as into the underground water system, due to rainwater rinsing and leaching. Crude oil typically contains approximately 82%–85% carbon, 10%–14% hydrogen, 0.01%–7% sulfur, 0.02%–2% nitrogen, and 0.1%–1% oxygen, as reported by several sources (Nametkin 1955). Petroleum can be categorized based on its composition into three types: light (density 0.65–0.87 g/cm^3), middle (density 0.87–0.91 g/cm^3), or heavy (0.91–1.05 g/cm^3) (Ryabov 2009). This characteristic is crucial since it directly affects the environmental impact of petroleum (Glazovskaya 1988).

Ecological damage caused by mining and processing incidents involving petroleum leaks is substantial. In such instances, the soil is most impacted as it has the capacity to amass significant quantities of contaminants owing to its vast adsorptive outer area. Petroleum pollution has a detrimental impact on soil biocenosis, causing significant alterations in the chemical composition, structure, and characteristics of soil. This pollution also diminishes soil fertility and its suitability for cultivation. Petroleum spills have the potential to transform soils into technogenic deserts characterized by a significant absence of biological activity. Soils contaminated with petroleum are not appropriate for agricultural or recreational use and might potentially contaminate surface and groundwater. This contamination leads to severe harmful effects such as delayed crop germination, decreased chlorophyll concentration in leaves, and sudden death of some crops grown in soil with petroleum

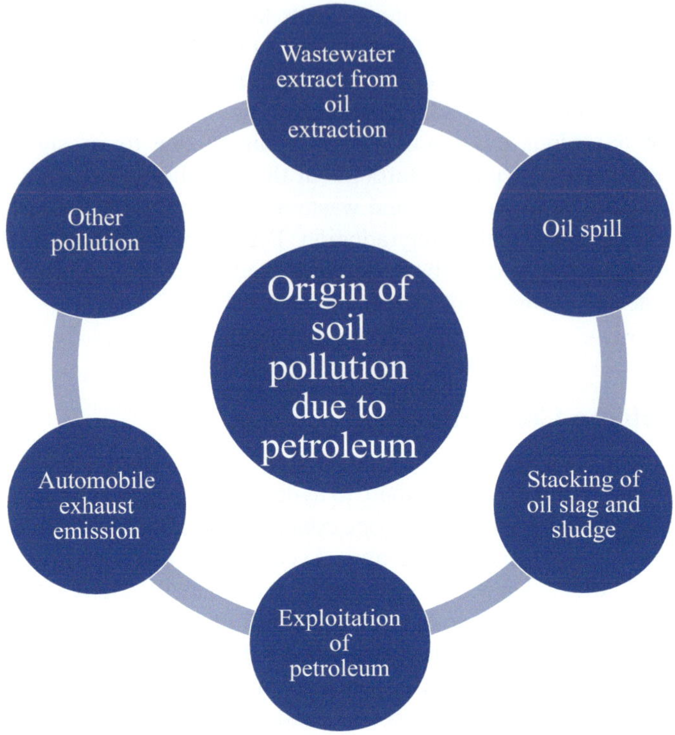

Fig. 5.2 Different sources of soil contamination by petroleum pollutants

contamination at higher levels (Farrington and Takada 2014). Some components of pollutants have the capability to infiltrate the human body via various means, such as inhalation, direct contact with the skin, or consuming food contaminated with petroleum. This can lead to severe skin diseases, visual and auditory impairment, problems in the gastrointestinal tract, and a significantly increased likelihood of leukemia. While hydrocarbons with low molecular weight can be easily degraded with respect to time, higher molecular weights are unable to be degraded and can persist in soil for a longer duration of time because of their hydrophobic nature. This can lead to more pollution in the ecosystem (Halanych et al. 2021; Rahimi Moazampour et al. 2023).

Remediating sites contaminated by petroleum can be arduous for many reasons, and these obstacles go beyond using bioremediation techniques. Several primary obstacles arise when addressing petroleum-contaminated sites:

1. The presence of petroleum contamination typically consists of an intricate combination of hydrocarbons, including volatile organic compounds (VOCs) and semi-volatile organic compounds (SVOCs), posing difficulties in the rehabilitation process.

2. Site heterogeneity refers to the variation in soil and subsurface conditions within a contaminated site. This variability poses difficulties in establishing consistent treatment throughout the site. Heterogeneity influences the spread and dispersion of pollutants, making the planning and execution of remediation methods more complex.
3. The persistence of contaminants refers to the ability of certain petroleum hydrocarbons to remain in the environment for long periods, which poses challenges in achieving total remediation. Soil and groundwater may still contain residual contamination, necessitating continuous monitoring and treatment.
4. The contamination can penetrate the soil, necessitating remediation procedures that address the uppermost layer and the layers beneath it. Penetrating pollution might provide difficulties in terms of accessibility and efficient treatment.
5. The existence of non-aqueous phase liquids (NAPLs), such as crude oil and gasoline, can result in long-lasting soil contaminants that also lead to infecting the groundwater. Removing these compounds can be difficult, and their existence might lead to lasting environmental consequences.
6. Technological Constraints: Certain remediation solutions may have limits in efficiently addressing specific petroleum contaminants or may need to be more suitable for particular site circumstances. Technological progress is consistently tackling these constraints. However, obstacles persist.
7. Ensuring adherence to regulatory standards and criteria can present difficulties during rehabilitation. Rigorous laws may require comprehensive documentation, monitoring, and reporting, increasing the complexity of repair activities.
8. The use of certain remediation technologies, such as excavation, carries the potential for spreading contaminants to nearby areas or releasing pollutants into the air during transportation and disposal.
9. Human and Ecological Health Concerns: Petroleum pollutants can adversely affect health and ecological receptors. Striking a balance between efficient cleanup and limiting possible threats to human health and ecosystems necessitates thoughtful deliberation.
10. Cost and Resource Constraints: The remediation process can require significant resources, and the financial expenses of certain technologies may be too high to be feasible. The availability of insufficient finance and resources may impact the process of identifying and implementing effective remedial techniques.

5.3 Role of Microbes in Bioremediation

Microorganisms play a crucial role in bioremediation. There are various types of microorganisms involved in this process. Microorganisms readily break down saturated hydrocarbons and aromatic hydrocarbons, which possess low molecular weight, but struggle to break down hydrocarbons of petroleum origin with enormous weight due to their ability to resist the degradation ability of microbes. Although bacteria capable of breaking down naphthalene, phenanthrene, and pyrene have been found, breaking down large polycyclic aromatic hydrocarbons is still tricky. The microbial degradation occurs in the following sequence: The order of compounds in terms of increasing molecular weight is as follows: N-alkanes, branched-chain alkanes, branched alkenes, low-molecular-weight n-alkyl aromatics, monoaromatics, cyclic alkanes, polynuclear aromatics, and asphaltenes (Chen et al. 2020). The primary concern impacting the planet is eradicating enduring environmental pollutants. PAHs have become prominent ecological pollutants due to their hydrophobic nature (Premnath et al. 2021). Microorganisms capable of breaking down alkanes and PAHs in a liquid medium containing inorganic salts have been identified and isolated within the specified period, as indicated in Table 5.1 (Heitkamp and Cerniglia 1987).

Comprehending the degradation processes is crucial for enhancing bioremediation techniques and creating effective microbial-driven treatment systems for settings contaminated with hydrocarbons. The process of microbial degradation of hydrocarbons encompasses a range of mechanisms, with distinct microbes playing pivotal roles in the breakdown of these chemical molecules. These methods demonstrate the range of microbial techniques used to break down hydrocarbons. Microbial populations frequently collaborate, each contributing to distinct stages of the degradation process, rendering bioremediation a versatile and viable method in many contexts. Table 5.2 represents organisms involved in involved in the removal of petroleum contamination from the soil with their specific substrates (Dell'Anno et al. 2021; Stepanova et al. 2022; Sui et al. 2021).

1. Aerobic hydrocarbon degradation refers to the process in which aerobic microorganisms break down hydrocarbons in the presence of oxygen. This degradation is facilitated by monooxygenases, dioxygenases, and hydroxylases, which oxidize the hydrocarbons and convert them into more straightforward, soluble molecules. *Pseudomonas aeruginosa* is frequently present in soil and water and is recognized for its capacity to break down a diverse array of hydrocarbons—certain species of *Mycobacterium* bacteria. Microbial organisms are recognized for their ability to break down hydrocarbons through aerobic degradation, and they are commonly found in soil (Leahy and Colwell 1990).

Table 5.1 Different types of PAHs are sequenced based on mineralization rate with their estimated half-life (in weeks) (Sui et al. 2021)

Sr. No.	Different PAHs (based on mineralization rate)	Estimated half-life (in weeks)
1.	Naphthalene	2.4–4.4
2.	Hexadecane	2.2–4.2
3.	Phenanthrene	4–18
4.	2-methyl Base naphthalene	14–20
5.	Pyrene	34–>90
6.	3-methycholanthrene	87–>200
7.	Benzo[a] pyrene	200–>300

Table 5.2 Name of microbes involved in the removal of petroleum contamination from the soil with their specific substrates (Dell'Anno et al. 2021; Stepanova et al. 2022; Sui et al. 2021)

Sr. No.	Name of Microbes	Bacteria/Fungi	Substrate
1.	Achromobacter sp.	Bacteria	PAHs
2.	Acinetobacter sp. WSD	Bacteria	PAHs
3.	Bacillus subtilis BMT4i (MTCC 9447)	Bacteria	PAHs
4.	Caulobacter sp. (T2A12002)	Bacteria	PAHs
5.	Enterobacter sp. (MM087)	Bacteria	PAHs
6.	Klebsiella pneumoniae AWD5	Bacteria	PAHs
7.	Mycobacterium vanbaalenii PYR-1	Bacteria	PAHs
8.	Raoultella planticola	Bacteria	PAHs
9.	Rhodococcus sp. P14	Bacteria	PAHs
10.	Pseudomonas sp. MPDS	Bacteria	PAHs
11.	Pseudoxanthomonas sp. DMVP2	Bacteria	PAHs
12.	Sphinogmonas sp.	Bacteria	PAHs
13.	Stenotrophomonas sp. IITR87	Bacteria	PAHs
14.	Streptomyces sp. (ERI-CPDA-1)	Bacteria	PAHs
15.	Aspergillus sp. RFC-1	Fungus	PAHs
16.	Nocardia sp. H17-1	Fungus	PAHs
17.	Penicillium sp. CHY-2	Fungus	PAHs
18.	Trichoderma sp.	Fungus	PAHs
19.	Fusarium sp.	Fungus	PAHs
20.	Achromobacter sp. HZ01	Bacteria	Alkanes
21.	Acinetobacter sp. (KC211013)	Bacteria	Alkanes
22.	Bacillus subtilis	Bacteria	Alkanes
23.	Pseudomonas sp. WJ6	Bacteria	Alkanes
24.	Rhodococcus sp.	Bacteria	Alkanes
25.	Alcanivorax sp. IO_7	Bacteria	Alkane
26.	Alcanivorax sp. 24	Bacteria	Alkanes
27.	Cupriavidus metallidurans CH34	Bacteria	Alkanes
28.	Cycloclasticus sp. strain BG-2	Bacteria	Alkanes
29.	Cycloclasticus sp. 78-ME	Bacteria	Alkanes
30.	Cycloclasticus sp. strain P1	Bacteria	Alkanes
31.	Halomonas sp. strain MCTG39a	Bacteria	Alkanes
32.	Halomonas pacifica strain Cnaph3	Bacteria	Alkanes
33.	Marinobacter hydrocarbonoclasticus SdK644	Bacteria	Alkanes
34.	Oleispira antarctica RB-8	Bacteria	Alkanes
35.	Pseudomonas aeruginosa N6P6	Bacteria	Alkanes
36.	Pseudomonas pseudoalcaligenes NP103	Bacteria	Alkanes
37.	Pseudomonas sp. sp48	Bacteria	Alkanes
38.	Pseudomonas aeruginosa GOM1	Bacteria	Alkanes
39.	Ralstonia pickettii	Bacteria	Alkanes
40.	Cladosporium resinae	Fungus	Alkanes
41.	Penicillium sp. CHY-2	Fungus	Alkanes
42.	Gordonia sp.	Actinomycetes	Alkanes
43.	Tsukamurella sp. MH1	Actinomycetes	Alkanes

2. Anaerobic hydrocarbon degradation refers to the process by which microorganisms break down hydrocarbons without oxygen. These microorganisms use electron acceptors such as nitrate, sulfate, or carbon dioxide instead of oxygen. As a result of this degradation, methane or other reduced compounds are produced. The Desulfobacteraceae family, which includes sulfate-reducing bacteria such as *Desulfobacteraceae*, can participate in the anaerobic breakdown of hydrocarbons. *Methanosaeta* spp., a specific type of archaea, is recognized for its role in the anaerobic breakdown of hydrocarbons (Widdel and Rabus 2001).
3. Alkane hydroxylation refers to introducing hydroxyl groups into alkanes, increasing their reactivity and making them easier to degrade. *Alcanivorax borkumensis* is present in maritime habitats and is recognized for its capacity to hydroxylate and break down alkanes (Van Beilen and Funhoff 2007).
4. Aromatic hydrocarbon ring cleavage is achieved through the action of dioxygenases or monooxygenases, which catalyze the breaking of aromatic rings in hydrocarbons. This process enables the degradation of various chemicals such as BTEX (benzene, toluene, ethylbenzene, and xylene). *Pseudomonas putida* is recognized for its capacity to metabolize aromatic substances, as it can enzymatically break down aromatic rings during degradation (Fuchs et al. 2011).
5. Surfactant Production: Certain microbes synthesize biosurfactants that increase the solubility of hydrophobic hydrocarbons in water, accelerating their microbial breakdown. Some specific types of *Bacillus* possess biosurfactants, which enhance the bioavailability of hydrocarbons (Antoniou et al. 2015; Marchut-Mikolajczyk et al. 2018).
6. Extracellular Polymeric Substances (EPS) have the ability to create biofilms, which improve the adhesion of cells to hydrocarbon surfaces and facilitate the breakdown process. Several types of bacteria, including *Marinobacter species, Acinetobacter species, Mycobacterium vanbaalenii* and *Mycobacterium smegmatis, Arthrobacter species, Sphingomonas species, and Alcanivorax species*, are present (Flemming and Wingender 2010).

5.4 Advantage of Microbes Over Other Techniques

The utilization of microbes has numerous advantages compared to other treatments (Azubuike et al. 2016), including.

1. Natural and Sustainable: Microbial bioremediation is an innate technique that utilizes the inherent capacities of microorganisms to decompose and disintegrate pollutants. It adheres to the principles of sustainable environmental management by employing biological agents instead of introducing extra chemicals.
2. Economical: Bioremediation is frequently a more cost-effective alternative to conventional approaches like excavation, disposal, or chemical treatments. Due to its small infrastructure and energy requirements, it is a financially feasible choice.

3. Reduces Environmental Disruption: Bioremediation can be performed on-site, eliminating the necessity for digging and transporting polluted soil. This reduces disturbance to the ecology and the surrounding environment.
4. Adaptability: Microorganisms can adjust to diverse environmental circumstances and pollutants. Microbial communities possess a high degree of adaptability, enabling them to remediate various forms of petroleum contaminants effectively.
5. Bioremediation improves soil quality by facilitating the decomposition of organic matter the release of nutrients and eliminating contaminants. It can aid in the reestablishment of the soil's inherent microbial community.
6. Targeted Degradation: Microorganisms have the ability to selectively and efficiently degrade hydrocarbons found in petroleum while having minimal impact on non-target molecules. This selectivity minimizes the likelihood of producing detrimental by-products.
7. Minimized Environmental Impact: Bioremediation has a lesser environmental impact than specific chemical techniques that may bring supplementary pollutants. It is a more ecologically conscious choice, in line with the principles of environmentally friendly and sustainable technologies.
8. Microbial communities can adapt to a polluted site's particular conditions, encompassing temperature fluctuations, pH, and nutrition availability. Bioremediation's capacity to adjust to various geographical and environmental conditions renders it appropriate for a broad spectrum of environments.
9. Microbial communities, once established, exhibit long-term stability by persisting in the environment and effectively degrading pollutants over an extended duration. The enduring stability over an extended time enhances the effectiveness of the repair process.
10. Community and Stakeholder Acceptance: Bioremediation is frequently perceived as more socially acceptable than specific chemical or physical approaches. Communities and stakeholders may prefer environmentally sustainable and minimally disruptive methods.

Bioremediation techniques encompass both in-situ and ex-situ technologies that can remediate contaminated environments using biological processes.

5.4.1 *In Situ Bioremediation*

In situ bioremediation, microorganisms are used to break down or convert pollutants in their original place, without digging or removing the polluted material. The efficacy of in situ bioremediation methods relies on multiple elements, such as the nature of the pollutants, site characteristics, and the composition of the microbial population. Site-specific evaluations and surveillance are commonly carried out to ascertain the optimal methodology (Sharma 2019). Below is a compilation of prevalent in situ bioremediation techniques:

1. Biostimulation refers to introducing nutrients, such as nitrogen, phosphorus, and carbon, to promote the growth of naturally occurring microbes that can improve the breakdown of contaminants. The introduction of oxygen enhances the process of aerobic biodegradation of pollutants.
2. Bioaugmentation involves the introduction of particular strains of microorganisms, such as bacteria or fungi, that can break down or convert specific pollutants. This results in an enhancement of the degradation process.
3. Phytoremediation is a process that utilizes plants and the microbes in their root zone to improve the breakdown, containment, or absorption of pollutants.
4. Bioventing is a technique used to improve the aerobic biodegradation of pollutants in the soil. It involves the injection of air into the subsurface.
5. Biofiltration is the process of directing polluted air or water through beds of microorganisms to facilitate deterioration.
6. Permeable reactive barriers (PRBs) are constructed by filling barriers with reactive substances and microorganisms to treat contaminated groundwater effectively as it passes through.
7. Co-metabolic bioremediation involves the utilization of microorganisms that break down pollutants as a secondary effect while metabolizing a primary substrate.
8. Intrinsic Bioremediation is a process that relies on the natural activity of microorganisms and the surrounding environmental circumstances to decrease the contaminants' levels gradually.
9. Hydrocarbon slurry bioreactors involve using a combination of polluted substances, water, and microorganisms to facilitate the process of biodegradation.
10. Electrokinetic Bioremediation involves the application of an electric field to polluted soil to facilitate the movement of nutrients and microorganisms, hence promoting biodegradation.
11. Microbial Fuel Cells (MFCs) are devices that harness the power of microbial activity to produce electricity while also facilitating the breakdown of pollutants.
12. Bio-extraction involves the stimulation of microorganisms to gather or alter pollutants, thereby reducing their mobility or toxicity.

5.4.2 Ex-Situ Bioremediation

Ex-situ bioremediation is extracting polluted environmental elements, such as soil or water, and transporting them to a different site for treatment. Diverse methodologies are utilized to enhance the bioremediation process. The efficacy of these approaches can vary based on the particular pollutants, environmental circumstances, and site attributes. Evaluating and observing the site is essential for establishing the effectiveness of ex-situ bioremediation endeavors (Hussain et al. 2022). Below is a compilation of conventional ex-situ bioremediation methods:

1. Biopiles include the excavation and placement of contaminated soil into heaps. Supplementing amendments such as nutrition, oxygen, and moisture increases microbial activity. The pile is regularly rotated to facilitate aeration and blended with the soil.
2. Composting involves blending polluted material with organic substances such as yard trash or agricultural wastes, similar to the process used in biopiles. Additionally, it stimulates the proliferation of microbes that decompose the pollutants.
3. Landfarming involves spreading contaminated soil onto a specially prepared bed and regularly tilling it. Microbial activity is enhanced by the introduction of air, regulation of moisture levels, and the supply of nutrients.
4. Aerated Static heaps: Aeration is accomplished like heaps, by directing air through the pile using pipes or other methods. Increases the amount of oxygen accessible for microbial decomposition.
5. Slurry bioreactors include the combination of contaminated soil with water and additional additions to create a mixture. Subsequently, the mixture undergoes treatment in a bioreactor, where the pollutants are broken down by microbial activity.
6. Solid-phase bioextraction is a method that utilizes plants and their accompanying microbes to extract and accumulate pollutants from the soil. The plants undergo harvesting, eliminating the pollutants in conjunction with the plant biomass.
7. Soil Vapour Extraction (SVE) and Bioventing: This method entails removing volatile contaminants from the soil by utilizing a vacuum. Bioventing enhances microbial activity in the unsaturated zone by introducing nutrients and air.
8. Ex-situ anaerobic bioremediation refers to using anaerobic microorganisms to break down and eliminate pollutants. It can be utilized to remediate chlorinated solvents and certain heavy metals.
9. Biological filters, also known as biofilters, direct polluted air through a bed of microorganisms that break down the contaminants. Frequently employed for the remediation of volatile organic compounds (VOCs) in air streams.
10. Rotary drum reactors involve the placement of contaminated soil in a drum that rotates, together with the addition of microbial inoculants and amendments. It offers a regulated setting for the breakdown of microorganisms.
11. Constructed wetlands employ wetland plants and their accompanying microorganisms to remediate polluted water, specifically wastewater containing organic contaminants.

Bioaugmentation and biostimulation are two interrelated techniques employed in bioremediation to amplify the microbial activity and efficacy of natural processes in breaking down contaminants (Lopes et al. 2022). Table 5.3 represents the major differences between bioaugmentation and biostimulation.

Bioaugmentation refers to deliberately introducing particular microorganisms, such as bacteria, fungi, or other microbes, into a polluted environment to improve the breakdown of harmful substances. The objective is to introduce microorganisms

Table 5.3 Difference between bioaugmentation and biostimulation

Sr No.	Basis	Bioaugmentation	Biostimulation
1.	Objective	Enhance the microbial community by introducing organisms that have the ability to break down the specific pollutants of interest.	Establish ideal circumstances to facilitate the proliferation and functionality of native microorganisms.
2.	Microbial origin	Addition of exogenous microorganisms possessing targeted metabolic capabilities.	Reliance on the indigenous microbial population present at a polluted location.
3.	Microbial Diversity	A limited range of particular microbial strains are specifically chosen to target the contaminants of concern.	Enhance the growth of different native microorganisms and increase the variety within the microbial community.
4.	Risk and Long-Term Sustainability	The introduction of non-native organisms carries the danger of ecological disruption, and the success of such introductions depends on the compatibility of the introduced strains with the existing ecosystem in the long term.	Regarded as less risky, as it interacts with the inherent microbial ecology, fostering sustainability.

with the metabolic capacity to degrade or convert particular pollutants that may be present in inadequate amounts or not naturally found in the environment. Bioaugmentation is frequently employed when the indigenous microbial community lacks the metabolic pathways to efficiently degrade specific contaminants. This technique can be utilized in both ex-situ and in-situ cleanup applications.

Biostimulation refers to the process of creating optimal environmental conditions or introducing certain compounds to enhance the development and activity of native microorganisms that are already present in the polluted area. The objective is to augment the inherent capacity of the indigenous microbial community to break down pollutants by supplying necessary nutrients, electron acceptors (such as oxygen or nitrate), or other elements that promote development. Biostimulation is commonly employed when the current microbial population possesses the capacity to break down pollutants but is hindered by factors like nutrient scarcity, oxygen deficiency, or other environmental circumstances. It is frequently utilized in both ex-situ and in-situ bioremediation endeavors.

Utilizing both bioaugmentation and biostimulation is an advantageous approach for improving bioremediation procedures, with the choice between the two relying on the site's unique conditions, the characteristics of the pollutants, and the project's objectives. These approaches are frequently combined to create a more comprehensive repair plan.

5.5 Case Studies

The case studies highlight the significance of considering individual site conditions, maintaining ongoing monitoring, and employing adaptive management to implement microbial bioremediation for soil contaminated with petroleum effectively. Several research is examining the practical implementation, acquired knowledge, and difficulties encountered and resolved in the use of microorganisms for the bioremediation of soil contaminated with petroleum:

5.5.1 The Exxon Valdez Oil Spill Occurred in Alaska, USA (1989) (Exxon Valdez | Oil Spills | Damage Assessment, Remediation, and Restoration Program n.d.)

- **Effective Implementation**: In response to the Exxon Valdez oil spill, a large-scale bioremediation initiative was carried out by combining nutrients with introducing microorganisms capable of breaking down oil. Indigenous microbes were stimulated by adding nutrients such as nitrogen and phosphorus.
- **Takeaways**: The outcome of the bioremediation process was impacted by factors such as the efficiency of the selected microbial strains and the meticulous evaluation of conditions particular to the location. It was essential to continuously monitor and modify the bioremediation method in response to ongoing evaluations.
- **Challenges and Solutions**: The challenges encompassed the unforgiving Arctic environment and the meticulous selection of bioremediation agents. The resolution entailed employing cryophilic microbes and consistently monitoring the ambient conditions.

5.5.2 Guanabara Bay, Located in Rio de Janeiro, Brazil, on the 18th of January, 2000 (Short 2003)

- **Effective Implementation**: Due to industrial activity, Guanabara Bay had substantial petroleum contamination. The contaminated sediments were treated via bioremediation, which involved the introduction of a consortium of bacteria capable of digesting hydrocarbons.
- **Takeaways**: The bioremediation project greatly benefited from the essential aspect of effectively communicating with stakeholders and the public, which was vital in garnering support. Enhancing success rates was achieved by comprehending the sediment properties and customizing the microbial community to suit the specific local conditions.

- **Challenges and Solutions**: The challenges encompassed the ever-changing characteristics of the bay environment and the requirement for ongoing surveillance. The solutions entailed employing adaptive management tactics and utilizing biodegradable nutrient carriers to maintain microbial activity.

5.5.3 Bemidji, Minnesota, USA (1994–1996) (U.S. Department of the Interior and U.S. Geological Survey 1998)

- **Effective Implementation**: Bemidji successfully executed a large-scale bioremediation initiative to address soil contamination caused by petroleum. The addition of oxygen and nutrients boosted Indigenous microbial activity.
- **Key Takeaway**: It was crucial to comprehend the microbial population specific to the site and adapt biostimulation tactics accordingly. Efficient contact with the community facilitated the resolution of complaints and fostered the development of trust.
- **Challenges and Solutions**: The challenges encompassed fluctuations in soil conditions and the necessity to maintain a balance in fertilizer addition. The solutions entailed regular monitoring, fine-tuning nutrient amounts, and implementing novel modification delivery techniques.

5.5.4 Belgium Refinery Site (Country & Territory Profiles Belgium a Summary of Oil Spill Response Arrangements & Resources Worldwide 2018)

- **Effective Implementation**: The remediation of petroleum-contaminated soil at a refinery site in Belgium was achieved by successfully using a combination bioaugmentation strategy using specialist microbial strains and biostimulation through nutrient addition.
- **Key Takeaways**: Thoroughly examining and comprehending the hydrogeological conditions were essential for successfully executing the project. Enhancing effectiveness was achieved by tailoring the microbial community to match the specific pollutants found at the site.
- **Challenges and Solutions**: The challenges encompassed the diversity of the underground layers and the requirement for effective nutrient distribution. The solutions entailed employing bioaugmentation with specifically modified strains and creating a customized nutrient delivery system.

5.6 Metadata Analysis Regarding Petroleum Contamination: Oil Spill

5.6.1 Introduction

Metadata is a type of data that offers information about other data, aiding in its description, organization, and management. Metadata encompasses many attributes such as file names, creation dates, authorship, keywords, tags, and other descriptors that offer contextual information and organization to the corresponding data. Metadata analysis is extracting essential knowledge, patterns, or trends from metadata. It helps with information retrieval, data management, and decision-making. The keyword "Oil spill" was utilized for metadata analysis on sciencedirect.com which is represented in Fig. 5.3.

5.6.2 Monitoring and Assessment

Monitoring and assessment play vital roles in bioremediation initiatives by evaluating the progress and efficacy of the treatment. When addressing petroleum-contaminated soil and employing microorganisms for bioremediation, diverse approaches can be utilized to evaluate the microbial activity and the physicochemical alterations at the location. Ongoing surveillance and evaluation during the bioremediation process enable modifications to treatment options and guarantee the efficacy of the remediation project. Systematic information gathering and analysis facilitate comprehension of the intricate interplay among microorganisms, pollutants, and environmental variables. The following are crucial elements of surveillance and assessment:

5.6.2.1 Biomarker Analysis

Biomarker analysis refers to quantifying distinct indications or markers that indicate microorganisms' existence, functioning, or reaction to petroleum hydrocarbons (Alimzhanova and Abdykarimov 2023; Garcia et al. 2019). Examples include

1. Phospholipid Fatty Acid (PLFA) Analysis: PLFAs are lipid molecules that comprise the membranes of bacteria, fungi, and other microorganisms. Alterations in PLFA profiles can serve as indicators of changes in the organization of microbial communities.
2. Dehydrogenase Activity: This enzyme participates in microbial respiration and can indicate the total microbial activity.

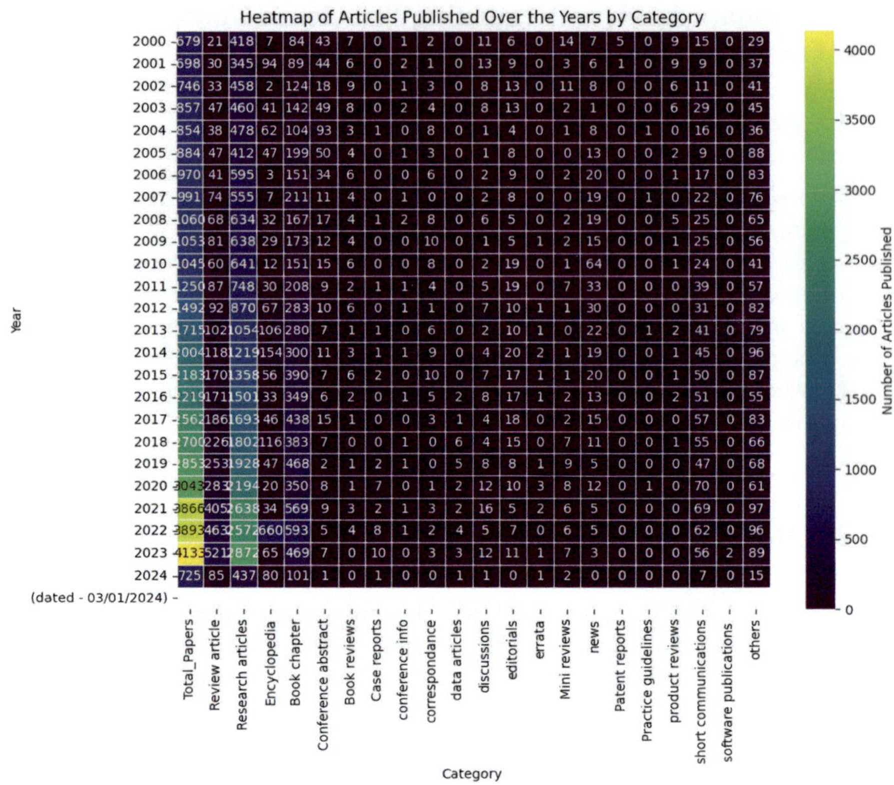

Fig. 5.3 Metadata analysis data of keyword 'Oil Spill'. (Data taken from ScienceDirect, from 2000 to 2024)

3. Genetic Markers: The presence of genes related to hydrocarbon-degrading microorganisms (such as alkane monooxygenase genes) can be detected using the polymerase chain reaction (PCR).

5.6.2.2 Microbial Dynamic Population

Monitoring the dynamics of microbial populations allows for the evaluation of how indigenous or introduced microorganisms respond to petroleum pollutants and their capacity to break them down (Hamamura et al. 2006). Some of the methods include-

1. Quantitative PCR (qPCR) measures the abundance of specific microbial genes linked to hydrocarbons' breakdown.
2. Denaturing Gradient Gel Electrophoresis (DGGE) or High-Throughput Sequencing: These methods are used to examine alterations in microbial communities' composition over time.

5.6.2.3 Physiochemical Parameters

The main objective of assessing physicochemical characteristics is to evaluate any alterations in these properties and gain a deeper understanding of the general circumstances at the remediation site and the impact of microbial activity (Pathak et al. 2011; Saikia et al. 2023). The monitored parameters:

1. pH: Variations in pH can impact microbial activity and the accessibility of nutrients.
2. Temperature: Microbial activity is frequently influenced by temperature.
3. Redox Potential: Indicates the oxidation-reduction state of the environment.
4. Monitoring the levels of nutrients such as nitrogen, phosphorus, and other substances that facilitate the growth of microorganisms.
5. Monitoring the levels of individual hydrocarbons consistently and over a period of time.

5.6.2.4 Soil Sampling and Analysis

Regular soil sampling is crucial for evaluating the spatial dispersion of pollutants and bioremediation efficiency throughout the site. The methodology involves the collection of soil samples from various depths and locations with the purpose of analyzing hydrocarbon concentrations, microbial populations, and physicochemical properties.

1. **Gas Emission Monitoring**: The main objective is to monitor gas emissions, specifically carbon dioxide (CO_2) and methane (CH_4), to gain insights into the metabolic processes of microorganisms. Gas chromatography or similar analytical methods measure gas concentrations in soil or the space above treatment units.
2. **Ecotoxicological assessments**: Ecotoxicological assessments are conducted to determine the ecological consequences of the bioremediation process on soil and aquatic habitats. The toxicity of soil or water samples is determined through bioassays, which involve the assessment of selected indicator species (Musa et al. 2022; Zawierucha et al. 2022).

5.6.3 Future Trends and Innovations

Future research in these areas is to enhance bioremediation systems' efficiency, sustainability, and applicability for soil contaminated by petroleum. Future advancements are expected to be propelled by continuous multidisciplinary partnerships involving microbiologists, geneticists, engineers, and environmental scientists (Dutta and Shityakov 2021). Possible forthcoming developments and domains of investigation in the field of bioremediation:

1. **Genetic Engineering for Enhanced Biodegradation**: Genetic engineering can boost biodegradation by developing microorganisms with an improved ability to break down particular petroleum hydrocarbons. Engineered bacteria or fungi can be created to produce new enzymes or pathways to enhance breakdown efficiency.
2. **Omics Technologies for Microbial Community Analysis**: Omics technologies, including metagenomics, metatranscriptomics, and metaproteomics, can be integrated to understand microbial communities participating in bioremediation processes thoroughly. This methodology enables scientists to pinpoint crucial microbial species and functional genes accountable for the breakdown of hydrocarbons.
3. **Synthetic Biology Approaches**: Synthetic biology techniques can be used to create artificial groups of microorganisms that are specifically designed to break down petroleum hydrocarbons. This technique can facilitate the development of microbial communities tailored explicitly to different environmental conditions and pollutants.
4. **Nanotechnology for Enhanced Delivery of Nutrients and Microorganisms**: Nanoparticles have the ability to enhance the transportation of nutrients, electron acceptors, or microorganisms to contaminated places. Microbial consortia can be nanoencapsulated to improve their viability and functionality in challenging environmental circumstances.
5. **Use of Microbial Electrochemical Systems (MES)**: Microbial Electrochemical Systems (MES) are utilized for in-situ bioremediation of contaminated settings by facilitating the contact between microorganisms and electrodes. These systems can improve microbial metabolism and facilitate the breakdown of petroleum hydrocarbons.
6. **Application of Artificial Intelligence (AI) and Machine Learning (ML)**: The utilization of Artificial Intelligence (AI) and Machine Learning (ML) involves the application of algorithms to examine intricate datasets produced from monitoring and assessment endeavors. These technologies can be utilized to forecast the most favorable conditions for bioremediation, ascertain the pivotal components that influence microbial activity, and refine treatment procedures.
7. **Phytoremediation-Microbial Interactions**: Combining plant-based phytoremediation and microbial augmentation techniques may result in synergistic results. Specific plant species can stimulate microbial activity in the rhizosphere, establishing a favorable setting for biodegradation.
8. **Bioinformatics Tools for Predictive Modeling**: Advancements in bioinformatics technologies can aid in developing prediction models to forecast bioremediation outcomes. These models use environmental parameters, microbial diversity, and pollutant properties to forecast the efficacy of particular bioremediation approaches.
9. **Ecological Engineering Approaches**: Ecological engineering approaches involve incorporating ecological engineering ideas into bioremediation solutions, which entails manipulating the ecosystem to optimize natural degrada-

tion processes. This strategy considers the interplay between bacteria, plants, and other species in the surrounding ecosystem.

10. **Remote Sensing and Monitoring Technologies**: The progress in remote sensing technologies, including satellite imaging and drone-based monitoring, enables the evaluation of contamination levels and the tracking of variations in vegetation and soil characteristics across extensive regions.

5.6.4 Regulatory Considerations

The utilization of microorganisms for the bioremediation of soil contaminated with petroleum is subject to a range of environmental rules and requirements for compliance. Regulatory factors are essential to guarantee that bioremediation initiatives are carried out in an environmentally conscientious way. Below are a few crucial factors to consider regarding regulations in the field of bioremediation:

1. **Permitting and Approval**: Obtaining the requisite permits and approvals from relevant regulatory bodies is imperative before commencing a bioremediation project. These permissions may encompass environmental impact studies, discharge permits, or other authorizations, contingent upon the project's geographical location and magnitude.
2. **Adherence to Environmental rules**: Bioremediation initiatives must conform to local, national, and international environmental rules and regulations. These may encompass legislation about water quality, air pollution, waste management, and land use.
3. **Monitoring and Reporting Requirements**: Regulatory authorities frequently mandate the submission of monitoring and reporting data at various stages of the bioremediation process. This may encompass periodic reports on the advancement of the remediation process, alterations in contaminant levels, and the treated soil's condition.
4. **Health and Safety Regulations**: Compliance with health and safety rules is necessary to protect workers engaged in the bioremediation process. This entails furnishing suitable personal protective equipment (PPE) and guaranteeing compliance with safety regulations.
5. **Waste Management**: Waste materials such as soil polluted with pollutants or biomass resulting from microbiological processes must be handled in compliance with waste disposal rules. It is crucial to adhere to appropriate disposal methods, such as treatment or disposal in authorized facilities.
6. **Community and Stakeholder Engagement**: Numerous regulatory systems emphasize the involvement and participation of the community and stakeholders. This entails ensuring that local communities and pertinent stakeholders are well-informed about the bioremediation project, addressing any concerns, and integrating their comments into the decision-making process.

7. **Risk Assessment and Mitigation**: A comprehensive risk assessment is frequently necessary before and throughout bioremediation endeavors. This evaluation aids in the identification of potential environmental and human health hazards linked to the remediation process, and it is crucial to apply suitable steps to mitigate these risks.
8. **Documentation and Record-keeping**: Thorough documentation and record-keeping are crucial for the bioremediation process, encompassing precise techniques, monitoring outcomes, and any variations from the initial plan. Thorough documentation must be kept to show adherence to regulatory obligations.
9. **Long-Term Monitoring and Post-Remediation Requirements**: Regulatory bodies may require long-term monitoring of the remediated site to verify that pollutants are maintained within permitted limits. After the repair process, there may be a need for periodic reporting and continuous assessments.
10. **Public Disclosure and Transparency**: Maintaining transparent communication with the public is frequently mandated by regulations. This includes disseminating information regarding the bioremediation project, its objectives, and strategies to mitigate adverse environmental effects.

Project managers and stakeholders engaged in bioremediation initiatives must prioritize staying updated on the precise environmental requirements relevant to their region. To ensure the successful and compliant execution of bioremediation activities, it is essential to collaborate closely with regulatory agencies, get the required permissions, and adhere to set rules. Furthermore, obtaining guidance from specialists and engaging in consultations with environmental experts can significantly assist in efficiently navigating the complex regulatory environment.

5.6.5 Conclusion (Metadata Analysis Regarding Petroleum Contamination: Oil Spill)

Using microorganisms for bioremediation in petroleum-contaminated soil has shown considerable promise as an ecologically sound and economically efficient method for reducing the effects of hydrocarbon pollutants. The critical discoveries from the study and practical applications can be summarized as follows:

1. Microorganisms are crucial in breaking down various petroleum hydrocarbons, such as complex molecules like polycyclic aromatic hydrocarbons (PAHs) and aliphatic hydrocarbons.
2. The adaptability of indigenous microbial communities refers to the impressive capacity of microorganisms native to an area to adjust and flourish in settings contaminated with petroleum. This adaptability plays a crucial role in the effectiveness of in-situ bioremediation methods.
3. Synergistic approaches involve the combination of bioaugmentation, biostimulation, and other new methods to improve the overall efficiency of bioremedia-

tion. This approach is particularly useful in tackling unique issues at a site and achieving better results.

4. Advances in monitoring technology, like omics studies and remote sensing, offer valuable insights into the dynamics of microbial communities, pathways for pollutant degradation, and the overall success of remediation.
5. Bioremediation is acknowledged for its sustainability and cost-effectiveness due to its utilization of natural processes and reduction of substantial site excavation. Furthermore, it frequently demonstrates cost-effectiveness in comparison to conventional cleanup techniques.

5.7 Conclusion

Microbial bioremediation of soil contaminated with petroleum is a leading approach to sustainable environmental management. The sector constantly evolves due to continuous study, technical breakthroughs, and a solid dedication to regulatory compliance. Amidst the difficulties caused by human activities, bioremediation presents a hopeful approach to remediate polluted areas, safeguard ecosystems, and guarantee a more pristine and healthier environment for future generations. Effectively utilizing microbial-based treatments for remediating petroleum-contaminated soil will require close collaboration among scientists, environmental practitioners, regulatory agencies, and industry stakeholders.

5.8 Suggestions

Expanding research to varied geographic regions and climates will facilitate the development of universally applicable bioremediation systems that consider regional variations in microbial communities and environmental circumstances, thereby enabling global application. The following suggestions are made to assist the use of microbial degradation in bioremediation:

I. Future studies should focus on comprehensively investigating the variety and functionality of microbial communities in various contexts to enhance the effectiveness of bioremediation technologies.
II. Exploring genetic engineering and synthetic biology methods may create customized microbes with incredible degrading skills and increased resistance to environmental conditions.
III. Using cutting-edge technologies, such as artificial intelligence and nanotechnology, in bioremediation processes can enhance treatment conditions and improve efficiency.

IV. Research should prioritize the development of reliable long-term monitoring procedures to evaluate the durability of remediation results and to analyze any potential ecological consequences on the surrounding ecosystem.

References

Abbaspour A, Zohrabi F, Dorostkar V, Faz A, Acosta JA (2020) Remediation of an oil-contaminated soil by two native plants treated with biochar and mycorrhizae. J Environ Manag 254:109755. https://doi.org/10.1016/J.JENVMAN.2019.109755

Adipah S (2018) Introduction of petroleum hydrocarbons contaminants and its human effects. J Environ Sci Public Health 3(1):1–9. http://www.fotunejournals.com/introduction-of-petroleum-hydrocarbons-contaminants-and-its-human-effects.html

Alimzhanova M, Abdykarimov B (2023) Analysis of Kazakhstan crude oil biomarkers by gas chromatography in combination with mass spectrometry. Separations 10(11):561. https://doi.org/10.3390/SEPARATIONS10110561

Ansari N, Hassanshahian M, Ravan H (2018) Study the microbial communities' changes in desert and farmland soil after crude oil pollution. Int J Environ Res 12(3):391–398. https://doi.org/10.1007/S41742-018-0099-6/METRICS

Antoniou E, Fodelianakis S, Korkakaki E, Kalogerakis N (2015) Biosurfactant production from marine hydrocarbon-degrading consortia and pure bacterial strains using crude oil as carbon source. Front Microbiol 6:133025. https://doi.org/10.3389/FMICB.2015.00274/BIBTEX

Apulu OG, Potravny IM, Sukhorukova IV (2021) Methods of justification and selection of technologies for remediation of oil-contaminated land. Ecol Ind Russia 25(6):38–43. https://doi.org/10.18412/1816-0395-2021-6-38-43

Azubuike CC, Chikere CB, Okpokwasili GC (2016) Bioremediation techniques–classification based on site of application: principles, advantages, limitations and prospects. World J Microbiol Biotechnol 32(11):1–18. https://doi.org/10.1007/S11274-016-2137-X

Broniatowski M, Binczycka M, Wójcik A, Flasiński M, Wydro P (2017) Polycyclic aromatic hydrocarbons in model bacterial membranes – Langmuir monolayer studies. Biochim Biophys Acta Biomembr 1859(12):2402–2412. https://doi.org/10.1016/J.BBAMEM.2017.09.017

Bulanova AV, Gretskova IV, Muratova OV (2005) Sorption properties of sorbents used for cleaning up soils from oil pollution. Vestn Samara Univ Nat Sci Ser 3:150–158

Chen YA, Grace Liu PW, Whang LM, Wu YJ, Cheng SS (2020) Effect of soil organic matter on petroleum hydrocarbon degradation in diesel/fuel oil-contaminated soil. J Biosci Bioeng 129(5):603–612. https://doi.org/10.1016/J.JBIOSC.2019.12.001

Country & territory profiles Belgium a summary of oil spill response arrangements & resources worldwide (2018). http://www.itopf.org/knowledge-resources/countries-territories-regions/

Crini G, Lichtfouse E (2018) Wastewater treatment: an overview. In: Crini G, Lichtfouse E (eds) Green adsorbents for pollutant removal: fundamentals and design. Springer International Publishing, pp 1–21. https://doi.org/10.1007/978-3-319-92111-2_1

Dell'Anno F, Rastelli E, Sansone C, Dell'Anno A, Brunet C, Ianora A (2021) Bacteria, fungi and microalgae for the bioremediation of marine sediments contaminated by petroleum hydrocarbons in the omics era. Microorganisms 9(8):1695. https://doi.org/10.3390/MICROORGANISMS9081695

Dutta K, Shityakov S (2021) New trends in bioremediation technologies toward environment-friendly society: a mini-review. Front Bioeng Biotechnol 9:666858. https://doi.org/10.3389/FBIOE.2021.666858/BIBTEX

Exxon Valdez | Oil Spills | Damage Assessment, Remediation, and Restoration Program (n.d.) Retrieved 6 Jan 2024, from https://darrp.noaa.gov/oil-spills/exxon-valdez

Farrington JW, Takada H (2014) Persistent organic pollutants (POPs), polycyclic aromatic hydrocarbons (PAHs), and plastics: examples of the status, trend, and cycling of organic chemicals of environmental concern in the ocean. Oceanography 27(1):196–213

Flemming HC, Wingender J (2010) The biofilm matrix. Nat Rev Microbiol 8(9):623–633. https://doi.org/10.1038/nrmicro2415

Fuchs G, Boll M, Heider J (2011) Microbial degradation of aromatic compounds – from one strategy to four. Nat Rev Microbiol 9(11):803–816. https://doi.org/10.1038/NRMICRO2652

Garcia MR, Cattani AP, da Cunha Lana P, Figueira RCL, Martins CC (2019) Petroleum biomarkers as tracers of low-level chronic oil contamination of coastal environments: a systematic approach in a subtropical mangrove. Environ Pollut 249:1060–1070. https://doi.org/10.1016/J.ENVPOL.2019.03.006

Glazovskaya MA (1988) Geokhimiya prirodnykh i tekhnogennykh landshaftov SSSR. Geochemistry of Natural and Technogenic Landscapes of the Soviet Union. Vysshaya Shkola, Moscow

Halanych KM, Ainsworth CH, Cordes EE, Dodge RE, Huettel M, Mendelssohn IA, Murawski SA, Paris-Limouzy CB, Schwing PT, Shaw RF (2021) Effects of petroleum by-products and dispersants on ecosystems. Oceanography 34(1):152–163

Hamamura N, Olson SH, Ward DM, Inskeep WP (2006) Microbial population dynamics associated with crude-oil biodegradation in diverse soils. Appl Environ Microbiol 72(9):6316. https://doi.org/10.1128/AEM.01015-06

Heitkamp MA, Cerniglia CE (1987) Effects of chemical structure and exposure on the microbial degradation of polycyclic aromatic hydrocarbons in freshwater and estuarine ecosystems. Environ Toxicol Chem 6(7):535–546. https://doi.org/10.1002/ETC.5620060706

Hisamuddin NH, Jalaludin J, Bakar SA, Latif MT (2022) The influence of environmental polycyclic aromatic hydrocarbons (PAHs) exposure on DNA damage among school children in urban traffic area, Malaysia. Int J Environ Res Public Health 19(4):2193. https://doi.org/10.3390/IJERPH19042193/S1

Hussain A, Rehman F, Rafeeq H, Waqas M, Asghar A, Afsheen N, Rahdar A, Bilal M, Iqbal HMN (2022) In-situ, ex-situ, and nano-remediation strategies to treat polluted soil, water, and air—a review. Chemosphere 289:133252. https://doi.org/10.1016/J.CHEMOSPHERE.2021.133252

Leahy JG, Colwell RR (1990) Microbial degradation of hydrocarbons in the environment. Microbiol Rev 54(3):305. https://doi.org/10.1128/MR.54.3.305-315.1990

Lopes PRM, Cruz VH, de Menezes AB, Gadanhoto BP, de Almeida Moreira BR, Mendes CR, Mazzeo DEC, Dilarri G, Montagnolli RN (2022) Microbial bioremediation of pesticides in agricultural soils: an integrative review on natural attenuation, bioaugmentation and biostimulation. Rev Environ Sci Biotechnol 21(4):851–876. https://doi.org/10.1007/S11157-022-09637-W

Marchut-Mikolajczyk O, Drozdzyński P, Pietrzyk D, Antczak T (2018) Biosurfactant production and hydrocarbon degradation activity of endophytic bacteria isolated from Chelidonium majus L. Microb Cell Factories 17(1):1–9. https://doi.org/10.1186/S12934-018-1017-5/FIGURES/1

Musa Z, Al-Asadi MK, Khwedem AA, al Kushimoto M, Zhang Z, Xiao H (2022) Ecotoxicity risk study of petroleum hydrocarbon contaminated soil. IOP Conf Ser: Earth Environ Sci 1035(1):012018. https://doi.org/10.1088/1755-1315/1035/1/012018

Nametkin SS (1955) Himija nefti. Izd. Akad. nauk SSSR

Pathak H, Bhatnagar K, Jaroli DP (2011) Physico-chemical properties of petroleum polluted soil collected from Transport Nagar (Jaipur). Indian J Fundam Appl Life Sci 1(3):84–89. http://www.cibtech.org/jls.htm

Peng W, Li X, Xiao S, Fan W (2018) Review of remediation technologies for sediments contaminated by heavy metals. J Soils Sediments 18(4):1701–1719. https://doi.org/10.1007/S11368-018-1921-7

Premnath N, Mohanrasu K, Guru Raj Rao R, Dinesh GH, Prakash GS, Ananthi V, Ponnuchamy K, Muthusamy G, Arun A (2021) A crucial review on polycyclic aromatic hydrocarbons –

environmental occurrence and strategies for microbial degradation. Chemosphere 280:130608. https://doi.org/10.1016/J.CHEMOSPHERE.2021.130608

Rahimi Moazampour S, Nabavi SMB, Mohammadi Roozbahani M, Khodadadi M (2023) Determination of total petroleum hydrocarbons and selected heavy metal (Pb, CO, V, Ni) concentration levels in surficial sediments of the Arvand River Estuary and their impact on benthic macroinvertebrates assemblages. Int J Environ Anal Chem 103(12):2841–2857. https://doi.org/10.1080/03067319.2021.1900143

Ryabov VD (2009) Khimiya nefti i gaza. Shelemina, NV, Ed

Saikia T, Bora J, Gogoi C (2023) Impact of crude oil contamination on soil physicochemical properties around the sildubi oil spill, Borhola, Assam. Int J Plant Soil Sci 35(22):241–253. https://doi.org/10.9734/IJPSS/2023/V35I224130

Sharma J (2019) Recent advances in biology and medicine advantages and limitations of in situ methods of bioremediation advantages and limitations of in situ methods of bioremediation. Recent Adv Biol Med 1. https://doi.org/10.18639/RABM.2019.955923

Short MKJ (2003) Guanabara bay oil spill 2000, Brazil – cetacean response. Int Oil Spill Conf Proc 2003(1):1035–1037. https://doi.org/10.7901/2169-3358-2003-1-1035

Stepanova AY, Gladkov EA, Osipova ES, Gladkova OV, Fernandes-Júnior PI, Stepanova AY, Gladkov A, Osipova ES, Gladkova OV, Tereshonok DV (2022) Bioremediation of soil from petroleum contamination. Processes 10(6):1224. https://doi.org/10.3390/PR10061224

Sui X, Wang X, Li Y, Ji H (2021) Remediation of petroleum-contaminated soils with microbial and microbial combined methods: advances, mechanisms, and challenges. Sustainability 13(16):9267. https://doi.org/10.3390/SU13169267

U.S. Department of the Interior, & U.S. Geological Survey (1998) Ground-water contamination by crude oil near Bemidji, Minnesota. USGS Fact Sheet, pp 84–98

Van Beilen JB, Funhoff EG (2007) Alkane hydroxylases involved in microbial alkane degradation. Appl Microbiol Biotechnol 74(1):13–21. https://doi.org/10.1007/S00253-006-0748-0

Wang S, Xu Y, Lin Z, Zhang J, Norbu N, Liu W (2017) The harm of petroleum-polluted soil and its remediation research. AIP Conf Proc 1864(1):20222. https://doi.org/10.1063/1.4993039/629189

Widdel F, Rabus R (2001) Anaerobic biodegradation of saturated and aromatic hydrocarbons. Curr Opin Biotechnol 12(3):259–276. https://doi.org/10.1016/S0958-1669(00)00209-3

Zawierucha I, Malina G, Herman B, Rychter P, Biczak R, Pawlowska B, Bandurska K, Barczynska R (2022) Ecotoxicity and bioremediation potential assessment of soil from oil refinery station area. J Environ Health Sci Eng 20(1):337–346. https://doi.org/10.1007/S40201-021-00780-0/FIGURES/4

Chapter 6
Mycoremediation Mechanisms of Petroleum- Contaminated Soil or Sediments

Anamita Sen, Biswajit Jena, Manoj Kumar Prajapati, Bipasha Datta, Sudip Sarkar, and Anupam Kumari

Abstract Fungi can degrade a wide range of contaminants due to their ability to synthesize extracellular enzymes that break down complex polysaccharides. Mycoremediation is a subset of bioremediation that involves fungi to degrade contaminants including petroleum pollutants. Despite the rich ecology, limited fungal species are associated with mycoremediation. The principal mechanisms of mycoremediation include oxidation, reduction, methylation, and demethylation processes that detoxify hazardous contaminants. *Trichoderma, Penicillium, Rhizopus, Aspergillus, Fusarium,* and *Curvularia* fungal genera are involved in the mycoremediation process. *Basidiomycetes*, wood rotting fungi play a crucial role in bioremediation of organic contaminants. *Trametes versicolor* and *Pleurotus* spp. have greater capacity of petroleum breakdown compared to weak competitors *Ganoderma lucidum*. *Polyporus* spp. are white rot fungi that produce enzymes such as manganese peroxidase, and lignin peroxidase and successfully degrade or uptake petro-

A. Sen
School of Agriculture Sciences and Forestry, Sardar Bhagwan Singh University, Balawala, Dehradun, Uttarakhand, India

B. Jena
Faculty of Agriculture and Allied Sciences, C V. Raman Global University, Bhubaneswar, Odisha, India

M. K. Prajapati
United University Rawatpur Jhalwa, Prayagraj, Uttar Pradesh, India

B. Datta
Department of Genetics and Plant Breeding, School of Crop Improvement, College of Post Graduate Studies in Agricultural Sciences, Central Agricultural University, Umiam, Meghalaya, India

S. Sarkar (✉)
ICAR-Research Complex for Eastern Region, Patna, Bihar, India

A. Kumari
Department of Plant Pathology, BPSAC, Purnea, Bihar Agricultural University, Sabour, Bihar, India

A. Parnian et al. (eds.), *Bioremediation of Petroleum Contaminated Soil and Sediment*, Environmental Contamination Remediation and Management,
https://doi.org/10.1007/978-3-031-95524-2_6

leum hydrocarbons, chlorinated insecticides, and polychlorinated biphenyl. *Trametes hirsuta* can cause 100% degradation of hydrocarbons whereas *Fusarium proliferatum* can degrade up to 95%. *Trametes versicolor* can degrade petrochemicals with laccase enzymes. Intrinsic bioremediation, biosparging, bioaugmentation, and bioventing are advanced and suitable in situ methods of mycoremediation that are cost-effective, as excavation of contaminants is avoided. Overall, the application of different fungi remediate petroleum contaminants in an eco-friendly manner with little or no effect on the ecosystem.

Keywords Fungi · Contaminants · Enzymes · Biodegradation · Eco-friendly

6.1 Introduction

Soil is an intricate nexus where living organisms and non-living factors interact. Humans, animals, and bacteria all benefited from healthy soil for growth and survival. However, over the last few decades, extensive exploitation has made soil unhealthy and unsuitable for supporting healthy plants. The principal barrier to recovering and safeguarding the life-supporting system in the soil is industrialization in addition to the extensive use of chemicals and release of petroleum byproducts. Pollutants have durable effects on the environment and eventually find their way into the agri-food chain because they are difficult to degrade into harmless forms. Exposing the environment to petrochemicals is detrimental to all living things because of their noxious, tenacious, and non-degradable nature (Hamba 2016). In most cases, toxic hydrocarbons act by disrupting proteins including enzymes either by removing an essential ion from biomolecules or by altering the active conformation resulting in the loss of a particular activity. Due to the formation of free radicals, the presence of heavy metalloids can cause oxidative stress and other types of cellular damage (Sharma and Dietz 2009). The unpleasant side effects of long-term exposure to these harmful and persistent substances cause ecological destruction. Since the mid-1980s, numerous attempts have been made in response to determine how to prevent petroleum pollutants from continuously deteriorating soil health and to improve soil health.

Several techniques, oxidation-reduction, chemical precipitation, filtration, dialysis, ion exchange, and electrochemical treatments eliminate petroleum and other contaminants from soil, but none are fully efficient. Furthermore, they are abandoned due to their high cost, restricted use with few opportunities, and incapacity to enhance the innate health of the soil. As a viable substitute that offers a cost and environmentally friendly remediation approach, bioremediation techniques have evolved (Ahemad and Khan 2011). Bioremediation is essentially the process of lowering contaminant levels to undetectable, nontoxic, or at least acceptable levels utilizing naturally occurring bio-resources of plants, microbes, and organic amendment. Microbial metabolism is a key component of soil bioremediation. Hydrocarbons are absorbed by microorganisms through redox reactions, intracellular accumulation, bioleaching, biosorption, and mineralization (Lloyd and Renshaw 2005). Certain fungi not only degrade but also produce trace amounts of biostatic compounds aiding in the bioremediation of complex organic and inorganic molecules such as hydrocarbons. Various methods are utilized for treating polluted soils such as adding manure and compost to the soil or by applying single organisms such as plants or animals in large quantities. In the meantime, consortia of bacteria, fungi, and plants are incorporated holistically (Singh 2006).

Overall, combinations of methods maximize the potential and efficacy of bioremediation and work in concert with one another. Fungi are known to break down a wide range of contaminants and toxicants, Fungi, akin to other microorganisms, have a promising future in the bioremediation of environmental pollutants. The process of detoxification of contaminated sites by employing one or more types of fungi is termed as mycoremediation. This method makes use of fungi to absorb harmful pollutants from soil and water. Fungi are an excellent option for the remediation of various pollutants owing to their fast growth, intricate hyphal network, ability to produce versatile extracellular ligninolytic enzymes, high surface area-to-volume ratio, resilience to heavy metals, adaptability to fluctuations in pH and temperature, and possession of metal-binding proteins (Khan et al. 2019). Fungi are widely distributed, representing a diverse range of groupings found in a range of habitats, and contribute significantly to recycling by releasing enzymes that decompose complex organic compounds into simpler chemical constituents that they or other organisms can use. They accordingly participate in the recycling of nutrients in soil and aid in the regulation of the C:N ratio. In addition to their functions in the environment as lignin and cellulose decomposers. Fungi have also drawn interest for their potential in bioremediation of inorganic pollutants due to their adaptability to extremes in pH, temperature, nutrient availability, and metal concentration. (Dhankhar and Hooda 2011). Mycotoxin-producing filamentous fungi have also been shown to be capable of breaking down a broad range of hazardous waste materials that are produced both artificially and naturally. Multiple advantages are associated with mycoremediation, including high efficiency, reduction of chemical usage, cheap operating costs, and fewer harmful chemical byproducts (Harms et al. 2011). The biotechnological application fungi in bioremediation allows for in-situ remediation of a variety of contaminants, including petroleum. Mycoremediation may also be applied in bioreactors which can be used ex-situ for removal of tar and polycyclic aromatic hydrocarbons from the soil (Tekere 2019).

Fungal cultures that have been specifically enhanced to accelerate decomposition are used in bioremediation. Saprophytic fungi are pivotal in the breakdown of organic compounds. Fungi release a variety of extracellular acids and enzymes that degrade natural polymers including lignin, pectin, cellulose, hemicellulose, chitin, and keratin (Lamar and White 2001). Many fungi provide energy to soil-dwelling bacteria that facilitate aggressive destruction of toxins through enzymatic breakdown of pollutants. Additionally, fungi eliminate heavy metals by translocation from the soil into their fruiting bodies; they contribute to revitalization of the rhizospheric ecosystem (Prasad et al. 2017).

6.2 Approaches in Bioremediation

Bioremediation is a naturally occurring process that uses living organisms to rid soil and groundwater of organic contaminants (Rhodes 2014). It deals with waste management strategies that remove pollutants from the environment efficiently and sustainably. Two types of mycoremediation techniques can be distinguished: in situ and ex-situ (Azubuike et al. 2016). The in-situ approach eliminates the possibility of toxicants spreading during excavation by treating contaminated soils at the exact location of polluted sites and facilitating the avoidance of excavation and the transportation of contaminants (Talley 2005). However, the ex-situ strategy includes eliminating contaminated soils and water from the site by moving it to another site where it will be treated to remove pollutants and toxicants.

6.2.1 In Situ Approaches

The process of transformation is entirely biological with fungal enzymes acting as a catalyst to break down and eliminate any harmful materials that may be trapped. Hydrocarbon pollutants are broken down during in situ bioremediation and utilized in bacterial growth. In general, the processes utilized to biodegrade contaminants include abiotic oxidation, hydrolysis, dispersion, sorption, dilution, volatilization, and penetration of nutrients including water or other electron acceptors. In situ, bioremediation is the most commonly used technique for the biodegradation of hazardous waste in saturated soils and groundwater. Utilizing microbial organisms to break down harmful compounds and purge contaminated environments is a safer, more effective, and less expensive approach (Tiwari and Singh 2014). Subdivisions of in situ bioremediation are seen below.

6.2.1.1 Bioventing

Bioventing involves forcing air through the soil to remove petroleum contaminants that are susceptible to oxygen from the unsaturated zone. Increased oxygen availability in the waste leads to the degradation of aerobiologically degradable pollutants, hence promoting the activity of indigenous microorganisms. It requires an adequate supply of oxygen or air because it is mostly an aerobic process. In some cases, hazardous soils with reduced oxygen concentrations can be recovered by adding (nitrogen and phosphorus) nutrients to support microbial biodegradation (Rockne and Reddy 2003). Subsurface vacuum extraction wells use air injection, as they pull air through the subsurface and introduce oxygen. Spatial distribution of nutrients and oxygen in the soil is influenced by soil texture. Bioventing provides sufficient oxygen for microbial development and the breakdown of large petroleum molecules by incorporating nutrients and moisture along with moderate airflow rates. During aerobic bioventing, air is pumped into contaminated soil above the water table through a vent well for the elimination of toxins. Depending on the types of toxicants and the texture of the soil, different sites have different rates of elimination.

Biodegradation rate is accelerated by homogenous oxygen distribution, the primary purpose is to enhance elimination of hydrocarbons such as oil, gasoline, and petroleum (Tiwari and Singh 2014). In anaerobic conditions, rapid degradation of resistant pollutants such as polycyclic aromatic hydrocarbons (PAHs) and chlorinated solvents occurs occasionally. Anaerobic bioventing is the technique of releasing substrates other than oxygen to promote microbial growth. In order to eliminate pollutants, methanogenesis, and fermentation are additional microbial processes used in anaerobic bioventing. Anaerobic conditions are generated by infusing nitrogen and an electron donor in place of air with the nitrogen taking the place of soil oxygen. The subsurface soil zone produces small amounts of electron donors in a reducing state which assists with microbial dechlorination. Anaerobic bioventing yields volatile and semi-volatile compounds which break down gradually. The soils around the treatment zone may absorb volatile compounds, which might lead to aerobic deterioration. Through the process of aerobic bioventing followed by anaerobic bioventing, it is feasible to treat both halogenated and nonhalogenated volatile organic substances as well as semi-volatile organic substances that remain in the treatment zone.

6.2.1.2 Biosparging/Air Sparging

In contrast to bioventing, which increases oxygen concentration in the unsaturated zone, biosparging elevates oxygen levels in the saturated zone by injecting oxygen or air, beneath the groundwater. Naturally occurring microbes, in biosparging promote the aerobic biological breakdown of hydrocarbon contaminants (Singh 2006). Airborne volatile contaminants can be eliminated from the saturated zone due to enhanced volatilization and desorption in the increased airstream. Due to their volatilization, light to mid-weight petroleum-based pollutants are removed by biosparging (Tiwari and Singh 2014). This method is recommended for cleaning underground storage tanks and aquifers.

6.2.1.3 Bioaugmentation

Addition of extrinsic microorganisms to the petroleum-contaminated area with particular metabolic activities in order to promote the biological breakdown or transformation of organic petroleum-based pollutants, bioaugmentation is applied in remediation areas where groundwater and soil are contaminated with chlorinated ethane such as tetrachloroethylene and trichloroethylene. In situ, microorganisms have the ability to efficiently break down these organic pollutants into harmless compounds like ethylene. Biotic and abiotic factors, including soil type, nutrient availability, moisture content, hydrolytic and osmotic pressure, pH, and existing microbial composition affect the survival and activity of microbes alone or in consortia, and may prevent inoculated specialized microbial populations from producing the desired transformation of contaminants. Microbes with a greater affinity for hydrocarbons and the capacity to endure in a variety of environments are typically favored for bioaugmentation (Mrozik and Piotrowska-Seget 2010).

6.2.2 *Ex Situ Approaches*

This technique involves the excavation or pumping of the original site followed by the application of ex-situ remedies to address contaminated soil or groundwater. Multiple methods facilitate the removal of a diverse array of harmful hydrocarbons and soil contaminants.

6.2.2.1 Slurry Phase

A large tank known as a bioreactor is used in this process to integrate the petroleum-contaminated site with water and other reagents. In the bioreactor, the favorable conditions encourage the best possible growth of microorganisms and amalgamation facilitates their continued interaction with the organic toxicants found in the soil. To improve the rate at which the indigenous microorganisms decompose harmful wastes, more oxygen and nutrients are added. The procedure ends with water moving away from the soil which is then inspected and reinserted into the surrounding area. The main factors constraining biodegradation which can be easily controlled in a bioreactor include temperature, pH, aeration, agitation, nutrient supply, augmentation, substrate availability, and inoculum concentrations (Purohit et al. 2018). Therefore, compared to other methods, it offers more advantages for quicker bioremediation.

6.2.2.2 Solid Phase

This method involves treating petroleum sediment-contaminated soil above ground in the solid phase which can be facilitated by soil bio-piling, land farming, and the composting process (Girma 2015). For decontamination, excavated material is placed in an aboveground enclosure or treatment chamber. An aeration system helps to ensure a controlled atmosphere for the best possible treatment. While these techniques are relatively simple to implement, they require time and space resources, and are not widely used.

6.2.2.3 Land Farming

Land farming, often called land treatment, is a low-tech method that involves excavating hydrocarbon-contaminated soil, spreading it over a ready-made bed, and then tilling it occasionally to allow the contaminants to break down naturally. To encourage the growth of native biodegradative microorganisms and promote the aerobic breakdown of petroleum-based pollutants, excavated soil is layered between clean soil and clay concrete, with the concrete layer on top and clean soil at the bottom. Generally, the procedure can only be used on soil that is between 10 and 35 centimeters deep. Land farmers use a variety of techniques, such as irrigation, tillage, and fertilizer application to encourage microbial activity for faster bioremediation (Felsot et al. 2003).

6.2.2.4 Compositing

Compositing is a biologically regulated method that involves combining materials, such as straw, to remediate harmful petroleum molecules in soil. The process of composting involves the high-temperature digestion of waste material (55–65 °C) facilitated by thermophilic microbes. The heat created in the compost pit over the complete degradation process rises the temperature of the material, which in turn causes waste to turn into more soluble and complex compounds to break down because of increased microbial metabolism (Blanca et al. 2007). In order to offer containment and protection from inclement weather, the soil is carried to a composting pit equipped with a temporary structure. To provide an additional carbon source during the composting process, organic materials such as sawdust, wood chips, vegetable and agricultural wastes, and more are added together with the compost components. The three most common methods for composting are:

(i) Aerated static piles are formed by stacking materials and then utilizing vacuum pumps or blowers to provide air.
(ii) Mechanically agitated in-vessel composting, which takes place in a reactor vessel following it has been completely mixed and aerated.

(iii) Windrow composting is a technique where soil and organic additives are stacked using mobile equipment into long, narrow, low piles known as windrows and then routinely mixed.

6.2.2.5 Bio-piling

This form of ex-situ bioremediation combines composting with land gardening. Control over moisture, heat, oxygen, pH, and nutrients is necessary for proper degradation. The irrigation system uses a vacuum to supply nutrients and air while it is buried beneath the earth. Plastic coating on soil serves to reduce evaporation and volatilization while promoting solar heating, thus aiding in runoff prevention. The breakdown of petroleum hydrocarbons in a bio pile treatment takes 20–90 days to complete. Usually, it makes use of the type of land farming that, through volatilization and leaching, serves to limit the physical losses of soil contaminants (Tiwari and Singh 2014). Another advantage is that bio piles make it simple to control the ambient conditions for native aerobic and anaerobic microorganisms.

6.3 Enzymes Involved in Mycoremediation

The presence of numerous enzymes in fungi renders them potent tools for the adaptation, detoxification, and breakdown of various hydrocarbons. The potential of various extracellular and intracellular enzymes as effective biodegradation agents has been evaluated and these enzymes catalyze a variety of contaminants as substrates in enzymatic processes. Recent advance in the field has significantly increased the rates of bioremediation utilizing more contemporary molecular biology techniques, recombinant enzyme systems, and genetically modified fungal strains (Kumar et al. 2013).

6.3.1 Extracellular Degradative Enzymes in Fungi

Extracellular enzymes are released by fungi to mobilize both organic and inorganic nutrients. These are additionally produced during the fermentation process which disintegrates the bonds between organic molecules and initiates their total breakdown or conversion into less hazardous and simpler forms. Fungal extracellular lignin-transforming enzymes are the cause of their effective biodegradation of a wide range of organic substances. Fungal-produced extracellular enzymes have a variety of roles in the biodegradation of complex hydrocarbons. These enzymes include peroxygenases, versatile peroxidase, and laccase. (Hofrichter et al. 2015).

6.3.1.1 Peroxidases

Peroxidases are a type of oxidoreductases that predominantly utilize H_2O_2 as a substrate however they can also effectively convert a variety of hydrocarbons such as oil and gasoline via the free radical mechanism. The action of these enzymes is non-specific and they can break down a wide range of substances. Heme and non-heme proteins are the two main categories into which peroxidases can be classified based on the presence of a heme cofactor.

6.3.1.2 Lignin Peroxidases (LiPs)

Lignin peroxidases, or LiPs, utilize H_2O_2 as a co-substrate to effectively degrade a variety of organic compounds, including complex hydrocarbons like phenolics and the substance lignin which is present in plant cell walls. (Dashtban et al. 2010). LiP's heme cofactor facilitates the transport of electrons during redox processes. The enzymatic activity of LiP helps fungi to transform and detoxify contaminants, promoting the remediation of polluted soils and water bodies. These enzymes are frequently observed in white rot fungi that belong to the genera *Phanerochaete*, *Trametes*, *Bjerkandera*, and *Phlebia* (Asemoloye et al. 2021).

6.3.1.3 Manganese Peroxidase (MnP)

MnPs are heme-containing proteins that fungi secrete outside their cells, predominantly produced by basidiomycetes fungi, notably those classified as white rot fungi such as *Trametes versicolor*, *Agrocybe praecox, Dichomitus squalens,* and *Agaricus bisporus* that decompose wood (Shourie and Vijayalakshmi 2022). Manganese peroxidase (MnP) plays a pivotal role in mycoremediation, a process that utilizes the unique metabolic capabilities of fungi to remediate and detoxify contaminated environments. Mycoremediation harnesses the enzymatic activities of fungi, with MnP being a key player in the degradation of various recalcitrant petroleum-based pollutants. This versatile enzyme is produced by certain fungi, particularly white-rot fungi, and is known for its ability to catalyze the oxidation of manganese ions and organic substrates. In the context of mycoremediation, MnP is crucial for the breakdown of complex organic pollutants, including lignin and other aromatic compounds which are often present in industrial effluents and environmental contaminants. Through its peroxidase activity, MnP generates highly reactive free radicals, facilitating the degradation of recalcitrant pollutants into simpler, less toxic compounds. The mycelial network of fungi further enhances the efficiency of mycoremediation by providing an extensive surface area for petroleum molecules absorption and transportation. The environmentally friendly and cost-effective nature of mycoremediation driven by enzymes like MnP, holds promise for sustainable solutions in addressing various oil sediment-related pollution challenges.

6.3.1.4 Versatile Peroxidase

Manganese-lignin peroxidase, also known as versatile peroxidase, has the ability to oxidize a wide range of common substrates that are catalyzed by other basidiomycetes peroxidases including hydrocarbons (Fernandez-Fueyo et al. 2015). As versatile peroxidases can oxidize a wide range of organic molecules, including aromatic compounds that are phenolic and non-phenolic, these are considered a unique class of ligninolytic enzymes. Furthermore, white rot fungi such as *Pleurotus ostreatus*, *Phlebia radiata, Pleurotus eryngii*, *Bjerkandera adusta,* and *Phanerochaete chrysosporium* are often found possessing these peroxidases (Baratto et al. 2015).

6.3.1.5 Laccases

Laccases classified as oxidoreductases are enzymes produced by both ascomycetes and basidiomycetes, belonging to the copper-containing oxidases family. They facilitate the oxidation of broad-spectrum substrates including phenols, polycyclic aromatic hydrocarbons, arylamines, anilines, nitroaromatics, chlorinated aromatic compounds thiols, and lignin. Laccase activity is intricately involved in diverse biochemical transformations. Numerous fungal taxa belonging to the Basidiomycota class, including *Trametes versicolor*, *Trametes villosa*, *Phanerochaete chrysosporium*, *Lenzites betulina*, *Pycnoporus cinnabarinus*, *Pleurotus ostreatus,* and *Phlebia radiate* have been reported to produce laccases (Shekher et al. 2011). *Trametes versicolor* can degrade toluene and benzene with laccase enzyme. Moreover, laccases have been found in Ascomycota fungi, including *Magnaporthe grisea*, *Fusarium* sp., *Gaeumannomyces graminis*, *Myceliophthora thermophile*, *Phoma* sp., *Botrytis cinerea*, and *Podospora anserine* (Yang et al. 2017).

6.3.2 Intracellular Enzymes in Fungi

Fungal cells can catabolize internally absorbed large hydrocarbons using their intracellular enzymes. This includes a group of mixed-function oxidases known as cytochrome P450s which exhibit the capacity to catalyze a variety of biodegradation reactions, including those involving epoxidation, nitro reduction, hydroxylation, dehalogenation deamination and desulfuration (Harms et al. 2011). Cytochrome P450 mono-oxygenases (CYPs) are a superfamily of intracellular enzymes found in various organisms including fungi. In the mycoremediation process, CYPs play a crucial role in the metabolism of xenobiotics by catalyzing oxidation reactions. These enzymes are involved in the breakdown of a wide range of organic compounds, including pollutants like petrochemicals and polycyclic aromatic hydrocarbons (PAHs). CYPs play a role in the adaptation of fungi to environmental stresses by metabolizing toxic compounds. Fungi are known for their ability to produce secondary metabolites involving cytochrome P450 monooxygenases including *Penicillium* spp., *Aspergillus* spp., *Fusarium* spp., and *Trichoderma* spp.

6.4 Fungal Diversity Involved in Mycoremediation

There are an estimated 2.2–3.8 million species of fungi in the world that are pathogens, mutualists, and decomposers that have a major impact on the environment (Hawksworth and Lucking 2017). Fungi have a remarkable degree of flexibility and adaptation when faced with challenging and unfavorable environmental conditions. Many fungal strains from the two major phyla Ascomycota and Basidiomycota have shown notable degradative activity, which has led to their recognition as efficient degraders of organic molecules. The formation and expansion of fungal communities are influenced by a number of biotic and abiotic variables, including moisture, temperature, aeration, carbon, nitrogen, and phosphorus availability as well as interspecific microbial competition. However, these fungi have been classified into the following groups according to their ecology and functionality.

6.4.1 Wood Rotting Fungi

The fungi that cause wood rot are members of the class Basidiomycetes (Li et al. 2022). Not only are these fungi useful for decaying wood, but they are also essential for the bioremediation of organic contaminants. They can use a range of enzymes that decompose complex substances including cellulose, hemicellulose, and lignin to reduce wood tissue into simpler forms. These fungal species degrade lignin extracellularly as a means of eliminating a variety of harmful environmental pollutants. Additionally, due to the extracellular enzymes in them, they are nonspecific and nonselective in their biodegradation and can tolerate relatively higher quantities of harmful hydrocarbon contaminants (McErlen et al. 2006). Compared to other fungal species, wood-rotting fungi are more adaptable and have a higher potential for degradation. The wood-degrading fungi are categorized as weak competitors, such as *Ganoderma applanatum* and *Dichomitus squalens,* and strong competitors, such as *Pleurotus* sp. and *Phanerochaete* sp., based on their capacity to colonize new areas. Based on their ability to decompose wood tissue, they are classified as either brown rot fungi or white rot fungi (Martínez et al. 2005).

6.4.1.1 White-Rot Fungi

The capacity of white-rot fungi to break down lignin in addition to cellulose and hemicellulose distinguishes them apart from other types of wood-rotting fungi and results in the white decay of the wood. With the exception of a few Ascomycota fungi that are exclusive to the Xylariaceae family, they are mostly composed of Basidiomycota fungi. These white-rot species were the first fungi used to investigate mycoremediation. They also produce lignin peroxidase, manganese peroxidase, laccase, and enzymes that produce H_2O_2. Some of these, like laccase result in ring breakage, while others generate strong oxidants that can lead to the degradation of petroleum substances (Anastasi et al. 2009). Extensive research has been performed on the biodegradation of complex substances by extracellular oxidative ligninolytic enzymes that generate CO_2 and H_2O more effectively than other microorganisms by breaking down large hydrocarbon compounds. Due to the various oxidative and reductive processes involved in biotransformation or the breakdown of petroleum compounds, its application is highly versatile across a range of matrices. Potent and non-specific, ligninolytic enzymes can be utilized to effectively eliminate a wide range of aromatic substances and xenobiotics from polluted soil. A number of white-rot fungi such as *Phanerochaete chrysosporium* are also known to degrade these substances, additional fungi which remove contaminants from soil include *Pleurotus ostreatus*, *Lentinula edodes*, *Bjerkandera adusta*, *Trametes versicolor*, and *Irpex lacteus*.

6.4.1.2 Brown-Rot Fungi

This species of wood-degrading fungi decompose cellulose and hemicellulose in wood, leaving a typically brownish residue of lignin. They are predominantly belong to the phylum Basidiomycota, specifically the class Agaricomycetes, and classified under the orders Agaricales, Gloeophyllales, Hymenochaetales, and Polyporales. They partially oxidize, depolymerize, and demethylate the lignin utilizing a nonenzymatic Fenton-type catalytic mechanism. Because of the partially converted lignin, the rotting wood has a distinctive dark brown color. The cellulose and hemicellulose are broken down by the brown-rot fungus via an oxidative mechanism that yields hydrogen peroxide. This process contributes to the formation of free hydroxyl (OH) radicals, which help break down and mineralize synthetic chemotherapeutants. Moreover, their ability to breakdown hydrocarbons and resist antimicrobial agents increases their ability to degrade petrochemicals. The cellulose that is rapidly depolymerized by brown rot fungi causes the degraded wood tissues to lose rigidity and disintegrate easily under pressure, especially under dry conditions.

6.4.2 Mycotoxin-Producing Fungi

Aspergillus, Penicillium, and *Fusarium* are the three genera of fungi that have been found to be the primary producers of mycotoxin which is crucial to public health. A broad range of toxins are produced by Aspergillus, especially citrinin, cyclopiazonic acid (CPA), ochratoxin A (OTA), patulin A, aflatoxins (B1, B2, G1, G2), and sterigmatocystin being the most commonly produced. These mycotoxins have a significant global impact on agriculture, health, economy, and environmental sustainability. Aflatoxin exposure results in cancer in both humans and animals. *Aspergillus flavus* is primarily responsible for the production of aflatoxin (Hedayati et al. 2007), and CPA (Georgianna et al. 2010). This ubiquitous fungus can thrive in conditions that hinder its ability to develop and can outcompete other organisms in its biological niche for nutrition. Along with 14 other species, *Aspergillus carbonarious*, *Aspergillus niger*, *Aspergillus acidus,* and *Aspergillus braciliensis* produce a total of 145 different secondary metabolites (Nielsen et al. 2009). Fumonisins and ochratoxin A provide the most severe health concerns among these. These toxins are frequently found in food, feed, and the environment but their ability to survive in a range of environmental conditions makes them crucial for bioremediation. Approximately 100 species of *Penicillium* produce mycotoxins, which are classified into several structural classes. Trichothecenes, zearalenone, and fumonisins are mycotoxins produced by the *Fusarium* genus. There are great opportunities for bioremediation with all of these mycotoxin manufacturers. *Fusarium graminearum* has excellent potential for degrading hydrocarbons. Similarly, it has been found that *Fusarium moniliforme* degrades organic hydrocarbons, environmental DNA, and nucleotides. Several studies have been carried out on the Aspergillus group specifically *Aspergillus fumigatus*, *A. gracilis*, *A. flavus*, *A. penicillioides,* and *A. restrictus* in relation to lead remediation (Tang et al. 2021). *A. versicolor* absorbs up to 45.0 mg g^{-1} of hydrocarbons, but *A. fumigatus* shows a maximum adsorption potential of 43.7 mg per g^{-1} (Bairagi et al. 2010).

6.4.3 Endophytic Fungi

Plants can be colonized by bacterial and fungal endophytes without affecting the host or the endophytes themselves (Rout 2014). They produce a range of compounds that are utilized in the fixation of nitrogen and the assimilation of methane (Roshchina 2012). Because of their saprophytic attributes, these endophytes remain viable in dead litter (Kubartová et al. 2009). As they may form stretched and widely dispersed mycelial networks, fungi are an excellent choice for bioremediation processes (Harms et al. 2011). Endophytic fungi employ metal sequestration or chelation systems to enhance biomass and resistance to organic compounds thereby facilitating the bioremediation process (Aly et al. 2011). Additionally, they produced a range of enzyme groups including cellulase, protease, and peroxidase, for the biotransformation of heavy metals in their reduced forms and the degradation of environmental chemicals, in addition to polyaromatic hydrocarbons, petrochemicals, herbicides, insecticides, and polyester polyurethane (Purohit et al. 2018). They increase the tolerance's capacity and flexibility to hydrocarbons and toxins. *Penicillium janthinellum* an endophyte of *Solanum lycopersicum* improves organic phytoextraction (Khan et al. 2014a, 2014b). *Trichoderma* spp. derived from Acacia plays a key role in reclamation of soil contaminated with oil sediments (Cao et al. 2008). Similarly, isolates of *Peyronellaea* induce host plant accumulation of contaminants (Shen et al. 2013).

6.4.4 Mycorrhizae

Mycorrhizae are versatile, mutually beneficial associations between plant roots and fungi. The associations are diverse including ectomycorrhiza, arbuscular mycorrhiza, orchid mycorrhiza, monotropoid mycorrhiza, arbutoid mycorrhiza, ericoid mycorrhiza, and ectendomycorrhiza (Finlay 2008). In addition to aiding nutrient availability, these interactions are essential in protecting plants from environmental hazards like petroleum toxicity. A key mechanism for plant adaptation and survival in highly contaminated environments is the modulation of petroleum sediment translocation, which is closely associated with the ability of plants to cope with heavy metal toxicity (Sharma and Dietz 2006). In return, the host plant facilitates the fungus by establishing an area that is contaminated and permitting enzymatic processes to break down different organic pollutants like PAHs, petroleum products, and chlorinated aromatic pesticides (John 2013). The complex relationship between fungi and plants exhibits the ways they can support each other in overcoming environmental stresses and detoxification of polluted regions. Certain potentially harmful soil components must be broken down by mycorrhizal fungus (Lehmann et al. 2011). Certain VAM fungi have been reported to be present in the stem and thallus despite usually being found in the rhizosphere. Fungal hyphae also absorb petroleum compounds, much like Arbuscular mycorrhiza functions. Therefore, mycorrhizal plants exhibit enhanced absorption of hydrocarbons whereas, VAM fungi aid in the immobilization of hydrocarbons in soil. Restoration of the ecosystem thus involves the removal of petroleum compounds from the soil via the Petroleum-Plant-Mycorrhizal connection. Plants utilize many strategies to keep the petroleum equilibrium both internally and externally in the presence of stress conditions, such as excessive concentrations of harmful compounds. This method involves a variety of gene types that aid in the remediation of toxic compounds and ultimately, in the detoxification of compounds from contaminated soil. Detoxification aims to prevent the intake of hazardous metals by immobilizing them in the root cell wall and chelating them using exudates from the roots (Table 6.1).

6.4.5 Aquatic Fungi

Microbial life is impacted by a wide range of activities that are supported by the complex ecosystem found in aquatic environments. Microbial communities perform a major ecological role in aquatic environments. A variety of biogeochemical processes depend on aquatic fungi, which are among the most prevalent species. These can have an obligatory or facultative nature and they can be found in both freshwater and marine habitats. An increasing number of human activities including oil spills and industrial petroleum discharge are disrupting the aquatic ecosystem. Basidiomycetous fungi that cause white rot are commonly linked to the production of ligninolytic enzymes which are produced by more than 80% of marine fungi that

Table 6.1 Petroleum pollutants removal efficiency by various fungi

Fungal strain	Pollutant Category	Organic pollutant	Maximum removal efficiency (%)	Time	References
Aspergillus sp.	Crude oil and petroleum	Naphthalene	97.40	7 days	Al-Hawash et al. (2019)
		Phenanthrene	84.90		
		Pyrene	90.70		
Aspergillus ustus		Crude oil (2%)	30.40	10 days	Benguenab and Chibani (2021)
Mucor sp.		Crude oil	65	21 days	Dirisu et al. (2018)
Penicillium sp.		Crude oil	55	21 days	Dirisu et al. (2018)
Purpureocillium lilacinum HM4		Crude oil (2%)	44.50	10 days	Benguenab and Chibani (2021)

are members of the Ascomycota group. Numerous research works on the application of marine fungi for bioremediation show encouraging results especially when it comes to bio-pulping, dye decolorization, and bio-bleaching (Bhardwaj et al. 2018). Lignocellulolytic substances and hydrocarbon-based industrial petroleum pollutants need the production of certain enzymes produced by aquatic fungi, such as Mn-peroxidase, laccase, and lignin peroxidase. Fungi can withstand harsh conditions including high pressure, high salinity, cold temperature, metal, and toxicity from oil spills. Various freshwater aquatic fungi, including *Aspergillus sclerotiorum*, *Mucor racemosus*, *Cladosporium cladosporioides*, *Phaeosphaeria spartinicola*, *Sordaria fimicola*, *Halosarpheia ratnagiriensis,* and *Flavodon flavus* have generated ligninolytic enzymes, which in turn have mineralized lignin-based compounds to carbon dioxide (Bonugli-Santos et al. 2010). Comparably, the biocatalytic capability of a few marine aquatic fungi has been explored, including *Penicillium raistrickii*, *A. sydowii*, and *Trichoderma sp*. (da Silva et al. 2013).

6.5 Mechanisms Involved in Mycoremediation

Fungi have a remarkable ability to connect with metal ions, involving processes such as expelling petroleum compounds, accumulating and forming complexes within the cell, and subsequently converting poisonous compounds into a non-toxic form. Various processes have developed to immobilize, mobilize, or alter pollutants making them inert or enabling the absorption of compounds. Fungi employ various mechanisms for bioremediation including:

(i) Exclusion: A permeable barrier is formed to keep pollutants from entering the specified sites.
(ii) Extrusion: The pollutants are driven out of the cells by active transport.

(iii) Fixation: This process involves binding pollutants to binding proteins or other components of the cell through intra- and extracellular sequestration, enzymatic detoxification, chelation, acid and organic base production-induced precipitation.
(iv) Biotransformation: This mechanism includes the conversion of highly toxic pollutants into less toxic forms through processes such as methylation, demethylation, oxidation, reduction, and volatilization.

6.5.1 *Immobilization*

The industrial utilization of fungal enzymes, especially in bioremediation techniques, has experienced consistent growth. Enzyme immobilization is a useful technique that may be used to enhance the stability of enzymes, increase their catalytic activity, and enable their reuse, therefore rendering biocatalysis economical for large-scale activities. Enzyme immobilization prevents denaturation and increases the catalytic efficiency by enhancing the enzymes' tolerance to temperature, pH, and organic solvents. The capacity of fungi to eliminate toxic materials through enzymes such as tyrosinases, laccases, peroxidases, dioxygenases, and polyphenol oxidases is increasingly being studied for their ability to remove harmful substances from the environment. Phenolic chemicals are frequently found in the environment as contaminants and can be efficiently eliminated by using tyrosinase-mediated dephenolization (Xu and Yang 2013). In addition to tyrosinases, immobilized laccases have been extensively researched for their ability to eliminate pollutants from the environment. This is due to their remarkable pollution removal capabilities and economic deterioration. For a wide range of uses, including the decolorization of dyes, the removal of PAHs, the breakdown of toxic compounds, and the modification of therapeutic compounds, peroxidases are often immobilized. To improve their shelf life and stability in relation to pH and temperature, the MnPs obtained from *Ganoderma lucidum* have been immobilized using gel polymers such as tetramethoxysilane and alginate (Iqbal and Asgher 2013). For the intent of biocatalysis and the application of multi-enzyme systems in the bioremediation of petroleum pollutants, the immobilization of intact fungal cells or fungal biomass presents notable advantages. Pellets, mycelium adhering to a cross-linked porous substrate, or encasing biomass in gel beads are a few ways to immobilize biomass. Aid with its adhesion and fixation to the substrate, the fungal cell secretes exopolysaccharides, which facilitates the immobilization process.

6.5.2 *Mobilization*

Contaminants can be mobilized by microorganisms via redox transformations, alkylation, methylation, leaching, and chelation by microbial metabolites and siderophores. Leaching occurs when microorganisms induce the soil environment to become acidic by their proton efflux, which triggers the release of free pollutants cations. Leaching is typically mediated in most fungi by the synthesis of low-molecular-weight organic acids, which, when broken down, yield protons and organic acid anions that bind with pollutants. *Trichoderma harzianum* is capable of enhancing the solubilization capability (Altomare et al. 1999). An alkyl group can be transferred as an alkyl carbocation, a free radical, a carbanion, or a carbene during the process of alkylation. This process is initiated by the synthesis and secretion of fungi that combined with pollutants, create non-toxic complexes (Verma et al. 2019). Yeast like *Saccharomyces* spp. can transform petroleum-origin PAHs to non-toxic simpler forms in soil.

6.5.3 Biosorption

Biosorption is a very efficient and non-active method for removing pollutants from petroleum sediments by utilizing macro-fungi. This technique is highly effective in treating industrial wastewater and may also extract pollutants from it. The biosorption mechanism has been classified into three categories depending on the location of the sorted petroleum pollutants from the effluent:

(i) Extracellular accumulation or precipitation
(ii) Sorption/precipitation on the cell surface
(iii) Intracellular accumulation

Based on cellular metabolism, absorption mechanisms may also be categorized as follows: (i) metabolism-independent and (ii) metabolism-dependent. The rapidity, reversibility, and lack of reliance on metabolic energy for the binding of contaminants to the biomass are characteristics of the sorption process, which remains unaltered by metabolism. Physicochemical interaction is the term used to describe how pollutants interact to enable the uptake of pollutants into microbial cells. (Pagnanelli et al. 2002). Other processes such as ion exchange, chemical sorption, and physical adsorption are included in the concept of biosorption (Fawzy et al. 2018). These procedures can be accomplished using both deceased and living biomass (Banerjee et al. 2018). Discarded mushroom compost and different mushroom parts such as mycelia and fruiting bodies, can also be used to make biosorbents. In functionalized fungi, the intracellular mineral scaffold acts as an internal pollutants container that facilitates petroleum biosorption.

It has been established through kinetic studies that the step most likely limiting the rate of biosorption is chemisorption. By effectively removing petroleum-based pollutants from soil and water, the fungus *Trichoderma harzianum* has shown great promise for promoting environmental sustainability. While these species have a limited capacity to extract pollutants, they exhibit notable resistance to *nicke*. However, the same species also exhibit a higher ability to remove the metal up to 46.30% (Cecchi et al. 2017). *A. fumigatus* and *A. flavus* can effectively remove 40.90% and 59.70% of toxic oil contaminants, respectively. The highly adsorptive substances found in *Agaricus bisporus* and *Lentinula edodes* mushrooms such as chitin and chitosan, aid in the removal of medication residues, including paracetamol and 17α-ethynyl estradiol. *L. edodes* show a noteworthy degree of porosity and a large degree of deacetylation in chitosan. *Shittake* is also shown to be more effective in adsorbing paracetamol and 17α-ethynyl estradiol, according to an evaluation of adsorption kinetics and isotherms (Menk et al. 2019). In order for effective removal of pollutants from contaminated water and soil, fungal chitosan and sodium tripolyphosphate are predominantly used in the production of chitosan nanoparticles.

6.5.4 Bioaccumulation

This process of sorption takes place in live cells. Depending on the metabolism of the cell, intracellular accumulation occurs when pollutants are transported across the cell membrane. There are three different ways that pollutants might bioaccumulate within an organism: passive diffusion, aided diffusion, and active transport (Banerjee et al. 2018). The filamentous fungal strains *Beauveria bassiana* and *A. fumigatus* have demonstrated an elevated capacity to extract many sediment oil-based pollutants from a combination of multiple sources. Both stains showed a substantial tolerance index towards all pollutants 500 mg/L (Gola et al. 2016). These fungal strains have been noted for their exceptional adaptation to diverse stresses, attributed to their ability to accumulate multiple pollutants across various concentrations. Mushrooms respond to petroleum sediment stress in their environment by producing stress-related chemicals, both proteinaceous and non-proteinaceous. The mushroom cap plays a role in absorbing metal ions by releasing substances such as glutathione, metallothionein and plastocyanin in response to stress. They also possess increased flexibility and an efficient method for metal accumulation (Damodaran et al. 2013).

6.5.5 Bioconversion

The process of recycling lignocellulosic wastes into more valuable products is termed bioconversion. Petroleum wastes that lack utility exhibit significant potential for bioconversion into useful byproducts. As they're capable of enzymatic activities that efficiently transform complex organic forms into simpler ones, certain types of fungi are essential to the bioconversion of wastes. These waste materials can also be effectively used to develop various products, such as infrastructure or road fillup material (Naraian et al. 2016). The substrate and wheat straw combination resulted in a significant increase in biological efficiency, as well as a decrease in the incidence of frameshift mutagens (Kulshreshtha et al. 2013). Moreover, phenolic compounds and antioxidant activity in the white-rot fungus are enhanced when surfactant is added to wheat straw during bioconversion. Furthermore, this novel approach presents a low-cost method of transforming lignocellulosic biomass into medicinal fungus (Zhao et al. 2020). Multiple species of a biofilm are produced by co-cultivating *A. phoenicis* and *Trichoderma reesei* and this biofilm may improve the β-glucosidase production by up to 2.5 times which is essential to break down plant biomass into individual sugar molecules (Xiros and Studer 2017).

6.5.6 *Bioreduction*

The bioreduction mechanism involves the reduction of the oxidation state of pollutants, resulting in a decrease in their toxicity. This reduction is facilitated by living organisms (Li et al. 2017). The toxicity potential of the petroleum substance is reduced when it is in its non-toxic form. *A. tubingensis* Ed8 is capable of producing glucose oxidase (GOX), which plays a crucial role in the reduction of non-toxic forms. This indicates that the GOX enzyme is essential for the transformation of pollutants. Furthermore, it has been observed that this transformation reaction is reliant on glucose, suggesting that the reduction of components requires a higher amount of energy (Romo-Rodríguez et al. 2015). The presence of oxalic acid, citric acid, and maleic acid in the solution resulted in a fall in pH which facilitated a more significant reduction of compounds by *Pisolithus* sp. (Shi et al. 2018). The formation of ROS (H_2O_2) is linked to the reduction of compounds through a multi-step process (Irazusta et al. 2018). A multistep electron transfer reaction is involved in this process and may depend on substrates or enzymes (Qamar et al. 2011). These extracellular activities can occasionally be triggered by functional groups or negative charges on the surface of fungal cells. (Cao et al. 2018; Dang et al. 2018).

6.5.7 *Mycoprecipitation*

Mycoprecipitation is a primary process employed by microorganisms to eliminate pollutants from wastewater (Maisa et al. 2018). It might be phrased as mycoprecipitation in regards to fungus, which is a part of bioprecipitation. Bioprecipitation is categorized into two different forms, intracellular and extracellular. Through the fungal cell wall, metal ions enter the cell during intracellular precipitation and react with certain anionic species to generate insoluble compounds. Petroleum compounds are produced outside of the cell and precipitates accumulate in extracellular precipitation. The fungus or the surrounding environment can provide the anionic species needed for this activity. The surface of the fungus acts as a catalyst to accelerate the process. The resultant precipitate mostly adheres to the inner surface of the cell during intracellular precipitation. In contrast, extracellular precipitation permits the precipitate to either accumulate on the fungal cell surface or diffuse across the surrounding media following precipitation (Liang et al. 2015). Phosphate ions have been found to cause the formation of uranium complex compounds, including chernikovite, meta-ankoleite, bassetite, and uramphite in the presence of fungus (Liang et al. 2016). *A. niger* synthesizes uranyl acetate hydrate, a kind of organo-uranyl compound, from low-grade uranium ore (Maisa et al. 2018). The bioprecipitate is found to include many forms of calcium carbonate ($CaCO_3$), including calcite, and strontianite (Li et al. 2015; Kumar et al. 2019). Petrochemical lead precipitation with aid of *Paecilomyces javanicus* involves the formation of plumbonacrite, cerussite ($PbCO_3$), and lead oxalate (PbC_2O_4) caused the formation by precipitating lead (Rhee et al. 2014).

6.5.8 Biovolatilization

Biovolatilization refers to the process by which pollutants and metalloids are released from water and soil into the environment through biological means. The process of biovolatilization is widely documented for the release of gasoline into the atmosphere by microorganisms such as bacteria, fungi, algae, and plants. Biovolatilization, a pivotal process in environmental biogeochemistry, is driven by microbial activity that converts hydrocarbons into volatile forms, ultimately releasing them into the atmosphere. This mechanism plays a vital role in the cycling of elements and can have significant implications for ecosystem health. Hydrocarbon transformation mechanisms in aquatic environments are illuminated by the microbial methylation and volatilization of hydrocarbon in polluted sediments (Li et al. 2021). The part soil microbes also play in the volatilization of sediment traped metals, delivering information about hazards related to the contaminated environment (Fai et al. 2020; Cui et al. 2019).

6.5.9 Biomethylation

Methylation is the process of adding methyl groups to a metal by enzymatic transfer, which results in the formation of several distinct compounds. Through oxidoreduction processes, redox transformations can enable microbes to mobilize various organic compounds and organometallic compounds. A widely distributed and crucial metabolic phase is the formation and transfer of the methyl group. Methyl groups are received by organic component C, O, N, and S atoms during the metabolic process. Hydrocarbons and metalloids have been identified in numerous investigations as acceptors of methyl groups, resulting in the production of methylated end products. However, the word biomethylation encompasses the creation of either a gaseous or non-gaseous substance containing methyl groups which can be formed by pollutants and metalloids. Methylation of adenine is ubiquitous, observed in bacteria, fungi, algae, and plants. The concept of methylation was initially suggested in the fungus *Scopulariopsis brevicaulis*. In the process of metal methylation, contaminated sediments are initially converted into a simpler form which is then further

Table 6.2 Mechanisms involved in the remediation of petroleum pollutants by fungal diversity

Fungal group	Fungal species	Target pollutant/ Degraded substances	Mechanisms	References
White rot fungi	*Pleurotus tuber-regium*	Crude oil	Degrade and utilize hydrocarbon as the source of carbon and energy	Ikhajiagbe et al. (2021)
Brown-rot fungi	*Daedalea dickinsii,*	Crude oil	Production of hydrocarbon byproducts	Rizqi and Purnomo (2017)
Endophytic fungi	*Coriolopsis gallica*	Diesel fuel hydrocarbons	Oxidation of laccase enzymes	Daâssi et al. (2021)

modified through the addition of a methyl group (–CH_3). It is proposed that in the *fungi Scopulariopsis brevicaulis and Penicillium notatum*, phenylstibonic acid is methylated to phenyldimethylstibine through the reduction of compounds (Hartmann et al. 2003; Andrewes et al. 2001). Fungi, including *Coprinellus radians, Coprinus comatus*, *Saccharomyces cerevisiae,* and *Candida albicans* are used in biomethylation (Table 6.2).

6.5.10 Biofiltration

Currently, bioreactors and biofiltration are shown to be efficient technical approaches for treating wastewater. Biofilters are made of porous carrier materials that hold either dead or living microorganisms such as fungi, bacteria, and algae, immobilized (Ali 2013). The petroleum pollutants detoxification or superficial removal from wastewater is the exclusive purpose of this technology. This is achieved by employing one or more fungal species that adhere to an inert surface. Numerous processes, including bioprecipitation, biosorption, redox reactions biovolatilization, and bioaccumulation are used in this system to remove pollutants. The presence of *A. niger* biomass in the biofilter can remove 89% Cr, 90% Ni and 91% at a pH of 6 and 10 min of aresidence period (Ali 2013).

6.6 Conclusions and Future Prospects

Mycoremediation emerges as an eco-friendly, cost-effective technique for tackling petroleum pollutants through the utilization of fungi or their metabolites. This method offers a blend of affordability, resilience, and effectiveness, making it a compelling option. Being inherently natural, it capitalizes on fungi's abundant biomass and can be easily adapted to various scenarios with minimal intervention, rendering it suitable for situations where conventional methods are unsuccessful.

The efficacy of mycoremediation depends on the specific fungal species and their compatibility with different contaminants. Given the diverse characteristics of fungi such as habitat preferences, morphology, and metabolic activities, this approach finds application across a wide range of scenarios, both in situ and ex-situ. Mycoremediation employs various methods to mitigate petroleum pollution, including precipitation, biosorption, sequestration, and biotransformation. In addition to the diversity and fungal habitat, the physical, chemical, and biological effects on fungal bioremediation as well as their potential significance are analyzed. Studying myco-nanotechnology from various ecologies for the bioremediation of contaminants is also growing more popular. The removal of toxins from polluted land surfaces and water bodies through the application of mycoremediation in combination with nano technology and other advanced technologies like next-generation sequencing (NGS), enzyme engineering, and microarray technology, therefore, has a promising future. It is critically necessary to develop a synergistic approach and favorable regulatory structure for mycoremediation that extends from laboratory scale to pilot scale testing using fungal species with acceptable traits.

The exploration of mycoremediation faces certain challenges, primarily due to its time-consuming nature and the inability to eliminate contaminants. Mycoremediation is thought to be greatly hindered by a lack of awareness of ecological tools and methods for sustaining fungal biomass, activity, and enzyme production into polluted matrices. To increase the biodegradation of a variety of petroleum contaminants, bioinformatics approaches must be explored for novel species of fungal and microbial consortia with better ecological adaptation. This will improve the bioremediation's application possibilities, competitiveness, and practicality. This synergistic approach serves as a special tool for the long-term security of natural resources, particularly in the context of the two intricate global issues of petroleum pollution and climate change. The efficiency of mycoremediation is limited in the field since it is still in its infancy at the laboratory/greenhouse level. The mycoremediation capacities of each species must thus be assessed in their natural habitat before the commercialization of this green technology. In order to gain an in-depth knowledge of the remediation processes, it is necessary to explore the genetic alteration and genealogy of microbes.

6.7 Suggestion

The following suggestions are made to enhance the use of mycoremediation techniques:

(i) The use of mycoremediation in conjunction with nanotechnology and other cutting-edge technologies like next-generation sequencing, enzyme engineering, and microarray technology to remove toxins from contaminated land surfaces has a bright future.

(ii) Creating a synergistic strategy for mycoremediation that goes beyond laboratory testing to pilot scale testing utilizing fungal species with desirable characteristics is imperative.
(iii) To enhance the biodegradation of various petroleum pollutants, bioinformatics techniques need to be used to find new fungal species and microbial consortia with improved ecological adaption.

References

Ahemad M, Khan MS (2011) Effects of insecticides on plant-growth-promoting activities of phosphate solubilizing *Rhizobacterium Klebsiella* sp. strain PS19. Pest Biochem Physiol 100(1):51–56. https://doi.org/10.1016/j.pestbp.2011.02.004

Al-Hawash AB, Zhang X, Ma F (2019) Removal and biodegradation of different petroleum hydrocarbons using the filamentous fungus Aspergillus sp. RFC-1. Microbiologyopen 8(1):e00619. https://doi.org/10.1002/mbo3.619

Ali SAA (2013) Removal of heavy metals from synthesis industrial wastewater using local isolated *Candida utilis* and *Aspergillus Niger* as bio-filter. Int J Biotechnol 2(5):83–90

Altomare C, Norvell WA, Björkman T, Harman GE (1999) Solubilization of phosphates and micronutrients by the plant-growth-promoting and biocontrol fungus *Trichoderma harzianum*. Rifai1295-22. Appl Environ Microbiol 65:2926–2933

Aly AH, Debbab A, Proksch P (2011) Fungal endophytes: unique plant inhabitants with great promises. Appl Microbiol Biotechnol 90:1829–1845

Anastasi A, Coppola T, Prigione V, Varese GC (2009) Pyrene degradation and detoxification in soil by a consortium of basidiomycetes isolated from compost: role of laccases and peroxidases. J of Hazardous Materials 165(1-3):1229-1233. https://doi.org/10.1016/j.jhazmat.2008.10.032

Andrewes P, Cullen WR, Polishchuk E, Reimer KJ (2001) Antimony biomethylation by the wood rotting fungus *Phaeolus schweinitzii*. Appl Organomet Chem 15(6):473–480

Asemoloye MD, Marchisio MA, Gupta VK, Pecoraro L (2021) Genome-based engineering of ligninolytic enzymes in fungi. Microb Cell Factories 20(1):1–18. https://doi.org/10.1186/s12934-021-01510-9

Azubuike CC, Chikere CB, Okpokwasili GC (2016) Bioremediation techniques-classification based on site of application: principles, advantages, limitations and prospects. World J Microbiol Biotechnol 32(11):180. https://doi.org/10.1007/s11274-016-2137-x

Bairagi H, Khan M, Ray L, Guha A (2010) Adsorption profile of lead on *Aspergillus versicolor*: a mechanistic probin. J Hazard Mater 186:756–764. https://doi.org/10.1016/j.jhazmat.2010.11.064

Banerjee A, Jhariya MK, Yadav DK, Raj A (2018) Micro-remediation of metals: a new frontier in bioremediation. In: Hussain CM (ed) Handbook of environmental materials management. Springer, New York, pp 1–25

Baratto MC, Juarez-Moreno K, Pogni R, Basosi R, Vazquez-Duhalt R (2015) EPR and LC-MS studies on the mechanism of industrial dye decolorization by versatile peroxidase from *Bjerkandera adusta*. Environ Sci Pollut Res Int 22(11):8683–8692

Benguenab A, Chibani A (2021) Biodegradation of petroleum hydrocarbons by filamentous fungi (*Aspergillus ustus and Purpureocillium lilacinum*) isolated from used engine oil contaminated soil. Acta Ecol Sin 41:416–423. https://doi.org/10.1016/j.chnaes.2020.10.008

Bhardwaj P, Sharma A, Capalash N, Sharma P (2018) Azo dye decolorization by marine bacterial and fungal species: a comparative study. Environ Sci Pollut Res 25(10):10153–10163

Blanca AL, Angus JB, Katarina S, Joe LR, Nicholas JR (2007) The influence of different temperature programmes on the bioremediation of polycyclic aromatic hydrocarbons (PAHs) in a coal-tar contaminated soil by in-vessel composting. J Hazard Mater 14:340–347

Bonugli-Santos RC, Durrant LR, Da Silva M, Sette LD (2010) Production of laccase, manganese peroxidase and lignin peroxidase by Brazilian marine-derived fungi. Enzym Microb Technol 46(1):32–37

Cao L, Jiang M, Zeng Z, Aixue D, Tan H, Liu Y (2008) *Trichoderma atroviride* F6 improves phytoextraction efficiency of mustard (*Brassica juncea* (L.) Coss. var. foliosa Bailey) in Cd, Ni contaminated soils. Chemosphere 71(9):1769–1773. https://doi.org/10.1016/j.chemosphere.2008.01.066

Cao F, Bourven I, Guibaud G, Rene ER, Lens PN, Pechaud Y, Hullebusch EDV (2018) Alteration of the characteristics of extracellular polymeric substances (EPS) extracted from the fungus *Phanerochaete chrysosporium* when exposed to sub-toxic concentrations of nickel (II). Int Biodeterior Biodegradation 129:179–188

Cecchi G, Marescotti P, Piazza SD, Zotti M (2017) Native fungi as metal remediators: silver myco-accumulation from metal contaminated waste rock dumps (Libiola mine, Italy). J Environ Sci Health B 52(3):191–195. https://doi.org/10.1080/03601234.2017.1261549

Cui Y, Gao Y, Zhuang X (2019) Microbial volatilization of selenium: insight into the factors influencing the process and its relevance for selenium cycling in ecosystems. Crit Rev Environ Sci Technol 49(9):775–795

da Silva NA, Birolli WG, Seleghim MHR, Porto ALM (2013) Biodegradation of the organophosphate pesticide profenofos by marine fungi. In: Patil YB, Rao P (eds) Applied bioremediation-active and aassive approaches. InTechWeb, Rijeka, pp 149–180. https://doi.org/10.5772/56372

Daâssi D, Nasraoui-Hajaji A, Bawasir S, Frikha F, Mechichi T (2021) Biodegradation of C20 carbon clusters from diesel fuel by *Coriolopsis gallica*: optimization, metabolic pathway, phytotoxicity. Biotech 11:1–21. https://doi.org/10.1007/s13205-021-02769-w

Damodaran D, Balakrishnan RM, Shetty VK (2013) The uptake mechanism of Cd(II), Cr(VI), Cu(II), Pb(II), and Zn(II) by mycelia and fruiting bodies of *Galerina vittiformis*. Biomed Res Int 2013:149120. https://doi.org/10.1155/2013/149120

Dang C, Yang Z, Liu W, Du P, Cui F, He K (2018) Role of extracellular polymeric substances in biosorption of Pb2+ by a high metal ion tolerant fungal strain *Aspergillus Niger* PTN31. J Environ Chem Eng 6(2):2733–2742

Dashtban M, Schraft H, Syed TA, Qin W (2010) Fungal biodegradation and enzymatic modification of lignin. Int J Biochem Mol Biol 1(1):36–50

Dhankhar R, Hooda A (2011) Fungal biosorption-an alternative to meet the challenges of heavy metal pollution in aqueous solutions. Environ Technol 32(5–6):467–491

Dirisu CG, Ogbulie JN, Orji JC, Eresia-Eke R, Olowu MD, Neighbour R (2018) Crude oil degradation by *Penicilium* and *Mucor* species isolated from fresh water swamp. Ann Ecol Environ Sci 2:76–81

Fai EY, Yan J, Chi L, Cao Y, Jia Z, Wang L, Li J (2020) Role of soil microbes in the volatilization of arsenic from arsenic-contaminated soils: a review. J Environ Sci 89:186–198

Fawzy M, Nasr M, Adel S, Helmi S (2018) Regression model, artificial neural network, and cost estimation for biosorption of Ni(II)-ions from aqueous solutions by *Potamogeton pectinatus*. Int J Phytoremediation 20(4):321–329. https://doi.org/10.1080/15226514.2017.1381941

Felsot AS, Racke KD, Hamilton DJ (2003) Disposal and degradation of pesticide waste. Rev Environ Contam Toxicol 177:123–200

Fernandez-Fueyo E, Linde D, Almendral D, López-Lucendo MF, Ruiz-Dueñas FJ, Martınez AT (2015) Description of the first fungal dyedecolorizing peroxidase oxidizing manganese (II). Appl Microbiol Biotechnol 99(21):8927–8942

Finlay RD (2008) Ecological aspects of mycorrhizal symbiosis: with special emphasis on the functional diversity of interactions involving the extraradical mycelium. J Exp Bot 59:1115–1126

Georgianna DR, Fedorova ND, Burroughs JL, Dolezal AL, Bok JW, Horowitz-Brown S, Woloshuk CP, Yu J, Keller NP, Payne GA (2010) Beyond aflatoxin: four distinct expression patterns and functional roles associated with *Aspergillus flavus* secondary metabolism gene clusters. Mol Plant Pathol 11(2):213–226. https://doi.org/10.1111/j.1364-3703.2009.00594.x

Girma G (2015) Microbial bioremediation of some heavy metals in soils: an updated review. Indian J Sci Res 6(1):147–161

Gola D, Deya P, Bhattacharya A, Mishra A, Malik A, Namburath M, Ahammad SZ (2016) Multiple heavy metal removal using an entomopathogenic fungi *Beauveria bassiana*. Bioresour Technol 218:388–396. https://doi.org/10.1016/j.biortech.2016.06.096

Hamba T (2016) Mycoremediation of heavy metals and hydrocarbons contaminated environment. Asian J Nat and Appl Sci 5(2)

Harms H, Schlosser D, Wick LY (2011) Untapped potential: exploiting fungi in bioremediation of hazardous chemicals. Nat Rev Microbiol 9(3):177–192. https://doi.org/10.1038/nrmicro2519

Hartmann LM, Craig PJ, Jenkins RO (2003) Influence of arsenic on antimony methylation by the aerobic yeast *Cryptococcus humicolus*. Arch Microbiol 180(5):347–352

Hawksworth DL, Lucking R (2017) Fungal diversity revisited: 2.2 to 3.8 million species. Microbiol Spect 5(4):5–4

Hedayati MT, Pasqualotto AC, Warn PA, Bowyer P, Denning DW (2007) *Aspergillus flavus*: human pathogen, allergen and mycotoxin producer. Microbiology 153(6):1677–1692. https://doi.org/10.1099/mic.0.2007/007641-0

Hofrichter M, Kellner H, Pecyna MJ, Ullrich R (2015) Fungal unspecific peroxygenases: heme-thiolate proteins that combine peroxidase and cytochrome P450 properties. Adv Exp Med Biol 851:341–368

Ikhajiagbe B, Enerijiofi KE, Umendu PO (2021) Mycorestoration of crude oil polluted soil using *Pleurotus tuber-regium*. Stud Univ Babes-Bolyai Biol 66(1):73–84

Iqbal HMN, Asgher M (2013) Decolorization applicability of sol–gel matrix immobilized manganese peroxidase produced from an indigenous white rot fungal strain *Ganoderma lucidum*. BMC Biotechnol 13(1):1–7

Irazusta V, Bernal AR, Estévez MC, De Figueroa LIC (2018) Proteomic and enzymatic response under Cr(VI) overload in yeast isolated from textile-dye industry effluent. Ecotoxicol Environ Saf 148:490–500. https://doi.org/10.1016/j.ecoenv.2017.10.076

John J (2013) Assessment of arbuscular mycorrhizal fungi in a green roof system. Dalhousie University, Halifax

Khan AL, Waqas M, Hussain J, Al-Harrasi A, Lee I-J (2014a) Fungal endophyte *Penicillium janthinellum* LK5 can reduce cadmium toxicity in *Solanum lycopersicum* (Sitiens and Rhe). Biol Fertil Soils 50:75–85

Khan SA, Uddin I, Moeez S, Ahmad A (2014b) Fungus-mediated preferential bioleaching of waste material such as fly-ash as a means of producing extracellular, protein capped, fluorescent and water soluble silica nanoparticles. PLoS One 9:e107597. https://doi.org/10.1371/journal.pone.0107597

Khan I, Aftab M, Shakir S, Ali M, Qayyum S, Rehman MU, Haleem KS, Touseef I (2019) Mycoremediation of heavy metal (Cd and Cr)–polluted soil through indigenous metallotolerant fungal isolates. Environ Monit Assess 191:585. https://doi.org/10.1007/s10661-019-7769-5

Kubartová A, Ranger J, Berthelin J, Beguiristain T (2009) Diversity and decomposing ability of saprophytic fungi from temperate forest litter. Microb Ecol 58:98–107

Kulshreshtha S, Mathur N, Bhatnagar P, Kulshreshtha S (2013) Cultivation of *Pleurotus citrinopileatus* on handmade paper and cardboard industrial wastes. Ind Crop Prod 41:340–346. https://doi.org/10.1016/j.indcrop.2012.04.053

Kumar S, Dagar VK, Khasa YP, Kuhad RC (2013) Genetically modified microorganisms (GMOs) for bioremediation. In: Kuhad RC, Singh A (eds) Biotechnology for environmental management and resource recovery. Springer, pp 191–218

Kumar V, Singh S, Singh G, Dwivedi SK (2019) Exploring the cadmium tolerance and removal capability of a filamentous fungus *Fusarium solani*. Gemicrobiol J 36(9):782–791. https://doi.org/10.1080/01490451.2019.1627443

Lamar RT, White RB (2001) Mycoremediation: commercial status and recent developments. In: von Fahnestock MF, Leeson A (eds) Magar VS. Proceedings Sixth International Symposium on In Situ and On-Site Bioremediation, San Diego, pp 263–278

Lehmann J, Rillig M, Thies J, Masiello C, Hockaday W, Crowley D (2011) Biochar effects on soil biota–a review. Soil Biol Biochem 43:1812–1836

Li Q, Csetenyi L, Paton GI, Gadd GM (2015) $CaCO_3$ and $SrCO_3$ bioprecipitation by fungi isolated from calcareous soil. Environ Microbiol 17(8):3082–3097. https://doi.org/10.1111/1462-2920.12954

Li X, Wang Y, Pan Y, Yu H, Zhang X et al (2017) Mechanisms of Cd and Cr removal and tolerance by macrofungus *Pleurotus ostreatus* HAU-2. J Hazard Mater 330:1–8. https://doi.org/10.1016/j.jhazmat.2017.01.047

Li H, Zhang Y, Liu G, Wang J (2021) Microbial methylation and volatilization of mercury in polluted sediments: mechanisms and implications for aquatic environments. Sci Total Environ 778:146228

Li T, Cui L, Song X, Cui X, Wei Y, Tang L, Mu Y, Xu Z (2022) Wood decay fungi: an analysis of worldwide research. J Soils Sediments 22:1688–1702

Liang X, Csetenyi L, Gadd GM (2015) Lead bioprecipitation by yeasts utilizing organic phosphorus substrates. Geomicrobiol J 33(3–4):294–307. https://doi.org/10.1080/01490451.2015.1051639

Liang X, Csetenyi L, Gadd GM (2016) Uranium bioprecipitation mediated by yeasts utilizing organic phosphorus substrates. Appl Microbiol Biotechnol 100:5141–5151. https://doi.org/10.1007/s00253-016-7327-9

Lloyd JR, Renshaw JC (2005) Bioremediation of radioactive waste: radionuclide-microbe interactions in laboratory and field-scale studies. Curr Opin Biotechnol 16:254–260

Maisa M, Amin I, Elaassy E, Mohamed G, El-Feky NA, Kawady MST, Abdelsattar MS (2018) Recovery of uranium from low-grade ore using microorganism isolated from uraniferous rock sample. Sep Sci Technol 53:1–6. https://doi.org/10.1080/01496395.2018.1437182

Martínez ÁT, Speranza M, Ruiz-Dueñas F, Ferreira P, Camarero S, Guillén F, Martínez M, Suárez AG, Río AJC (2005) Biodegradation of lignocellulosics: microbial, chemical, and enzymatic aspects of the fungal attack of lignin. Int Microbiol 8:195–204

McErlen C, Marchant R, Banat IM (2006) An evaluation of soil colonisation potential of selected fungi and their production of ligninolytic enzymes for use in soil bioremediation applications. Anton Leeuw Int J G 90:147–158

Menk JDJ, Nascimento AISD, Leite FG, Oliveira RAD, Jozala AF, Junior JMDO, Chaud MV, Denise Grotto D (2019) Biosorption of pharmaceutical products by mushroom stem waste. Chemosphere 237:124515. https://doi.org/10.1016/j.chemosphere.2019.124515

Mrozik A, Piotrowska-Seget Z (2010) Bioaugmentation as a strategy for cleaning up of soils contaminated with aromatic compounds. Microbiol Res 165(5):363–375

Naraian R, Singh MP, Ram S (2016) Supplementation of basal substrate to boost up substrate strength and oyster mushroom yield: an overview of substrates and supplements. Int J Curr Microbiol App 5:543–553

Nielsen KF, Mogensen JM, Johansen M, Larsen TO, Frisvad JC (2009) Review of secondary metabolites and mycotoxins from the *Aspergillus Niger* group. Anal Bioanal Chem 395(5):1225–1242. https://doi.org/10.1007/s00216-009-3081-5

Pagnanelli F, Toro L, Veglio F (2002) Olive mill solid residues as heavy metal sorbent material: a preliminary study. Waste Manag 22:901–907

Prasad R, Bhola D, Akdi K, Cruz C, Sairam KVSS, Tuteja N, Varma A (2017) Introduction to mycorrhiza: historical development. In: Varma A, Prasad R, Tuteja N (eds) Mycorrhiza. Springer, Cham, pp 1–7

Purohit J, Chattopadhyay A, Biswas MK, Singh NK (2018) Springer international publishing AG, part of springer nature 91. In: Prasad R (ed) Mycoremediation and environmental sustainability, fungal biology. https://doi.org/10.1007/978-3-319-77386-5_4

Qamar M, Gondal MA, Yamani ZH (2011) Synthesis of nanostructured NiO and its application in laser-induced photocatalytic reduction of Cr(VI) from water. J Mol Catal A Chem 341:83–88. https://doi.org/10.1016/j.molcata.2011.03.029

Rhee YJ, Hillier S, Pendlowski H, Gadd GM (2014) Fungal transformation of metallic lead to pyromorphite in liquid medium. Chemosphere 113:17–21. https://doi.org/10.1016/j.chemosphere.2014.03.085

Rhodes CJ (2014) Mycoremediation (bioremediation with fungi)-growing mushrooms to clean the earth. Chem Speciat Bioavailab 26(3):196–198

Rizqi HD, Purnomo AS (2017) The ability of brown-rot fungus Daedalea dickinsii to decolorize and transform methylene blue dye. World J Microbiol Biotechnol 33:1–9. https://doi.org/10.1007/s11274-017-2256-z

Rockne K, Reddy K (2003) Bioremediation of contaminated sites. Proceedings international e-conference on modern trends in foundation engineering: geotechnical challenges and solutions

Romo-Rodríguez P, Acevedo-Aguilar FJ, Lopez-Torresm A, Wrobel K, Wrobel K, Gutiérrez-Corona JF (2015) Cr (VI) reduction by gluconolactone and hydrogen peroxide, the reaction products of fungal glucose oxidase: cooperative interaction with organic acids in the biotransformation of Cr(VI). Chemosphere 134:563–570

Roshchina VV (2012) The excretory function of higher plants. Springer Science and Business Media, Cham

Rout ME (2014) The plant microbiome. In: Advances in botanical research. Elsevier, Amsterdam, pp 279–309

Sharma SS, Dietz KJ (2006) The significance of amino acids and amino acid-derived molecules in plant responses and adaptation to heavy metal stress. J Exp Bot 57:711–726

Sharma SS, Dietz KJ (2009) The relationship between metal toxicity and cellular redox imbalance. Trends Plant Sci 14(1):43–50. https://doi.org/10.1016/j.tplants.2008.10.007

Shekher R, Sehgal S, Kamthania M, Kumar A (2011) Laccase: microbial sources, production, purification, and potential biotechnological applications. Enzyme Res 2011:217861

Shen M, Liu L, Li D-W, Zhou W-N, Zhou Z-P, Zhang C-F, Luo Y-Y, Wang H-B, Li H-Y (2013) The effect of endophytic *Peyronellaea* from heavy metal-contaminated and uncontaminated sites on maize growth, heavy metal absorption and accumulation. Fungal Ecol 6:539–545

Shi L, Xue J, Liu B, Dong P, Wen Z, Shen Z, Chen Y (2018) Hydrogen ions and organic acids secreted by ectomycorrhizal fungi, *Pisolithus* sp1, are involved in the efficient removal of hexavalent chromium from waste water. Ecotoxicol Environ Saf 161:430–436. https://doi.org/10.1016/j.ecoenv.2018.06.004

Shourie A, Vijayalakshmi U (2022) Fungal diversity and its role in mycoremediation. Geomicrobiol J 39:3–5. https://doi.org/10.1080/01490451.2022.2032883

Singh H (2006) Mycoremediation—fungal bioremediation. Wiley, p 573

Talley J (2005) Introduction of recalcitrant compounds. In: Jaferey W, Talley L (eds) Bioremediation of recalcitrant compounds. CRC, Boca Raton, pp 1–9

Tang A, Lu Y, Li Q, Zhang X, Cheng N, Liu H, Liu Y (2021) Simultaneous leaching of multiple heavy metals from a soil column by extracellular polymeric substances of *Aspergillus tubingensis* F12. Chemosphere 263:127883. https://doi.org/10.1016/j.chemosphere.2020.127883

Tekere M (2019) Microbial bioremediation and different bioreactors designs applied. In: Biotechnol Bioeng. IntechOpen. https://doi.org/10.5772/intechopen.83661

Tiwari G, Singh SP (2014) Application of bioremediation on solid waste management. J Bioremed Biodegr 5:248. https://doi.org/10.4172/2155-6199.1000248

Verma A, Kore R, Corbin DR, Shiflett MB (2019) Metal recovery using oxalate chemistry: a technical review. Ind Eng Chem Res 58:15381–15393

Xiros C, Studer MH (2017) A multispecies fungal biofilm approach to enhance the celluloyltic efficiency of membrane reactors for consolidated bioprocessing of plant biomass. Front Microbiol 8:1930. https://doi.org/10.3389/fmicb.2017.01930

Xu DY, Yang Z (2013) Cross-linked tyrosinase aggregates for elimination of phenolic compounds from wastewater. Chemosphere 92(4):391–398

Yang J, Li W, Ng TB, Deng X, Lin J, Ye X (2017) Laccases: production, expression regulation, and applications in pharmaceutical biodegradation. Front Microbiol 8:832

Zhao W, Huang P, Zhu Z, Chen C, Xu X (2020) Production of phenolic compounds and antioxidant activity via bioconversion of wheat straw by *Inonotus obliquus* under submerged fermentation with the aid of a surfactant. Wiley. https://doi.org/10.1002/jsfa.10710

Chapter 7
Sustainable Bioremediation Techniques for Soil Petroleum Hydrocarbons

Ana Paula Honrado Pinto, Carolina S. Marques, Elisabete P. Carreiro, and Pedro C. Barrulas

Abstract Soil pollution with organic xenobiotics is an emerging environmental issue of global importance. This chapter reviews and investigates the critical impacts resulting from the presence of petroleum hydrocarbons in soils and their relationship with their physicochemical characteristics, describing suitable ecologically justified remediation processes and exploring synergies present in plant-microbe integrated approaches.

Keywords Eco-friendly · Hydrocarbons · Petroleum · Remediation · Soil

7.1 Introduction

Soil pollution is one of the biggest and most challenging problems of our time. Organic xenobiotics are defined as any chemical compounds possessing carbon-hydrogen or carbon-carbon chemical bonds "that do not necessarily occur naturally

---

A. P. H. Pinto (✉)
Mediterranean Institute for Agriculture, Environment and Development, MED & Global Change and Sustainability Institute, CHANGE, Institute for Advanced Studies and Research, Évora, Portugal

Chemistry and Biochemistry Department, School of Sciences and Technology, Évora, Portugal
e-mail: app@uevora.pt

C. S. Marques · E. P. Carreiro
LAQV-REQUIMTE, School of Sciences and Technology, University of Évora, Évora, Portugal

P. C. Barrulas
Chemistry and Biochemistry Department, School of Sciences and Technology, Évora University, Évora, Portugal

University of Évora, HERCULES Laboratory, Évora, Portugal

A. Parnian et al. (eds.), *Bioremediation of Petroleum Contaminated Soil and Sediment*, Environmental Contamination Remediation and Management,
https://doi.org/10.1007/978-3-031-95524-2_7

in the environment or whose naturally occurring concentrations are significantly different from those caused by anthropogenic activity" (Štefanac et al. 2021). Pesticides, pharmaceuticals, additives, and industrial and domestic synthetic compounds are developed to improve everyday life at the cost of ecosystem sustainability. Despite the implementation of severe global regulation regarding discharge of contaminants to the environment, waste disposal cannot be controlled, including household personal care products, residual pharmaceuticals, and agriculture xenobiotics (de Oliveira et al. 2020). Residual pollution is difficult to remove, as proper identification and quantification of complex molecular structures, physical and chemical properties, and health consequences are costly to determine.

Among organic xenobiotics (Štefanac et al. 2021) petroleum hydrocarbons (PHs) are of greatest concern, due to their neurotoxic, carcinogenic, and teratogenic effects (Webb et al. 2018; Kuppusamy et al. 2020).

PHs are stable compounds constituted primarily of carbon and hydrogen atoms, derived from crude oil and petroleum products (Hoang et al. 2021a). Oxygen, nitrogen, and sulfur, are also present in trace amounts in the structure of PHs. To fully recognize the impact of the PHs on the planet it is crucial to understand their sources, composition, and properties. Evaluation will naturally drive the creation and implementation of useful strategies to mitigate and remediate pollution.

PHs can be classified as saturated, unsaturated, and heteroaromatic compounds (Fig. 7.1), depending on their chemical structure and properties (Dike et al. 2021; Haider et al. 2021; Hoang et al. 2021a; Dike et al. 2022).

Among the three main fractions (Fig. 7.1) of PHs classification, saturated hydrocarbons (paraffins or alkanes) are considered the most abundant in crude oil. All

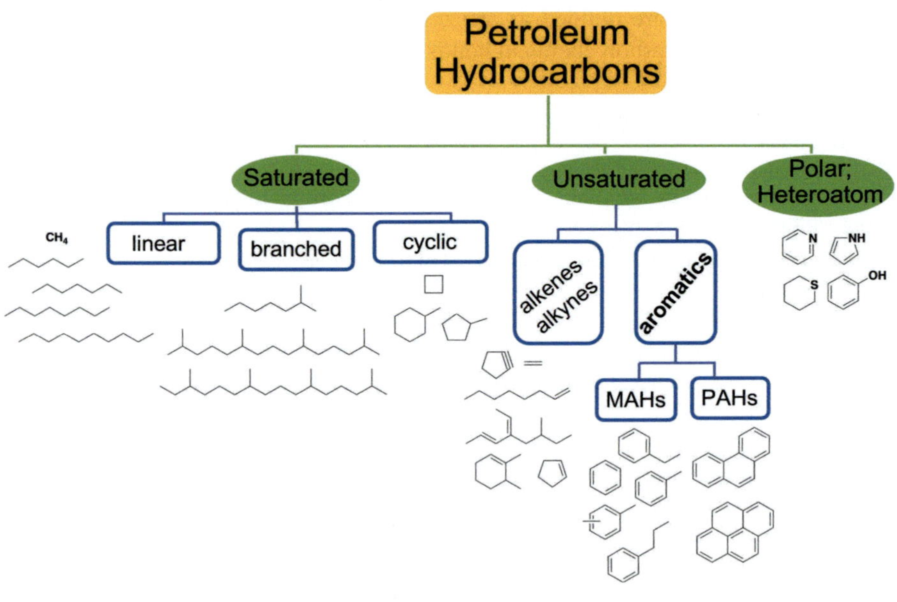

Fig. 7.1 Classification of Petroleum Hydrocarbons (PHs) and some selected examples

compounds containing C-H and C-C single bonds in linear, branched, or cyclic non-aromatic chains belong to this fraction (Fig. 7.1). Methane (CH_4), hexane (C_6H_{14}), heptane (C_7H_{16}), octane (C_8H_{18}), or decane ($C_{10}H_{22}$) are examples of linear saturated hydrocarbons. Isooctane (C_8H_{18}), 2,6,10,14-tetramethylpentadecane or pristane ($C_{19}H_{40}$) and 2,6,10,14-tetramethyl hexadecane or phytane ($C_{20}H_{42}$) are examples of branched saturated hydrocarbons. Regarding cycloalkanes with cyclic non-aromatic saturated hydrocarbons, cyclobutane (C_4H_8), methylcyclohexane (C_7H_{14}) and methyl cyclopentane (C_6H_{12}) are reported to be present (Stauffer et al. 2008; Atef et al. 2017; Macaya et al. 2019; Kuppusamy et al. 2020).

Within the category of saturated hydrocarbons, are *n*-alkanes (the linear saturated hydrocarbons and the fastest to be decomposed) which may be subdivided according to their molecular weight (MW): gaseous alkanes, low molecular weight (LMW) alkanes (C8-C16 containing carbons), medium molecular weight (MMW) alkanes (possessing C17-C28 containing carbons) and the high molecular weight (HMW) alkanes (with up to 28 carbons on their scaffold) (Abbasian et al. 2015).

Unsaturated PHs are essentially hydrocarbons with double or triple C-C bonds, where alkenes or olefins, alkynes, and arenes (or aromatics) group together. Among the alkenes, example compound types are aliphatic linear (such as ethylene (C_2H_4) and octene (C_8H_6)), branched (including 4-ethylidene-6-methyloct-2-ene ($C_{11}H_{20}$)), cyclic non-aromatic PHs (cyclopentene (C_5H_8), 2,3-dimethyl-cyclohexene (C_8H_{14}) and cyclopentyne (C_5H_6)) are within this category (Fig. 7.1) (Haider et al. 2021; Dike et al. 2022).

All PHs containing at least one benzene ring fall under the sub-category of aromatic hydrocarbons (Infante and Bingham 2012). Within this sub-category, we can find monocyclic aromatic hydrocarbons (MAHs) and polycyclic aromatic hydrocarbons (PAHs). The difference relies upon the number of benzene rings in the structures. Propylbenzene (C_9H_{12}), with one aromatic ring, is a MAH. Benzene (C_6H_6), toluene (C_7H_8), ethylbenzene (C_8H_{10}) and xylene (*ortho, meta,* and *para*, C_8H_{10}) are further examples of MAHs. This composite set is commonly described as BTEX in literature, a group of volatile organic pollutants associated with severe health risks (Yu et al. 2022).

Compounds possessing two or more benzene rings fit into the PAHs sub-division. Phenanthrene ($C_{14}H_{10}$) and pyrene ($C_{16}H_{10}$) are examples of PAHs with several fused benzene rings in their scaffold (Fig. 7.1) (Abdel-Shafy and Mansour 2016).

Finally, the last division covers all the polar PHs like asphaltenes (Fakher et al. 2020), resins (Mildenberg et al. 2007), and other non-hydrocarbon derivatives containing heteroatoms like oxygen, nitrogen, and sulfur. Pyridine (C_5H_5N), pyrrole (C_4H_5N), tetrahydro-2H-thiopyran ($C_5H_{10}S$), and phenol (C_6H_6O) are some examples of this class of PHs (Fig. 7.1).

Asphaltene, a complex solid mixture of high molecular weight PHs of many different structures emulsified in crude oil (Fig. 7.2a), is a cause of concern and difficult to remove from industrial plants, clogging the pipes or wells (Groenzin and Mullins 2001; Abdel-Raouf 2012).

Its accumulation due to high density and MW can cause severe operation and production problems and of course, an environmental problem if a leak occurs.

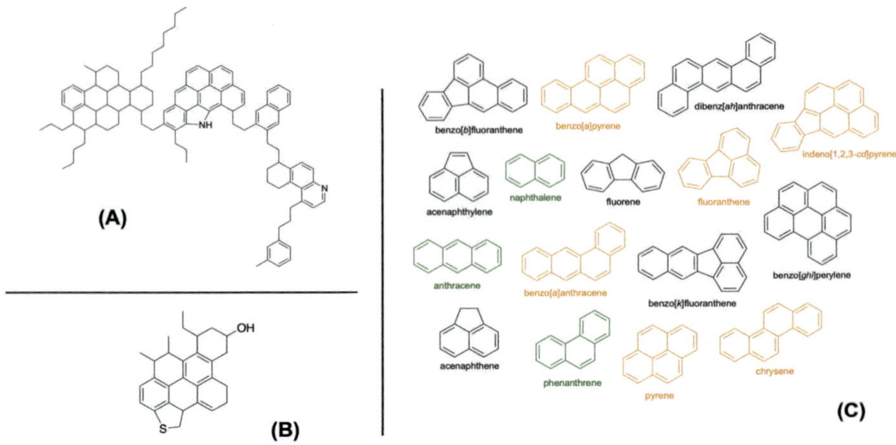

Fig. 7.2 Examples of molecular structures of asphaltene (**a**), resin (**b**), and some examples of PAHs (**c**): in green LMW PAHs; in orange HMW PAHs

Among characteristics used to identify asphaltene the presence of heteroatoms (like N, O, and S), high polarity, non-solubility in LMW alkanes (such as *n*-pentane or *n*-heptane), and solubility in generally non-polar aromatics such as toluene or xylene, are some examples (Seifried et al. 2013; Fakher et al. 2020).

Resins are very similar mixtures of complex aromatic and heteroaromatic rings (fused or not) formed structures (Fig. 7.2b), with attached alkyl chains, but having lower MW (light oils) in comparison to asphaltenes (heavy oils) (Abdel-Raouf 2012; Cheshkova et al. 2019; Fakher et al. 2020).

Difficult to fully characterize at a molecular level, hydrocarbon resins, obtained industrially via thermal or catalytic polymerization processes from petroleum distillates, are involved in the manufacture of important products used in daily life; plastics, paints, printing inks, and sealants, among others (Mildenberg et al. 2007). According to León and co-workers (León et al. 2002), resins take part in the stabilization of the highly polar asphaltene fraction in the crude oil, being involved in the bonding capacity to those complex mixtures, increasing their solubility (Miadonye and Evans 2010).

Within the fraction of the unsaturated PHs (Fig. 7.1), the distinctive characteristics of the PAHs (such as poor aqueous solubility, high boiling, and melting points, vapor pressure, and challenge regarding degradation, among others) target them in the class of high priority pollutants of natural resources (Abdel-Shafy and Mansour 2016).

The number of aromatic rings in PAHs scaffolds enables grouping them into two subclasses: LMW PAHs (with two or three benzene rings) and HMW PAHs (with to four benzene rings). Naphthalene (two fused benzene rings), phenanthrene, and anthracene (three fused benzene rings with different conformations) are examples of LMW PAHs (Fig. 7.2c). Four-fused rings pyrene, benzo[*a*]anthracene, chrysene, and fluoranthene, five-fused rings benzo[*a*]pyrene and six-fused rings indeno[1,2,3-cd]pyrene are some examples of HMW PAHs (Fig. 7.2c) (Kuppusamy et al. 2016a, b, 2017).

The release of PHs into the environment has two major points of origin: from incomplete combustion of natural geological sources (like volcanic activities or forest fires) and human anthropogenic activities, principally waste incineration, fossil fuels, industry, and agriculture (Kumar et al. 2021; Premnath et al. 2021).

The United States Environmental Protection Agency (US EPA) has recognizes 16 PAHs as priority pollutants: naphthalene, acenaphthylene, acenaphthene, fluorene, phenanthrene, anthracene, fluoranthene, pyrene, benzo[*a*]anthracene, chrysene, benzo[*b*]fluoranthene, benzo[*k*]fluoranthene, benzo[*a*]pyrene, dibenz[*ah*]anthracene, indeno[1,2,3-*cd*]pyrene and benzo[*ghi*]perylene, which chemical structures can be found in Fig. 7.2c (Kumar et al. 2021; Zhang et al. 2021a).

PAHs were flagged by the International Agency for Research on Cancer (IARC) regarding their potential carcinogenic, teratogenic, and mutagenic effects (Bauer et al. 2018; Kumar et al. 2021; Zhang et al. 2021a).

As a consequence of their source, PHs can be found in several environments; and consequently in aquatic life, soil, sediments and air. Due to naturally occurring progress in civilization and global energy demand from economic and demographic growth, there is an expected increase of general PH contamination. Several efforts by the scientific community to reverse this trend or to remediate the contaminated environments have been made in the last few years. Great attention has been focused on PAHs, the main causes of soil contamination and the most harmful to human health (Sayara and Sánchez 2020).

In subsequent sections, special attention will be given to the environmental impact of soil pollution by PHs, pointing out their unique characteristics as contaminants (*viz.* aqueous solubility, molecular structure, toxicity, resistance to degradation), and describing the most suitable ecologically remediation processes to detoxify the contaminated soils, exploring the plant-microbe synergies present in integrated approaches.

7.2 PHs and Environmental Impact

The increase in the world's population, allied with the economic expansion economy and subsequent energy requirement, still targets petroleum products (heavy oil) as vital. Obviously, this carries severe consequences of environmental pollution, mainly in the soil ecosystem, caused by the release of naturally or industrially modified PHs.

According to the US Energy Information Administration database, the standard of crude oil production was estimated to be 80.62 million barrels per day, in 2019. USA, Russia, China, Saudi Arabia, and Iran, among other countries, are among the world's top ten oil producers. Almost 68% of the crude oil production worldwide has its origin in those countries (Kuppusamy et al. 2020).

Approximately 90% of the global environment contamination by PHs is estimated to be linked to human activity. 1.7–8.8 Million tonnes of PHs are released to the environment per year in industrial and domestic wastewater, urban overspill, and oil leaks (Ambaye et al. 2022a). The discharge of an average of 209.000 tonnes of PHs per year to marine ecosystems is tremendously disturbing (Premnath et al. 2021).

Accidental spillages and leakages from tankers, old refinery plants, and underground tanks cause alarm, regarding contamination by the discharge of PHs into the ocean, surface soil, and marine ecosystems (Varjani 2017). According to the International Tanker Owners Pollution Federation (ITOPF), about 15,000 tonnes of oil was released into the environment in 2022, whereby more than 14,000 tonnes were attributed to three large tanker incidents in Asia and Africa (ITOPF 2023).

Figure 7.3 presents the average amounts of pollution of oil spills since 1970 and the representation of some of the greatest spills recorded until the moment (ITOPF 2023).

PAHs have been tagged by regulatory institutions worldwide regarding their priority as ubiquitous pollutants, due to their natural properties, including their persistence, toxicity, and potential carcinogenic for human exposure (Adeniji et al. 2019; Patel et al. 2020). Although naturally occurring events; volcanic activity, forest fires, and geological incidents contribute to the discharge of PHs into the environment, human activity is the main source (Hussain et al. 2018a).

Figure 7.4 illustrates some of the main causes of the release of PHs to the environment and the ecosystems affected by PHs exposure. Domestic incineration of wood and refuse, oil/gas burners in kitchens, and other domestic heating sources contribute to an increase in the concentration of PHs in the atmosphere (Patel et al. 2020).

The global distribution of PHs is interconnected with atmospheric conditions, landscape and, ecosystem conditions. It is safe to assume that industrially developed and heavily populated urban areas contribute significantly to PHs contamination (Ravindra et al. 2008; Hussain et al. 2018a).

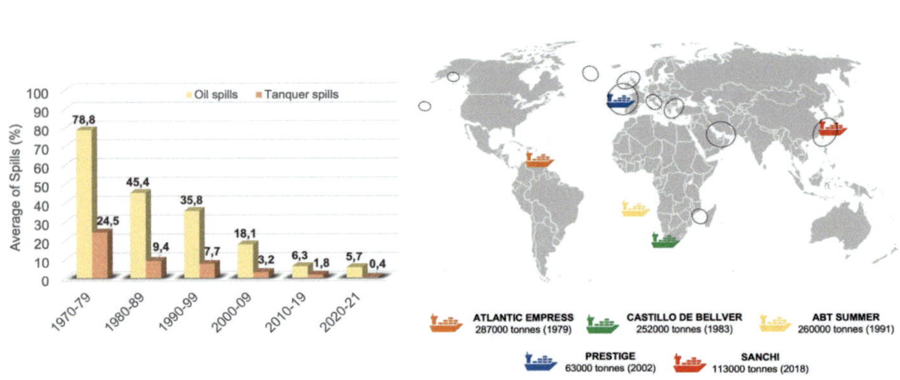

Fig. 7.3 Global oil spills in recent decades. (ITOPF 2023)

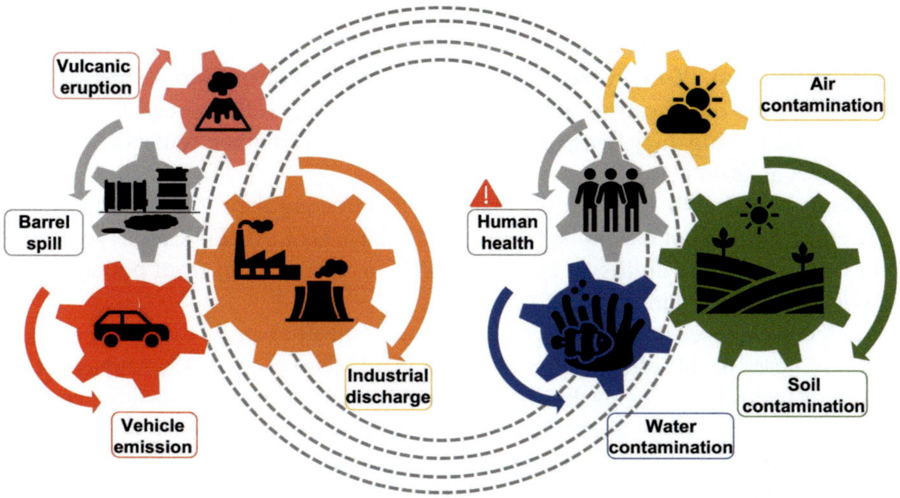

Fig. 7.4 Some of the main sources of PHs in the environment

The most remote and preserved areas could be affected by semi-volatile PHs contamination due to their airborne long-distance carriage (Motelay-Massei et al. 2007).

Despite air emissions from stationary and mobile causes, and due to intrinsic Earth dynamics, ultimately PHs tend to accumulate in soils, serving as the reservoir of atmospheric PHs (Cheruyiot et al. 2015). The solubility of PHS in non-polar solvents (hydrophobic nature), makes them easy to retain in soils with high organic matter (Włóka et al. 2015).

Soil pollution of PHs was detected in many different countries. Canada has 50% of the soil contaminated with PHs; Russia, on its side has>100.000 ha of surface area contaminated and, in Azerbaijan, it is estimated that about 11.143 ha is polluted with PHs from refineries (Dike et al. 2022).

The group of Premnath labeled diesel combustion (54%), cake oven (23%), gasoline and heavy oil combustion (14%), and coal/biomass burning (9%) PAHs in soils (Premnath et al. 2021).

Wang and co-workers conducted an interesting study to determine and compare the concentration of PAHs in house foundations, open spaces, busy streets, and soil close to residential spots in America (Wang et al. 2008). Data revealed that PAHs concentration in Detroit soils was 7.843 μg/kg and 5.100 μg/kg in New Orleans. Also, regarding the samples collected in New Orleans, the PAHs concentration found in the different soil samples were: 7.189 μg/kg on a busy street, 2.404 μg/kg on an open space, and 2.712 μg/kg on a house foundation (5–10 m farther then streets and buildings) and 5.237 μg/kg on a residential street. In summary, soil contamination with PHs is a timeless problem involving severe consequences to the environment and ecosystem residents (plants, wildlife, microorganisms) as well as on human health. Human exposure to PHs is correlated with several sources and may occur by multiple routes such as inhalation, skin contact, and ingestion (Kim et al. 2013). Direct inhalation by tobacco (Adesina et al. 2021), fireplaces (Sarigiannis et al. 2015), vehicles (Zhang et al. 2021b) or industrial hazards (Tsai et al. 2001) are major causes of human exposure to PHs. Additionally, the preparation of food using high temperatures or smoking is another route of exposure (Bansal et al. 2017).

Plants absorb PHs from contaminated soils, affecting their growth with drastic consequences in plant metabolic processes (Ambaye et al. 2022a). Crops rye, wheat, and lentils, have the ability to synthesize PHs or absorb them from polluted soils (Abdel-Shafy and Mansour 2016).

The consequences of PHs exposure to human health can be divided into short and long-term health effects (Ramesh et al. 2011; Kim et al. 2013; Ambaye et al. 2022a). Their toxicity levels depend on mode, level and time of exposure, chemical structure, composition, and properties of the contaminant (Ossai et al. 2020). Genotoxicity (ability to induce non-transmissible DNA damage), carcinogenicity (capacity for cancer induction), teratogenicity (induction of malformation of embryo or foetus), and human reproductive system abnormalities make PHs a global concern (Patel et al. 2020).Many studies reported some symptoms found in patients who were exposed. Eye irritation, cough, nausea, skin irritations, diarrhea, fatigue, and inflammation are commonly reported. With longer term exposure skin and lung cancer, haematotoxicity, hepatotoxicity, ocular toxicity, and immunotoxicity result (Ramesh et al. 2011; Kim et al. 2013; Abdel-Shafy and Mansour 2016; Ambaye et al. 2022a).

Dibenz[*ah*]anthracene is a five-fused aromatic ring hydrocarbon (Fig. 7.2c), found in tobacco smoke, certain food products, and gasoline exhaust. Inhalation or absorption through the skin causes accumulation in several organs. Short-term health effects of skin and eye irritation occur. Long-term exposure causes headaches, nausea, vertigo, and carcinogenicity (Bhattacharya et al. 2014).

Naphthalene, (Fig. 7.2c), is a product of petroleum or coal tar and can be found in moth repellents, air fresheners, and other household products. This PAH can induce hemolytic anemia in humans, causing the breakdown of red blood cells if inhaled or ingested in large amounts (Schreiner 2003).

Soil is a complex matrix of organic matter, minerals, and aqueous phase colloids. The dispersion and concentration of PHs within depend on characteristics of the soil itself and the PHs pollutants (physicochemical properties; solubility, molecular weight and polarity). Global contamination by PHs, aligning with the impact on the environment and ecosystems and consequent damages to human health has been, in the last two decades, a worldwide concern. Mechanisms for the removal or degradation of PHs in soils (natural and human intervention), have been and will continue to be broadly studied and investigated (Sakshi et al. 2019; Kumar et al. 2021; Ambaye et al. 2022a). Physical, chemical, and biological remediation techniques will be discussed throughout this chapter.

7.2.1 *Importance of PHs Molecular Weight and Aqueous Solubility*

Physicochemical properties of PHs: molecular weight (MW), framework, and aqueous solubility, contribute significantly to their accumulation and remediation in soils and other ecosystems. Due to stability and subsequent resistance to degradation, PHs are considered as ubiquitous pollutants of natural reserves, and a cause of worldwide concern (Andersson et al. 2003).

PAHs, petroleum-derived hydrocarbons possessing at least two fused or condensed aromatic rings in their scaffold, harmful to health (Cheruyiot et al. 2015). Their high stability is associated with π-electrons which are delocalized across the ring structure (resonance), making them unable to suffer a nucleophilic attack (Haritash and Kaushik 2009).

PAHs molecular weight ranges from 166–328 g/mol (Panwar and Mathur 2022). PHs exist in the atmosphere as gaseous phase or particle phase, depending on the same physic-chemical characteristic. Generally, LMW PHs exist in gaseous phase and HMW PHs in the particulate matter (Akyüz and Çabuk 2010). The consequence of HMW, results in poor solubility in water (high lipophilicity) and poor vapor pressure (melting and boiling points increase) (Varjani and Upasani 2017; Patel et al. 2020). In the other hand, PHs are generally highly soluble in organic solvents showing a strong affinity for solid phases (An et al. 2017).

Bioremediation is dependent on water solubility of PHs (Jabbar et al. 2022). The solubility of PHs in water can be measured according to their chemical structure and their octanol-water partition coefficients, and is dependent on pH, temperature, ionic strength and other dissolved organic chemicals (Gupte et al. 2016).

The octanol-water partition coefficient (K_{ow}) has a significant influence in soil contamination by PHs, because it is expressed as the solubility of an organic compound in water (Jonker 2016). High values of K_{ow} indicate high affinity to soil absorption and consequently low aqueous solubility. For example, the log K_{ow} value of naphthalene is 3.37 (Fig. 7.2c), lower than 6.50 that represents the log K_{ow} value for indeno[1,2,3-*cd*]pyrene (Fig. 7.2c), meaning that naphthalene is more water soluble (Gupte et al. 2016).

C10-C25 carbon chain PHs are easily bioremediated by microorganisms. However with carbon chain and complexity of the PHs structure increases, bioremediation finds resistance. PAHs for instance, due to their properties present high resistance to degradation (Hussain et al. 2018a).

PHs can be grouped according volatility. The first section represents the most volatile PHs and encompasses the C6 to C10 carbon chains PHs. For PHs with C10 to C16 chains, they are catalogued as second section, representing the semi-volatile PHs. PHs with C16 to C35 chains are considered the non-volatile third section. Lastly, HMW PHs (*viz.* PAHs) with more than C35 carbon chains are considered the fourth section, with low volatility and water solubility, and with concerning adverse effects on human health (Varjani 2017).

7.3 Plant-Assisted Remediation Technologies for Petroleum Hydrocarbons (PHs)

Soil remediation and water pollutants for sustainable ecosystem processes and functions are some of the major challenges that our society faces today.

Environmental pollution with PHs impacts soil, water, and air quality and consequently all living beings health (Guarino et al. 2017; Ajona and Vasanthi 2021; Ruley et al. 2022).

In ecosystems, specifically in soils, PHs are subject to attenuation or degraded by physical, chemical, and biological processes, such as abiotic loss (*viz.* photooxidation induced by ultraviolet light; volatilization), sorption by soil organic and clay constituents, uptake/metabolization by plants or microbial degradation (Agnello et al. 2016; Zahed et al. 2021; Eze et al. 2022; Zhuang et al. 2022).

The rate at which these processes occur and consequently the extent to which PHsare degraded or retained in soil particulate matter depends on several factors, from which emerge: (a) soil properties such as: clay or mineral content, organic matter, temperature, moisture, redox potential, nutrient availability, fulvic and humic colloid contents, diversity and abundance of degrading microorganisms; and (b) chemical and physical characteristics of the PHs: structural complexity, molecular weight (HMW hydrocarbons are more recalcitrant), bioavailability, (eco)toxicity and potential for (bio)degradation (Guarino et al. 2017; Feng et al. 2021;Ambaye et al. 2022a, 2023).

Plant and microbial uptake/metabolization are largely defined by the aqueous solubility of PHs (*viz.* PAHs), which is reduced by each additional ring in the structure (Sect. 7.2) (Hoang et al. 2021b; Kariyawasam et al. 2022; Panwar and Mathur 2022). In fact, PAHs persistence in natural ecosystems can be strengthened by increasing the number of aromatic rings present. One of the reasons pointed out for this trend in environmental persistence is the fact that HMW PAHs promote lower water solubility and consequently lower bioavailability (details in Sect. 7.2) (Kariyawasam et al. 2022; Panwar and Mathur 2022).

N-Alkanes (Fig. 7.1) of intermediate carbon chain length and MW *viz.* C10-C25 are easily biodegraded. However, alkanes with shorter carbon chains and high aqueous solubility may become more toxic to PAHs degraders (Martin et al. 2014; Varjani and Upasani 2017; Ossai et al. 2020). Long chain alkanes *viz.* C25-C40, are compounds that are not easily biodegraded due to their high hydrophobic nature and reduced water solubility, which also severely affects their bioavailability (Varjani and Upasani 2017). Furthermore, branched chain alkanes and cycloalkanes (Fig. 7.1) are slower biodegraded comparatively to linear alkanes (Martin et al. 2014; Varjani and Upasani 2017; Ossai et al. 2020). Highly condensed aromatics, cycloparaffin structures and asphaltic materials (Fig. 7.1) exhibit the higher resistance to biodegradation due to HMW and boiling points (Martin et al. 2014; Varjani and Upasani 2017; Ossai et al. 2020; Haider et al. 2021; Panwar and Mathur 2022).

HMW PHs have high chemical stability and therefore become more recalcitrant in the environment, while LMW PHs can be rapidly attenuated in soil due to their mobility/availability and higher potential for (bio)degradation (low K_{ow}, high water solubility and volatility) (Haider et al. 2021; Hoang et al. 2021b; Kariyawasam et al. 2022; Panwar and Mathur 2022). Commonly, naphthalene (Fig. 7.2c) does not persist and promptly degrades into simpler and non-toxic compounds (Haider et al. 2021). However, other persistent PAHs may be absorbed to different soil materials like clays, iron oxides, and humic substances (Haider et al. 2021). Soil to crop bioconcentration factors decrease with log K_{ow} increasing (Singh et al. 2019; Haider et al. 2021).

Sorption of PHs to soil organic or inorganic particles may be due to different processes, such as hydrogen or covalent bonds and electrostatic/hydrophobic interactions. The absorbed PHs become less accessible or bioavailable for microbes, thus limiting degradation rate and transport of these compounds into the soil (Hoang et al. 2021a). The sorption process is highly dependent on the organic matter content (both quantity and nature), clay content and of the hydrocarbons hydrophobicity (Hoang et al. 2021a).

Physical, chemical, thermal, biological, and combined/integrated approaches can be applied to remediate the pollution caused by PHs (Ossai et al. 2020; Haider et al. 2021; Tucci et al. 2021; Ambaye et al. 2022a, b; Lv et al. 2022; Michael-Igolima et al. 2022). Conventional techniques commonly require high energy inputs and operation costs *viz.* solvent extraction/soil washing, covering, supercritical fluid extraction, thermal desorption, pyrolysis, vitrification, excavation and landfilling, chemical oxidation (use of oxidants including permanganate, sodium persulfate, and ozone in soil remediation), incineration. The conventionally most used soil remediation options for the treatment of soils contaminated with PHs are shown in Fig. 7.5.

Ex-situ conventional physicochemical remediation technologies are aggravated by excavation and transport costs (Ossai et al. 2020; Michael-Igolima et al. 2022).

A growing need to develop 'green technologies' that reduce both economic and environmental costs has encouraged the effort to develop environmentally sustainable, and cost-effective remediation technologies, particularly based on tolerance and bioremediation abilities of plants and microorganisms such as: natural attenuation, biostimulation, bioaugmentation, vermicomposting, bioslurry, bioventing, biosparging, landfarming, composting/biopiles, mycoremediation, nanoremediation, and phytoremediation (Ossai et al. 2020; Ali et al. 2022; Ambaye et al. 2022a). It should be noted however that, in the remediation of aged soils with HMW PHs, or in soils with accumulation of recalcitrant and hazardous metabolites, or even if it is necessary to accelerate remediation processes, diverse integrated approaches can be applied such as: physical-chemical, physical-biological, chemical-biological, biological-biological and physical-chemical-biological (Hussain et al. 2018b; Pinto et al. 2021; Krzyszczak et al. 2022; Lv et al. 2022). Indeed, most single remediation methods are limited and cannot be used for all contaminants type (Michael-Igolima et al. 2022). Combined remediation approaches allow integrating the resources of each technology, promoting a synergy capable of overcoming the individual weaknesses of each remediation process. The poor environmental compatibility, high cost of implementation and poor public acceptability associated with conventional

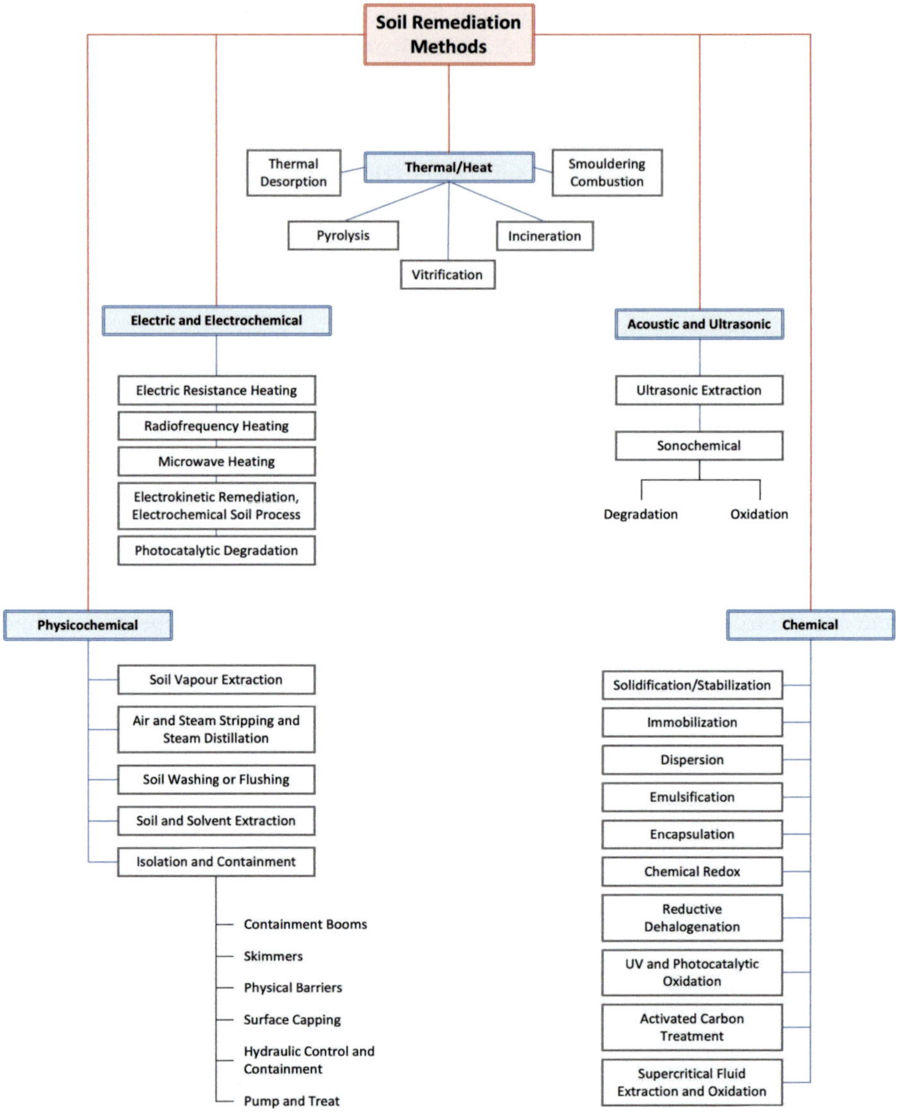

Fig. 7.5 Soil remediation options conventionally employed for the treatment of soils contaminated with PHs (Adapted from Pinto et al. 2018)

remediation technologies have increased the need to employ combined biological remediation methods.

Effectively, plant and microbe-assisted remediation processes offers extensive ecological and cost-associated benefits, with the overall efficiency and performance of these approaches being improved by genetically engineered microbes and plants.

7.3.1 Phytoremediation

Remediation methods that use living organisms such as plants and microorganisms, to remove, degrade or attenuate contaminants in soil are included in biological methods (Michael-Igolima et al. 2022). These methods encompass processes that are proposed as ecologically correct approaches with sustainable costs, capable of restoring soils, waters and sediments contaminated by organic compounds (PHs included) and metals over large areas with wide applicability (Ambaye et al. 2022a; Jabbar et al. 2022). Phytoremediation includes a set of different eco-friendly rehabilitation technologies that use plants and soil-associated microorganisms to uptake, accumulate/contain, transform/ metabolize, immobilize, or biodegrade organic and/ or inorganic pollutants, present in different environmental matrices, through physical, chemical, and biological processes (Pinto et al. 2016, 2021; Abdullah et al. 2020; Jabbar et al. 2022). The implementation of these biological technologies has received attention as an innovative, environmentally friendly, and cost-effective rehabilitation strategy, able to improve the native microflora and the soil's physical, chemical, and biological properties (Wang et al. 2017; Pinto et al. 2018; Jabbar et al. 2022). In addition, phytoremediation is effective for organic and inorganic pollutants, and is suitable to be used for mixed types of pollutants through different mechanisms (*viz.* phytoextraction, phytostabilization, phytovolatilization, phytodegradation and rhizodegradation) (Table 7.1), to extract or detoxify the pollutants. It is effective for contaminated soils with large volumes and highly dispersed contaminants at low-to-moderate concentrations and it can be conducted in situ with the maintenance of soil structure and texture. It is also ecologically accurate and aesthetically accepted by the public with pleasant panoramic views (Pinto et al. 2016, 2018, 2021; Abdullah et al. 2020; Ossai et al. 2020).

Phytoremediation has many advantages over other relevant treatments such as conventional approaches, to restore soils contaminated with PHs (Abdullah et al. 2020).

There are several benefits of including plants in bioremediation strategies, as plants act as filters in natural environments, improving soil structure and balancing hydrological parameters, in addition to improving aeration and, consequently, soil health and fertility (Agnello et al. 2016; Ajona and Vasanthi 2021). Different factors affect the accumulation, distribution and transformation/metabolization of organic compounds by plants (Agnello et al. 2016; Pinto et al. 2016, 2018, 2021; Abdullah et al. 2020). Indeed, phytoremediation performance can be affected by soil parameters as texture and structure, type and content of organic colloids, availability of water and oxygen, temperature, nutrients, and environmental parameters such as sunlight and climate (Pinto et al. 2018, 2021;Robichaud et al. 2019; Abdullah et al. 2020; Ossai et al. 2020).

Soil texture has consequences on phytoremediation performance by influencing the contaminant bioavailability, which is lower in soils with high clay contents since clay can bind to molecules more readily compared with silt or sand (Robichaud et al. 2019; Abdullah et al. 2020; Ossai et al. 2020). Moreover, water plays an

Table 7.1 Overview of most employed phytoremediation techniques. (Adapted from Pinto et al. 2016)

Technique	Description	References
Phytostabilization	Immobilization/stabilization of pollutants in the root zone reducing their mobility and bioavailability thereby preventing their migration. Phytostabilization cannot generally be applied as a permanent solution, as contaminants remain in the soil, although with reduced mobility. However, it can be a step in a strategy for managing toxic contaminants. Plants most used in phytostabilisation require extensive root systems able to uptake or concentrate the contaminants within the rhizospheric zone, on the plant cell walls and in the root cells	Pinto et al. (2016, 2018, 2021), Khashij et al. (2018), Hussain et al. (2018b), Ossai et al. (2020), Saravanan et al. (2020), Steliga and Kluk (2020), Haider et al. (2021), Jabbar et al. (2022), Kafle et al. (2022), and Lv et al. (2022)
Phytoextraction	Phytoextraction refers to the uptake of pollutants by plant roots and their translocation and subsequent accumulation in aboveground tissues. Remarkably efficient to remove xenobiotics from environment through an effective uptake and assimilation of these compounds in their own cells. This technology can also be known as phytoaccumulation or phytosequestration.	Abhilash et al. (2012), Mahar et al. (2016), Pinto et al. (2016, 2018, 2021), Hussain et al. (2018a, b, c), Haider et al. (2021), Jabbar et al. (2022), Kafle et al. (2022), and Lv et al. (2022)
Rhyzodegradation	It involves indirect processes that promote increased biodegradation driven by microbes residing in the plant's rhizosphere. This increase in microbial activity in cooperation with plant roots can promote a synergy able of significantly increasing the biodegradation of organic compounds. A powerful phytotechnology in the reclamation of soils contaminated with of PAHs. Diverse approaches have been implemented with the objective of increasing the efficiency of these processes namely the addition to the soil of inorganic fertilizers, nutrients, bulking agents, compost and biochar.	Abhilash et al. (2009, 2012), Abbasian et al. (2015), Pinto et al. (2016, 2021), Agrawal and Shahi (2017), Khan et al. (2017), Baoune et al. (2018), Sarma et al. (2019), Haider et al. (2021), Jabbar et al. (2022), Kafle et al. (2022), and Lv et al. (2022)

(continued)

Table 7.1 (continued)

Technique	Description	References
Phytodegradation	PHs may be transformed and broken down by several active metabolic processes into harmless by-products that occur within the plant tissues or by external processes driven by plants in their root vicinity. The last depends on plant metabolic activities and is not influenced by rhizospheric microorganisms. Phytotechnology suitable just for cleaning environments contaminated with organic compounds (viz. PAHs).	Dhankher et al. (2012), Pinto et al. (2016, 2018, 2021), Hussain et al. (2018b), Jeevanantham et al. (2019), Prabakaran et al. (2019), Sarma et al. (2019), Haider et al. (2021), Jabbar et al. (2022), Kafle et al. (2022), and Lv et al. (2022)
Phytovolatilization	Plants uptake pollutants through their roots and transform them by plant metabolism into volatile forms that are transferred to the atmosphere through the process of transpiration. LMW volatile hydrocarbons such as benzene, toluene, ethylbenzene and xylene (BTEX), naphthalene and trichloroethylene can experience phytovolatilization along the transpiration stream.	Pinto et al. (2016, 2018, 2021), Hussain et al. (2018b), Jeevanantham et al. (2019), Ossai et al. (2020), Haider et al. (2021), Jabbar et al. (2022), Kafle et al. (2022), and Lv et al. (2022)

important role in transporting nutrients to plants. In low humid soils there will be a loss of microbial activity and plants dehydration (Abdullah et al. 2020). Finally, adequate soil nutrients availability is required to support plants growth and their associated microorganisms, most notably during phytoremediation efforts when the plant-microbe community is already under stress from the contaminant (Jeevanantham et al. 2019; Abdullah et al. 2020).

The toxicity of hydrocarbons can negatively impact seed germination and nutrients translocation, disturbing the plant metabolic activity and photosynthesis activity, inducing oxidative stress, and inhibiting plant physiology and morphology that ultimately reduce plant yield (Pinto et al. 2016, 2018, 2021;Haider et al. 2021).

Photosynthesis is the most essential metabolic pathway for plants. Although plant growth is controlled by a great number of physiological, biochemical, and molecular processes, photosynthesis is a decisive route, which defines plant growth and biomass production (Jajoo et al. 2014; Pinto et al. 2016, 2018; Haider et al. 2021).

Different environmental stress conditions may reduce the photosynthetic ability of plants due to their influence on one or more events associated with the photosynthetic phenomenon. Stress can influence energy consumption, pigment level, chloroplasts, and photosystem II (PS II) structure, stomatal opening and closing, and even enzymatic synthesis associated with the biochemical efficiency of different photosynthetic reactions, especially CO_2 assimilation (Jajoo et al. 2014; Haider et al. 2021).

Contamination by PHs (*viz.* PAHs) in the rhizosphere has a significant influence on the photosynthesis-related principal and secondary processes with adverse effects on plant growth (Jajoo et al. 2014; Bashir et al. 2020; Haider et al. 2021). Due to the aggregation of hydrophobic PHs in thylakoid membranes (Duxbury et al. 1997; Haider et al. 2021; Jajoo et al. 2014), it is acceptable that they have an influence on the ordered arrangement of chloroplasts and interfere with electron transport and photosynthesis (Kummerova et al. 2008; Jajoo et al. 2014), resulting in disruption of electron transfer at PS II donor and acceptor sites (Kummerova et al. 2008; Jajoo et al. 2014). In addition, they can covalently bind into cells to proteins, DNA, and RNA, interfering with growth, metabolism, and enzymatic activity (Jajoo et al. 2014; Bao et al. 2020; Haider et al. 2021).

The antioxidant mechanism aids in protecting plant cells from reactive oxygen species (ROS) generated under stress conditions (Jajoo et al. 2014; Saleem et al. 2020; Haider et al. 2021). ROS is a term that covers molecules such as singlet oxygen ($^1O_2$), superoxide (O_2^-), hydrogen peroxide (H_2O_2), and the hydroxyl radical ($^{\bullet}OH$), among other reactive species, produced during metabolism, all of which impact cell function (Noctor et al. 2018; Kapoor et al. 2019).

When plants grow in PH contaminated soils, a higher ROS concentration can be registered, which causes lipidic peroxidation, inhibition of enzymatic activities, protein oxidation, deterioration of nucleic acids, and ultimately the plant death (Achuba and Ja-Anni 2018; Rusin et al. 2018; Kapoor et al. 2019; Haider et al. 2021). Consequently, cells develop a balanced system to counteract the effect of ROS, an antioxidative defense system including both enzymatic antioxidants like superoxide dismutase (SOD), catalase (CAT), guaiacol peroxidase (GPX), ascorbate peroxidase (APX), monodehydroascorbate reductase (MDHAR), dehydroascorbate reductase (DHAR), and glutathione reductase (GR) and non-enzymatic antioxidants such as ascorbate (AsA), glutathione (GSH), carotenoids, tocopherols, and phenolics which cooperatively reduce oxidative state (Noctor et al. 2018; Kapoor et al. 2019; Haider et al. 2021).

In plants, ROS generation diverges widely with PHs (*viz.* PAHs) concentration in soils, as well as with other environmental stress factors, plant species and its genotypes (Hasanuzzaman et al. 2020; Haider et al. 2021).

PHs polluted soils are typically deficient in nutrients (*viz.* nitrogen) and oxygen availability, and cultivated plants undergo different stresses, with frequent alterations in their detoxification mechanisms (Hussain et al. 2018b; Sarma et al. 2019; Haider et al. 2021). In addition, the ability of plants and microorganisms to uptake water and nutrients from the soil may be reduced due to hydrophobic properties of hydrocarbons (Khan et al. 2013; Haider et al. 2021; Parus et al. 2023).

PHs readily absorbed by plants can accumulate leading to toxic levels within the plant tissues, unless effective means of detoxification are present.

Plants adapt themselves to stress evolving detoxifying mechanisms in their cells, a competence known as the "green liver" model by Sandermann (1994), comparing the detoxification processes that occur in plants with those that occur in the mammalian liver.

The green liver model reports three distinct phases: conversion or transformation of the xenobiotic (phase I), conjugation with plant molecules (phase II), and active sequestration/ compartmentalization of the conjugate in apoplast or vacuoles (phase III) (Feng et al. 2017; Pinto et al. 2021).

Contrary to what happens in humans and animals, plants do not have an excretion system analogous to the mammalian one, so phase III in plants is the step where transformed xenobiotics accumulate in cells, sequestered in the vacuole or on the cell walls (Hussain et al. 2018b; Kong et al. 2018; Sarma et al. 2019; Pinto et al. 2021).

During phase I, termed by transformation/functionalization or activation, hydrophobic PHs are transformed into more hydrophilic and less toxic metabolites by oxidation, reduction and/or hydrolysis reactions. Phase I is considered rate-limiting in the metabolism of xenobiotics in plants and crucial reactions in this step are catalyzed by P450 monooxygenases, carboxylesterases, peroxidases, and laccases (Coleman et al. 1997; Hussain et al. 2018b; Pinto et al. 2018, 2021; Sarma et al. 2019; Ossai et al. 2020).

In the course of phase II (conjugation), detoxification reactions catalyzed by glucosyl, malonyl and glutathione transferases turn the metabolites produced in Phase I in water soluble conjugates, either nontoxic or less toxic than the parent compound (Hussain et al. 2018b; Khan et al. 2013; Ossai et al. 2020; Pinto et al. 2021; Coleman et al. 1997).

Ultimately, at phase III (compartmentalization), a set of other reactions as cleavage, rearrangement, and secondary conjugation, produce soluble conjugates that can be translocated and accumulated in vacuoles (or in the apoplast) or assimilated into cell wall fractions as lignin, pectin, and cellulose (Ibañez et al. 2016; Hussain et al. 2018b; Pinto et al. 2018; Ossai et al. 2020; Rai et al. 2020).

In soil, there are different dynamics between plants and PHs, depending on age and plant species, PHs physicochemical characteristics and concentration, as well as root morphology, chemical, physical and biological characteristics of the soil and density of microorganisms occupying the plant by-root zone, able to assist in remediation processes (Feng et al. 2017; Hussain et al. 2018b; Pinto et al. 2021).

For efficient phytoremediation of organic soil contaminants, the soil must provide the proper environment for both plants and associated microorganisms. Soil properties such as pH, texture, humic and fulvic content, cation exchange capacity and structure can affect not only plant growth and development, but also the colonization and metabolic activity of the microbiome, consequently affecting the biodegradation of organic pollutants (Khan et al. 2013; Haider et al. 2021).

Characteristics of PHs (*viz.* PAHs) such as solubility, polarity and hydrophobicity, determine the probability of these compounds being absorbed by plants. After their uptake, PHs may be sequestered into roots or translocated through xylem within plant tissues, and later stored in the vacuoles or volatilized to the atmosphere. Moreover, translocation of PAHs in plant shoots is limited as PAHs tend to accumulate in the epidermis of the shoot tissues due to their nonpolar nature (Abdullah et al. 2020; Kariyawasam et al. 2022).

Most PHs cannot move appreciably into plants from the soil due to their log K_{ow} values higher than 4, which provides evidence for constituent sorption in the soil matrix (Khan et al. 2013; Abdullah et al. 2020).

As described in Sect. 7.2.1 the log value of the octanol-water partition coefficient (log K_{ow}) gives information on the hydrophobicity or lipophilicity of the organic compound and its ability to cross lipidic bilayers (Yang et al. 2017; Hussain et al. 2018b; Pinto et al. 2021). Organic compounds with high log K_{ow} (>3 to 8) have low solubility in aqueous solutions, which makes it difficult to penetrate plant roots, due to strong sorption by soil particles and root surface, thus limiting the plant's ability to degrade PHs (Hussain et al. 2018b; Kong et al. 2018; Haider et al. 2021; Pinto et al. 2021).

Organic compounds with medium log K_{ow} value (0.5–3) are less hydrophobic, and more easily absorbed by plants. Lower log K_{ow} values (<0.5) imply high solubility in aqueous solution, low lipophilicity and ultimately low propensity to cross lipidic bilayers (Hussain et al. 2018b; Kong et al. 2018; Pinto et al. 2021).

The interface between microbes and plant roots in the rhizosphere is a vital habitat for diverse microbes, comprising protozoa, fungi, and bacteria and is believed to have great influence in the growth and survival of plants. Microbes can synthesize and detect signaling molecules able of triggering microbial populations to form biofilms around the root surface (Pinto et al. 2018, 2021; Haider et al. 2021).

Furthermore, microbes in the rhizosphere may produce biosurfactants, organic ligands via microbial degradation of soil organic matter, exudates, determining hydrocarbons availability and modifying bioavailability and uptake by plants (Pinto et al. 2018; Haider et al. 2021; Pinto et al. 2021). The rhizosphere of most healthy plants is characterized by intense microbial activity. In fact, in the rhizosphere, soil, plant roots, microorganisms and other soil organisms can interact in complex ways promoting mechanisms that stimulate the biodegradation of organic compounds.

The rhizosphere microbiome benefits plants through a mutualistic/synergistic plant-microbe partnership. Microbiomes comprise a wide range of prokaryotic and eukaryotic organisms that favor plants by providing essential nutrients and organic substrates such as amino acids, vitamins, and cytokines for healthy plant growth, while plant roots provide suitable habitat where microbial degradation can be driven (Hussain et al. 2018b; Sarma et al. 2019; Lv et al. 2022).

Microorganisms can transform pollutants into plant-absorbable forms, reduce their toxicity, and even promote increased plant tolerance (Hussain et al. 2018b; Sarma et al. 2019; Lv et al. 2022).

Exudates released by plants (sugars, sterols, alcohols, carbohydrates, nucleotides, growth factors, flavanones, enzymes, organic acids, and other compounds synthesized by plants) can act as chemotactic factors that direct microbial chemotaxis towards the food source (Thomas et al. 2017; Pinto et al. 2018, 2021).

Plant roots can effectively alter the bioavailability of PHs in soils due to the release of exudates. Indeed, LMW organic acids from root exudates may contribute to increase PHs bioavailability, either by providing competing sorption sites on soil particles, or by boosting microbial growth and activity in the soil (Hussain et al. 2018c; Hoang et al. 2021a).

Bioactive compounds produced by plants and microorganisms can therefore be essential in the desorption process of contaminants in the soil matrix, with positive effects on their bioavailability (Truskewycz et al. 2019; Hoang et al. 2021a).

It has also been observed that several compounds released from roots act as inducers of microbial genes involved in the degradation of PHs. Root exudates appear to support the development of a high diversity of bacteria containing known hydrocarbon degrading genes (Cébron et al. 2011; Khan et al. 2013; Hussain et al. 2018b; Sarma et al. 2019).

Different plant species have different skills to manage with PHs. The ability to exploit additional soil depth through large root surface area and the presence of bulky roots are some of the advantages revealed by grasses. On the other hand, symbiotic nitrogen fixation is typically associated with Leguminosae, providing the soil with a nitrogen source (Haider et al. 2021; Lv et al. 2022).

The screening and cultivation of plants that have efficient strategies to tolerate and deal with PHs is one of the main goals in the phytoremediation of contaminated soils (Ossai et al. 2020; Pinto et al. 2021; Lv et al. 2022).

Phytoremediation includes a set of plant-based remediation processes, distinct strategies used by plants to deal with organic xenobiotics such as PHs. The most important strategies are described in Table 1. All the processes aim at converting the contaminants into less toxic and less persistent forms in the environment (Pinto et al. 2018, 2021; Ossai et al. 2020).

Although phytoremediation includes ecologically and economically attractive techniques to remediate soils contaminated with PHs, there are several shortcomings that must be overcome: i) restoring contaminated environments is time-consuming ii) contaminants which are highly absorbed to soil particles may not be treatable due to their limited bioaccessibility or bioavailability; ii) many plant species are sensitive to PHs (*viz.* PAHs) and even plants with tolerance, do not adequately develop the intended degradation of PHs; iii) in most PHs polluted soils, the microbial population is decreased, and consequently, there are not enough microorganisms to promote degradation or to improve plant growth and development. These gaps can be mitigated through soil amendments or by improving soil properties with the addition of fertilizers/surfactants and microbes' inoculation. The latter is able of promoting plant biomass and triggering the release of exudates in the rhizosphere, useful to significantly accelerate the hydrocarbons removal rate (Khan et al. 2013; Guarino et al. 2017; Hussain et al. 2018b; Pinto et al. 2018, 2021; Varjani and Upasani 2019; Abdullah et al. 2020; Behera et al. 2022; Curiel-Alegre et al. 2022; Gupta et al. 2022; Jabbar et al. 2022; Nwankwegu et al. 2022).

Further future research is indispensable in the field of genetic engineering to improve the phytoremediation abilities of transgenic plants and understand the mechanisms involved, in order to make these processes more efficient, faster and cost-effective.

7.4 Assisted Phytoremediation—Additives for the Remediation of PHs Contaminated Soils

In many scenarios, plant species cannot perform properly and need supports to enhance phytoremediation results. Such aids include soil organic amendments like biochar (Dike et al. 2021; Haider et al. 2021; Kafle et al. 2022), rhizospheric and endophytic bacteria (Hussain et al. 2018b; Pinto et al. 2018; Kafle et al. 2022), arbuscular mycorrhizal fungi (AMF) (Hussain et al. 2018b; Pinto et al. 2018) or even transgenic plants (Kawahigashi 2009; Mishra et al. 2020; Kafle et al. 2022).

Organic amendments are materials derived from biomass or living beings which, due to their unique structural properties, chemical composition, and surface functional groups, have been widely used for environmental remediation (Gómez-Sagasti et al. 2018; Hoang et al. 2021a).

Natural biodegradation of PHs in contaminated soils is often limited due to low microbial density and activity, nutrients deficiency, and oxygen status, among other factors (Hoang et al. 2021a, b). Soil aeration is of essential importance in the bioremediation process because the initial degradation of PHs requires the activity of oxidative enzymes (dioxygenases and monooxygenases), thus making oxygen a limiting factor for biodegradation in many processes (Varjani 2017; Hoang et al. 2021a). Conversely, incorporation of organic amendments can stimulate the capability of naturally existing microorganisms to biodegrade PHs in soils (Hoang et al. 2021a). In fact, increasing the organic matter content by adding organic amendments in the contaminated soil could improve soil fertility, soil aggregate stability, and other beneficial properties, such as soil porosity, water availability, and water infiltration and percolation, promoting microbial degradation of PHs in the impacted soils (Wang et al. 2016; Hoang et al. 2021a; Jabbar et al. 2022).

Soil organic matter (SOM) plays a key role in the formation of stable soil aggregates as it helps to form cationic bridges that bind mineral particles to organic polymers. In addition, increased biological activity derived from the decomposition of the added organic matter, together with the release of microorganisms' agglutinants (e.g., exo-polysaccharides), may further contribute to increasing the stability of soil aggregates (Qurashi and Sabri 2012; Hoang et al. 2021a). Improved structural stability, in turn, results in reduced bulk density and increased porosity, hydraulic conductivity, and water availability of the impacted soil. Similarly, the addition of organic additives can also improve the balance between macropores (among aggregates) and micropores (within aggregates), which can contribute to increasing permeability and water availability (Gul et al. 2015; Hoang et al. 2021a).

Within the context of remediation of soils contaminated by PHs, a wide range of organic amendments, including crop residues, manures, biosolids, and biochar can be used due to their positive effects on soil health and plant growth (Gul et al. 2015; Hoang et al. 2021a). Amendments not only favorably alter soil's physical (like bulk density), chemical (like cation exchange capacity), and biological (like microbial activity) properties but also supply essential nutrients able to strengthen plant growth (Hoang et al. 2021a).

Organic amendments also provide abundant labile carbon and nutrient sources to increase and stimulate soil microbial biomass and activity in PHs-contaminated soils, thereby increasing the natural attenuation potential of the soils (Wang et al. 2016; Hoang et al. 2021a).

Sorption of PHs contaminants to soil particles can occur through various mechanisms such as hydrogen bonding, covalent bonding, and hydrophobic interactions. Typically, absorbed PHs contaminants by mineral particles are less accessible or available to microbes, thereby limiting the degradation and transport of the compounds in soil (Khan et al. 2018; Hoang et al. 2021a).

SOM (both quantity and nature) has been recognized as one of the critical factors in determining the sorption of PHs in soil because SOM may increase the density of exchange sites in the impacted soil, leading to enhanced surface sorption and complexation, thereby having an effect on PHs mobility and bioavailability (Farzadkia et al. 2019; Hoang et al. 2021a).

The sorption process increases with the increase of organic matter amount, clay content, and hydrophobicity of PHs. Generally, LMW PHs contaminants, when entering the soil, are relatively mobile and degradable due to their low hydrophobicity (i.e., a low K_{ow} value), high water solubility, and high volatility. In contrast, HMW PHs contaminants are more persistent and resistant to degradation (Harmsen and Rietra 2018; Hoang et al. 2021a). On the other hand, the presence of dissolved organic matter (DOM) can increase the solubility and mass transfer of the PHs to the aqueous phase. A variety of organic amendments can produce DOM resulting in increased mobilization/availability of PHs and consequent biodegradation, due to the lability of the sorption processes involved (Cai et al. 2017; Hoang et al. 2021a, b).

Despite the wide diversity of the amendments/additives, amendments can be broadly grouped into (1) stimulation of plants and microorganisms capable of biodegradation (biostimulation), and (2) addition of plant growth-promoting and/or hydrocarbon-degrading and/or biosurfactant-producing microorganisms (bioaugmentation) (Hoang et al. 2021b; Kariyawasam et al. 2022). The former involves the addition of nutrients, soil moisture, and aeration for the proliferation of indigenous microbes (Mohan et al. 2006). Incorporating nutrients promotes co-metabolism and enhances PHs dissipation (Premnath et al. 2021; Kariyawasam et al. 2022).

Biostimulation requires modifying the environment via nutrients addition (N, P, K) and carbon sources, electron acceptors (oxygen), and other compounds (*viz.* biosurfactants) into the polluted media to activate autochthonous hydrocarbon-degrading microorganisms and increase PHs attenuation (Abena et al. 2019; Hoang et al. 2021b; Nwankwegu et al. 2022). This approach optimizes the vital processes and conditions for the remediation of polluted sites. Although the adoption of biostimulation in the recovery of polluted environment has attracted a lot of attention, it presents challenges. For example, the introduction of inorganic fertilizers is limited by their high cost. Secondly, there is always a shortage of chemical formulations and lack of knowledge about the application rate due to the complexity of sites and heterogeneity of pollutants (Odoh et al. 2019; Nwankwegu et al. 2022). Furthermore, biostimulation can lead to increased cases of plant and soil toxicity, as well as certainty of eutrophication and algal blooms due to the abundant supply of nitrogen and phosphorus, particularly in aquatic ecosystems (Nwankwegu et al. 2016, 2022; Xu et al. 2021).

7.4.1 *Biochar-Mediated Remediation of PHs*

A lot of research has recently been focused on rich carbon sources *i.e.*, biochar for the remediation of PHs. Indeed, biochar represents a promising biostimulation agent for the remediation of hydrocarbon-contaminated soils (Sarma et al. 2019; Dike et al. 2021; Haider et al. 2021; Hoang et al. 2021b).

Biochar is basically a carbon-rich, fine-grained, and porous material produced by pyrolysis (300 to 500 °C), under oxygen-limited conditions, which has been targeted for the management of soil as a conditioner (Sarma et al. 2019; Haider et al. 2021; Hoang et al. 2021b). Biochar can originate from a variety of feedstock materials, including agricultural and forestry by-products, animal/poultry wastes, and sewage sludge or slurries (Wiszniewska et al. 2016; Hoang et al. 2021b). The addition of biochar to degraded soils can not only improve plant growth by enhancing soil fertility, nutrient retention, and oxygen supply but also promote microbial growth and activity, which in turn stimulate bioremediation efficiency (Barati et al. 2018; Sarma et al. 2019; Zhen et al. 2019; Dike et al. 2021; Hoang et al. 2021b).

The addition of biochar in soils can contribute in different ways to stimulate the biodegradation of PHs (Table 7.2). Its potential for remediation is well documented owing to its high surface area, nano-, and micro-porosity, and surface functional groups, acting not only as a soil pH regulator, but also improving cation exchange capacity (CEC), water-holding capacity, surface sorption capacity, and soil revitalization (Tang et al. 2013; Sarma et al. 2019; Hoang et al. 2021a). The latter essentially results from the improvement in the soil's physicochemical and biological characteristics and the supply of a large number of essential nutrients, together with the modification of soil enzymes and microbial density in the rhizosphere, factors that favor plant growth and microbial degradation of PHs (Haider et al. 2021; Hoang et al. 2021a).

Biochar is known to be one of the most effective amendments to increase some soil capabilities, inhibiting/decreasing the movement and influencing the fate and actions of soil pollutants such as PHs (Haider et al. 2021; Krzyszczak et al. 2022; Meki et al. 2022).

Different mechanisms are attributed to increased sorption of hydrocarbon compounds onto biochar after incorporating it into PH contaminated soils. Firstly, the sorption of PH contaminants on the carbonized fraction of biochar is achieved by electrostatic and van der Waals interactions (physical sorption) (Hoang et al. 2021a; Krzyszczak et al. 2022; Meki et al. 2022), or via π–π interactions, hydrogen bonding, and coordination bonding (chemical sorption) (Dike et al. 2021; Hoang et al. 2021a). Secondly, the partition of PHs contaminants on biochar due to hydrophobic interactions is closely associated with the solubility and hydrophobicity of PHs. Thirdly, pollutants may bind strongly to biochar through micropore filling (Varjani

Table 7.2 Biochar expedites PAHs biodegradation in soil by biostimulation strategy (last 5 years)

Biostimulation strategy	Mechanism	Reference
Biochar	Biochar addition induced an enrichment of PAHs degraders as shown by molecular analysis.	Kong et al. (2018)
Biochar and root exudates	Biochar and root exudates synergistically promoted high-ring PAHs dissipation and enhanced microbial biomass and activity.	Li et al. (2019a)
Biochar and oxalic acid (OA)	Biochar and OA stimulated the functional genes participating in PAHs degradation.	Li et al. (2019b)
Biochar and biosurfactants, and N treatment	Biochar increased PAHs dissipation primarily by adsorption. Overall, the findings of this study affirm the positivity of the application of integrated biochar, biosurfactant, and N treatment in soil remediation.	Wei et al. (2020)
Biochar and biosurfactant (rhamnolipid)	Alleviated the toxicity of *Spartina anglica* via improving plant growth by increasing shoot height, root vitality, and total chlorophyll in PAHs contaminated soils.	Zhen et al. (2019)
Biochar and/or mycorrhizae inoculation	Better growth of plants through applying biochar and/or mycorrhizae inoculation led to an increase in the degradation rate of PAHs.	Abbaspour et al. (2020)
Biochar and mushroom residue and corn straw	PAHs degradation in soils increased with added biochar, mushroom residue, and corn straw.	Bao et al. (2020)
Biochar and rice straw feedstock	Biochar plays a key role in stimulating microbial abundance (bacteria) and diversity (bacteria and fungi) in soil. Accordingly, biochars pyrolyzed at a relatively high heating temperature with rice straw feedstock containing relatively high mineral nutrients are recommended for application to heavily PAHs contaminated soils.	Zhang et al. (2020)
Biochar	Enhancing microbial biomass and biodegradation of PAHs.	Dike et al. (2021)
Biochar and bioaugmentation	Biochar co-application with bioaugmentation was more effective in hydrocarbon removal than biochar co-application with either phytoremediation or biostimulation.	Dike et al. (2022)

et al. 2017; Dai et al. 2019; Hoang et al. 2021a). Under various conditions, one of the above-mentioned mechanisms predominates in the sorption process (Varjani et al. 2017; Dai et al. 2019; Hoang et al. 2021a).

In practice, these phenomena translate into the fact that the incorporation of biochar results in the sorption of soil pollutants by different mechanisms; ion exchange processes, electrostatic attraction, complexation, precipitation, or transformation of soil pollutants from toxic to non-toxic forms, which alter the mobility and bioavailability of pollutants, thereby improving soil quality (Varjani et al. 2017; Dai et al. 2019; Haider et al. 2021; Hoang et al. 2021a; Zahed et al. 2021).

Moreover, biochar can strongly sorb residual PHs and metabolites to provide a slow release of PHs source for continued microbial degradation (Hoang et al. 2021b).

Biochar has unique sorption properties and therefore, their combination with organic compost and inorganic fertilizers expedites the activity of plant-microbe systems in the PHs remediation process (Sarma et al. 2019; Hoang et al. 2021a). The porous structure and high surface area of biochar contribute to the efficient HMW PHs (*viz.* PAHs) removal in soils amended with biochar (Dike et al. 2021).

Furthermore, biochar can be a suitable habitat for soil microbes and can also protect microbes from predators. The surface and pores in biochar can support the growth of soil microbes, due to the presence of dissolved organic carbon (DOC) and nutrients released, as well as a surface charge on the biochar (which allows the immobilization of microbial cells, chemical ions, and other compounds) (Sarma et al. 2019; Zhang et al. 2020; Dike et al. 2021).

Zhang et al. (2020) suggested that one of the reasons for a higher PAHs biodegradation in soil amended with biochar and inorganic fertilizers is the increase in nutrients and water transfer from the soil to the biochar pores. This net effect contributes to the creation of a habitat that supports microbial growth and activity.

Although biochar has been associated with nutrient supply to soil microbes, it cannot replace the addition of nitrogen-enriched compounds, especially in biochar with a high C/N ratio. Soils contaminated by PHs are characterized by nitrogen deficiency since PHs are mainly composed of C and H and the addition of biochar (usually high C/N biomass) to soil can further increase the soil C/N ratio and cause nitrogen immobilization. Co-applying contaminated soils amended with biochar with fertilizer/inorganic nutrients could address this problem (Varjani et al. 2017; Dai et al. 2019; Sarma et al. 2019; Haider et al. 2021; Hoang et al. 2021a; Zahed et al. 2021; Curiel-Alegre et al. 2022; Dike et al. 2022).

In conclusion, due to the porous nature of biochar, this can act as a matrix for organic contaminants, and thus its sole application to the soil could result in the reduction of the organic contaminant's bioavailability and biodegradation (Kuppusamy et al. 2016c; Dike et al. 2022). However, the co-application of biochar with fertilizer/inorganic nutrients and/or with hydrocarbon-degrading microorganisms (bacteria, fungi, yeast, and algae) has demonstrated to promote the metabolic activity in the biochar-assisted bioaugmentation treatment significantly increasing the degradation of PHs (Dai et al. 2019; Haider et al. 2021; Hoang et al. 2021a; Zahed et al. 2021; Curiel-Alegre et al. 2022; Dike et al. 2022; Krzyszczak et al. 2022; Lv et al. 2022; Wu et al. 2022).

One of the drawbacks of the use of biochar alone in soils is the fact that it can decrease the efficiency of removing PHs due to its adsorption, making them unavailable for microbial metabolism (Xiong et al. 2017; Dike et al. 2022). Moreover, different studies have shown that co-applying biochar with a surfactant can be useful in mitigating this challenge because a surfactant enhances the desorption of the hydrocarbon from the biochar (Kang et al. 2019; Feng et al. 2021; Hoang et al. 2021b; Dike et al. 2022; Kariyawasam et al. 2022; Krzyszczak et al. 2022). Based on this, it is anticipated that co-applying biochar with surfactant should result in higher hydrocarbon removal.

7.4.2 (Bio)Surfactants-Mediated Remediation

The bioremediation efficiency of PHs in contaminated soils may be dependent on the characteristics of pollutants (*viz.* bioavailability and toxicity), the spatial distribution of microbes, and their metabolic ability. The main limiting factor in bioremediation of soils contaminated with PHs is the possible slow desorption of these compounds from the solid phase to the aqueous phase, which may reduce the respective bioavailability (Feng et al. 2021; Kariyawasam et al. 2022). In fact, PHs (*viz.* PAHs) with low aqueous solubility tend to remain bounded to soils and sediment particles, having to undergo desorption in order to be solubilized and degraded (Kariyawasam et al. 2022). Therefore, surfactant-enhanced bioremediation is a promising remediation technique as it could improve the bioaccessibility of the contaminant to microorganisms (Feng et al. 2021; Hoang et al. 2021b). Surfactants may increase the freely dissolved hydrocarbons in the aqueous phase and intensify the PHs biodegradation processes (Feng et al. 2021; Hoang et al. 2021b).

Surfactants are a group of amphiphilic compounds, which include in their molecular structures a hydrophilic and a hydrophobic portion, allowing the formation of micelles in the aqueous phase when the surfactant concentration reaches the critical micellar concentration (CMC) (Hoang et al. 2021b). The distinct structure of surfactants can enhance the water solubility of soil pollutants, especially in cases of hydrophobic organic compounds such as PHs (*viz.* PAHs), because hydrocarbons get solubilized in the hydrophobic cores of micelles produced by surfactants, thereby facilitating their assimilation by microorganisms (Hoang et al. 2021b; Kariyawasam et al. 2022). Micelle-bound PHs can be directly taken up by microbes or diffused into the aqueous media to make them available for microbial uptake (Li and Chen 2009). Surfactants enhance the enzymatic activity, microbial mobility, and interfacial uptake of PHs by increasing the cell surface hydrophobicity of microbes. These may act as dispersants, solubilizing agents, or emulsifiers (Liu et al. 2018; Kariyawasam et al. 2022).

Surfactants can be simply separated into chemical surfactants and biosurfactants. The latter have been widely applied due to their better environmental compatibility and antibacterial ability. Their processing and purification costs are the most significant drawbacks. However, researchers are focused on lowering the cost by the greener production of biosurfactants in-situ, using renewable feedstock such as glucose, glycerol, and waste cooling oils as substrates (Li and Chen 2009; Mulligan 2021; Kariyawasam et al. 2022).

Biosurfactants are LMW amphiphilic molecules produced by biological sources consisting of a hydrophilic and a hydrophobic portion, with a wide range of chemical structures, such as mycolic acid, glycolipids, lipopeptides, polysaccharide-protein complexes, phospholipids, fatty acids, etc. (Li et al. 2021; Pinto et al. 2021; Kariyawasam et al. 2022). These molecules can decrease the surface tension, CMC, and interfacial tension, therefore affecting the distribution of the pollutants among the solid and aqueous phases (Pinto et al. 2018; Li et al. 2021; Pinto et al. 2021; Kariyawasam et al. 2022).

Biosurfactants comprise a promising choice to improve the bioremediation of PHs due to their lower toxicity, better environmental compatibility, higher biodegradability, higher selectivity, and effectiveness under several extreme environments compared to chemical surfactants (Li et al. 2021; Mulligan 2021; Kariyawasam et al. 2022).

Studies have shown that biosurfactants can be produced by a wide range of microorganisms or extracted from plant and animal metabolites (Table 7.3) (Mukherjee et al. 2006; Geetha et al. 2018; Marcelino da Silva and Maranho 2019).

Microbial surfactants are widely classified based on their chemical structure/composition as LMW biosurfactants, comprising glycolipids and lipopeptides and HMW biosurfactants, as polysaccharides and lipoproteins (Mukherjee et al. 2006; Geetha et al. 2018; Marcelino da Silva and Maranho 2019; Carolin et al. 2021; Markande et al. 2021; Ng et al. 2022; Sarubbo et al. 2022). The most important groups of glycolipids are rhamnolipids, sophorolipids, trehalolipids and mannosylerythritol lipids, on the other hand, the most important groups of lipopeptides are surfactants, fengycins and iturins (Table 7.3). The LMW compounds have an emulsifying capacity that can be applied in the bioremediation of hydrophobic compounds in different environments (Carolin et al. 2021; Kadiri et al. 2022).

Production of biosurfactants by microorganisms varies with the microorganism as well as with the nutrient source. Microbial communities, including *Pseudomonas*, *Bacillus*, *Enterobacter*, *Actinobactor*, *Rhodococcus*, and *Candida* sp., produce biosurfactants (Mulligan 2021; Kariyawasam et al. 2022). Among bacterial species, *Pseudomonas aeruginosa* is one of the most notable candidates involved in the production of biosurfactants. Furthermore, *Enterobacteriaceae* family, *Microbacterium* genus are able to degrade both HMW and LMW PHs and could be efficiently used without further purification (Cazals et al. 2020; Kariyawasam et al. 2022). As an example, *Pseudomonas* species produces a glycolipid-based biosurfactant known as rhamnolipid while a lipoprotein-based biosurfactant known as surfactin is produced by *Bacillus subtilis* (Table 7.3) (Cao et al. 2022; Nwankwegu et al. 2022).

Surfactin is a widely studied lipopeptide biosurfactant used for the remediation of PHs contaminated soils and is separated into three families, surfactin, iturin, and fengycin, (with different structures and functions). For instance, the surfactin family has excellent surface activity, yet the iturin and fengycin families dominate antifungal activity (Goswami and Deka 2019; Wang et al. 2022).

Biosurfactants increase the rate of microbial degradation of hydrocarbons through two mechanisms first, by enhancing the substrate bioavailability through emulsification and, second, by reducing surface and interfacial tensions, facilitating the involvement of bacterial cells with hydrophobic substrates, due to the reduction of the hydrophobicity of the bacterial cell surface (Posada-Baquero et al. 2019; Sarma et al. 2019; Gaur et al. 2022; Jagaba et al. 2022).

Apart from their efficiency in increasing the bioavailability of contaminants, (bio)surfactants are currently used as the preferred source of energy by microorganisms (Ali et al. 2022; Liu et al. 2018, 2022). Both surfactants and biosurfactants can act as rich energy sources by microbes (Li et al. 2017; Ali et al. 2022). Microorganisms use the fatty acid lipophilic fraction of surfactants as a privileged carbon source for

Table 7.3 Important classes of biosurfactants and respective producer microorganisms

Class	Microorganisms
Glycolipids (Rhamnolipds)	*Pseudomonas aeruginosa*
	Pseudomonas cepacia
	Lysinibacillus sphaericus
	Serratia rubidaea
	Pseudomonas putida
	Pseudomonas fluorescens
	Bacillus sp.
	Trametes versicolor
	Pseudomonas balearica
Glycolipids (Trehalolipids)	*Nocardia farcinica*
	Rhodococcus erythropolis
	Candida bombicola
	Mycobacterium sp.
Glycolipids (Mannosylerythritol lipids)	*Pseudozyma aphidis*
Glycolipids (Sophorolipids)	*Candida sphaerica*
	Starmerella bombicola
	Cutaneotrichosporon mucoides
	Torulopsis bombicola
	Torulopsis apícola
	Candida glabrata
	Candida lipolytica
Phospholipids	*Vibrio* sp. LQ2
	Thiobacillus thiooxidans
	Klebsiella pneumoniae
	Corynebacterium alkanolyticum
	Staphylococcus hominis
Lipopeptides (Surfactin)	*Bacillus amyloliquefaciens*
	Bacillus subtilis
	Enterobacter sp.
	Micromonospora marina
Lipopeptides (Lichenysin)	*Bacillus licheniformis*
Lipopeptides (Fengycin)	*Bacillus mojavensis*
	Bacillus amyloliquefaciens
	Bacillus megaterium
	Bacillus subtilis
Lipopeptides (Iturin)	*Bacillus amyloliquefacien*
	Bacillus megaterium
	Aspergillus carbonarius
Lipopeptides (Kurstakin)	*Bacillus thuringiensis*
	Bacillus mojavensis
Lipopeptides (Polymyxin)	*Paenibacillus polymyxa*

(continued)

Table 7.3 (continued)

Class	Microorganisms
Lipopeptides (Pseudofactin II)	*Pseudomonas* sp.
Polymeric Surfactants (Liposan)	*Candida lipolytica*
Polymeric Surfactants (Emulsan)	*Acinetobacter Iwoffii* *Acinetobacter calcoaceticus* *Acinetobacter radioresistens*

References: Geetha et al. 2018; Marcelino da Silva and Maranho 2019; Carolin et al. 2021; Markande et al. 2021; Kadiri et al. 2022; Ng et al. 2022; Sarubbo et al. 2022.

their growth. The biosurfactant alkyl polyglycoside, is an effective source for enhancing the solubility of PAHs and has been used by microorganisms as a preferential carbon source (Li et al. 2017; do Vale et al. 2020; Ali et al. 2022).

Compatibility of surfactants and biosurfactants with microorganisms is necessary for the safe and efficient bioremediation of contaminated sites. However, there are some synthetic surfactants that effectively enhance the contaminant's solubility but unfortunately are toxic to microorganisms (sodium dodecyl sulfate and triton X-100 are some examples) (Ali et al. 2022; Kariyawasam et al. 2022). These negative effects can be reduced by applying biosurfactants instead of synthetic ones, increasing microbial tolerance using other additives capable of reducing the susceptibility of cells to surfactants, or even selecting appropriate combinations of surfactants and microorganisms (Kariyawasam et al. 2022).

The combination of more than one approach could serve as an effective tool for the efficient reduction of PHs contamination from diverse ecosystems.

7.5 Plant-Microbe Systems—A Synergistic Approach

Many reports have mentioned the use of plants to remediate organic pollutants. Nonetheless, the synergistic use of plants with specific microbes is recommended for consolidated efficiency (Hussain et al. 2018b; Sarma et al. 2019). In fact, the application of plant-microbe systems for the removal of organic pollutants in the soil has emerged as a synergistic ecological approach with high potential for the remediation of soils contaminated with PHs. This led to the identification of a plethora of plants, algae, bacteria, and fungi, able to metabolize hydrocarbons into non-hazardous by-products through their respective physiological and molecular processes (Sarma et al. 2019). Over the past decade, high-throughput sequencing technology and microbial biotechnology have been developed to aid in the identification and screening of a multitude of hydrocarbon-using/degrading microorganisms from environments contaminated with PHs (Hussain et al. 2018b; Mishra et al. 2021; Gaur et al. 2022).

The rhizosphere has been considered a hotspot of microbiome benefitting plants through mutualistic plant-microbe associations. Indeed, deep microbial activity has been reported in the rhizosphere, where microbes flourish and are attracted to root exudates. The exudates released by the plants contain specific phytochemicals (sugars, alcohols, carbohydrates, among others), which drive microbial chemotaxis (Hussain et al. 2018b; Sarma et al. 2019; Gaur et al. 2022).

Rhizospheric microbiomes encompasse a wide range of prokaryotic and eukaryotic organisms, comprising consortia of rhizobacteria, yeast, and fungi, which interact synergically and symbiotically with the roots, using plant exudates and boosting plant growth and the removal and detoxification of pollutants (Hussain et al. 2018b; Sarma et al. 2019; Abdullah et al. 2020). Therefore, the combined use of plants and suitable microorganisms is potentially more promising for the remediation of soils contaminated with PHs than the use of plants itself.

Bacteria are reported as the most effective PHs degrading microbial community (Hussain et al. 2018b; Abena et al. 2019; Wu et al. 2019). In plant-bacterial interactions, bacteria may proliferate around plant roots (as rhizospheric bacteria) and in internal tissues (as endophytes) without being pathogenic to their host plant (Hussain et al. 2018b; Pinto et al. 2018). Plants support bacterial growth by providing the right habitat and food; in turn, bacteria help their host plants overcome stress caused by pollutants, and enhance their growth, thereby increasing aerial and root biomass, which in turn supports microbial populations and pollutant degradation in the rhizosphere and endosphere. This synergistic relationship enhances the role of each participant in this symbiosis (Hussain et al. 2018b; Pinto et al. 2018).

Bacterial associations with plants can improve the phytoremediation of organic pollutants and stabilize them into less harmful metabolites or even mineralize the complex organic structures into carbon dioxide and water (Hussain et al. 2018b; Kariyawasam et al. 2022; Nwankwegu et al. 2022).

Acinetobacter, *Dietzia*, *Achromobacter*, *Burkholderia*, *Marinobacter*, *Arthrobacter*, *Alteromonas*, *Mycobacterium*, *Enterobacter*, *Sphingomonas*, *Stenotrophomonas*, *Streptobacillus*, *Pseudomonas*, *Staphylococcus*, and *Streptococcus*, bacterial genera use PHs as an energy source for their microbial activities (Poi et al. 2017; Behera et al. 2021; Behera et al. 2022; Gaur et al. 2022; Sun et al. 2022). In the presence of carbon sources, the process of enzyme production is triggered in degraders and these enzymes quickly catalyze the efficient degradation of the hydrocarbon molecules (Gaur et al. 2022).

Pseudomonas are PH degrading bacterial species with great potential as they can improve the bioremediation rate by different mechanisms, such as the production of rhamnolipid biosurfactants, and act as plant growth promoting rhizobacteria (PGPR), which increases the phytoremediation rate (Gaur et al. 2022; Kariyawasam et al. 2022; Nwankwegu et al. 2022; Raklami et al. 2022; Muthukumar et al. 2023).

The use of bacteria having both pollutants degrading as well as plant growth-promoting abilities has shown better results compared to the use of bacteria having only pollutant-degrading or plant growth-promoting activities (Khan et al. 2013; Raklami et al. 2022; Muthukumar et al. 2023).

These PGP (Plant Growth Promoter) bacteria may develop symbiotic/mutualistic associations with plants and may be found as free-living rhizospheric or endophytic bacteria (Khan et al. 2013; Pinto et al. 2018, 2021).

PGP bacteria can enhance plant development by acting as biofertilizers, consequently increasing the bioavailability of essential nutrients (*viz.* N_2 fixation and phosphate and iron solubilization); as phytostimulants (producing plant growth regulators and hormones, such as indole acetic acid–IAA; cytokinins and other auxins); as stress controllers (by decreasing ethylene production through the synthesis of 1-aminocyclopropane-1-carboxylic acid deaminase-ACCD); and as plant defense inducers against phytopathogens (by producing siderophores, antibiotics, or fungicidal compounds) (Khan et al. 2013; Balseiro-Romero et al. 2017; Pinto et al. 2018; Sarma et al. 2019).

The ability of fungi to form extended mycelial networks also makes them very suitable for bioremediation processes (Pinto et al. 2018, 2021; Li et al. 2021). Fungi can effectively exploit soil microsites that are not accessible to most plant roots due to the small diameter of mycelia and can compete with other microorganisms for water and nutrients, protecting roots from direct contact with toxic compounds. The application of filamentous fungi can thus be considered a promising alternative or a valuable complement in situations of bacterial malfunction (Deng and Cao 2017; Pinto et al. 2018; Li et al. 2021).

Fungi such as *Aspergillus*, *Drechslera*, *Candida*, *Curvularia*, *Fusarium*, *Mucor*, *Lasiodiplodia*, *Rhizopus*, *Penicillium*, *Trichoderma*, have been reported to be involved in the degradation of hydrocarbons (aliphatic and aromatic) (Al-Hawash et al. 2019; Abdullah et al. 2020; Li et al. 2021; Gaur et al. 2022; Kariyawasam et al. 2022).

PHs biodegradation can significantly be increased using associations of degrading microorganisms such as fungi from genera *Achremonium*, *Absidia*, *Aspergillus*, and *Penicillium* and bacteria *Alcaligenes*, *Bacillus*, *Flayobacterium*, *Mycobacterium*, *Pseudomonas*, and *Sphingobium*. Effectively, considering the complex nature of PHs, mixed microbial strains are more effective than individual strains (Gupta et al. 2022; Jabbar et al. 2022).

Strategies to accelerate hydrocarbons biodegradation in the soil can be achieved by enhancing the growth and metabolic activity of native microorganisms, optimizing factors such as nutrients, organic matter, electron acceptors, oxygenation, temperature, pH, possible addition of biosurfactants (biostimulation) and/or employing selected organisms (bioaugmentation) with a remarkable catabolic diversity to degrade, transform or accumulate a broad spectrum of organic compounds (bioaugmentation) (Guarino et al. 2017; Hussain et al. 2018b; Varjani and Upasani 2019; Behera et al. 2022; Curiel-Alegre et al. 2022; Gupta et al. 2022; Jabbar et al. 2022; Nwankwegu et al. 2022). Several inorganic nutrients, surfactants, and carbon-rich materials have been used as sources of substrates for the remediation of different PHs contaminated sites (Ali et al. 2022; Gupta et al. 2022).

Nutrients nitrogen and phosphorus-based, enhance microbial metabolic activities and as basic components of cellular proteins and nucleic acids, play a vital role in microbial cell division and growth. Moreover, a balanced ratio of C:N:P (12:1:0.1) in soils promotes an increase in microbial activity and contributes to fast detoxification of contaminants (Ali et al. 2022; Behera et al. 2022; Jabbar et al. 2022).

Bioremediation strategies are grouped into two main groups: biostimulation and bioaugmentation, which can be used alone or together in order to achieve, accelerate, and improve decontamination processes. The simultaneous application of both approaches resulted in a further improvement in plant growth, soil microbial activity, and consequently, in PHs biodegradation (Marcelino da Silva and Maranho 2019; Hoang et al. 2021b; Behera et al. 2022; Sattar et al. 2022).

7.5.1 Bioaugmentation

Bioaugmentation has been reported as an effective microbiological approach capable of accelerating bioremediation efficiency through the application of specialized microbial communities or consortia (Behera et al. 2021; Gupta et al. 2022). Autochthonous, allochthonous, or genetically engineered microbial strains with specific catabolic activity, isolated strains, or enriched groups can be used for this purpose (Marcelino da Silva and Maranho 2019; Xu et al. 2021; Nwankwegu et al. 2022; Muthukumar et al. 2023). Depending on the origin of the microorganisms, three types of approaches can be distinguished: (1) allochthonous bioaugmentation, which inoculates foreign species that were recruited in a different geographic region from the one to be remedied (non-indigenous); (2) autochthonous bioaugmentation, which uses an indigenous consortium, enriched from the contaminated area, and which will subsequently be reintroduced in the same site; or (3) genetically modified organisms bioaugmentation, which involves the use of genetically engineered microbial strains that have been equipped with genes that encode the enzymes responsible for the desired degradation functions (Marcelino da Silva and Maranho 2019; Behera et al. 2021; Hoang et al. 2021b; Nwankwegu et al. 2022; Sattar et al. 2022; Muthukumar et al. 2023). However, the use of foreign microorganisms often results in low biodegradation efficiency due to the inability to adapt to the environment of the polluted area (Abena et al. 2019; Xu et al. 2021; Nwankwegu et al. 2022; Sattar et al. 2022). Effectively, indigenous microbial communities have the potential to perform more effective biodegradation due to their genetic mutation, synergistic effect, resistivity, and better adaptability to environmental toxicity (Abena et al. 2019; Behera et al. 2021; Xu et al. 2021; Nwankwegu et al. 2022).

It is often desirable to use mixed multi-microbe (consortium) systems to biodegrade complex mixtures of PHs relative to individual strains, due to the interactive synergies between consortium members (Poi et al. 2017; Wu et al. 2019; Behera et al. 2021). One of the main advantages of treatment with consortiums is the presence of a wide range of enzymatic and degradative pathways, capable of degrading the different polluting intermediates, promoting the process of co-metabolism and commensalism, essential for the biodegradation of complex hydrocarbons (Abena et al. 2019; Varjani and Upasani 2019; Wu et al. 2019; Behera et al. 2021; Nwankwegu et al. 2022). Moreover, the selection of an efficient microbial culture should take into consideration the following microbial features: microbial stability (survivability) over a wide range of environmental conditions (stress-tolerant microorganisms), ability to withstand high concentrations of the pollutants without exhibiting signs of cytotoxicity, fast growth, and easily culturable (Nwankwegu et al. 2022).

Bioaugmentation includes the introduction of adapted microorganisms able to survive, rapidly flourishing, and becoming the most dominant microorganisms suited to degrade the contaminants. Introduced microorganisms may come from a consortium already proven capable of degrading contaminants or may be isolated from the originally contaminated environment and subsequently adapted to be reintroduced (Poi et al. 2017; Li et al. 2021; Gupta et al. 2022).

A successful bioaugmentation strategy requires knowledge about the type and concentration of the contaminant, the characteristics of the soil to be cleaned (pH, moisture content, nutrient, and organic matter availability), and environmental factors such as the climate of the region, with the possibility of success in large-scale application varying between regions (Guarino et al. 2017; Feng et al. 2021; Nwankwegu et al. 2022).

Regional variability may be linked to other potential background factors including ecosystem heterogeneity, environment-specific microbial diversity, characteristics of pollutants, and the environmental acceptability of foreign microorganisms that, unfortunately, are not considered and identified in most cases. Inherently, the existence of different strategies on the part of microbes, such as modification of the cell membrane, surfactant compounds production, use of efflux pumps, and maintenance of biological functions, may favor the detoxification of compounds inside microbial cells (Guarino et al. 2017; Feng et al. 2021; Nwankwegu et al. 2022).

Several successful studies have provided evidence of the success of bioaugmentation considered an ecologically correct and cost-effective technology capable of total mineralization of contaminants into innocuous by-products (Table 7.4) (Guarino et al. 2017; Feng et al. 2021; Nwankwegu et al. 2022).

Operating conditions for enhanced degradation during bioaugmentation include: ease of microbial adaptability, genetic stability, microbial motility, as well as mass transfer mechanisms through the polluted system. These operational conditions can be optimized by immobilizing the microbial cells of the bioaugmentation system. The process of cell entrapment, adsorption, and encapsulation are immobilization techniques of utmost importance used to make the microbial system introduced in the polluted environment more robust, reducing the inadequacies (environmental disturbances) associated with the initial cellular adaptation and the environmental toxicity of the polluted environment for the microbial species (Ali et al. 2022; Nwankwegu et al. 2022). Immobilized materials provide large surface areas, and high transfer volumes with long retention time for adsorption of PHs and protect the exogenous microbes from competition with indigenous microorganisms, resulting in improved bioremediation. Both natural and synthetic materials have been used for the immobilization of microorganisms. Polymeric materials such as agarose, polyacrylamide gel, K-carrageenan alginate, clay, anthracite, porous glass, and ceramic are suitable immobilizing agents for cell immobilization in a bioaugmentation system (Ali et al. 2022; Nwankwegu et al. 2022). Immobilization media are also suitable attachment surfaces for optimized microbial actions, improving

Table 7.4 The effectiveness of bioaugmentation to restore PHs contaminated soils

Host plant	Inoculum	Contaminants	Main results	References
L. purpureus	The mixture of novel strains *Pseudomonas aeruginosa* PP3 and *Pseudomonas aeruginosa* PP4 were used. Four different bioaugmentation systems (BS 1–4) were carried out in this experiment labelled as BS 1 (Crude oil contaminated soil), BS 2 (BS 1 + bacterial consortia), BS 3 (Motor oil sludge contaminated soil), and BS 4 (BS 3 + bacterial consortia).	Crude oil/motor oil contaminated soil	The highest PHs removal and biodegradation rates was recorded for the BS 4 inoculum. The bioaugmented soil was subjected to the plant cultivation and the results revealed that the *L. purpureus* showed the greatest growth in the BS 4 system.	Muthukumar et al. (2023)
Lolium perenne cv. Pinia	*Enterobacter ludwigii* ZCR5	Hydrocarbons and heavy metals	A higher efficiency of phytoremediation in the ZCR5-bioaugmented soil than in the controls, although the authors indicate the need to study bacteria-plant interactions during assisted phytoremediation at a more detailed level.	Pacwa-Plociniczak et al. (2023)
Medicago sativa L. and *S. nigrum* L.	*Bacillus* sp.; *Micrococcus* sp. and *Sphingobacterium multivorum*	*PAHs and multi-metals*	Bioaugmentation removed 81.4% and 95.4% of PAHs in heavier-contaminated soils and lighter-contaminated soils. The fresh weights of *S. nigrum* L. and *M. sativa* L. were increased by the addition of microbial agents.	Cao et al. (2022)

(continued)

Table 7.4 (continued)

Host plant	Inoculum	Contaminants	Main results	References
Medicago sativa L	Growth-promoting and hydrocarbon-degrading bacterial consortium dominated by Alphaproteobacteria (*Acidocella aminolytica* and *Acidobacterium capsulatum)*	Petroleum hydrocarbons	The inoculation of *M. sativa* with the consortium resulted in 66% increase in plant biomass and achieved a 91% reduction in diesel fuel hydrocarbon concentrations in the soil within 60 days.	Eze et al. (2022)
Mangroves (*Avicennia marina*)	*Sphingomonas* sp.-LK11; *Rhodococcus corynebacterioides*-NZ1; *Bacillus subtilis*-EP1; *Bacillus safensis*-SH10	Diesel	Mangrove seedlings' associations with bacteria could reprogram host growth, physiology, and ability to degrade diesel.	Khan et al. (2022)
Vigna radiata	*Ochrobactrum* sp., *Pseudomonas aeruginosa,* and *Stenotrophomonas maltophila*	Petroleum sludge	Hydrocarbon degrading bacterial consortium degraded 93.14 ± 1.75% petroleum sludge in 42 days. Petroleum sludge-polluted agricultural soil has been restored (confirmed by seed germination).	Varjani et al. (2021)
Alfalfa (*Medicago sativa* L.)	Bacillus cereus (Acc KR232400), Bacillus altitudinis (Acc KF859970), Commomonas (Delftia sp.) (Acc KF859971), and Stenotrophomonas maltophilia (Acc KF859973).	Oily sludge mixed with soil in 30:70 ratio	Consortium degraded more than 63% of total PHs.	Shahzad et al. (2020)
Distichlis spicata (Saltgrass)	A consortium consists of five bacterial strains (*Pseudomonas* sp. PFS1 and BSS3B2, *Serratia proteamaculans* S1BD1, *Alcaligenes* sp. PKS1, and *Bacillus* sp. PSS2) with different plant growth-promoting traits were used	Co-contamination with PHs (15,000 mg kg^{-1} dry soil) and heavy metals	Bacterial inoculation significantly enhanced plant biomass in the contaminated soil compared to non-inoculated treatment	Xia et al. (2020)

contact between microorganisms and pollutants. However, studies on the positive influence of microbial cell immobilization on microbial activities under PHs contamination are limited, and therefore, future studies to explore the effects on the success of bioremediation processes in detail are mandatory.

7.6 Conclusions

The remediation of PHs-polluted soil includes physical, chemical, and biological processes, and their associations. There is a wide variety of techniques available for soil remediation, although none is the best for all pollutants and for the site-specificity of each contaminated area. However, some innovative bioremediation strategies can provide environmentally safe and economically viable options for the degradation of PHs in contaminated soils.

Widely used methods for the bioremediation of PHs-contaminated soils include phytoremediation, biostimulation (providing suitable environment and nutrients), and bioaugmentation (introducing efficient PHs degrading microbes), alone or combined in integrated approaches.

Phytoremediation efficiency can be augmented by optimizing plant-microbe synergies, through the inoculation of seeds/plants with specific endophytic and rhizospheric microbes. Furthermore, microbial strains having adequate features such as plant growth-promoting abilities, contaminant degrading competencies, and/or surfactant-producing skills can fairly increase remediation efficiency.

Inoculation with efficient single strains or consortia of many bacterial strains is a form of bioaugmentation that proved to be an effective approach for the biodegradation of PHs in contaminated soils, as it improves the effectiveness of contaminant clean-up and reduces the time and cost of clean-up.

An inoculation of mixtures of ecologically diverse microbes instead of single strains represent a highly successful strategy, due to the different and extensive metabolic capabilities of bacterial consortia. Moreover, evidence of fungal-bacterial synergism on PHs degradation warns that future research efforts should focus on the regulation of cooperative catabolism between bacteria and fungi, which seems a promising topic to support more effective bioaugmentation.

PHs removal from the soil matrix is usually hampered by the low bioavailability and high Freundlich adsorption coefficient of long-chain hydrocarbons. To overcome this issue, the addition of surfactants to facilitate emulsification and desorption of hydrophobic pollutants has become one of the most widely used solutions. Surfactants can be simply separated into chemical surfactants and biosurfactants, with biosurfactants being widely applied due to their greater ability to regulate healthy microbial ecosystems, environmental compatibility, and antibacterial ability.

The use of biochar, plants, and microbes is recommended as it offers a practical, sustainable, eco-friendly, and cost-effective solution for the remediation of PHs.

In conclusion, this review demonstrated that bioremediation is a productive and effective solution to remediate, clean, manage, and rehabilitate polluted environments. Different remediation technologies are nowadays in use, but no single approach can be applied to all environmental conditions.

A full and exhaustive characterization of the contaminated environment; the composition, properties, and nature of the contaminants, transport, and distribution in soil; better understanding of the degradation mechanisms involved; interactions and relationships with microbes; extrinsic factors affecting the remediation process and the potential impact of the remediation measures to adopt, are relevant to determine the best remediation strategy to be used.

Future research on joint strategies combining microbial remediation technology and physicochemical technologies may be effective to accelerate the removal of PHs pollutants and should be strongly encouraged. In addition, the use of genetically modified plants and microorganisms can further exploit the plant-microbe cooperation and provide time-efficient remediation options. However, the release of genetically modified microbes and plants into the environment encompasses potential risks. These can be mitigated by adopting safe and advanced environmental biotechnological techniques and appropriate technical regulatory guidelines, provided by environmental protection agencies and other regulatory frameworks.

7.7 Suggestions

The following suggestions are put forward to emphasize sustainable bioremediation techniques:

(i) Dveloping environmentally friendly technologies and viable strategies for remediation of PHs contamination in soils is key for restoring soil ecosystem function, mitigating adverse effects on global climate change, and protecting biodiversity.

(ii) The concerted use of plants and competent microorganisms (*viz.* bacteria, fungi) aided by adequate biostimulation measures, can be a promising strategy for the remediation of PHs contaminated soils as compared to bioaugmentation and phytoremediation alone.

(iii) Evaluating all parameters of the remediation process that considerably impact the decision-making process is crucial to increase the effectiveness of remediating soils contaminated with PHs.

References

Abbasian F, Lockington R, Mallavarapu M, Naidu R (2015) A comprehensive review of aliphatic hydrocarbon biodegradation by bacteria. Appl Biochem Biotechnol 176(3):670–699. https://doi.org/10.1007/s12010-015-1603-5

Abbaspour A, Zohrabi F, Dorostkar V et al (2020) Remediation of an oil-contaminated soil by two native plants treated with biochar and mycorrhizae. J Environ Manag 254(15):109755

Abdel-Raouf ME-S (ed) (2012) Crude oil emulsions-composition stability and characterization. InTech. https://doi.org/10.5772/2677

Abdel-Shafy HI, Mansour MSM (2016) A review on polycyclic aromatic hydrocarbons: source, environmental impact, effect on human health and remediation. Egypt J Pet 25(1):107–123. https://doi.org/10.1016/j.ejpe.2015.03.011

Abdullah SRS, Al-Baldawi IA, Almansoory AF et al (2020) Plant-assisted remediation of hydrocarbons in water and soil: application, mechanisms, challenges and opportunities. Chemosphere 247(125):932

Abena MTB, Li T, Shah MN, Zhong W (2019) Biodegradation of total petroleum hydrocarbons (TPH) in highly contaminated soils by natural attenuation and bioaugmentation. Chemosphere 234:864–874

Abhilash P, Jamil S, Singh N (2009) Transgenic plants for enhanced biodegradation and phytoremediation of organic xenobiotics. Biotechnol Adv 27(4):474–488

Abhilash P, Powell J, Singh H, Singh B (2012) Plant–microbe interactions: novel applications for exploitation in multipurpose remediation technologies. Trends Biotechnol 30(8):416–420. https://doi.org/10.1016/j.tibtech.2012.04.004

Achuba FI, Ja-Anni MO (2018) Effect of abattoir waste water on metabolic and antioxidant profiles of cowpea seedlings grown in crude oil contaminated soil. Int J Recycl Org Waste Agric 7(1):59–66

Adeniji AO, Okoh OO, Okoh AI (2019) Levels of polycyclic aromatic hydrocarbons in the water and sediment of buffalo river estuary, South Africa and their health risk assessment. Arch Environ Contam Toxicol 76(4):657–669. https://doi.org/10.1007/s00244-019-00617-w

Adesina OA, Nwogu AS, Sonibare JA (2021) Indoor levels of polycyclic aromatic hydrocarbons (PAHs) from environment tobacco smoke of public bars. Ecotoxicol Environ Saf 208(111):604. https://doi.org/10.1016/j.ecoenv.2020.111604

Agnello AC, Bagard M, van Hullebusch ED et al (2016) Comparative bioremediation of heavy metals and petroleum hydrocarbons co-contaminated soil by natural attenuation, phytoremediation, bioaugmentation and bioaugmentation-assisted phytoremediation. Sci Total Environ 563–564:693–703

Agrawal N, Shahi S (2017) Degradation of polycyclic aromatic hydrocarbon (pyrene) using novel fungal strain Coriolopsis byrsina strain APC5. Int Biodeterior Biodegrad 122:69–81. https://doi.org/10.1016/j.ibiod.2017.04.024

Ajona M, Vasanthi P (2021) Bioremediation of petroleum contaminated soils–a review. Mater Today Proc 45(7):7117–7122

Akyüz M, Çabuk H (2010) Gas-particle partitioning and seasonal variation of polycyclic aromatic hydrocarbons in the atmosphere of Zonguldak, Turkey. Sci Total Environ 408(22):5550–5558. https://doi.org/10.1016/j.scitotenv.2010.07.063

Al-Hawash AB, Zhang X, Ma F (2019) Removal and biodegradation of different petroleum hydrocarbons using the filamentous fungus Aspergillus sp. RFC-1. Microbiology 8(1):e00619. https://doi.org/10.1002/mbo3.619

Ali M, Song X, Ding D et al (2022) Bioremediation of PAHs and heavy metals co-contaminated soils: challenges and enhancement strategies. Environ Pollut 295(118):686. https://doi.org/10.1016/j.envpol.2021.118686

Ambaye TG, Chebbi A, Formicola F et al (2022a) Remediation of soil polluted with petroleum hydrocarbons and its reuse for agriculture: recent progress, challenges, and perspectives. Chemosphere 293(133):572. https://doi.org/10.1016/j.chemosphere.2022.133572

Ambaye TG, Chebbi A, Formicola F et al (2022b) Ex-situ bioremediation of petroleum hydrocarbon contaminated soil using mixed stimulants: response and dynamics of bacterial community and phytotoxicity. J Environ Chem Eng 10(6):108814. https://doi.org/10.1016/j.jece.2022.108814

Ambaye TG, Vaccari M, Franzetti A et al (2023) Microbial electrochemical bioremediation of petroleum hydrocarbons (PHCs) pollution: recent advances and outlook. Chem Eng J 452(3):139372. https://doi.org/10.1016/j.cej.2022.139372

An X, Xiao B, Di X, Dong H, Tang H (2017) Research progress on aging of organic pollutants in geosorbents: a review. Acta Geochimica 36(1):27–43. https://doi.org/10.1007/s11631-016-0129-z

Andersson T, Hartonen K, Hyötyläinena T, Riekkola M-L (2003) Stability of polycyclic aromatic hydrocarbons in pressurised hot water. Analyst 128(2):150–155. https://doi.org/10.1039/B211447J

Atef N, Kukkadapu G, Mohamed SY et al (2017) A comprehensive iso-octane combustion model with improved thermochemistry and chemical kinetics. Combust Flame 178:111–134. https://doi.org/10.1016/j.combustflame.2016.12.029

Balseiro-Romero M, Gkorezis P, Kidd PS et al (2017) Use of plant growth promoting bacterial strains to improve Cytisus striatus and Lupinus luteus development for potential application in phytoremediation. Sci Total Environ 581–582:676–688

Bansal V, Kumar P, Kwon EE, Kim K-H (2017) Review of the quantification techniques for polycyclic aromatic hydrocarbons (PAHs) in food products. Crit Rev Food Sci Nutr 57(15):3297–3312. https://doi.org/10.1080/10408398.2015.1116970

Bao H, Wang J, Zhang H et al (2020) Effects of biochar and organic substrates on biodegradation of polycyclic aromatic hydrocarbons and microbial community structure in PAHs-contaminated soils. J Hazard Mater 385(121):595. https://doi.org/10.1016/j.jhazmat.2019.121595

Baoune H, El Hadj-Khelil A, Pucci G et al (2018) Petroleum degradation by endophytic *Streptomyces* spp. isolated from plants grown in contaminated soil of southern Algeria. Ecotoxicol Environ Saf 147:602–609. https://doi.org/10.1016/j.ecoenv.2017.09.013

Barati M, Bakhtiari F, Mowla D, Safarzadeh S (2018) Comparison of the effects of poultry manure and its biochar on barley growth in petroleum-contaminated soils. Int J Phytoremediation 20(2):98–103

Bashir MA, Naveed M, Ahmad Z et al (2020) Combined application of biochar and sulfur regulated growth, physiological, antioxidant responses and Cr removal capacity of maize (Zea mays L.) in tannery polluted soils. J Environ Manag 259:110051. https://doi.org/10.1016/j.jenvman.2019.110051

Bauer AK, Velmurugan K, Plöttner S et al (2018) Environmentally prevalent polycyclic aromatic hydrocarbons can elicit co-carcinogenic properties in an in vitro murine lung epithelial cell model. Arch Toxicol 92(3):1311–1322. https://doi.org/10.1007/s00204-017-2124-5

Behera ID, Manoranjan N, Biswas S et al (2021) Enhanced biodegradation of total petroleum hydrocarbons by implementing a novel two-step bioaugmentation strategy using indigenous bacterial consortium. J Environ Manag 292(112):746. https://doi.org/10.1016/j.jenvman.2021.112746

Behera ID, Nayak M, Mishra A et al (2022) Strategic implementation of integrated bioaugmentation and biostimulation for efficient mitigation of petroleum hydrocarbon pollutants from terrestrial and aquatic environment. Mar Pollut Bull 177(113):492. https://doi.org/10.1016/j.marpolbul.2022.113492

Bhattacharya S, Chakraborty P, Roy SS (2014) Dibenz[a,h]anthracene. In: Wexler, reference module in biomedical sciences, encyclopedia of toxicology, 3rd edn. Academic, Oxford, pp 62–66. https://doi.org/10.1016/B978-0-12-386454-3.00303-1

Cai D, Yang X, Wang S, Chao Y et al (2017) Effects of dissolved organic matter derived from forest leaf litter on biodegradation of phenanthrene in aqueous phase. J Hazard Mater 324(Part B):516–525

Cao X, Cui X, Xie M, Zhao R et al (2022) Amendments and bioaugmentation enhanced phytoremediation and micro-ecology for PAHs and heavy metals co-contaminated soils'. J Hazard Mater 426(128):096

Carolin F, Kumar P, Ngueagni P (2021) A review on new aspects of lipopeptide biosurfactant: types, production, properties and its application in the bioremediation process. J Hazard Mater 407(5):124827

Cazals F, Huguenot D, Crampon M et al (2020) Production of biosurfactant using the endemic bacterial community of a PAHs contaminated soil, and its potential use for PAHs remobilization. Sci Total Environ 709(136):143. https://doi.org/10.1016/j.scitotenv.2019.136143

Cébron A, Louvel B, Faure P et al (2011) Root exudates modify bacterial diversity of phenanthrene degraders in PAH-polluted soil but not phenanthrene degradation rates. Environ Microbiol 13(3):722–736. https://doi.org/10.1111/j.1462-2920.2010.02376.x

Cheruyiot NK, Lee W-J, Mwangi JK et al (2015) An overview: polycyclic aromatic hydrocarbon emissions from the stationary and mobile sources and in the ambient air. Aerosol Air Qual Res 15(7):2730–2762. https://doi.org/10.4209/aaqr.2015.11.0627

Cheshkova TV, Sergun VP, Kovalenko EY et al (2019) Resins and asphaltenes of light and heavy oils: their composition and structure. Energy Fuel 33(9):7971–7982. https://doi.org/10.1021/acs.energyfuels.9b00285

Coleman J, Blake-Kalff M, Davies E (1997) Detoxification of xenobiotics by plants: chemical modification and vacuolar compartmentation. Trends Plant Sci 2(4):144–151. https://doi.org/10.1016/S1360-1385(97)01019-4

Curiel-Alegre S, Velasco-Arroyo B, Rumbo C et al (2022) Evaluation of biostimulation, bioaugmentation, and organic amendments application on the bioremediation of recalcitrant hydrocarbons of soil. Chemosphere 307(1):135638

Dai Y, Zhang N, Xing C et al (2019) The adsorption, regeneration and engineering applications of engineering applications of biochar for removal organic pollutants: a review. Chemosphere 223:12–27. https://doi.org/10.1016/j.chemosphere.2019.01.161

de Oliveira M, Frihling BEF, Velasques J et al (2020) Pharmaceuticals residues and xenobiotics contaminants: occurrence, analytical techniques and sustainable alternatives for wastewater treatment. Sci Total Environ 705(135):568. https://doi.org/10.1016/j.scitotenv.2019.135568

Deng Z, Cao L (2017) Fungal endophytes and their interactions with plants in phytoremediation: a review. Chemosphere 168:1100–1106

Dhankher O, Pilon-Smits E, Meagher R, Doty S (2012, 2012) Biotechnological approaches for phytoremediation. Plant Biotechnol Agric:309–328

Dike CC, Shahsavaria E, Surapanenic A et al (2021) Can biochar be an effective and reliable biostimulating agent for the remediation of hydrocarbon-contaminated soils? Environ Int 154(106):553. https://doi.org/10.1016/j.envint.2021.106553

Dike CC, Hakeem IG, Rani A et al (2022) The co-application of biochar with bioremediation for the removal of petroleum hydrocarbons from contaminated soil. Sci Total Environ 849(157):753. https://doi.org/10.1016/j.scitotenv.2022.157753

do Vale TO, de Magalhães RS, de Almeida PF et al (2020) The impact of alkyl polyglycoside surfactant on oil yields and its potential effect on the biogenic souring during enhanced oil recovery (EOR). Fuel 280(118):512

Duxbury CL, Dixon DG, Greenberg BM (1997) Effects of simulated solar radiation on the bioaccumulation of polycyclic aromatic hydrocarbons by the duckweed Lemna gibba. Environ Toxicol Chem 16:1739–1748

Eze MO, Thiel V, Hose GC et al (2022) Enhancing rhizoremediation of petroleum hydrocarbons through bioaugmentation with a plant growth-promoting bacterial consortium. Chemosphere 289(133):143

Fakher S, Ahdaya M, Elturki M, Imqam A (2020) Critical review of asphaltene properties and factors impacting its stability in crude oil. J Pet Explor Prod Technol 10(3):1183–1200. https://doi.org/10.1007/s13202-019-00811-5

Farzadkia M, Esrafili A, Gholami M, Koolivand A (2019) Effect of immature and mature compost addition on petroleum contaminated soils composting: kinetics. J Environ Health Sci Eng 17:839–846

Feng N-X, Yu J, Zhao H-M et al (2017) Efficient phytoremediation of organic contaminants in soils using plant–endophyte partnerships. Sci Total Environ 583:352–368. https://doi.org/10.1016/j.scitotenv.2017.01.075

Feng L, Jiang X, Huang Y et al (2021) Petroleum hydrocarbon-contaminated soil bioremediation assisted by isolated bacterial consortium and sophorolipid. Environ Pollut 273(116):476. https://doi.org/10.1016/j.envpol.2021.116476

Gaur VK, Gautam K, Sharma P et al (2022) Sustainable strategies for combating hydrocarbon pollution: special emphasis on mobil oil bioremediation. Sci Total Environ 832(155):083. https://doi.org/10.1016/j.scitotenv.2022.155083

Geetha SJ, Banat IM, Joshi SJ (2018) Biosurfactants: production and potential applications in microbial enhanced oil recovery (MEOR). Biocatal Agric Biotechnol 14:23–32. https://doi.org/10.1016/j.bcab.2018.01.010

Gómez-Sagasti MT, Hernández A, Artetxe U et al (2018) How valuable are organic amendments as tools for the phytomanagement of degraded soils? The knowns, known unknowns, and unknowns. Front Sustain Food Syst 2:68

Goswami M, Deka S (2019) Biosurfactant production by a rhizosphere bacteria *Bacillus altitudinis* MS16 and its promising emulsification and antifungal activity. Colloids Surf B 178:285–296. https://doi.org/10.1016/j.colsurfb.2019.03.003

Groenzin H, Mullins OC (2001) Molecular size and structure of asphaltenes'. Pet Sci Technol 19(1–2):219–230. https://doi.org/10.1081/LFT-100001236

Guarino C, Spada V, Sciarrillo R (2017) Assessment of three approaches of bioremediation (Natural attenuation, landfarming and bioagumentation—assistited landfarming) for a petroleum hydrocarbons contaminated soil. Chemosphere 170:10–16. https://doi.org/10.1016/j.chemosphere.2016.11.165

Gul S, Whalen JK, Thomas BW et al (2015) Physico-chemical properties and microbial responses in biochar-amended soils: mechanisms and future directions. Agric Ecosyst Environ 206:46–59

Gupta PK, Mustapha HI, Singh B, Sharma YC (2022) Bioremediation of petroleum contaminated soil-water resources using neat biodiesel: a review. Sustain Energy Technol Assess 53(102):703

Gupte A, Tripathi A, Patel H et al (2016) Bioremediation of polycyclic aromatic hydrocarbon (PAHs): a perspective. Open Biotechnol J 10(1):363–378. https://doi.org/10.2174/1874070701610010363

Haider FU, Ejaz M, Cheema SA et al (2021) Phytotoxicity of petroleum hydrocarbons: sources, impacts and remediation strategies. Environ Res 197(111):031. https://doi.org/10.1016/j.envres.2021.111031

Haritash AK, Kaushik CP (2009) Biodegradation aspects of Polycyclic Aromatic Hydrocarbons (PAHs): a review. J Hazard Mater 169(1–3):1–15. https://doi.org/10.1016/j.jhazmat.2009.03.137

Harmsen J, Rietra RP (2018) 25 years monitoring of PAHs and petroleum hydrocarbons biodegradation in soil. Chemosphere 207:229–238

Hasanuzzaman M, Bhuyan MB, Zulfiqar F et al (2020) Reactive oxygen species and antioxidant defense in plants under abiotic stress: revisiting the crucial role of a universal defense regulator. Antioxidants 9(8):681. https://doi.org/10.3390/antiox9080681

Hoang SA, Sarkar B, Seshadri B et al (2021a) Mitigation of petroleum-hydrocarbon-contaminated hazardous soils using organic amendments: a review. J Hazard Mater 416(125):702. https://doi.org/10.1016/j.jhazmat.2021.125702

Hoang SA, Lamb D, Seshadri B et al (2021b) Rhizoremediation as a green technology for the remediation of petroleum hydrocarbon-contaminated soils. J Hazard Mater 401(123):282. https://doi.org/10.1016/j.jhazmat.2020.123282

Hussain K, Hoque RR, Balachandran S et al (2018a) Monitoring and risk analysis of PAHs in the environment. In: Hussain C (ed) Handbook of environmental materials management. Springer, Cham. https://doi.org/10.1007/978-3-319-58538-3_29-2

Hussain I, Aleti G, Naidu R et al (2018b) Microbe and plant assisted-remediation of organic xenobiotics and its enhancement by genetically modified organisms and recombinant technology: a review. Sci Total Environ 628–629:1582–1599. https://doi.org/10.1016/j.scitotenv.2018.02.037

Hussain I, Puschenreiter M, Gerhard S et al (2018c) Rhizoremediation of petroleum hydrocarbon-contaminated soils: improvement opportunities and field applications. Environ Exp Bot 147:202–219

Ibañez S, Talano M, Ontañon O et al (2016) Transgenic plants and hairy roots: exploiting the potential of plant species to remediate contaminants. New Biotechnol 33:625–635. https://doi.org/10.1016/j.nbt.2015.11.008

Infante PF, Bingham E (2012) Aromatic hydrocarbons-benzene and other alkylbenzenes. Patty's Toxicol 2:1–68. https://doi.org/10.1002/0471435139.tox051.pub2

ITOPF (2023) International tanker owners pollution federation–oil tanker spill statistics 2022. https://www.itopf.org/knowledge-resources/data-statistics/statistics/. Accessed 13 Jan 2023.

Jabbar NM, Alardhi SM, Mohammed AK et al (2022) Challenges in the implementation of bioremediation processes in petroleum-contaminated soils: a review. Environ Nanotechnol Monit Manage 18(100):694. https://doi.org/10.1016/j.enmm.2022.100694

Jagaba AH, Kutty SRM, Lawal IM et al (2022) Diverse sustainable materials for the treatment of petroleum sludge and remediation of contaminated sites: a review. Clean Waste Syst 2(100):010. https://doi.org/10.1016/j.clwas.2022.100010

Jajoo A, Mekala NR, Tomar RS et al (2014) Inhibitory effects of polycyclic aromatic hydrocarbons (PAHs) on photosynthetic performance are not related to their aromaticity. J Photochem Photobiol B Biol 137:151–155. https://doi.org/10.1016/j.jphotobiol.2014.03.011

Jeevanantham S, Saravanan A, Hemavathy RV et al (2019) Removal of toxic pollutants from water environment by phytoremediation: a survey on application and future prospects. Environ Technol Innov 13:264–276. https://doi.org/10.1016/j.eti.2018.12.007

Jonker MTO (2016) Determining octanol–water partition coefficients for extremely hydrophobic chemicals by combining "slow stirring" and solid-phase microextraction. Environ Toxicol Chem 35(6):1371–1377. https://doi.org/10.1002/etc.3300

Kadiri F, Zahid B, Thoume A et al (2022) Characterization and valorization of a biosurfactant produced by *Aeromonas salmonicida* isolated from the marchika lagoon of Nador. Biocatal Agric Biotechnol 45(102):523. https://doi.org/10.1016/j.bcab.2022.102523

Kafle A, Timilsina A, Gautam A et al (2022) Phytoremediation: mechanisms, plant selection and enhancement by natural and synthetic agents. Environ Adv 8(100):203. https://doi.org/10.1016/j.envadv.2022.100203

Kang S, Kim G, Choe JK (2019) Effect of using powdered biochar and surfactant on desorption and biodegradability of phenanthrene sorbed to biochar. J Hazard Mater 371:253–260. https://doi.org/10.1016/j.jhazmat.2019.02.104

Kapoor D, Singh S, Kumar V et al (2019) Antioxidant enzymes regulation in plants in reference to reactive oxygen species (ROS) and reactive nitrogen species (RNS). Plant Gene 19(100):182. https://doi.org/10.1016/j.plgene.2019.100182

Kariyawasam T, Doran GS, Howitt JA, Prenzler PD (2022) Polycyclic aromatic hydrocarbon contamination in soils and sediments: sustainable approaches for extraction and remediation. Chemosphere 291(132):981

Kawahigashi H (2009) Transgenic plants for phytoremediation of herbicides. Curr Opin Biotechnol 20(2):225–230. https://doi.org/10.1016/j.copbio.2009.01.010

Khan S, Afzal M, Iqbal S, Khan QM (2013) Plant–bacteria partnerships for the remediation of hydrocarbon contaminated soils. Chemosphere 90:1317–1332

Khan A, Tanveer S, Alia S et al (2017) Role of nutrients in bacterial biosurfactant production and effect of biosurfactant production on petroleum hydrocarbon biodegradation. Ecol Eng 104(A):158–164. https://doi.org/10.1016/j.ecoleng.2017.04.023

Khan MAI, Biswas B, Smith E et al (2018) Toxicity assessment of fresh and weathered petroleum hydrocarbons in contaminated soil-a review. Chemosphere 212:755–767. https://doi.org/10.1016/j.chemosphere.2018.08.094

Khan A, Numan M, Bilal S et al (2022) Mangrove's rhizospheric engineering with bacterial inoculation improve degradation of diesel contamination. J Hazard Mater 423(A):127046. https://doi.org/10.1016/j.jhazmat.2021.127046

Khashij S, Karimi B, Makhdoumi P (2018) Phytoremediation with Festuca arundinacea: a Mini Review. Int J Health Life Sci 4(2):86625. https://doi.org/10.5812/ijhls.86625

Kim KH, Jahan SA, Kabir E, Brown RJC (2013) A review of airborne polycyclic aromatic hydrocarbons (PAHs) and their human health effects. Environ Int 60:71–80. https://doi.org/10.1016/j.envint.2013.07.019

Kong F-X, Sun G-D, Liu Z-P (2018) Degradation of polycyclic aromatic hydrocarbons in soil mesocosms by microbial/plant bioaugmentation: performance and mechanism. Chemosphere 198:83–91. https://doi.org/10.1016/j.chemosphere.2018.01.097

Krzyszczak A, Dybowski MP, Zarzycki R et al (2022) Long-term physical and chemical aging of biochar affected the amount and bioavailability of PAHs and their derivatives. J Hazard Mater 440(129):795. https://doi.org/10.1016/j.jhazmat.2022.129795

Kumar M, Bolan NS, Hoang SA et al (2021) Remediation of soils and sediments polluted with polycyclic aromatic hydrocarbons: to immobilize, mobilize, or degrade? J Hazard Mater 420(126):534. https://doi.org/10.1016/j.jhazmat.2021.126534

Kummerova M, Váňová L, Krulová J, Zezulka Š (2008) The use of physiological characteristics for comparison of organic compounds phytotoxicity. Chemosphere 71(11):2050–2059. https://doi.org/10.1016/j.chemosphere.2008.01.060

Kuppusamy S, Thavamani P, Megharaj M, Naidu R (2016a) Biodegradation of polycyclic aromatic hydrocarbons (PAHs) by novel bacterial consortia tolerant to diverse physical settings–Assessments in liquid- and slurry-phase systems. Int Biodeterior Biodegrad 108:149–157. https://doi.org/10.1016/j.ibiod.2015.12.013

Kuppusamy S, Thavamani P, Megharaj M et al (2016b) Kinetics of PAH degradation by a new acid-metal-tolerant Trabulsiella isolated from the MGP site soil and identification of its potential to fix nitrogen and solubilize phosphorous. J Hazard Mater 307:99–107. https://doi.org/10.1016/j.jhazmat.2015.12.068

Kuppusamy S, Thavamani P, Megharaj M et al (2016c) Agronomic and remedial benefits and risks of applying biochar to soil: current knowledge and future research directions. Environ Int 87:1–12. https://doi.org/10.1016/j.envint.2015.10.018

Kuppusamy S, Thavamani P, Venkateswarlu K et al (2017) Remediation approaches for polycyclic aromatic hydrocarbons (PAHs) contaminated soils: technological constraints, emerging trends and future directions. Chemosphere 168:944–968. https://doi.org/10.1016/j.chemosphere.2016.10.115

Kuppusamy S, Maddela NR, Megharaj M, Venkateswarlu K (2020) Methodologies for analysis and identification of total petroleum hydrocarbons. In: Total petroleum hydrocarbons. Springer, Cham. https://doi.org/10.1007/978-3-030-24035-6_2

León O, Contreras E, Rogel E et al (2002) Adsorption of native resins on asphaltene particles: a correlation between adsorption and activity. Langmuir 18(13):5106–5112. https://doi.org/10.1021/la011394q

Li J-L, Chen B-H (2009) Surfactant-mediated biodegradation of polycyclic aromatic hydrocarbons. Materials 2(1):76–94. https://doi.org/10.3390/ma2010076

Li G, Lan G, Liu Y et al (2017) Evaluation of biodegradability and biotoxicity of surfactants in soil. RSC Adv 7(49):31018–31026. https://doi.org/10.1039/c7ra02105d

Li X, Song Y, Wang F et al (2019a) Combined effects of maize straw biochar and oxalic acid on the dissipation of polycyclic aromatic hydrocarbons and microbial community structures in soil: a mechanistic study. J Hazard Mater 364(15):325–331. https://doi.org/10.1016/j.jhazmat.2018.10.041

Li X, Song Y, Yao S et al (2019b) Can biochar and oxalic acid alleviate the toxicity stress caused by polycyclic aromatic hydrocarbons in soil microbial communities? Sci Total Environ 695(10):133879. https://doi.org/10.1016/j.scitotenv.2019.133879

Li Q, Li J, Jiang L et al (2021) Diversity and structure of phenanthrene degrading bacterial communities associated with fungal bioremediation in petroleum contaminated soil. J Hazard Mater 403(123):895

Liu G, Zhong H, Yang X et al (2018) Advances in applications of rhamnolipids biosurfactant in environmental remediation: a review. Biotechnol Bioeng 115(4):796–814. https://doi.org/10.1002/bit.26517

Liu X, He L, Zhang X et al (2022) Bioremediation of petroleum-contaminated saline soil by Acinetobacter baumannii and Talaromyces sp. and functional potential analysis using metagenomic sequencing. Environ Pollut 311:119970. https://doi.org/10.1016/j.envpol.2022.119970

Lv Y, Bao J, Zhu L (2022) A comprehensive review of recent and perspective technologies and challenges for the remediation of oil-contaminated sites. Energy Rep 8:7976–7988. https://doi.org/10.1016/j.egyr.2022.06.034

Macaya CC, Durán RE, Hernández L et al (2019) Bioremediation of petroleum. Ref Mod Life Sci 12:1–6. https://doi.org/10.1016/b978-0-12-809633-8.20810-8

Mahar A, Wang P, Ali A et al (2016) Challenges and opportunities in the phytoremediation of heavy metals contaminated soils: a review. Ecotoxicol Environ Saf 126:111–121. https://doi.org/10.1016/j.ecoenv.2015.12.023

Marcelino da Silva B, Maranho LT (2019) Petroleum-contaminated sites: decision framework for selecting remediation technologies. J Hazard Mater 378(120):722. https://doi.org/10.1016/j.jhazmat.2019.05.115

Markande A, Patel D, Varjani S (2021) A review on biosurfactants: properties, applications and current developments. Bioresour Technol 330(124):963

Martin BC, George SJ, Price CA et al (2014) The role of root exuded low molecular weight organic anions in facilitating petroleum hydrocarbon degradation: current knowledge and future directions. Sci Total Environ 472:642–653. https://doi.org/10.1016/j.scitotenv.2013.11.050

Meki K, Liu Q, Wu S, Yuan Y (2022) Plant- and microbe-assisted biochar amendment technology for petroleum hydrocarbon remediation in saline-sodic soils: a review. Pedosphere 32(1):211–221. https://doi.org/10.1016/S1002-0160(21)60041-3

Miadonye A, Evans L (2010) The solubility of asphaltenes in different hydrocarbon liquids. Pet Sci Technol 28(14):1407–1414. https://doi.org/10.1080/10916460902936960

Michael-Igolima U, Abbey SJ, Ifelebuegu AO (2022) A systematic review on the effectiveness of remediation methods for oil contaminated soils. Environ Adv 9(100):319. https://doi.org/10.1016/j.envadv.2022.100319

Mildenberg R, Zander M, Collin G (2007) Hydrocarbon resins. Wiley VCH Verlagsgesellschaft mbH., ISBN:9783527286171, pp 1–179. https://doi.org/10.1002/9783527614653

Mishra SK, Kumar PR, Singh RK (2020) 3–Transgenic plants in phytoremediation of organic pollutants. In: Pandey VC, Singh V (eds) Bioremediation of pollutants. Elsevier, pp 39–56

Mishra B, Varjani S, Kumar G et al (2021) Microbial approaches for remediation of pollutants: innovations, future outlook, and challenges. Energy Environ 32(6):1029–1058. https://doi.org/10.1177/0958305X19896781

Mohan SV, Kisa T, Ohkuma T et al (2006) Bioremediation technologies for treatment of PAH-contaminated soil and strategies to enhance process efficiency. Rev Environ Sci Biotechnol 5:347–374. https://doi.org/10.1007/s11157-006-0004-1

Motelay-Massei A, Ollivon D, Garban B et al (2007) Fluxes of polycyclic aromatic hydrocarbons in the Seine estuary, France: mass balance and role of atmospheric deposition. Hydrobiologia 588(1):145–157. https://doi.org/10.1007/s10750-007-0659-9

Mukherjee S, Das P, Sen R (2006) Towards commercial production of microbial surfactants. Trends Biotechnol 24:509–515. https://doi.org/10.1016/j.tibtech.2006.09.005

Mulligan CN (2021) Sustainable remediation of contaminated soil using biosurfactants. Front Bioeng Biotechnol 9(635):196

Muthukumar B, Surya S, Sivakumar K et al (2023) Influence of bioaugmentation in crude oil contaminated soil by *Pseudomonas* species on the removal of total petroleum hydrocarbon. Chemosphere 310(136):826. https://doi.org/10.1016/j.chemosphere.2022.136826

Ng Y, Lim H, Khoo K et al (2022) Recent advances of biosurfactant for waste and pollution bioremediation: substitutions of petroleum-based surfactants. Environ Res 212(A):113126. https://doi.org/10.1016/j.envres.2022.113126

Noctor G, Reichheld J-P, Foyer CH (2018) ROS-related redox regulation and signaling in plants. Semin Cell Dev Biol 80:3–12

Nwankwegu AS, Orji MU, Onwosi CO (2016) Studies on organic and in-organic biostimulants in bioremediation of diesel-contaminated arable soil. Chemosphere 162:148–156. https://doi.org/10.1016/j.chemosphere.2016.07.074

Nwankwegu AS, Zhang L, Xie D et al (2022) Bioaugmentation as a green technology for hydrocarbon pollution remediation. Problems and prospects. J Environ Manag 304(114):313. https://doi.org/10.1016/j.jenvman.2021.114313

Odoh CK, Zabbey N, Sam K, Eze CN (2019) Status, progress and challenges of phytoremediation-An African scenario. J Environ Manag 237:365–378. https://doi.org/10.1016/j.jenvman.2019.02.090

Ossai IC, Ahmed A, Hassan A, Hamid FS (2020) Remediation of soil and water contaminated with petroleum hydrocarbon: a review. Environ Technol Innov 17(100):526. https://doi.org/10.1016/j.eti.2019.100526

Pacwa-Plociniczak M, Byrski A, Chlebek D et al (2023) A deeper insight into the phytoremediation of soil polluted with petroleum hydrocarbons supported by the Enterobacter ludwigii ZCR5 strain. Appl Soil Ecol 181(104):651

Panwar R, Mathur J (2022) Mitigation of polycyclic aromatic hydrocarbon contaminated soil using microbes and nanoparticles: a review. Pedosphere 1–16. https://doi.org/10.1016/j.pedsph.2022.06.032

Parus A, Ciesielski T, Woźniak-Karczewska M et al (2023) Basic principles for biosurfactant-assisted (bio)remediation of soils contaminated by heavy metals and petroleum hydrocarbons–a critical evaluation of the performance of rhamnolipids. J Hazard Mater 443(Part A):130171

Patel AB, Shaikh S, Jain KR et al (2020) Polycyclic aromatic hydrocarbons: sources, toxicity, and remediation approaches. Front Microbiol 11(562):813. https://doi.org/10.3389/fmicb.2020.562813

Pinto AP, de Varennes A, Lopes ME, Teixeira DM (2016) Biological approaches for remediation of metal-contaminated sites, in phytoremediation. Manag Environ Contam 3(3):65–112. https://doi.org/10.1007/978-3-319-40148-additives5_3

Pinto AP, de Varennes A, Dias CMB, Lopes ME (2018) Microbial-assisted phytoremediation: a convenient use of plant and microbes to clean up soils. In: Ansari A, Gill S, Gill R, Lanza R, G., Newman, L. (eds) Phytoremediation. Springer, Cham, pp 21–87. https://doi.org/10.1007/978-3-319-99651-6_2

Pinto AP, Lopes ME, Dordio A, Castanheiro JEF (2021) Microbe and plant-assisted remediation of organic xenobiotics. In: Prasad MNV (ed) Handbook of assisted and amendment: enhanced sustainable remediation technology. Wiley, pp 437–475. https://doi.org/10.1002/9781119670391.ch22

Poi G, Aburto-Medina A, Mok PM et al (2017) Large scale bioaugmentation of soil contaminated with petroleum hydrocarbons using a mixed microbial consortium. Ecol Eng 102:64–71. https://doi.org/10.1016/j.ecoleng.2017.01.048

Posada-Baquero R, Martín ML, Ortega-Calvo J-J (2019) Implementing standardized desorption extraction into bioavailability-oriented bioremediation of PAH-polluted soils. Sci Total Environ 696(134):011

Prabakaran K, Li J, Anandkumar A et al (2019) Managing environmental contamination through phytoremediation by invasive plants: a review. Ecol Eng 138:28–37. https://doi.org/10.1016/j.ecoleng.2019.07.002

Premnath N, Mohanrasub K, Rao GR et al (2021) A crucial review on polycyclic aromatic Hydrocarbons – Environmental occurrence and strategies for microbial degradation. Chemosphere 280(130):608. https://doi.org/10.1016/j.chemosphere.2021.130608

Qurashi AW, Sabri AN (2012) Bacterial exopolysaccharide and biofilm formation stimulate chickpea growth and soil aggregation under salt stress. Braz J Microbiol 43:1183–1191. https://doi.org/10.1590/S1517-83822012000300046

Rai PK, Kim K-H, Lee SS, Lee J-H (2020) Molecular mechanisms in phytoremediation of environmental contaminants and prospects of engineered transgenic plants/microbes. Sci Total Environ 705(135):858

Raklami A, Meddich A, Oufdou K, Baslam M (2022) Plants—microorganisms-based bioremediation for heavy metal cleanup: recent developments, phytoremediation techniques, regula-

tion mechanisms, and molecular responses. Int J Mol Sci 23(9):5031. https://doi.org/10.3390/ijms23095031
Ramesh A, Archibong AE, Hood DB et al (2011) Global environmental distribution and human health effects of polycyclic aromatic hydrocarbons. In: Global contamination trends of persistent organic chemicals, Chapter 5. CRC Press, Boca Raton, pp 97–126. https://doi.org/10.1201/b11098-10
Ravindra K, Sokhi R, Van Grieken R (2008) Atmospheric polycyclic aromatic hydrocarbons: source attribution, emission factors and regulation. Atmos Environ 42(13):2895–2921. https://doi.org/10.1016/j.atmosenv.2007.12.010
Robichaud K, Stewart K, Labrecque M et al (2019) An ecological microsystem to treat waste oil contaminated soil: using phytoremediation assisted by fungi and local compost, on a mixed-contaminant site, in a cold climate. Sci Total Environ 672:732–742. https://doi.org/10.1016/j.scitotenv.2019.03.447
Ruley JA, Amoding A, Tumuhairwe JB, Basamba TA (2022) Rhizoremediation of petroleum hydrocarbon-contaminated soils: a systematic review of mutualism between phytoremediation species and soil living microorganisms. In: Bhat RA, Tonelli FM, Dar GH, Hakeem K (eds) Phytoremediation: biotechnological strategies for promoting invigorating environs. Academic, Amsterdam, pp 263–296. https://doi.org/10.1016/B978-0-323-89874-4.00008-X
Rusin M, Gospodarek J, Barczyk G, Nadgórska-Socha A (2018) Antioxidant responses of Triticum aestivum plants to petroleum-derived substances. Ecotoxicology 27(10):1353–1367. https://doi.org/10.1007/s10646-018-1988-3
Sakshi, Singh SK, Haritash AK (2019) Polycyclic aromatic hydrocarbons: soil pollution and remediation. Int J Environ Sci Technol 16(10):6489–6512. https://doi.org/10.1007/s13762-019-02414-3
Saleem MH, Kamran M, Zhou Y et al (2020) Appraising growth, oxidative stress and copper phytoextraction potential of flax (Linum usitatissimum L.) grown in soil differentially spiked with copper. J Environ Manag 257:109994. https://doi.org/10.1016/j.jenvman.2019.109994
Saravanan A, Jeevanantham S, Narayanan V et al (2020) Rhizoremediation–a promising tool for the removal of soil contaminants: a review. J Environ Chem Eng 8(2):103543
Sarigiannis DA, Karakitsios SP, Zikopoulos D et al (2015) Lung cancer risk from PAHs emitted from biomass combustion. Environ Res 137:147–156. https://doi.org/10.1016/j.envres.2014.12.009
Sarma H, Nava AR, Prasad MNV (2019) Mechanistic understanding and future prospect of microbe-enhanced phytoremediation of polycyclic aromatic hydrocarbons in soil. Environ Technol Innov 13:318–330
Sarubbo L, Silva M, Durval I et al (2022) Biosurfactants: production, properties, applications, trends, and general perspectives. Biochem Eng J 181(108):377. https://doi.org/10.1016/j.bej.2022.108377
Sattar S, Hussain R, Shah SM et al (2022) Composition, impacts, and removal of liquid petroleum waste through bioremediation as an alternative clean-up technology: a review. Heliyon 8(10):e11101. https://doi.org/10.1016/j.heliyon.2022.e11101
Sayara T, Sánchez A (2020) Bioremediation of PAH-contaminated soils: process enhancement through composting/compost. Appl Sci 10(11):3684. https://doi.org/10.3390/app10113684
Schreiner CA (2003) Genetic toxicity of naphthalene: a review. J Toxicol Environ Health–Part B: Crit Rev 6(2):161–183. https://doi.org/10.1080/10937400306472
Seifried CM, Crawshaw J, Boek ES (2013) Kinetics of asphaltene aggregation in crude oil studied by confocal laser-scanning microscopy. Energy Fuel 27(4):1865–1872. https://doi.org/10.1021/ef301594j
Shahzad A, Siddiqui S, Bano A et al (2020) Hydrocarbon degradation in oily sludge by bacterial consortium assisted with alfalfa (*Medicago sativa* L.) and maize (*Zea mays* L.). Arab J Geosci 13:879. https://doi.org/10.1007/s12517-020-05902-w
Singh A, Kumar A, Yadav S, Singh IK (2019) Reactive oxygen species-mediated signaling during abiotic stress. Plant Gene 18(100):173

Stauffer E, Dolan JA, Newman R (2008) Interpretation of data obtained from neat ignitable liquids. Fire Debris Anal 9:295–354. https://doi.org/10.1016/b978-012663971-1.50013-0

Štefanac T, Grgas D, Landeka Dragičević T (2021) Xenobiotics-division and methods of detection: a review. J Xenobiot 11(4):130–141. https://doi.org/10.3390/jox11040009

Steliga T, Kluk D (2020) Application of Festuca arundinacea in phytoremediation of soils contaminated with Pb, Ni, Cd and petroleum hydrocarbons. Ecotoxicol Environ Saf 194(110):409

Sun S, Su Y, Chen S, Cui W, Zhao C, Liu Q (2022) Bioremediation of oil-contaminated soil: exploring the potential of endogenous hydrocarbon degrader Enterobacter sp. SAVR S-1. Appl Soil Ecol 173:104387. https://doi.org/10.1016/j.apsoil.2022.104387

Tang J, Zhu W, Kookana R, Katayama A (2013) Characteristics of biochar and its application in remediation of contaminated soil. J Biosci Bioeng 116(6):653–659. https://doi.org/10.1016/j.jbiosc.2013.05.035

Thomas JC, Russell DK, Rugh CL (2017) Polyaromatic hydrocarbon phytoremediation stimulated by root exudates. Int J Environ Bioremediat Biodegrad 5(1):1–7. https://doi.org/10.12691/ijebb-5-1-1

Truskewycz A, Gundry TD, Khudur LS et al (2019) Petroleum hydrocarbon contamination in terrestrial ecosystems—fate and microbial responses. Molecules 24:3400. https://doi.org/10.3390/molecules24183400

Tsai PJ, Shieh H-Y, Lee W-J, Lai S-O (2001) Health-risk assessment for workers exposed to polycyclic aromatic hydrocarbons (PAHs) in a carbon black manufacturing industry. Sci Total Environ 278(1–3):137–150. https://doi.org/10.1016/S0048-9697(01)00643-X

Tucci M, Viggi CC, Núñez AE et al (2021) Empowering electroactive microorganisms for soil remediation: challenges in the bioelectrochemical removal of petroleum hydrocarbons. Chem Eng J 419(130):008. https://doi.org/10.1016/j.cej.2021.130008

Varjani SJ (2017) Microbial degradation of petroleum hydrocarbons. Bioresour Technol 223:277–286. https://doi.org/10.1016/j.biortech.2016.10.037

Varjani SJ, Upasani VN (2017) A new look on factors affecting microbial degradation of petroleum hydrocarbon pollutants. Int Biodeterior Biodegrad 120:71–83. https://doi.org/10.1016/j.ibiod.2017.02.006

Varjani S, Upasani VN (2019) Influence of abiotic factors, natural attenuation, bioaugmentation and nutrient supplementation on bioremediation of petroleum crude contaminated agricultural soil. J Environ Manag 245:358–366. https://doi.org/10.1016/j.jenvman.2019.05.070

Varjani SJ, Gnansounou E, Pandey A (2017) Comprehensive review on toxicity of persistent organic pollutants from petroleum refinery waste and their degradation by microorganisms. Chemosphere 188:280–291

Varjani S, Pandey A, Upasani V (2021) Petroleum sludge polluted soil remediation: integrated approach involving novel bacterial consortium and nutrient application. Sci Total Environ 763(1):142934

Wang G, Zhang Q, Ma P et al (2008) Sources and distribution of polycyclic aromatic hydrocarbons in urban soils: case studies of Detroit and New Orleans. Soil Sediment Contam 17(6):547–563. https://doi.org/10.1080/15320380802425055

Wang S, Wang X, Zhang C et al (2016) Bioremediation of oil sludge contaminated soil by landfarming with added cotton stalks. Int Biodeterior Biodegradation 106:150–156. https://doi.org/10.1016/j.ibiod.2015.10.014

Wang L, Ji B, Hu Y et al (2017) A review on in situ phytoremediation of mine tailings. Chemosphere 184:594–600. https://doi.org/10.1016/j.chemosphere.2017.06.025

Wang Y, Wu S, Wang H et al (2022) Optimization of conditions for a surfactant-producing strain and application to petroleum hydrocarbon-contaminated soil bioremediation. Colloids Surf B: Biointerfaces 213(112):428. https://doi.org/10.1016/j.colsurfb.2022.112428

Webb E, Moon J, Dyrszka L et al (2018) Neurodevelopmental and neurological effects of chemicals associated with unconventional oil and natural gas operations and their potential effects on infants and children. Rev Environ Health 33(1):3–29. https://doi.org/10.1515/reveh-2017-0008

Wei Z, Wang J, Gaston L et al (2020) Remediation of crude oil-contaminated coastal marsh soil: integrated effect of biochar, rhamnolipid biosurfactant and nitrogen application. J Hazard Mater 396(5):122595. https://doi.org/10.1016/j.jhazmat.2020.122595

Wiszniewska A, Hanus-Fajerska E, Muszyńska E, Ciarkowska K (2016) Natural organic amendments for improved phytoremediation of polluted soils: a review of recent progress. Pedosphere 26(1):1–12. https://doi.org/10.1016/S1002-0160(15)60017-0

Włóka D, Kacprzak M, Grobelak A et al (2015) The Impact of PAHs Contamination on the Physicochemical Properties and Microbiological Activity of Industrial Soils. Polycycl Aromat Compd 35(5):372–386. https://doi.org/10.1080/10406638.2014.918887

Wu M, Wu J, Zhang X, Ye X (2019) Effect of bioaugmentation and biostimulation on hydrocarbon degradation and microbial community composition in petroleum-contaminated loessal soil. Chemosphere 237(124):456

Wu M, Liu Z, Gao H et al (2022) Assessment of bioremediation potential of petroleum-contaminated soils from the shanbei oilfield of China revealed by qPCR and high throughput sequencing. Chemosphere 308(Part 3):136446

Xia M, Chakraborty R, Terry N, Singh R, Fu D (2020) Promotion of saltgrass growth in a saline petroleum hydrocarbons contaminated soil using a plant growth promoting bacterial consortium. Int Biodeterior Biodegradation 146(104):808. https://doi.org/10.1016/j.ibiod.2019.104808

Xiong B, Zhang Y, Hou Y et al (2017) Enhanced biodegradation of PAHs in historically contaminated soil by M. gilvum inoculated biochar. Chemosphere 182:316–324. https://doi.org/10.1016/j.chemosphere.2017.05.020

Xu J, Xu L, Qiao X et al (2021) Stimulated biodegradation of all alkanes in soil. Chemosphere 278(130):444. https://doi.org/10.1016/j.chemosphere.2021.130444

Yang CY, Chang ML, Wu SC, Shih YH (2017) Partition uptake of a brominated diphenyl ether by the edible plant root of white radish (Raphanus sativus L.). Environ Pollut 223:178–184. https://doi.org/10.1016/j.envpol.2017.01.009

Yu B, Yuan Z, Yu Z, Xue-song F (2022) BTEX in the environment: an update on sources, fate, distribution, pretreatment, analysis, and removal techniques. Chem Eng J 435(P1):134825. https://doi.org/10.1016/j.cej.2022.134825

Zahed MA, Salehi S, Madadi R, Hejabi F (2021) Biochar as a sustainable product for remediation of petroleum contaminated soil. Curr Res Green Sustain Chem 4(100):055. https://doi.org/10.1016/j.crgsc.2021.100055

Zhang G, He L, Guo X et al (2020) Mechanism of biochar as a biostimulation strategy to remove polycyclic aromatic hydrocarbons from heavily contaminated soil in a coking plant. Geoderma 375(114):497. https://doi.org/10.1016/j.geoderma.2020.114497

Zhang T, Liu F, Yu X et al (2021a) Risk assessment and ecotoxicological diagnosis of soil from a chemical industry park in Nanjing, China. Ecotoxicology 30(7):1303–1314. https://doi.org/10.1007/s10646-020-02320-2

Zhang X, Yang L, Zhang H et al (2021b) Assessing approaches of human inhalation exposure to polycyclic aromatic hydrocarbons: a review. Int J Environ Res Public Health 18(6):3124. https://doi.org/10.3390/ijerph18063124

Zhen M, Chen H, Liu Q et al (2019) Combination of rhamnolipid and biochar in assisting phytoremediation of petroleum hydrocarbon contaminated soil using Spartina anglica. J Environ Sci (China) 85:107–118. https://doi.org/10.1016/j.jes.2019.05.013

Zhuang X, Wang Y, Wang H et al (2022) Comparison of the efficiency and microbial mechanisms of chemical- and bio-surfactants in remediation of petroleum hydrocarbon. Environ Pollut 314(120):198. https://doi.org/10.1016/j.envpol.2022.120198

Chapter 8
Plant Growth in Petroleum Contaminated Soils: An Appraisal of Phytoremediation

Abin Sebastian, Eilin Maria Martin, Malavika M. Nair, Rinnu Joseph, P. M. Sumayya, and Vandana K. Manoj

Abstract Petroleum contamination in soils forms an impermeable layer that prevents water entry and gas exchange. Petroleum hydrocarbons cause alteration in plant available nutrients and chemical composition of the soil respectively. Phytoremediation of petroleum-contaminated sites is a low-cost feasible approach. The physiological activity of plants and the colonization of microbes in their rhizosphere help to alter and remove petroleum products in the soil. Rhizodegradation as well as the accumulation of various hydrocarbons in plants determines the rate of decontamination of petroleum products in the course of phytoremediation. However, plants that grow in petroleum-contaminated sites have metabolic alterations. Understanding the phytotoxicity of petroleum products is important. It is imperative to identify local plant species for phytoremediation of petroleum-contaminated soil. The present chapter discusses plant growth and ideal plants for phytoremediation in petroleum-contaminated sites.

Keywords Soil · Phytotoxicity · Phytoremediation · Microbial · Diversity

8.1 Introduction

Petroleum is a yellow-black liquid mixture of hydrocarbons such as alkanes, cycloalkanes, and aromatic hydrocarbons. It is formed when large amounts of dead organisms, mostly zooplankton and algae, are buried under sedimentary rocks, and exposed to prolonged heat and pressure. Deposits of petroleum are found in porous rock formations. Petroleum is also found in oil sands. Crude oil containing, petrol is mainly retrieved via drilling after studying structural geology, analysis of sedimentary basin, and characterization of the oil reservoir. Recent developments in

A. Sebastian (✉) · E. M. Martin · M. M. Nair · R. Joseph · P. M. Sumayya · V. K. Manoj
Department of Botany, St. George's College, Erattupetta, India
e-mail: abin@sgcaruvithura.ac.in

A. Parnian et al. (eds.), *Bioremediation of Petroleum Contaminated Soil and Sediment*, Environmental Contamination Remediation and Management,
https://doi.org/10.1007/978-3-031-95524-2_8

technologies have also led to the exploitation of unconventional petroleum reserves such as oil sands and oil shale. Once extracted, the crude oil is refined and separated via distillation into products such as gasoline (petrol), diesel and kerosene (Ashoori et al. 2017). Hydrocarbon chains of varying lengths are present in petroleum. Petroleum can be used for a variety of purposes through processing it with other chemical processes. For example, petroleum is a raw material for many chemical products, including pharmaceuticals, solvents, fertilizers, pesticides, and plastics.

The word petroleum covers all liquid, gaseous, and solid hydrocarbons found in crude oil. The lighter hydrocarbons such as methane, ethane, propane, and butane are gases, while pentane and the heavier hydrocarbons are liquids or solids. The proportions of gas, liquid, and solid in an oil reservoir depend on geological conditions, and the phase profile of the petroleum mixture respectively. An oil well produces mainly crude oil with some dissolved natural gas. The low pressure on the surface leads to the release of gases out of the liquid phase which can be recovered as an associated gas or solution gas (Juwarkar and Jambhulkar 2008). Sometimes petroleum gas may contain heavier hydrocarbons such as pentane, hexane, and heptane in the gaseous state. These hydrocarbons condense to form natural gas condensate which is often abbreviated as condensate. The organic compounds in crude oil contain elements such as nitrogen, oxygen, and sulfur, as well as trace amounts of metals such as iron, nickel, copper, and vanadium. The chemical composition of petroleum is heterogeneous.

The release of petroleum products into the environment is a serious threat to ecosystems (Fig. 8.1). The Exxon Valdez oil spill has shows that small spills have large impacts on ecosystems (Prabowo and Bae 2019). Crude oil and refined fuel from tanker ship accidents have destroyed natural ecosystems and human life in many parts of the world including Alaska, the Gulf of Mexico, and the Galapagos Islands. Oil spills at sea are more damaging than those on land because they can

Fig. 8.1 Mangroves are one of the responsive ecosystems to petroleum contamination in the ocean. Studies show that more than 1.94 million hectares of mangroves are affected by petroleum contamination. The oil coating on the aerial and submerged roots is the main threat to oil pollution near mangroves

spread hundreds of nautical miles in a thin layer of oil slick which can cover beaches with oil. Oil films in water can kill seabirds, mammals, shellfish, and other organisms. Oil spills on land can be more easily contained if a temporary earthen dam can be bulldozed quickly before most of the oil escapes. Heterocyclic nitrogen compounds such as pyridine, picoline, and quinoline are reported as contaminants associated with crude oil, and, have been found in oil shale or coal processing facilities. Naturally occurring bacteria such as *Micrococcus* spp., *Arthrobacter* spp., and *Rhodococcus* spp. have been shown to degrade these contaminants (Maila and Cloete 2002). The presence of petroleum in the environment is not necessarily due to human causes. Phenomena such as seeps and tar pits cause the release of petroleum into the environment without human involvement.

Crude oil contamination directly affects the growth and development of plants (da Silva Correa et al. 2022). It causes changes in the physiology, anatomy, and morphology of plants. Studies on the effects of oil contamination on plant growth led to the development of techniques such as bioremediation and biomonitoring (Maila and Cloete 2002; Khandare and Govindwar 2015; Wang et al. 2022). Phytoremediation is the use of plants to reduce the concentration or toxic effects of pollutants in the environment. Phytoremediation is widely recognized as a cost-effective environmental restoration technology. It is an alternative to engineering procedures that are usually more destructive to soils. However, phytoremediation is limited to the root zone of plants (Khandare and Govindwar 2015). Secondly, phytoremediation of contaminated sites should not exceed one decade to reach acceptable levels of contaminants. This technology has limited application where concentrations of contaminants are toxic to plants. Plant roots transport pollutants from the soil into the plant. The root zone of the plant is the key factor determining the rate of removal of contaminants during phytoremediation. Advantages of phytoremediation include cost effectiveness, suitability for remediation of large areas of soil, environmentally friendly nature, aesthetic nature, and low maintenance respectively.

Phytoremediation of petroleum often depends on the stimulation of petroleum hydrocarbon-degrading microbes in the rhizosphere (Allamin et al. 2020). Soil physico-chemical properties such as pH, moisture content, and mineral content were found to change in the course of phytoremediation (Devatha et al. 2019). Plants that grow in petroleum-contaminated sites are prone to toxicity due to oxidative stress. However, plants have been found to develop tolerance against petroleum through the operation of detoxification pathways. Prior selection of suitable native plants is critical before initiating time-consuming and costly greenhouse studies. Several studies have been conducted with plants such as *Bromus tectorum*, *Festuca arundinacea*, *Pteris vittata*, *Epipremnum aureum*, *Mucuna bracteata*, and *Imperata cylindrica* to explore their ability to remove petroleum from the environment. The utilization of local plants for phytoremediation helps to avoid difficulties in adaptation to growing plants in local climate and soil conditions. Studies with *Bassia scoparia*, *Salicornia persica*, and castor bean show that phytoremediation of petroleum-contaminated sites is a promising method (Moubasher et al. 2015).

Selection of plants that can grow well in the local environment is critical for the successful phytoremediation of petroleum-contaminated sites.

8.2 Plant Growth Environment in Petroleum Contaminated Sites

Petroleum contamination of soil affects the germination, growth, and health of plants. Therefore, remediation and removal of petroleum from the soil is necessary for a sustainable ecosystem. The soil conditions influence plant-microbe interactions. Therefore, an ideal soil for phytoremediation should have proper characteristics to ensure plant growth and biomass productivity. The efficiency of phytoremediation depends on factors such as soil type, water status, aeration, pH, and salinity of soil (Fig. 8.2). Petroleum-derived substances such as diesel oil, petrol, and engine oil can negatively affect the biochemical processes of soil and reduce soil fertility. Some of these derivatives are mixtures of different hydrocarbons serving as a carbon source for some microorganisms. However, contamination due to petroleum-derived substances can negatively impact the microbial properties of soil. It will affect the activity of enzymes such as phosphatase, dehydrogenase, and urease, and the occurrence of soil invertebrates (Allamin et al. 2020; Asif et al. 2022; Gospodarek et al. 2021; Silva Correa et al. 2022). Among the various enzymes, dehydrogenases are the best enzymes to determine soil microbial activity. They also reflect the degree of organic matter decomposition and nutrient availability in the soil. Petroleum products significantly modify the activity of soil enzymes analyzed even after a lapse of 5 years from the incident of contamination. Dehydrogenase seems to be a good indicator of soil contamination with petroleum products where the activity of this enzyme increases in the presence of diesel and decreased in the presence of petrol respectively (Kaczyńska et al. 2015). Crude oil in the field also results in a slight decrease in urease activity. The effect of phytoremediation using

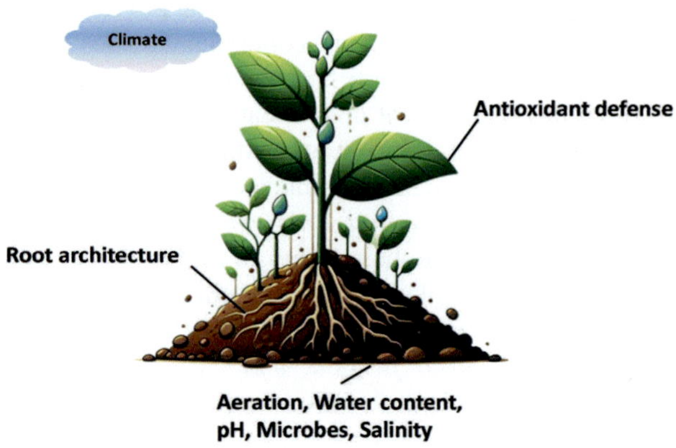

Fig. 8.2 Soil and plant-related factors affecting phytoremediation of petroleum products. The effectiveness of phytoremediation can vary depending on the specific site conditions. Therefore a holistic approach considering multiple environmental factors is often necessary for successful remediation of petroleum products

ZB-01 biopreparation on the activity of selected enzymes in the soil was seen to be variable (Gospodarek et al. 2020). The activity of antioxidant enzymes was found to decrease in plants growing in diesel-contaminated sites in the presence of ZB-01 biopreparation. The enzyme activities in plants used for phytoremediation of petroleum also depend on the type of enzymes, the time elapsed after the contamination, and the use of the preparation respectively (Mocek-Płóciniak et al. 2023).

The biological balance of the soil affected by the toxicity of petroleum compounds has been assessed by ecotoxicological assays using soil invertebrates alongside plants and bacteria (Zawierucha et al. 2022). Soil invertebrates including nematodes, springtails and arachnids are also used as bio-indicators of petroleum pollution. Petroleum is harmful to soil invertebrates because these substances permeate into the body, accumulate in fat tissues, and are released into their blood (Savin et al. 2015). This causes damage to their nervous and reproductive systems by precipitating body fluids. The toxic potential of hydrocarbons of petroleum results from their ability to bind to nonpolar components of biological membranes thereby causing disintegration of biological membranes. As a result species structure of soil invertebrates is found to be altered in the polluted environment. Many microorganisms transform petroleum-derived hydrocarbons into nontoxic compounds (Ajona and Vasanthi 2021). Natural microbial activity is used in the process of bioremediation, which can be considered an efficient method of soil purification from petroleum products. Studies have shown the usefulness of organisms belonging to the groups Hypogastruidae, Isotomidae, and Entomobryidae in assessing soil pollution with petroleum products and monitoring the progress of soil pollution with crude oil (Gospodarek et al. 2021). The biomass of earthworms was directly proportional to the amount of oil applied. The usage of chainsaws leads to the emission of polycyclic aromatic hydrocarbons in forest soil and affects the activities of enzymes in the soil which affects the survival and reproduction of earthworms. Hence it is clear that crude oil contamination also affects the earthworm population in the soil (Ekperusi et al. 2020).

Crude oil contamination can affect the physical and chemical properties of soil (Klamerus-Iwan et al. 2015). Crude oil-contaminated soils are hydrophobic compared with pristine sites. Petroleum pollution reduces the pores in the soil and thereby reduces soil aeration and water infiltration which will badly affect plant growth. It reported that crude oil contamination causes about a 4.0% decrease in air porosity which also increases the water resistance of soil aggregates (Adeniran et al. 2023). Aeration porosity is the pore spaces in the soil filled with air. Air pores are necessary for plant and microbial growth. The presence of air pores is necessary for the phytoremediation of crude oil-contaminated places because oxygen is required for aerobic respiration as well as aerobic biodegradation. Soil moisture content is another important factor when phytoremediation of petroleum products is considered. Soil moisture helps in hydrocarbon degradation through several direct and indirect mechanisms (Stepanova et al. 2022; Mocek-Płóciniak et al. 2023). For example, moisture content in the soil helps the movement of salts and nutrients for plant growth which is critical for phytoremediation.

Petroleum pollution can destroy marshland soil (Asif et al. 2022). The Momoge National Natural Reserve (MNNR) is one of the largest inland wetland and hydrosphere ecosystems in China (Wang et al. 2010). There are abundant crude oil resources beneath the MNNR with huge economic importance. Studies in MNNR showed that the presence of oil in the soil reduces soil fertility by reducing bioavailable phosphorous, and increasing soil pH (Zheng et al. 2022). It also reduces the water content in the soil and thereby reduces marsh water use efficiency in the MNNR where the prevalence of the semi-arid conditions leads to the water supply being relatively limited. The results of phytoremediation by planting *Curcuma angustifolia* in MNNR highlight that restoring and remediating the petroleum hydrocarbons-contaminated wetlands requires a longer time. However, the growth of *Curcuma angustifolia* resulted in the degradation of crude oil in the marsh soil ecosystem. Crude oil contamination alkalinizes the marsh soils due to the hydrophobic nature of the crude oil. The presence of oil increased salinity and pH via the accumulation of exchangeable bases such as Ca^{2+}, and Na^{+}, and a reduction in effective cation exchange capacity respectively.

8.3 Phytotoxicity of Petroleum Products

Petroleum contamination harms plant growth and development. Toxic petroleum hydrocarbons present in agricultural soil minimize crop development and metabolic process associated with plant growth. Extraction and exploration of petroleum hydrocarbons to satisfy the rising fossil fuel demand plays havoc with the ecosystem. Petroleum products deteriorate the physiochemical and biochemical properties of soil which in turn restrict plant growth and productivity (Daâssi and Almaghribi 2022). The major changes in soil properties due to petroleum-derived compounds are shortages of oxygen, water, and essential nutrients such as phosphorus and nitrogen respectively (Asif et al. 2022; Mitter et al. 2021).

Studies indicate that the toxicity of petroleum hydrocarbons significantly alters the physiochemical properties of soils, microbial populations, and nutrient uptake in plants. Microbes are the key components of terrestrial ecosystems that act by fostering nutrient cycling, improving fertility, and contributing to the regulation of the energy flow from decomposer food webs to higher food webs (Zhang et al. 2021). Studies showed that the composition of the soil microbiome helps plants recover from high kerosene doses (Shelyakin et al. 2022). It is generally assumed that there is a two-way positive relationship between plants and bacteria during the phytoremediation of oil pollution. For example, plants stimulate microbiological activity and plant-associated microorganisms. Plant-microbial interactions reduce environmental harshness for plants, and improve their growth during phytoremediation (Zhang et al. 2021; Lednev et al. 2023). The latter not only supported the relationship between microbiome and vegetation recovery after hydrocarbon fuel pollution but also lead to thinking about the priority role of each of the components of the ecosystem in this process. When petroleum hydrocarbons are taken by plants, they enter the food chain and directly or indirectly affect ecosystems. Decline in microbial population during petroleum hydrocarbon contamination is associated with an increase in the C/N ratio, and the resulting nitrogen deficiency causes retardation of growth and development of plants. Studies with mangroves showed that crude oil contamination significantly reduced the availability of macronutrients phosphorus and potassium and resulted in nutrient stress in cultivars with frequent changes in defense mechanisms (Orji 2013).

The pollutants present in petroleum-derived substances decrease the activity of the microbes in the soil. Hydrocarbons in the petroleum serve as a carbon source for some microorganisms. Microbes can convert petroleum-derived hydrocarbon into nontoxic components. This natural microbial activity is used in the process of bioremediation which is an efficient method of soil decontamination. However, environmental pollution by hydrocarbon is a major ecological problem because hydrocarbon fuels (such as kerosene, petrol, and diesel) are known to be phytotoxic. Research shows that petroleum hydrocarbon inhibits seed germination, seedling development, and photosynthesis activity in plants (Haider et al. 2021). Soil and air are the two well-known exposure mediums of petroleum hydrocarbons in plants. Roots are closely in contact with the pollutants in their surrounding soil, and are considered an important factor that significantly affects the uptake and translocation of petroleum hydrocarbons. The majority of petroleum hydrocarbons are absorbed by roots translocated to aerial parts. However, the uptake of organic pollutants by plants from polluted soils is influenced by a series of abiotic and biotic factors (Begum et al. 2022). The abiotic components in the soil such as iron oxides and dead organic matter content influenced the plant uptake of petroleum hydrocarbons in the contaminated soil. Biotic factors such as transpiration, lipid-protein content, and density of root biomass in the rhizosphere also affect the accumulation of petroleum hydrocarbons in plants. Roots are closely in contact with the phytotoxic petroleum pollutants in the surrounding soil. Root changes both the composition of tissues and water content in response to petroleum pollutants. Inhibition of root growth and hypocotyl elongation may reflect the toxicity of petroleum hydrocarbons in low concentrations. The degradation of petroleum products and biodiesel increases soil acidity due to the breakdown of fatty acid methyl ester molecules (Badzinski et al. 2021; Haider et al. 2021). Therefore, plants that grow in biodiesel-contaminated sites confront toxicity associated with low soil pH too.

Incomplete combustion of fossil fuels produces polycyclic aromatic hydrocarbons. These compounds are sufficient to disturb and inhibit the growth of plants. Recent studies show that threshold levels of hydrocarbon fuel pollution cause a significant impact on vegetation (Lednev et al. 2023). Toxicity due to petroleum hydrocarbons significantly decreases the growth of root hairs and trichomes in plants. Late flowering, chlorosis, and the appearance of white spots that later developed into necrotic lesions have also noticed in plants exposed to petroleum hydrocarbons (Alkio et al. 2005). Additionally, total area, leaf number, stem girth, plant height, total biomass, and grain yield have also been affected in crop plants exposed to petroleum hydrocarbons. Kerosene has been found to affect the growth of plant species including herbs, shrubs, and moss layers of forest and bog communities

(Lednev et al. 2023). Plants have shown immediate reactions towards petroleum products such as necrosis of leaves. Raised bogs, large erect woody shrubs, and mosses are more resistant to the effect of kerosene. In the latter study mosses were seen less affected by petroleum when compared to other plants. However herbaceous plants and shrubs were affected by petroleum pollutants and completely disappeared from the contaminated lands. The phytotoxic effect of kerosene was mainly manifested by tissue necrosis and the partial defoliation of shoots. Some plants had chemical burns in the presence of kerosene. Ferns are very susceptible to kerosene, and can be used as a pollution indicator. Ferns were found to be disappearing from forested area after exposure to low levels of hydrocarbon pollution.

Plants grown in soil polluted with petroleum hydrocarbons show various anomalies in the growth and agronomic indices (Fig. 8.3). The main symptoms shown by plants grown in contaminated soil are a reduction in chlorophyll contents, starch metabolism, and antioxidant enzyme activity (Skrypnik et al. 2021). Lipoproteins of plants incorporate petroleum products, and decrease the germination rate of different plant species. Accumulation of hydrocarbons also causes leaf deformation due to necrosis of tissue and cells. Petroleum hydrocarbons cause molecular level alterations in plants such as enzymatic dysfunction. Hydrocarbons in petroleum interfere with the arrangement of chloroplasts and the transport of electrons in photosynthesis (Chai et al. 2023). Petroleum hydrocarbons can covalently bind with proteins, DNA, and RNA, due to hydrophilicity and electrophilicity respectively.

Fig. 8.3 *Chamaecostus cuspidatus*es before (**a**) and after treatment with petroleum (**b**). The plant is found in tropical climates, and the plant can be found growing densely in gardens of southern India, where it is nicknamed the insulin plant. The plant is a perennial, erect, spreading plant up to two feet tall, with the tallest stem drooping and resting on the ground. Chlorosis was one of the responses of *Chamaecostus* to petroleum. However, petroleum treatment is mainly found to affect water transport processes because of the collapse of tissues near the root

However, the phytotoxicity of petroleum is mainly due to oxidative stress in plants which decreases chlorophyll content and biomass. Petroleum hydrocarbons affect both the primary and secondary photosynthesis process. In petroleum hydrocarbons-polluted soils, the concentration of malondialdehyde (MDA) in leaves studied increased indicating petroleum hydrocarbons-induce oxidative stress in plants (Skrypnik et al. 2021). Also, the concentration of organic acids and amino acids has been seen to be significantly reduced in plants cultivated in petroleum-contaminated soils. Photosynthesis is one of the most important metabolic processes in plants. It is extremely stress-sensitive and closely associated with biomass. Photosynthesis activity is significantly affected and reduced due to the presence of petroleum hydrocarbons. Some petroleum compounds, such as anthracene, impede photosynthetic activities in green algae due to the reduction in quantum efficiency of PSII and PSI respectively (Kottuparambil and Park 2019). Other petroleum compounds such as phenanthrene and pyrene significantly reduce photosynthetic activity through a decrease in conductance of stomata and photosynthesis pigment synthesis respectively.

8.4 Petroleum Tolerance in Plants

The release of crude oil and its constituents can have harmful effects on ecological health. In early times, one of the assumptions used in the derivation of the safety guidelines was that plants were unable to take up Petroleum hydrocarbons. Nowadays studies suggest that plants are indeed able to accumulate petroleum hydrocarbons in their tissues (Hunt et al. 2019). Various plant species, including some edible plants, were shown to uptake petroleum hydrocarbons from the contaminated soil and aqueous media in both laboratory and field studies. Petroleum hydrocarbons are chemical compounds either found within or originate from crude oil. The compounds such as benzene, toluene, and xylene (which are present in gasoline) can affect plant life. Crude oil (petroleum) hydrocarbons are one of the most common groups of persistent organic pollutants. Petroleum hydrocarbons are known to be toxic to many living organisms due to their carcinogenic properties. The low rate of decomposition of oil and oil products in the environment triggers the accumulation and gradual increase of petroleum hydrocarbons in the environment including soil (Lednev et al. 2023; Lv et al. 2022). After getting into soil, crude oil products destroy soil structure, upset the air balance, and alter the soil physicochemical properties that can inhibit microbial proliferation, disrupt soil enzymatic activity, and harm terrestrial plant growth and development. Firstly, the absorption of toxic petroleum molecules by plants can modify the permeability and structure of their plasma membrane, alter the shape and size of the parenchyma tissues, reduce the intercellular space in the cortex of the stem and roots, and inhibit the mitotic activity of the root meristem. Secondly, insufficient aeration caused by air displacement from the pore spaces between the soil particles by crude oil lead to stress and low water availability to plants (Skrypnik et al. 2021). Moreover, oil pollution minimizes the percentage of organic matter available to plants and reduces the amount of mineral nutrients such as potassium, sulfates, and nitrates.

The response of plants to oil pollution can manifest itself at various levels (Table 8.1). Detailed information about the oil contamination on photosynthetic activity is scarce. For example, it has recently been reported that crude oil pollution reduces overall photosynthetic activity and chlorophyll content in plants. Notable symptoms observed in plants growing on oil-polluted soil also include a decrease in the activity of starch metabolizing enzymes and a decrease in the content of total carbohydrates, total proteins, and total amino acids respectively (Skrypnik et al. 2021). The most important disorder resulting from the effect of pollutants, including crude oil and petroleum hydrocarbons on plants is oxidative stress which leads to the formation of reactive oxygen species (ROS) such as superoxide radicals, H_2O_2, and hydroxyl radicals. In one respect, ROS destroy cell-membrane complexes, disrupt transport processes and intercellular reactions, and thereby inhibit plant growth activity. Patowary et al. (2017b) determined that hydrocarbons present at sites contaminated with crude oil are transferred to roots, shoots, and grains of rice crops (*Oryza sativa*) grown at those sites. In this study, the soil was artificially contaminated with crude oil, and it was followed by the planting of rice seedlings. After 4 months of cultivation in contaminated soil, rice plants were harvested and washed thoroughly with tap water followed by the separation of the root, shoot, grains, and husks. The hydrocarbons in plant samples were extracted with dichloromethane (TPH) and analyzed with GC-MS. The abundance of hydrocarbon-degrading genes was higher in phenanthrene-contaminated soils planted with ryegrass (Li et al. 2011). The plants were found to favor the activities of bacterial degraders belonging to the Pseudomonadales, Actinobacteria, Caulobacterales, Rhizobiales, and Xantomonadales. Similarly, the presence of ryegrass was shown to stimulate the expression of bacterial dioxygenase genes such as nidA3, Pdoa, nahAC, and phnAC which hydroxylate aromatic rings of hydrocarbons (Guo et al. 2017; Li et al. 2011).

The tolerance of plants to petroleum was studied in many plant species. The physiological response of two petroleum petroleum-tolerant plant species *Mirabilis jalapa* and *Orychophragamus violace* was investigated in petroleum-contaminated soil (Peng et al. 2009). The study was conducted on soil contaminated with 10.0,

Table 8.1 Spatial distribution of petroleum accumulation in various plant species

Plant species	Hydrocarbon	Plant part	Amendment	References
Scirpus grossus	Diesel	Stem and root	None	Al-baldawi et al. (2015)
Rhizophora mangle	Bunker oil	Whole plant	None	Suprayogi and Murray (1999)
Centrosema pubescence	Crude oil	No accumulation	Chicken manure	Nwaichi et al. (2011)
Lycopersican esculentum	Crude oil	Stem and root	None	Morelos-Moreno et al. (2019)
Elusine indica	Crude oil	No accumulation	None	Lu et al. (2010)
Zea mays	Crude oil	Whole plant	Tween 80	Liao et al. (2016)

20.0, and 40.0 g petroleum per kg soil for 120d. Petroleum degradation rate, germination rate, free proline and superoxide dismutase and peroxidase activities of *M. jalapa* were higher than that of *O. violoace* under petroleum stress. However, the decrease in the soluble proteins, plant height, chlorophyll, and fresh weight of roots was greater in *O. violence* as compared to *M. jalapa*. Pearson correlation coefficient analysis showed that the higher petroleum tolerance of *M. jalapa* was correlated with the higher level of free proline and antioxidant enzyme activities. Segura et al. (2009) found that polycyclic aromatic hydrocarbon is more rapidly degraded in the rhizosphere than in the bulk soil. Stephanova et al. (2022) suggested that the combined use of plants and biologically active bacteria could facilitate the removal of petroleum products. Plant roots can release simple sugars, amino acids, enzymes, aliphatic, and aromatics that increase microbial counts which in turn promote degradation of petroleum hydrocarbons.

Applying ornamental plants for the remediation of petroleum hydrocarbon-contaminated soils is also a practical way to remove petroleum contaminants from soil. Lu et al. (2010) studied petroleum uptake by *Eleusine indica* which is an agricultural and environmental weed. Seeds were planted in pots containing soil spiked with crude oil consisting of 12.5% asphaltene, 57.2% aliphatic hydrocarbons, and 13.7% aromatic hydrocarbons. The pots were incubated in a greenhouse under controlled light conditions of 14.0 hours of light and 10.0 hrs of darkness. Fertilizer containing NPK was applied to all the treatment groups. After 5 months of growth, the plants were harvested and washed with tap water and separated into above-ground and below-ground biomass. Petroleum was extracted from plant tissue by soxhlet extraction using dichloromethane, and the petroleum hydrocarbons were analyzed by gas chromatography-mass spectrometry. No detectable concentration of saturated hydrocarbons was found in the plant material. Also, hydrocarbons were detected in the root tissues but not in the shoot. Accumulation of hydrocarbons in the roots ranged between 0.11 and 438.0 mg/kg. The accumulation of total measured petroleum hydrocarbons in the plant tissue was 1.4% of that present in the soil. The smaller compounds were most likely volatilized or metabolized by plants, and these compounds were virtually not present in plant tissue. Patowary et al. (2017a) found that hydrocarbons present at sites contaminated with crude oil are transferred to roots, shoots, and grains of rice crops. In this study, the soil was artificially contaminated with crude oil at concentrations up to 15,000.0 mg/kg and it was followed by planting of rice seedlings. After 4 months of cultivation in contaminated soil, rice plants were harvested and washed thoroughly with tap water, followed by the separation of the roots, shoots, grains, and husks. It was concluded that the uptake of hydrocarbons by the plants gradually increased as the concentration of oil increased in the soil. This study demonstrated that rice plants grown in the soil contaminated with crude oil can uptake petroleum hydrocarbons which could pose health hazards to humans and grazing animals. Analysis of the biochemical parameters of the plants such as photosynthetic pigments, malondialdehyde, hydrogen peroxide, proline, non-protein and protein thiols, ascorbic acid, dehydroascorbic acid and 2,3,-diketogulonic acids and antioxidant enzyme activity were performed. The results showed that regulation of redox–homeostasis in plants in the conditions of oil pollution is based on the activation of the ant oxidative system, which includes several low molecular weight compounds and antioxidant enzymes.

Proline accumulation in plants can regulate the osmotic potential of cells, stabilize the cell structure, and remove excess reactive oxygen species that improve the resistance of plants to environmental stress. The studies in rye shoots showed that proline increased with an increase in the oil concentration in the soil (Skrypnik et al. 2021). Additionally, studies reveal a decrease in proline levels in the leaves of the plant in the course of petroleum exposure. Phenolic compounds in rye varieties increased with an increase in the oil concentration of the soil. Plants are known to enhance the synthesis of polyphenols such as phenolic acid and flavonoids under abiotic stress conditions to resist adverse environmental factors. It is known that plant resistance to free radicals is also associated with an increase in the activities of antioxidant enzymes after exposure to pollutants. Non-enzymatic antioxidants such as non-protein thiols including glutathione and cysteine are involved in several detoxification processes in the cell. Ascorbic acid is an important component of the Asada-Halliwell cycle of plant anti-oxidant defense. The increase in the content of ascorbic acid in shoots of rye varieties indicates that ascorbic acid plays an important role in cellular ROS homeostasis regulation during growth in crude oil-contaminated soil (Peng et al. 2009).

De novo transcription analysis of plants grown in oil-contaminated soil showed a total of 2.4 billion reads that belonged to 68,191 trinity genes. Out of the total 7275 unique genes that were identified as significantly differentially expressed, 6536 were assigned protein identity whilst 739 were classified as unknown. The most prominent plant species was *Sarracenia purpurea* as expected with 41.0% of all unique differentially expressed genes. The rhizosphere microbes, especially bacteria are thought to be major players in organic contaminant degradation during rhizoremediation (Bell et al. 2014). Recent plant-microbe metatranscriptomic studies confirmed that the hydrocarbon degradation genes expressed in the root-rhizosphere environment were mostly linked to bacteria (Correa-García et al. 2018; Peng et al. 2009).

8.5 Diversity of Petroleum-Tolerant Plants

Plants play a major role in the remediation of petroleum-contaminated soils (Liu et al. 2012). Assessment of plants grown at petroleum-contaminated sites is important to identify plants that tolerate petroleum contamination. Phytoremediation is one of the efficient and environmentally friendly means of restoring contaminated soil. Phytoremediation takes advantage of the ability of plants to accumulate elements and compounds from the environment. Some plants are called hyperaccumulators that can accumulate relatively high concentrations of contaminants. However, phytoremediation very often works best where contaminant levels are low because high concentrations may limit plant growth (Panchenko et al. 2023). The use of plants to clean up the environment also helps to prevent groundwater pollution.

Some common examples of plants that are used in the phytoremediation of petroleum compounds are *Phargmites australis, Thlaspi caerulcens,* and *Festuca arundinacea.* These plants have potential for phytoremediation of petroleum hydrocarbon-contaminated soils as they grow in soils with high petroleum hydrocarbon concentrations. The screening of plants at contaminated sites has shown potential plants for phytoremediation of petroleum hydrocarbons. Examples of plants that are used for the phytoremediation of oil-contaminated sites are given in Table 8.2. Plant growth has been shown to encourage organic reduction of contaminants principally by providing the optimal environment for microbial proliferation in the root zone. Studies found 6 families and 11 species that dominate the sludge farm can accumulate petroleum (Jagaba et al. 2022). These plants also exhibited an inherent genetic diversity that may give them a competitive advantage in becoming established under unfavorable conditions. Some examples of plant species sampled at the contaminated sites include *Paspulum vaginatum* (Graminea), *Caryatia trifoila* (Vitaceae), *Melochia corchorifila* (Malvanceae).

Most of the plants in the family Graminae can survive in a contaminated site with high concentrations of hydrocarbon (Liu et al. 2012). The hyper-accumulating plants have a relatively large biomass concentration of petroleum in the tissue than petroleum in the soil. Petroleum hyper-accumulating plants should germinate and grow in contaminated sites. The petroleum tolerance of plants is achieved with the help of metabolic changes associated with hormonal and enzyme activities. The plants were reported to tolerate petroleum concentrations in the soil ranging from ranging from 224.0–177.0 mg kg^{-1}, 595.0 mg kg^{-1} and 264.0–7268.0 mg kg^{-1}, respectively (Idris et al. 2016; Tang and Angela 2019). Plants belonging to the family Graminae have fibrous root systems with large surface areas and provide more contact with contaminants in the soil for phytoremediation. This feature also allows the colonization of petroleum-tolerant microbes in the rhizosphere which is imperative in enhancing phytoremediation of petroleum hydrocarbons. Plants that can grow under low water and aeration are ideal for the phytoremediation of petroleum because contamination of hydrocarbons often affects the infiltration of water and air in the soil. Also, a high antioxidant enzyme activity is seen in petroleum-tolerant plant species because volatile compounds in petroleum cause membrane damage.

The biodegradation of petroleum contaminants in the soil by plants is limited by factors such as adsorption, mass transfer and bioavailability, and bioaccessibility as well as the fate and behavior of the hydrocarbon. However, the presence of microbes in the rhizosphere may help to overcome the above limiting factors of phytoremediation. So the synergetic effect of the plant-microbe association is an important factor in the phytoremediation of petroleum-contaminated soils. The petroleum-contaminated soils had a high C/N ratio (Wang et al. 2022). Therefore, leguminous plants having symbiotic association with nitrogen-fixing microbes are ideal for phytoremediation of petroleum-contaminated sites. Studies showed that leguminous plants namely, *Cajanus cajan* and *Lablab purpureus* are effective in the removal of hydrocarbons from engine oil-contaminated sites. Local plant species are ideal for the phytoremediation of petroleum products because these plants adapt well to local climate and soil conditions resulting in proper growth in polluted sites. Studies with

Table 8.2 Plants with higher levels of petroleum tolerance

Plant name	Family	References
Epipremnum aureum	Araceae	Tangand and Angela (2019)
Imperata cylindrical	Poaceae	Khalid et al. (2021)
Pteris vittata	Pteridaceae.	Tangand and Angela (2019)
Mucuna bracteata	Leguminosae	Tangand and Law (2019)
Agropyron smithii	Poaceae	Besalatpour et al. (2010)
Festuca arundinacea	Poaceae	Besalatpour et al. (2010)
Bassia scoparia	Amaranthaceae	Moubasher et al. (2015)
Cynodon dactylon	Poaceae	Kathi (2022)
Lemna gibba	Araceae	Ekperusi et al. (2020)
Medicago sativa	Leguminosae	Rafique et al. (2023)
Panicum coloratum	Poaceae	Anukwa et al. (2020)
Lolium perenne	Poaceae	Wyszkowska et al. (2019)
Lolium multiflorum	Poaceae	Alarcon et al. (2006)

local plant species for the removal of petroleum in a contaminated site in Malaysia showed that *Pteris vittata* efficiently removes petroleum from the contaminated site (Idris et al. 2016). Also, *Epipremnum aureum* which is an indoor ornamental plant had a higher potential for rhizofilteration of crude oil components such as Co-60 and Cs-137. The transport index was 13.8 for Co-60 and 35.6 for Cs-137 respectively. Plants such as *Mucuna bracteata* and *Imperata cylindrica* are also reported as potential plants for the phytoremediation of crude oil (Tang and Angela 2019). Shrubs belonging to the Asteraceae and Cyperaceae families were dominant in a site contaminated with crude oil (Panchenko et al. 2018). These plants had extensively branched and deep root systems as a characteristic feature that helped in the phytoremediation of petroleum products.

8.6 Conclusion

Phytoremediation is a promising approach to clean up petroleum-contaminated soils. The harsh environmental conditions in the petroleum contaminated sites such as lack of aeration retard plant growth. Activities of soil enzymes and ecotoxicological assays are feasible to monitor petroleum contamination in the soil. Plants that are exposed to petroleum often show symptoms such as a decrease in photosynthesis. However, petroleum-tolerant plants resist petroleum toxicity with the help of compatible solutes and antioxidant enzyme activity. The breakdown pathways of petroleum products in the plants need to be explored in detail. The synergetic effects of microbial association with plants are crucial for the phytoremediation of petroleum-contaminated sites. The selection of native plants is a critical factor that influences the success of phytoremediation of petroleum hydrocarbons in the soil.

8.7 Suggestions

In order to make effective use of plants in phytoremediation the following suggestions are made:

I. Phytoremediation of petroleum-contaminated soil is a sustainable practice that helps to remove petroleum in the soil
II. The environmental factors affecting the phytodegradation of petroleum must be characterized for successful phytoremediation
III. It is critical to study the tolerance mechanisms of various plant species against petroleum products which will help to develop transgenic plants with petroleum hyperaccumulation capacity
IV. A study on plant diversity in petroleum-contaminated soil is also feasible to adopt successful phytoremediation practices

References

Adeniran MA, Oladunjoye MA, Doro KO (2023) Soil and groundwater contamination by crude oil spillage: a review and implications for remediation projects in Nigeria. Front Environ Sci 11. https://doi.org/10.3389/fenvs.2023.1137496

Ajona M, Vasanthi P (2021) Bioremediation of petroleum contaminated soils—a review. Mater Today Proc 45:7117–7122. https://doi.org/10.1016/j.matpr.2021.01.949

Alarcon A, Davies FT, Autenrieth RL, Reed DW, Zuberer DA (2006) Phytoremediation of petroleum hydrocarbons with a Loliummultiflorum–Glomusintraradices–inorganic fertilization system: influence on plant growth, antioxidant activity, microbial respiration, and hydrocarbon degradation. HortScience: A Publication of the American Society for Horticultural Science 41(4):1059C–1059. 10.21273/hortsci.41.4.1059c

Al-Baldawi IA, Abdullah SRS, Anuar N, Suja F, Mushrifah I (2015) Phytodegradation of total petroleum hydrocarbon (TPH) in diesel-contaminated water using Scirpusgrossus. Ecol Eng 74:463–473. https://doi.org/10.1016/j.ecoleng.2014.11.007

Alkio M, Tabuchi TM, Wang X, Colón-Carmona A (2005) Stress responses to polycyclic aromatic hydrocarbons in Arabidopsis include growth inhibition and hypersensitive response-like symptoms. J Exp Bot 56(421):2983–2994. https://doi.org/10.1093/jxb/eri295

Allamin IA, Halmi MIE, Yasid NA et al (2020) Rhizodegradation of petroleum oily sludge-contaminated soil using *Cajanuscajan* increases the diversity of soil microbial community. Sci Rep 10:4094. https://doi.org/10.1038/s41598-020-60668-1

Anukwa FA, Onuoha EM, Nkang A, Nkereuwem J (2020) Phytoremediation potential of Zea mays L. and Panicumcoloratum L. on hydrocarbon polluted soils. Int J Botan 17(1):1–7. https://doi.org/10.3923/ijb.2021.1.7

Ashoori I, Sharifi M, Masoumi M, Salehi MM (2017) The relationship between SARA fractions and crude oil stability. Egypt J Pet 26:209–213. https://doi.org/10.1016/j.ejpe.2016.04.002

Asif Z, Chen Z, An C, Dong J (2022) Environmental impacts and challenges associated with oil spills on shorelines. J Mar Sci Eng 10(6):762. https://doi.org/10.3390/jmse10060762

Badzinski C, Ferraz Ramos R, Godoi B, Daroit DJ (2021) Soil acidification and impacts over microbial indicators during attenuation of soybean biodiesel (B100) as compared to a diesel-biodiesel blend (B8). Fuel (London, England) 289(119989):119989. https://doi.org/10.1016/j.fuel.2020.119989

Begum M, Sarmah B, GoswamiKandali G, Kalita S, Ojha I, Bhagawati R, Talukdar L (2022) Persistant organic pollutants in soil and its phytoremediation. In: Biodegradation technology of organic and inorganic pollutants. IntechOpen

Bell TH, El-Din Hassan S, Lauron-Moreau A, Al-Otaibi F, Hijri M, Yergeau E, St-Arnaud M (2014) Linkage between bacterial and fungal rhizosphere communities in hydrocarbon-contaminated soils is related to plant phylogeny. ISME J 8(2):331–343. https://doi.org/10.1038/ismej.2013.149

Besalatpour AA, Hajabbasi MA, Khoshgoftarmanesh AH (2010) Reclamation of a petroleum-contaminated calcareous soil using phytostimulation. Soil Sediment Contamin 19(5):547–559. https://doi.org/10.1080/15320383.2010.499920

Chai X, Wang M, Fu X, Zhang W, Huang Y, Germaine KJ, Wang J (2023) Shift of combined eco-toxicity index in petroleum polluted soils during a bacterial remediation. Front Environ Sci 11. https://doi.org/10.3389/fenvs.2023.1141562

Correa-García S, Pande P, Séguin A, St-Arnaud M, Yergeau E (2018) Rhizoremediation of petroleum hydrocarbons: a model system for plant microbiome manipulation. Microb Biotechnol 11(5):819–832. https://doi.org/10.1111/1751-7915.13303

da Silva Correa H, Blum CT, Galvão F, Maranho LT (2022) Effects of oil contamination on plant growth and development: a review. Environ Sci Pollut Res Int 29:43501–43515. https://doi.org/10.1007/s11356-022-19939-9

Daâssi D, QabilAlmaghribi F (2022) Petroleum-contaminated soil: environmental occurrence and remediation strategies. 3 Biotech 12(6). https://doi.org/10.1007/s13205-022-03198-z

Devatha CP, Vishnu Vishal A, Chandra P, Rao J (2019) Investigation of physical and chemical characteristics on soil due to crude oil contamination and its remediation. Appl Water Sci 9:89. https://doi.org/10.1007/s13201-019-0970-4

Ekperusi AO, Nwachukwu EO, Sikoki FD (2020) Assessing and modelling the efficacy of Lemnapaucicostata for the phytoremediation of petroleum hydrocarbons in crude oil-contaminated wetlands. Sci Rep 10(1). https://doi.org/10.1038/s41598-020-65389-z

Gospodarek J, Rusin M, Nadgórska-Socha A (2020) The long-term effect of petroleum-derived substances and their bioremediation on the host plant (Viciafaba L.) and a herbivore (Sitona spp.). Agronomy (Basel, Switzerland) 10(8):1066. https://doi.org/10.3390/agronomy10081066

Gospodarek J, Rusin M, Barczyk G, Nadgórska-Socha A (2021) The effect of petroleum-derived substances and their bioremediation on soil enzymatic activity and soil invertebrates. Agronomy (Basel, Switzerland) 11(1):80. https://doi.org/10.3390/agronomy11010080

Guo M, Gong Z, Miao R, Rookes J, Cahill D, Zhuang J (2017) Microbial mechanisms controlling the rhizosphere effect of ryegrass on degradation of polycyclic aromatic hydrocarbons in an aged-contaminated agricultural soil. Soil Biol Biochem 113:130–142. https://doi.org/10.1016/j.soilbio.2017.06.006

Haider FU, Ejaz M, Cheema SA, Khan MI, Zhao B, Liqun C, Salim MA, Naveed M, Khan N, Núñez-Delgado A, Mustafa A (2021) Phytotoxicity of petroleum hydrocarbons: sources, impacts and remediation strategies. Environ Res 197(111031):111031. https://doi.org/10.1016/j.envres.2021.111031

Hunt LJ, Duca D, Dan T, Knopper LD (2019) Petroleum hydrocarbon (PHC) uptake in plants: a literature review. Environ Pollut (Barking, Essex: 1987) 245:472–484. https://doi.org/10.1016/j.envpol.2018.11.012

Idris M, Malaysia UK, Abdullah SRS, Titah HS, Latif MT, Abasa AR, Husin AK, Hanima RF, Ayub R (2016) Screening and identification of plants at a petroleum contaminated site in Malaysia for phytoremediation. J Environ Sci Manag 19(1):27–36. https://doi.org/10.47125/jesam/2016_1/04

Jagaba AH, Kutty SRM, Lawal IM, Aminu N, Noor A, Al-dhawi BNS, Usman AK, Batari A, Abubakar S, Birniwa AH, Umaru I, Yakubu AS (2022) Diverse sustainable materials for the treatment of petroleum sludge and remediation of contaminated sites: a review. Clean Waste Syst 2(100010):100010. https://doi.org/10.1016/j.clwas.2022.100010

Juwarkar AA, Jambhulkar HP (2008) Phytoremediation of coal mine spoil dump through integrated biotechnological approach. Bioresour Technol 99:4732–4741. https://doi.org/10.1016/j.biortech.2007.09.060

Kaczyńska G, Borowik A, Wyszkowska J (2015) Soil dehydrogenases as an indicator of contamination of the environment with petroleum products. Water Air Soil Pollut 226(11). https://doi.org/10.1007/s11270-015-2642-9

Kathi S (2022) Phytoremediation of heavy metals and petroleum hydrocarbons using Cynodondactylon (L.)Pers. In: Cost effective technologies for solid waste and wastewater treatment. Elsevier, pp 135–145

Khalid FE, Ahmad SA, Zakaria NN, Shaharuddin NA, Sabri S, Azmi AA, Khalil KA, Verasoundarapandian G, Gomez-Fuentes C, Zulkharnain A (2021) Application of cogon grass (Imperatacylindrica) as biosorbent in diesel-filter system for oil spill removal. Agronomy (Basel, Switzerland) 11(11):2273. https://doi.org/10.3390/agronomy11112273

Khandare RV, Govindwar SP (2015) Phytoremediation of textile dyes and effluents: current scenario and future prospects. Biotechnol Adv 33:1697–1714. https://doi.org/10.1016/j.biotechadv.2015.09.003

Klamerus-Iwan A, Błońska E, Lasota J, Kalandyk A, Waligórski P (2015) Influence of oil contamination on physical and biological properties of forest soil after chainsaw use. Water Air Soil Pollut 226(11). https://doi.org/10.1007/s11270-015-2649-2

Kottuparambil S, Park J (2019) Anthracenephytotoxicity in the freshwater flagellate alga Euglena agilisCarter. Sci Rep 9(1). https://doi.org/10.1038/s41598-019-51451-y

Lednev SA, Semenkov IN, Koroleva TV (2023) Phytotoxic effects of kerosene on plants of forest and bog phytocenoses of southern taiga. Forests 14(5):873. https://doi.org/10.3390/f14050873
Li JH, Yu XZ, Wu SC, Wang XR, Wang SH, Tam NFY, Wong MH (2011) Responses of bioaugmented Ryegrass to pah soil contamination. Int J Phytoremediation 13(5):441–455. https://doi.org/10.1080/15226510903353104
Liao C, Xu W, Lu G, Deng F, Liang X, Guo C, Dang Z (2016) Biosurfactant-enhanced phytoremediation of soils contaminated by crude oil using maize (Zea mays. L). Ecol Eng 92:10–17. https://doi.org/10.1016/j.ecoleng.2016.03.041
Liu R, Jadeja RN, Zhou Q, Liu Z (2012) Treatment and remediation of petroleum-contaminated soils using selective ornamental plants. Environ Eng Sci 29(6):494–501. https://doi.org/10.1089/ees.2010.0490
Lu M, Zhang Z, Sun S, Wei X, Wang Q, Su Y (2010) The use of goosegrass (Eleusineindica) to remediate soil contaminated with petroleum. Water Air Soil Pollut 209(1–4):181–189. https://doi.org/10.1007/s11270-009-0190-x
Lv Y, Bao J, Zhu L (2022) A comprehensive review of recent and perspective technologies and challenges for the remediation of oil-contaminated sites. Energy Rep 8:7976–7988. https://doi.org/10.1016/j.egyr.2022.06.034
Maila MM, Cloete TM (2002) Germination of Lepidiumsativum as a method to evaluate polycyclic aromatic hydrocarbons (PAHs) removal from contaminated soil. Int Biodeterior Biodegrad 50:107–113. https://doi.org/10.1016/S0964-8305(02)00059-8
Mitter EK, Germida JJ, de Freitas JR (2021) Impact of diesel and biodiesel contamination on soil microbial community activity and structure. Sci Rep 11(1). https://doi.org/10.1038/s41598-021-89637-y
Mocek-Płóciniak A, Mencel J, Zakrzewski W, Roszkowski S (2023) Phytoremediation as an effective remedy for removing trace elements from ecosystems. Plants 12(8):1653. https://doi.org/10.3390/plants12081653
Morelos-Moreno A, Martel-Valles JF, Morales-Díaz AB, Foroughbakhch-Pournavab R, Morales I, Benavides-Mendoza A (2019) Response of tomato plants to diesel fuel, gasoline and benzene. Terra Latinoamericana: OrganoCientifico de La Sociedad Mexicana de La Ciencia Del Suelo, AC 37(4):425. https://doi.org/10.28940/terra.v37i4.543
Moubasher HA, Hegazy AK, Mohamed NH, Moustafa YM, Kabiel HF, Hamad AA (2015) Phytoremediation of soils polluted with crude petroleum oil using Bassiascoparia and its associated rhizosphere microorganisms. Int Biodeterior Biodegradation 98:113–120. https://doi.org/10.1016/j.ibiod.2014.11.019
Nwaichi EO, Osuji LC, Onyeike EN (2011) Evaluation and decontamination of crude oil-polluted soils using*Centrosemapubescen*Benth and amendment-support options. Int J Phytoremediation 13(4):373–382. https://doi.org/10.1080/15226514.2010.495144
Orji (2013) Bioremediation of petroleum hydrocarbon-polluted mangrove swamps using nutrient formula produced from water hyacint (eicchorniacrassipes). Am J Environ Sci 9(4):348–366. https://doi.org/10.3844/ajessp.2013.348.366
Panchenko L, Muratova A, Dubrovskaya E, Golubev S, Turkovskaya O (2018) Dynamics of natural revegetation of hydrocarbon-contaminated soil and remediation potential of indigenous plant species in the steppe zone of the southern Volga Uplands. Environ Sci Pollut Res Int 25(4):3260–3274. https://doi.org/10.1007/s11356-017-0710-y
Panchenko L, Muratova A, Dubrovskaya E, Golubev S, Turkovskaya O (2023) Natural and technical phytoremediation of oil-contaminated soil. Life (Basel, Switzerland) 13(1):177. https://doi.org/10.3390/life13010177
Patowary K, Patowary R, Kalita MC, Deka S (2017a) Characterization of biosurfactant produced during degradation of hydrocarbons using crude oil as sole source of carbon. Front Microbiol 8. https://doi.org/10.3389/fmicb.2017.00279
Patowary R, Patowary K, Devi A, Kalita MC, Deka S (2017b) Uptake of total petroleum hydrocarbon (TPH) and polycyclic aromatic hydrocarbons (PAHs) by Oryza sativa L. grown in

soil contaminated with crude oil. Bull Environ Contam Toxicol 98(1):120–126. https://doi.org/10.1007/s00128-016-1990-5

Peng S, Zhou Q, Cai Z, Zhang Z (2009) Phytoremediation of petroleum contaminated soils by Mirabilis Jalapa L. in a greenhouse plot experiment. J Hazard Mater 168(2–3):1490–1496. https://doi.org/10.1016/j.jhazmat.2009.03.036

Prabowo AR, Bae DM (2019) Environmental risk of maritime territory subjected to accidental phenomena: correlation of oil spill and ship grounding in the Exxon Valdez's case. Results Eng 4:100035. https://doi.org/10.1016/j.rineng.2019.100035

Rafique HM, Khan MY, Asghar HN, Ahmad Zahir Z, Nadeem SM, Sohaib M, Alotaibi F, Al-Barakah FNI (2023) Converging alfalfa (*Medicago sativa* L.) and petroleum hydrocarbon acclimated ACC-deaminase containing bacteria for phytoremediation of petroleum hydrocarbon contaminated soil. Int J Phytoremediation 25(6):717–727. https://doi.org/10.1080/15226514.2022.2104214

Savin MC, Wolf DC, Davis KJ, Gbur EE, Thoma GJ (2015) Nematodes as bioindicators of ecosystem recovery during phytoremediation of crude oil contaminated soil. Int J Phytoremediation 17(2):182–190. https://doi.org/10.1080/15226514.2013.876964

Segura MF, Hanniford D, Menendez S, Reavie L, Zou X, Alvarez-Diaz S, Zakrzewski J, Blochin E, Rose A, Bogunovic D, Polsky D, Wei J, Lee P, Belitskaya-Levy I, Bhardwaj N, Osman I, Hernando E (2009) Aberrant miR-182 expression promotes melanoma metastasis by repressing FOXO3 and microphthalmia-associated transcription factor. Proc Natl Acad Sci USA 106(6):1814–1819. https://doi.org/10.1073/pnas.0808263106

Shelyakin PV, Semenkov IN, Tutukina MN, Nikolaeva DD, Sharapova AV, Sarana YV, Lednev SA, Smolenkov AD, Gelfand MS, Krechetov PP, Koroleva TV (2022) The influence of kerosene on microbiomes of diverse soils. Life (Basel, Switzerland) 12(2):221. https://doi.org/10.3390/life12020221

Skrypnik L, Maslennikov P, Novikova A, Kozhikin M (2021) Effect of crude oil on growth, oxidative stress and response of antioxidative system of two rye (Secalecereale L.) varieties. Plants 10(1):157. https://doi.org/10.3390/plants10010157

Stepanova AY, Gladkov EA, Osipova ES, Gladkova OV, Tereshonok DV (2022) Bioremediation of soil from petroleum contamination. Processes (Basel, Switzerland) 10(6):1224. https://doi.org/10.3390/pr10061224

Suprayogi B, Murray F (1999) A field experiment of the physical and chemical effects of two oils on mangroves. Environ Exp Bot 42(3):221–229. https://doi.org/10.1016/s0098-8472(99)00037-4

Tang KHD, Angela J (2019) Phytoremediation of crude oil-contaminated soil with local plant species. IOP Conf Ser Mater Sci Eng 495:012054. https://doi.org/10.1088/1757-899x/495/1/012054

Tang KHD, Law YWE (2019) Phytoremediation of soil contaminated with crude oil using Mucunabracteata. Res Ecol 1(1):20–30. https://doi.org/10.30564/re.v1i1.739

Wang X, Feng J, Zhao J (2010) Effects of crude oil residuals on soil chemical properties in oil sites, Momoge Wetland, China. Environ Monit Assess 161(1–4):271–280. https://doi.org/10.1007/s10661-008-0744-1

Wang A, Fu W, Feng Y, Liu Z, Song D (2022) Synergetic effects of microbial-phytoremediation reshape microbial communities and improve degradation of petroleum contaminants. J Hazard Mater 429(128396):128396. https://doi.org/10.1016/j.jhazmat.2022.128396

Wyszkowska J, Borowik A, Kucharski J (2019) The resistance of Loliumperenne L. × hybridum, Poapratensis, Festucarubra, F. arundinacea, Phleumpratense and Dactylisglomerata to soil pollution by diesel oil and petroleum. Plant Soil Environ 65(6):307–312. https://doi.org/10.17221/42/2019-pse

Zawierucha I, Malina G, Herman B, Rychter P, Biczak R, Pawlowska B, Bandurska K, Barczynska R (2022) Ecotoxicity and bioremediation potential assessment of soil from oil refinery station area. J Environ Health Sci Eng 20(1):337–346. https://doi.org/10.1007/s40201-021-00780-0

Zhang K, Maltais-Landry G, Liao H-L (2021) How soil biota regulate C cycling and soil C pools in diversified crop rotations. Soil Biol Biochem 156(108219):108219. https://doi.org/10.1016/j.soilbio.2021.108219

Zheng X, Wang H, Tao Y, Kou X, He C, Wang Z (2022) Community diversity of soil meso-fauna indicates the impacts of oil exploitation on wetlands. Ecol Indic 144(109451):109451. https://doi.org/10.1016/j.ecolind.2022.109451

Chapter 9
Biostimulation of Petroleum Contaminated Soil Using Organic Amendments

Suganthi Rajendran, Poornima Ramesh, and Rakesh Subramanian Srirangarayan

Abstract Soil serves as a sink for pollutants and a source of nutrients. The complexity and persistence of hydrocarbon contamination in the environment, which threatens livelihoods and food security, is gaining considerable attention. Due to their water-repellent properties, these hydrocarbons disrupt the normal functioning of plant systems, resulting in nutrient and moisture deficiencies. Additionally, they pose a risk to the well-being of humans, animals, and the environment. Phytoremediation is an effective and eco-friendly method of remediation. However, the success of remediation is influenced by the bioavailability of the contaminants. To promote rapid degradation, organic supplements such as biochar, compost, manure, pressmud, moringa seed powder, chicken waste, palm oil waste, and cow dung have been added to the soil. These amendments enhance microbial growth in the soil, facilitating successful degradation. Organic additions provide a continuous source of organic carbon for microorganisms and act as a soil conditioner, improving the physicochemical properties of the soil while supplying nutrients that support plant growth. By reducing the water's surface tension, these amendments aid in the extraction of petroleum compounds, affecting the biosorption of hydrocarbons and maintaining soil health. A crucial factor in the sequestration of petroleum hydrocarbons is the type of organic amendment, its dosage, the type and concentration of petroleum hydrocarbons, as well as other environmental factors such as soil characteristics, moisture content, and temperature. In this chapter, we will explore phytoremediation, emphasizing the significance of applying organic amendments to enhance the phytoremediation process of soil contaminated by petroleum.

S. Rajendran (✉)
Department of Environmental Sciences, JSA College of Agriculture and Technology, Cuddalore, India

P. Ramesh
Department of Environmental Sciences, Vanavarayar Institute of Agriculture, Coimbatore, India

R. S. Srirangarayan
Department of Crop Physiology, Tamil Nadu Agricultural University, Coimbatore, India

A. Parnian et al. (eds.), *Bioremediation of Petroleum Contaminated Soil and Sediment*, Environmental Contamination Remediation and Management,
https://doi.org/10.1007/978-3-031-95524-2_9

Keywords Hydrocarbon · Phytoremediation · Biostimulation · Restoration · Environment

9.1 Introduction

Crude oil, as the predominant source of global energy, exerts significant influence on the environment. The designation "petroleum" pertains to a composite of hydrocarbons that includes crude oil, diesel, various oils, and additional derivatives that are extracted from crude oil (Kuppusamy et al. 2020). The release of petroleum contaminants into the soil has the potential to adversely affect ecosystems, groundwater, soil integrity, and other environmental components. The two principal classifications of hydrocarbon molecules constituting total petroleum hydrocarbons (TPH) are extractable petroleum hydrocarbons and volatile petroleum hydrocarbons. The former category encompasses shorter-chain hydrocarbons (C6-C10), inclusive of BTEX (Benzene, Toluene, Ethyl Benzene, and Xylene), while the latter includes longer-chain hydrocarbons (C10-C40) and polycyclic aromatic hydrocarbons (Cheng et al. 2023; Kuppusamy et al. 2020; Fatima et al. 2018). Hydrocarbon contamination of soil resulting from extraction, processing, and transportation activities in oil-producing and importing nations, driven by the demand for petroleum products, adversely impacts the ecosystem. Due to their lipophilic nature, these hydrocarbons' insolubility obstructs the absorption of nutrients and water. The presence of petroleum molecules adversely affects both the physical and biochemical properties of soil, posing a threat to all forms of life. Empirical studies have demonstrated that petroleum exhibits acute physiological toxicity to flora, culminating in the deterioration of plant communities and a concomitant decline in biodiversity. The immobilization and accumulation of petroleum hydrocarbons have emerged as a critical environmental challenge, presently jeopardizing ecosystem integrity and human health (Hasan et al. 2024).

Petroleum contaminants are classified into three distinct categories: crude oil, gasoline, diesel fuel, and polycyclic aromatic hydrocarbons. Crude oil signifies the unrefined petroleum product comprising a heterogeneous mixture of hydrocarbons that incorporates volatile organic compounds (VOCs) and heavy metals. Refined petroleum derivatives, such as gasoline and diesel fuel, contain benzene, toluene, ethylbenzene, and xylene (BTEX), which are volatile and possess the ability to migrate through soil (Hasan et al. 2024; Ptaszek et al. 2020). Certain petroleum products are characterized by the presence of polycyclic aromatic hydrocarbons (PAHs), which are persistent organic pollutants recognized for their environmental durability and potential toxicological effects. Petroleum contaminants alter the soil's physical characteristics, including structure, texture, density, porosity, and water retention capacity (Patel et al. 2020; Wang et al. 2016). These contaminants undergo various chemical transformations within the soil matrix, resulting in the generation of metabolites that may exhibit toxicity comparable to or different from that of the parent compounds. The proliferation and functionality of soil microorganisms integral to nutrient cycling and organic matter degradation can be hindered by the presence of petroleum contamination. The infiltration of these contaminants into the soil reaches aquifers, compromising potential drinking water resources (Daassi and Qabil Almaghribi 2022).

Confronting the challenge of petroleum contamination in soil necessitates a comprehensive and contextually relevant methodology, which must take into account the extent of contamination, the properties of the soil, and the possible impacts on surrounding ecosystems and groundwater resources (Ambaye et al. 2022). Efficient remediation involves a suite of techniques meticulously adapted to the particular conditions of the site.

The utilization of flora to eradicate or mitigate pollution, referred to as phytoremediation, has emerged as an effective strategy for the restoration of soil affected by petroleum hydrocarbons. This methodology leverages the inherent capabilities of select plant species to withstand and metabolize petroleum pollutants, thereby facilitating the gradual remediation of the soil. Numerous types of plants have been identified for their ability to clean up soil tainted by oil using phytoremediation methods. These species exhibit significant tolerance to petroleum contamination and possess the capability to translocate or degrade the pollutants, ultimately enhancing the ecological quality of impacted regions (Montreemuk et al. 2023; Nedjimi 2021).

9.2 Phytoremediation

Phytoremediation encompasses the strategic utilization of flora and their symbiotic microorganisms to eliminate, alleviate, or transform deleterious contaminants into innocuous substances within the environment (Table 9.1). Empirical investigations have demonstrated its efficacy and reliability in remediating an array of pollutants, including heavy metals, pesticides, cyanide, and petroleum derivatives. The efficiency of phytoremediation is contingent upon various factors, including the species of plants employed, the properties of the soil, environmental conditions, climatic variables, and the concentration of the contaminants present. This technique is regarded as one of the safest and most dependable remediation methods, particularly in contrast to alternative physicochemical approaches that may result in the emission of greenhouse gases into the atmosphere. Bioremediation is predominantly dependent on the indigenous microorganisms that inhabit the contaminated environment. Both in-situ and ex-situ methodologies have been implemented to diminish the concentration of hydrocarbons present (Hegazy et al. 2023; Sharma et al. 2023; Wang et al. 2011). Selection of the remediation approach is profoundly influenced by soil pH, the presence of clay minerals, organic matter content, temperature, moisture levels, and nutrient availability. The strategies of biostimulation and bioaugmentation have been proposed to enhance the effectiveness of contaminant extraction. Biostimulation is defined as the provision of sufficient aeration, nutrients, and conducive environmental conditions to augment the metabolic

Table 9.1 Efficiency of plants in phytoremediation of petroleum-contaminated soils

Number	Plant	Removal Efficiency	References
1	*Medicago sativa* *Lolium perenne*	72–90%	Panchenko et al. (2023)
2	*Pteris vittata,* *Epipremnum aureum,* *Mucuna bracteata* *Imperata cylindrica*	36% 50.4% 30.9% 39.5%	Tang and Angela (2019)
3	*Jatropha curcas* *Vetiveria zizanioides*	78.8% 51.1%	Nero (2021)
4	*Oryza longistaminata,* *Sorghum arundinaceum,* *Tithonia diversifolia* *Hyparrhenia rufa*	58% 58% 56% 73%	Ruley et al. (2020)
5	*Leptochloa fusca* *Brachiaria mutica*	80%	Fatima et al. (2018)
6	*Lolium perenne*	31.1. %	Ptaszek et al. (2020)
7	*Phragmites australis* *Chenopodium marale*	90% 20%	Aloud et al. (2023)
8	*Mimosa pudica* *Gazania rigen* *Zinnia profusion* *Ipomoea quamoclit*	49%	Ikeura et al. (2016)
9	*Zea mays*	80%	Wojtowicz et al. (2023)
10	*Vetiveria zizanioides* *Cymbopogon nardus*	90%	Cheng et al. (2023)
11	*Vinca rosea*	19.5%	Hegazy et al. (2023)
12	*Melilotus officinalis*	49%	Steliga and Kluk (2021)
13	*Lolium perenne*	45%	Tao et al. (2016)

activity of indigenous microorganisms for the remediation of pollutants from the contaminated substrate, while bioaugmentation entails the introduction of non-native or allochthonous microorganisms alongside native populations to facilitate the effective removal of contaminants from the affected area. This is particularly relevant in zones contaminated by petroleum, often supplemented with hydrocarbon-degrading bacterial strains (Myazin et al. 2021; Sharma 2020; Goswami et al. 2018).

Phytoremediation includes various processes, such as phytodegradation, rhizodegradation, rhizofiltration, phytostabilization, phytovolatilization, and phytoextraction. The chemistry of the rhizosphere and endosphere, involving plants and microorganisms, is critical for plant metabolism, growth, and productivity, which in turn significantly impacts the remediation of hydrocarbons within the soil matrix. The limitation of utilizing organic pollutants as an energy source has constrained plant metabolic processes; however, this challenge has been addressed through the implementation of an integrated plant-microbe treatment system (Sharma et al. 2023; Sabreena et al. 2022; Nedjimi 2021).

9.3 Soil Organic Matter and Bioavailability

Once petroleum hydrocarbon contaminants infiltrate the soil environment from both natural and human-induced origins, they undergo a process of weathering involving a combination of physicochemical and biological mechanisms. The extent of weathering experienced by various petroleum hydrocarbon components primarily hinges on their individual physicochemical properties, concentration levels before degradation, and their structure and chemical composition (Balseiro-Romero et al. 2018). Furthermore, this process is intricately linked to soil attributes, encompassing both physicochemical aspects (including porosity, particle size, permeability, and organic matter content) and biological characteristics (such as the abundance and activity of indigenous microbial populations). Moreover, environmental factors like humidity, temperature variations, and precipitation patterns play a pivotal role in influencing the weathering process (Truskewycz et al. 2019). The fate of petroleum hydrocarbons embedded in the soil matrix is shaped by a combination of abiotic and biotic factors.

9.4 Dissolution and Volatilization

Gasoline products tend to evaporate rapidly owing to their highly volatile nature, low carbon number and low boiling point (Balseiro-Romero et al. 2018; Rivett et al. 2011). Simple aromatic hydrocarbons such as benzene, toluene, ethylene, and xylene evaporate completely in warmer conditions unless constrained by environmental interactions such as the entry of leachate into aquifers (Simantiraki et al. 2013; Truskewycz et al. 2019). In contrast to the latter, it was reported that merely a negligible share of petroleum hydrocarbons volatilize and break out into the atmosphere (Balseiro-Romero et al. 2018). Long-term mobilization of contaminants is greatly influenced by volatilization and shifts in viscosity and density (Fine et al. 1997). Furthermore, the generation of vapor plumes in the saturation zone occurs due to the downward movement of petroleum hydrocarbon in the soil matrix which tends to escape into the atmosphere and soil pores (Rivett et al. 2011). The transport of hydrocarbons into the atmosphere occurs by diffusion or advection primarily dictated by the characteristics of soil pores. Hydrocarbons with low molecular weight are easily soluble in water which enhances their soil permeability (Cipullo et al. 2018). The transport of petroleum hydrocarbon in the soil matrix follows first-order kinetics (Rahbeh and Mohtar 2007). These dissolved compounds navigate through the soil matrix via different mechanisms entailing advection, diffusion, and dispersion. Dispersion, specifically, involves the random movement of hydrocarbon due to pore water present in the soil system. Furthermore, solubility and mobility of the high molecular weight hydrocarbon can be augmented by addition of alcohol and acetone (Truskewycz et al. 2019).

9.5 Desorption, Sorption and Degradation

Biotic degradation by microbes and abiotic degradation by sunlight is pivotal for the natural breakdown of petroleum hydrocarbon and its residues present (Serrano et al. 2008). It involves convoluted interactions between the contaminant and the environmental components involving both biotic and abiotic units (Hoang et al. 2021). Sorption governs the fate and transport of hydrocarbon contaminants in the soil through bonding and complex interactions (Rivett et al. 2011; Pierzynski et al. 2005). Bioavailability of hydrocarbons to the microbes tends to reduce under sorption which hinders its degradation and mobilization (Khan et al. 2018). The sorption process may be influenced by various properties of hydrocarbon and soil, especially organic carbon (Providenti et al. 1993; Liu et al. 2013).

Typically, low molecular weight (LMW) petroleum hydrocarbon contaminants exhibit higher mobility and susceptibility to degradation upon entering the soil, attributed to their low hydrophobicity, elevated water-soluble nature, and enhanced volatility whereas high molecular weight hydrocarbons are resistant to degradation leading to their persistent nature (Harmsen and Rietra 2018). Over time petroleum hydrocarbon degradation diminishes due to the decrease in bioavailable fractions of these compounds. The primary mechanism for removing petroleum hydrocarbon contaminants is microbial degradation (Hussain et al. 2017). However, microbial degradation of hydrocarbons is impeded by the sorption process in the soil. Nevertheless, a long retention time of hydrocarbons exhibits lesser toxicity due to reduced availability; the aging process can also diminish the toxicity of contaminants to receptors by reducing their bioavailability (Ouvrard et al. 2013).

Sorbed petroleum hydrocarbon degradation cannot be dismissed outright since microbes access and degrade hydrocarbons (Chen et al. 2010). Microorganisms employ extracellular enzyme secretion or cellular adaptations for their uptake as well as degradation (Balaji et al. 2014; Xu et al. 2013). Additionally, soil-sequestered petroleum hydrocarbon either by partitioning or diffusion ultimately releases contaminants later facilitating microbial remediation (Cornelissen et al. 2000). Root exudates increase bioavailability of hydrocarbons in the soil (Martin et al. 2014). Organic acids of low molecular weight present in root exudates trade with the hydrocarbon for sorptive sites present in the soil matrix, rendering sorbed hydrocarbon available for degradation and organic acid kindles microbial growth thereby acting as a biostimulant in hydrocarbon removal (Hussain et al. 2017). Bioactive molecules from plants and microorganisms play a crucial role in releasing bound petroleum hydrocarbon from soil (Truskewycz et al. 2019).

9.6 Intervention of Organic Amendment in Remediating Petroleum Hydrocarbons

Organic amendments are defined as substances originating from biological processes, encompassing both plant-derived biomass and biomass from living organisms, predominantly consisting of carbon, which possess the capacity to enhance soil quality and consequently augment agricultural productivity, rendering their utilization in agricultural practices prevalent (Hamid et al. 2020). Owing to their unique physical and chemical properties, there has been a marked increase in scholarly inquiry and innovation aimed at their application for environmental remediation in the previous decade (Gomez-Sagasti et al. 2018; Park et al. 2011). Within the domain of remediation efforts targeting soil contamination by petroleum hydrocarbons, a wide array of organic amendments can be applied, given their advantageous effects on the physical, chemical, and biological attributes of soil, as well as on plant growth and yield. Organic amendments are considered safe, improving the soil's physico-chemical attributes and making them a potential soil conditioner. Organic amendments such as organic manure, compost, and biochar enrich the soil and enhance the autochthonous microbial activity; collectively leading to rapid degradation of hydrocarbons (Aliku et al. 2021; Hoang et al. 2021; Ruley et al. 2020; Tao et al. 2016).

Remediating soil befouled with petroleum hydrocarbons involves strategies to restrict availability to their receptors by chemical or biological immobilization (Cao et al. 2016). On the other hand, Phytoextraction uses plants to gather contaminants in harvestable tissues and chemical washing can be used to mobilize contaminants to address localized hydrocarbon contamination in urban settings (Huguenot et al. 2015). Furthermore, pollutants containing petroleum hydrocarbons may be biodegraded by microorganisms and plants, changing their concentration and bioavailability (Dubrovskaya et al. 2017; Garcés Mejía et al. 2018). Key remediation techniques for resolving soils polluted with petroleum hydrocarbons include immobilization, mobilization, and biodegradation (Hussain et al. 2017; Lamichhane et al. 2016). According to Farzadkia et al. (2019), organic amendments improve the degree of surface area, surface charge density, and exchange sites in affected soil. This leads to improved surface adsorption and complexation, which in turn reduces the mobility of petroleum hydrocarbons. Sigmund et al. (2018) found that adding 10% (w/w) compost and 5% (w/w) biochar raised the sorption coefficients for PAHs in soils by ten and up to a hundred times, respectively. Furthermore, by altering sorptive sites in soil particles and increasing their availability for microbial breakdown, organic amendments can transfer petroleum hydrocarbons from solid to aqueous phases (Cai et al. 2017). In cleaning oil-contaminated soil, adding sewage sludge and cow dung (10%, w/w) led to 82% and 94% biodegradation, respectively (Agamuthu et al. 2013). Organic amendments should be carefully chosen to complement remediation goals because of their dual implications (Table 9.2).

Table 9.2 Effect of organic amendments in remediating petroleum hydrocarbons

Number	Contaminant	Organic Amendment	Effect	References
1	Pyrene	Compost	Increased plant biomass and concentration of iso–butyric acid and butyric acid in the rhizosphere region	Wang et al. (2012)
2	Petroleum hydrocarbon	Meat waste and green waste compost	Reduced bioavailability of petroleum hydrocarbon	Wu et al. (2013)
3	PAH	Poultry dung	Reduced bioavailability of PAH for crop uptake	Nwaichi et al. (2015)
4	Pyrene and phenanthrene	Sewage sludge biochar	Increased sorption of pyrene and phenanthrene. Clay minerals and dissolved organic carbon dwindled the sorption efficiency of biochar	Oleszczuk (2016)
5	Alkanes and PAH	Compost	The abundance of catabolic enzymes	Bastida et al. (2016)
6	PAH	Spent mushroom substrate	Enhanced PAH–degrading microbes	Di Gregorio et al. (2016)
7	Petroleum hydrocarbon	Grass straw compost	Enhanced degradation of petroleum hydrocarbon since the addition of compost increased the nutrient level in the soil	Nwankwegu et al. (2016)
8	Petroleum hydrocarbon	Cotton stalks	Increased removal efficiency of petroleum hydrocarbon by increased physicochemical properties and microbial diversity of soil	Wang et al. (2016)
9	Monoaromatic petroleum hydrocarbon	Wood chip biochar	Adsorption inclined up to a factor of 36	Bushnaf et al. (2017)
10	Phenanthrene	Softwood chip	Increased soil adsorption and reduced leaching of pollutants	Trinh et al. (2017)
11	Acenaphthene, Naphthalene, pyrene and phenanthrene	Pin needly biochar	Sorption of PAH to biochar thereby dwindling the bioavailability and crop uptake of PAH	Zhu et al. (2017)
12	PAHs	Spent mushroom compost	Enhanced the growth of guinea grass	Asemoloye et al. (2017)
13	Petroleum hydrocarbon	Poultry manure biochar	Improved microbial respiration in the rhizosphere of oat and barley. Enhanced the degradation of Petroleum Hydrocarbon	Barati et al. (2017)

(continued)

Table 9.2 (continued)

Number	Contaminant	Organic Amendment	Effect	References
14	PAH	Compost	Dissipation of PAH was relatively greater in non–mature compost than the mature compost	Brimo et al. (2018)
15	PAH	Wheat straw and sawdust biochar	The addition of biochar reduced the availability of PAH content	Kong et al. (2018)
16	Petroleum hydrocarbon	Poultry manure and Poultry manure biochar	Reduced toxicity of petroleum hydrocarbon to barley	Barati et al. (2018)
17	PAH	Silver grass biochar and compost	Increased sorption of PAH by tenfold with the addition of compost and by a hundred times with the addition of biochar	Sigmund et al. (2018)
18	Anthracene, naphthalene and benzo[a]pyrene	Vermicompost	Inclined sorption of PAH and the efficiency increased with increasing benzene rings.	Dores-Silva et al. (2018)
19	Petroleum hydrocarbon	Wheat straw	Increased respiration of soil microbes and mineralization of organic carbon	Alotaibi et al. (2018)
20	Petroleum hydrocarbon	Sawdust	Increased microbial activity and higher degradation rate of Petroleum Hydrocarbon	Alvim and Pontes (2018)
21	Petroleum hydrocarbon and PAH	Sawdust and rice straw	Increase in the abundance of hydrocarbon-degrading taxa	Huang et al. (2019)
22	Petroleum hydrocarbon	Horse manure, ramial chipped wood, and brewer-spent grain	Reduction of TPH was relatively higher in horse manure than other inorganic fertilizer soil	Robichaud et al. (2019)
23	PAH	Pea straw	Enhanced degradation of PAH (66%)	Koshlaf et al. (2019)
24	Petroleum hydrocarbon	Sugarcane residue biochar	Increased dissipation of petroleum hydrocarbon by irreversible sorption	Wei et al. (2020)
25	Benz[a]anthracene	Lignin, cow manure, mushroom culture waste, and straw	All amendments except cow manure increased the degradation of the pollutant	Zhu et al. (2020)
26	Petroleum hydrocarbon	Pea straw	Increased number of 16s rRNA and alkB genes	Koshlaf et al. (2020)
27	Petroleum hydrocarbon	Sewage sludge biochar, fruit waste biochar, and cow dung	Increased diversity and abundance of soil microorganisms	Aziz et al. (2020)

9.6.1 Biowastes

Organic amendments derived from bio-wastes encompass a diverse array of materials readily available for use, including crop residues, green manures, animal manures, biosolids, and composts from yard waste (Quilty and Cattle 2011). These amendments play a significant role in the amelioration of hydrocarbon-contaminated soil (Sarkar et al. 2005; Shahsavari et al. 2013). Crop residues and green manures aid as a substantial source of carbon and necessary nutrients in restoring degraded soil which accounts for over 3.8 billion metric tons if managed effectively (Thangarajan et al. 2013). Reintroducing crop residues into soils contaminated with petroleum hydrocarbons may be a sustainable way to nourish these soils with nitrogen (Che et al. 2018; Chen et al. 2014; Lal 2005). Crop residues have a critical role in petroleum hydrocarbon remediation, as several research has demonstrated (Shahsavari et al. 2013; Wang et al. 2016). By sorption, these amendments immobilize petroleum hydrocarbons, reducing their toxicity and promoting the growth of microorganisms and plants (Medina et al. 2015; Nunes et al. 2020). Comparably, a considerable degree of animal manure holds the potential to be used for environmental cleanup. Globally, there is a significant daily production of animal feces, estimated at approximately 12 million tons (Faostat 2016). Besides supplying macronutrients such as nitrogen, phosphorus, potassium, sulfur, and micronutrients, animal manure contains organic fractions such as humic and fulvic substances, recognized as the most active compounds within it, influencing both soil characteristics and availability of petroleum hydrocarbon (Ezenne et al. 2014). Incorporating animal manure into petroleum hydrocarbon-contaminated soil has been shown to enhance soil dissolved organic matter, ascribed to a reduction in the soil's clay fraction sorption capacity (Wei et al. 2014). Dissolved Organic Matter serves as a better desorbing agent for various petroleum hydrocarbon compounds, including aliphatic and aromatic hydrocarbons (Chen et al. 2019; Kopinke et al. 2011; Chen and Yuan 2012; Zhang and He 2013). Furthermore, meats from animal byproducts can be used as a source of nutrients and carbon for the cleanup of petroleum hydrocarbons (Liu et al. 2019). The composition and application of manure were governed by various factors such as handling and storage processes (Miller et al. 2003). As such, their effects on impacted soils could differ (Olawale et al. 2020). For instance, in petroleum hydrocarbon-contaminated soil, Adesodun and Mbagwu (2008) found that the addition of poultry manure increased the rate of petroleum hydrocarbon reduction in comparison to treatments using cow dung and pig waste. Alterations in the carbon-nitrogen-phosphorus ratios of these animal manures were shown to be the cause of their varying performances. These changes in turn had distinct impacts on the hydrocarbon-degrading microbes in the impacted soil (Adesodun and Mbagwu 2008). The restoration of soils contaminated by petroleum hydrocarbons also makes use of a variety of other organic amendments, such as composts, biosolids, and wastes from paper mills (Dindar et al. 2017; Wang et al. 2012). A feasible means of cleaning up soil contaminated by petroleum hydrocarbons can be achieved by the humic materials included in these amendments (Dindar et al. 2017). To aid in soil rehabilitation, adding biosolids or slurries to contaminated soil with petroleum hydrocarbons can change the bioavailability of those hydrocarbons or increase microbial activity (Rivera-Espinoza and Dendooven 2004). For example, the incorporation of digestates from biomethanation has been shown to promote petroleum hydrocarbon biodegradation (Gielnik et al. 2019). Similarly, waste from the sugarcane and paper industries alters the soil properties and enhances the microbial activity which has been found to stimulate the biodegradation of petroleum hydrocarbon contaminants (Babaei et al. 2020). In soils contaminated with petroleum hydrocarbons, a variety of organic wastes, such as food scraps, leaves, cow dung, and sludge from sewage systems, can be decomposed and used (Ren et al. 2018a, b). Because of the generation of metal-humic complexes, the addition of composts frequently results in changes in the availability of petroleum hydrocarbons (Wu et al. 2013). Incorporation of compost or any manures can elevate soil organic matter content, thereby reducing the availability of petroleum hydrocarbon in contaminated soil (Puglisi et al. 2007). Furthermore, humic materials in composts are thought to have a cation exchange capacity similar to clay minerals in soil, which helps the soil hold onto vital nutrients (Kim et al. 2017).

9.6.2 Biochar

Biochar, a carbon-rich, porous substance produced by pyrolyzing different organic wastes in an oxygen-limited environment, is another important organic amendment that can be used in hydrocarbon-contaminated soil reclamation (Zhu et al. 2017). Due to its large surface area, functional groups, and nano- and micro-porosity, it has a well-documented remediation potential (Li et al. 2019). Biochar is frequently used as a petroleum hydrocarbon adsorbent in contaminated soils due to its porous character and potent adsorption capacity (Jia et al. 2020; Wei et al. 2020). Moreover, biochar can alter the physicochemical characteristics of soil and supply an abundance of vital nutrients that promote plant development and the microbial breakdown of petroleum hydrocarbons (Gul et al. 2015). Thus, it is acknowledged that adding biochar to contaminated soils is a successful method of cleaning up soils containing petroleum hydrocarbons (Lamichhane et al. 2016). As a way to cleanse an urban soil clouded with PAHs, Khan et al. (2015) applied biochars made from four distinct feedstocks—soybean straw, sewage sludge, peanut shell, and rice straw—that were heated to 500 °C at concentrations of 2% and 5% (w/w). The introduction of biochars reduced the bioaccumulation of hydrocarbons in turnip plants by diminishing the availability of hydrocarbon in soil (*Brassica rapa* L.), with the most effective outcome observed at the 5% biochar application rate. Due to its least polar surface and most significant amount of aromatic functional groups (C–H), peanut shell biochar showed the greatest efficiency in dropping bioaccessible PAHs among various organic amendments (Khan et al. 2015). In a long-term field trial, soil incorporated with wheat straw biochar generated at 650 °C at rates of 30 and 45 tons per hectare resulted in a significant reduction of hydrocarbons by 25% and 22%, respectively (Oleszczuk et al. 2016). There was a gradual decline in the strength of interactions between biochar and PAHs over time, with no discernible risk associated with the levels of contaminants in the soil treated with biochar.

Besides adsorption, hydrocarbon degradation was enhanced by the addition of biochar which proved to be an efficient biostimulant by enhancing soil microbial activity. Qin et al. (2013) examined the effects of adding 2% w/w rice straw biochar that was heated to 500 °C on the makeup of microorganisms and the final petroleum hydrocarbon biodegradation in a soil that had been plagued with petroleum hydrocarbons at a rate of 16,300 mg kg^{-1} over 180 days. Biochar was added on the 80th day, which significantly aided in the elimination of petroleum hydrocarbons by increasing the sorption of metabolites that were harmful to soil microorganisms. The microbial population was stimulated and petroleum hydrocarbon degradation ensued as a result of the biochar's reduction in the bioavailability and ecotoxicity of petroleum hydrocarbons. Likewise, Kong et al. (2018) employed sawdust and wheat straw biochars to hydrocarbon-contaminated soil and observed the biostimulation effect of biochars which created a more favorable environment for PAH degraders, leading to accelerated biodegradation of PAHs (Kong et al. 2018).

9.6.3 *Limitations*

Despite the aforementioned benefits, the use of organic amendments may present certain drawbacks. Incorporating crop residues containing substantial amounts of carbon-rich compounds might lead to the reduced availability of inorganic nitrogen by immobilization in soils (Chen et al. 2014). Correspondingly, soil application of manures and sludge resulted in undesirable impacts including: (i) elevated concentrations of aerobic and anaerobic degradation byproducts, such as nitrogen compounds such as ammoniacal nitrogen, nitrate nitrogen, and emissions of greenhouse gases namely methane and carbon dioxide (ii) introduction of undesirable substances, including inorganic contaminants (such as heavy metals) or organic pollutants (including antibiotics, PAHs, microplastics), as well as infectious agents, (iii) application of organic amendments resulted in the excess runoff with nitrogen and phosphorus (Park et al. 2011; Scotti et al. 2015; Thangarajan et al. 2013). Moreover, based on the composition of the feedstock, the incorporation of compost or biochar produced from metal-containing feedstock exhibited the presence of heavy metal concentration in soil. For instance, manures produced from animal waste in China showed high levels of cadmium (10 mg kg^{-1}) (Wang et al. 2014). Alteration of soil properties such as pH and electrical conductivity should be considered (Mohan et al. 2014; Scotti et al. 2015). Research investigating the consequences of organic amendment application on soil properties is necessary to achieve optimal remediation outcomes. Despite the co-generation of syngas and bioenergy along with biochar, the economic feasibility of biochar's pyrolysis system remains uncertain. In some cases, biochar production has been impacted negatively owing to high power consumption during pyrolysis (Brassard et al. 2018). Life cycle assessment of biochar is essential to exploit significant advantages for the remediation of various contaminants.

9.6.4 Seed Powder

Hydrocarbons from oil spills during manufacture and transportation contaminate soil, endangering ecosystem functions and the environment's ability to function normally. In addition to raising concerns about the nation's food security, it has an impact on food production in developing nations. To help overcome the constraints imposed by oil spills, seed powder is utilized as a biostimulant to promote phytoremediation. *Moringa olifera* (MSP) and *Asimina triloba* (PSP) seeds contain saponin which is often employed in water purification and reported to improve the availability of hydrocarbons and they have not yet dealt with the soil remediation research (Aliku et al. 2021).

Phytoremediation studies on soil hydrocarbon using seed powder from *Moringa olifera* and *Asimina triloba* and their combined mixture (PSP + MSP) as organic stimulants in crop production of *Vigna ungiculata* as Pawpaw seed powder of dosage 150 g m^{-2} has the efficiency of 78.2% TPH removal rate and closely tagged by PSP + MSP at the rate of 150 g m^{-2} of 77.4% removal rate. Application of PSP seed powder at 150 gm^{-2} favors crop productivity and phytoremediation of crude oil contaminated soil even though the fresh pod yield of cowpea was reported to be 2416.67 kg ha^{-1} (250 g m^{-2} PSP) which was tailed by 150 g m^{-2} PSP (2173.34 kg ha^{-1}) (Aliku et al. 2021).

Another investigation of hydrocarbon-contaminated soil with naphthalene and acenaphthalene using *Helianthus annus* seed husk (125 g) and poultry droppings (125 g) mixture reported a bacterial population of $184 \pm 4.0 \times 10^9$ CFU/g soil with the dehydrogenase enzyme activity of 0.74 ± 0.06 μg TPF/g soil/hour. The phytotoxicity of soil was reduced by amendments and the treated soils enhanced the germination rate of the seeds within 40–60 days in *Zea mays* and *Abelmoschus esculentus* (Boopathy et al. 2021). *Dacryodes edulis* (African pear tree) seeds contain an array of bioactive compounds such as alkaloids, saponins, flavonoids, steroids, tannin, phenol, phytosterol, terpenes, cardiac glycoside, anthocyanins, oxalates, phytate, and coumarins. The *Dacryodes edulis* seed powder was investigated for the remediation of polycyclic aromatic hydrocarbons and heavy metals in the leaves of Telfairia occidentalis. The studies showed that *Dacryodes edulis* seed powder (30–100 g dosage) reduced the levels of naphthalene, pyrene, chrysene, acenaphthylene, phenanthrene, anthracene, and nickel (Etim et al. 2019).

9.6.5 Compost

Biostimulation has frequently been shown to be more successful in places that are persistently polluted. The addition of organic amendments accelerated the biodegradation of total petroleum hydrocarbons (TPH) by 75% in just 56 days. The highest percentage of biodegradation (93%) has been observed in sterilized soil that was inoculated with optimized water hyacinth compost, followed by polluted soil amended with a blend of spent mushroom and water hyacinth composts, which had an 89% TPH biodegradation rate. The analysis of residual PAHs, including naphthalene, acenaphthylene, acenaphthene, fluorene, phenanthrene, anthracene, fluoranthene, pyrene, benzo[a]anthracene, chrysene, benzo[b]fluoranthene, benzo[k]fluoranthene, benzo[a]pyrene, dibenzo[a,h]anthracene, benzo[g,h,i]perylene, and indeno [1,2,3-cd]pyrene were significantly degraded in the combined treatment with the combination of spent mushroom and water hyacinth composts. This suggests the potential of this amendment for environmental remediation. (Udume et al. 2023). The addition of chopped Bermuda grass hay/ sawdust/vermiculite led to rapid biodegradation from 10 percent to 82 percent in 12 weeks (Sui et al. 2021).

9.7 Microbial Inoculants

Table 9.3 presents a unique, cost-effective, and potential biotechnological technique for using soil-based microbial-phytoremediation systems (Sarma et al. 2019). As oxygen acts as an electron acceptor and because oxidation, reduction, hydroxylation, and dehydrogenation mediate degradation, bacteria primarily break down hydrocarbons via aerobic pathways (Wang et al. 2011). Enzymes including cytochrome P450, peroxidase, hydroxylase, dehydrogenase, dioxygenase, and monooxygenase aid in the biodegradation of hydrocarbons (Huang et al. 2019). It was discovered that immobilized cells had a good utilization of hydrocarbons in the liquid medium when additions such as sawdust, styrofoam, and wheat bran were used as carriers for the incubation of hydrocarbon-degrading bacteria. Petroleum,

Table 9.3 Microbial inoculants in Hydrocarbon degradation

Number	Microbial strain and source	Hydrocarbon and Degradation rate	References
1	*Aspergillus sp.* RFC-1	Crude oil (84.6%), Naphthalene (50.3%), Phenanthrene and Pyrene (55.1%)	Al-Hawash et al. (2019)
2	*Stenotrophomonas sp.* IITR87	Phenanthrene (99%), Pyrene (98%) and Benzo-pyrene (50%)	Tiwari et al. (2017)
3	*Achromobacter sp.* HZ01	Total n-alkanes (96.6%)	Deng et al. (2014)
4	*Enterobacter sp.* (MM087)	Phenanthrene (80.2%) and Pyrene (59.7%)	Darma et al. (2016)
5	*Pseudomonas sp.* WJ6	N-dodecane (C12) (46.65%) and N-dodecane (C22,C32 and C40) (23.62%)	Xiao et al. (2013)
6	*Bacillus subtilis* BMT4i (MTCC 9447)	Benzo[a]Pyrene (84.66%)	Lily et al. (2009)
7	*Caulobacter sp.* (T2A12002)	Pyrene (36%)	Al-Thukair and Malik (2016)

phenanthrene, pyrene, and fluoranthene are the main sources of soil contamination in the Philippines; Acidovorax, Rhodoferax, Hydrogenophaga, and Polaromonas genera contain the main species spotted in contaminated soil tied to polycyclic aromatic hydrocarbons (Sui et al. 2021).

Hydrocarbons are broken down into smaller organic compounds by plants and microbes that produce certain enzymes such as laccase, peroxidase, dehalogenase, and nitroreductase. Certain contaminants are absorbed on the surface of the roots and become trapped there by the dual lipid layer of the cell's plasma membrane and the hemicellulose from the plant cell wall (Abdullah et al. 2020). Under Phytoextraction, a portion of the hydrocarbons gets transported to shoot and other aerial parts of the plant (Kiamarsi et al. 2020) and released through phytovolatilization into the atmosphere (Rodriguez et al. 2014). Following the enrichment of native or foreign microorganisms, the introduction of new microorganisms can more effectively metabolize pollutants and have a greater tolerance to toxicity (Barathi and Vasudevan 2003). In a bioremediation experiment with the duration of 12 weeks of peanut hull powder immobilized with bacterial cells, the TPH concentration in soil was 11,400 mg kg^{-1} and had a TPH removal efficiency of 61% (Xu and Lu 2010). Within the microbial-phytoremediation system, microbes break down pollutants like petroleum hydrocarbons into molecules that plants may readily absorb and use. Additionally, the roots of plants can serve as ideal habitats for microorganisms to grow and thrive (Wojtera-Kwiczor et al. 2014; Kiamarsi et al. 2020).

Biosolids

Biosolids, which are rich in nutrients and are generated during wastewater treatment in effluent treatment facilities, can be applied to agricultural fields to improve crop growth and production as well as soil nutrient levels. When biosolids were added to soil polluted by crude oil, the amount of hydrocarbons in the soil decreased. Biosolids are known to increase the microbial activity in the rhizosphere. They increase the porosity of the soil, which results in a lower bulk density and a greater ability to store water, both of which enhance the structure of the soil. As seen in pearl millet, biosolids enhanced the morphological and growth characteristics of the plant, including root biomass and root surface area (Markway et al. 2008). Biosolids enrich the soil with trace minerals, nitrogen, and other nutrients.

9.8 Factors Influencing Assisted Remediation

By providing the right nutrients, biostimulation can accelerate the breakdown of pollutants originating from petroleum and increase the metabolic activity of microorganisms, according to several studies. This can assist in overcoming the constraints that environmental factors such as temperature, oxygen, moisture content, pH, and nutrients place on native microorganisms (Wu et al. 2019). Soil parameters: soil moisture content, temperature, pH, salinity, availability of nutrients (particularly nitrogen and phosphorus), and redox potential, influence the bioremediation and phytoremediation processes (Waqas et al. 2014). It is necessary for plant seeds to be able to absorb water from the soil, germinate, and then anchor and stabilize the plant as it grows. Microbes must be able to live biologically in soil in adverse conditions in order to be a useful remediation agent (Afzal et al. 2011).

9.8.1 Soil pH

Remediation evaluations show that *Thalaspi caerulescens* absorbed more Cd from soils with high concentrations, low pH (between 5 and 6), and coarser textural particles. The transport of some heavy metal ions (Zn, Pb, Cu, and Cd) may be impeded by soil conditioners and inclusions such as zeolite or charcoal, which alter the soil's textural and structural characteristics (Wen et al. 2016). Additionally, Waqas et al. (2014) documented the impact of biochar, compost, and sewage sludges on the remediation of some PAHs. According to Marchal et al. (2014), the main reason for this was that soil amendments change the bioavailability of PAHs and have an impact on the rhizosphere of plants.

9.8.2 Soil Texture

The texture of the soil is responsible for root growth, water-holding capacity, nutrient dynamics, and soil-water and gaseous exchange (Abdel-Moghny et al. 2012). Soil texture can influence phytoremediation by controlling the bioavailability of plant nutrients and contaminants since clay can bind molecules, decreasing the bioavailability of pollutants in soils with higher clay concentrations (Abdel-Moghny et al. 2012; Figueiredo et al. 2016).

According to Falciglia et al. (2011), soil textural behavior (coarse, medium, fine sand, silt, and clay size aggregate fractions) was examined when diesel fuel that had been artificially contaminated was thermally treated at various temperatures. At 175 °C, when the pollutant concentration was 100 mg kg^{-1}, the fine sandy soil exhibited the highest desorption. Schist with the largest fine fraction degrades pollutants more quickly, whereas granite and limestone with particular clay minerals require the longest time to remediate (Carvalho et al. 2015). Soils with lower silt and clay content break down hydrocarbons at higher rates. This may develop due to porosity and aeration that depend on soil texture. Microbial activity suffers in clay soils that are prone to water logging. Clay soils that are prone to waterlogging have a negative impact on microbial activity and root respiration, which significantly slows down the breakdown of hydrocarbons.

9.8.3 Soil Structure

A broad spectrum of soil management techniques, including tillage, crop rotation, irrigation, drainage, and contaminants like petroleum hydrocarbons, can alter the dynamic quality of soil structure. Soil pores control the plant's physiological functions such as germination and emergence of seeds and crop growth and soil biodiversity. Soil organic matter for building aggregates influences the fate of PAHs, as PAHs strongly bind to soil organic matter. Amendments addition including vermiculite, sawdust, wheat bran, hay, and other grass residues augment the binding of PAHs with soil particles consequently increasing aeration which in turn favors microbial activity and hydrocarbon degradation (Razafimbelo et al. 2013; Xiao et al. 2014). Growing roots produce organic matter into the soil, stabilizing aggregates; in salinity, however, aggregates become fragmented, causing structural degradation, changing mycorrhizal colonization, and changing the distribution of pollutants (Tejeda-Agredano et al. 2013; Liu et al. 2013).

9.8.4 Soil Organic Matter

Organic carbon and matter present in the soil have multifaceted characteristics and functions that regulate the soil's essential services. Previous research has proved that the contamination of soil with hydrocarbons resulted in the reduced bioavailability of the contaminants caused by variations in soil physico-chemical and biological properties (Chakraborty et al. 2012). The degradation rate of hydrocarbons was found to be inversely proportional to the hydrocarbon concentration in sandy and loamy soil (Phillips et al. 2012). Hydrocarbons of higher molecular weight were found to be less degradable when compared with low molecular compounds. Low molecular compounds ensure complete degradation (Moubasher et al. 2015).

Air flow rate or gas diffusion in the soil facilitates the exclusion of contaminants such as benzene, toluene, Xylene, and other hydrocarbons from the soil which aligns with the fact that complete and rapid remediation was observed in sandy soil with sufficient gaseous exchange between soil particles (Albergaria et al. 2010). Gaseous diffusion especially oxygen deficiency resulted in the reduced nutrient transfer and persistence of the contaminant in the environment. Modification of soil aggregates such as crushing result in enhanced microbial respiration and assisted mineralization of carbon present in the soil (Razafimbelo et al. 2013). Carvalho et al. (2015) reported higher ($\geq$99.3%) remediation rates in bio venting technique confirming the important role of oxygen in remediation.

9.8.5 Soil Moisture

Moisture content in the soil is an important factor governing the degradation of hydrocarbon through different processes such as evaporation, transpiration, percolation, soil-gas exchange, and transportation of salts and nutrients. Wang et al. (2013) ascertained the fact that when marsh soils are contaminated with hydrocarbon, the soil pH tends to shift towards alkaline pH, affects soil health and fertility, and modifies the physical properties of soil. Larger molecules of hydrocarbon adsorb on the soil surface resulting in crust formation and small particles affect infiltration and gas diffusion by trapping soil micro and macro pores which altogether delays the degradation process (Hajabbasi 2016).

9.8.6 Additional Factors

The breakdown of petroleum hydrocarbons is either directly or indirectly impacted by the temporal traits of the soil. The temperature range of 9–50 °C has been found to promote microbial and plant growth, leading to better degradation of both organic and inorganic pollutants (Chunyan et al. 2023). Contrary to the findings above, it was found that increasing salinity from increased evaporation had a negative impact on earthworms and plant growth (Thavamani et al. 2015).

9.9 Conclusion

Phytoremediation is becoming a viable option for removing different types of toxins from water and soil. In this context, cleaning up soil tainted by petroleum is acknowledged as a very difficult task. Hydrocarbon-contaminated soil is subjected to phytoremediation with the help of several organic amendments, which greatly improve the soil's ability to remove hydrocarbons. Compared to phytoremediation, the use of conventional physical and chemical techniques is substantially more costly. Furthermore, adding organic amendments has no detrimental effects on the characteristics of the soil or how well plants grow. Organic amendments increase the soil's physical and chemical characteristics, microbial activity, organic carbon, and nutrient availability.

9.10 Suggestions

In order to enhance bioremediation approaches the following suggestions are made:

I. Future research should be focused on improving the bioavailability of the pollutant without deteriorating the environment
II. Remediation methods of thermal means or any pretreatment methods can be tested to modify the properties of petroleum hydrocarbons
III. A better understanding of the influencing properties and behavior of hydrocarbon will facilitate efficient remediation
IV. Bioremediation of hydrocarbons should be considered essential in a circular economy

References

Abdel-Moghny T, Mohamed RS, El-Sayed E, Mohammed Aly S, Snousy MG (2012) Effect of soil texture on remediation of hydrocarbons-contaminated soil at El-Minia district, Upper Egypt. International Scholarly Research Notices

Abdullah SRS, Al-Baldawi IA, Almansoory AF, Purwanti IF, Al-Sbani NH, Sharuddin SSN (2020) Plant-assisted remediation of hydrocarbons in water and soil: application, mechanisms, challenges and opportunities. Chemosphere 247:125932

Adesodun JK, Mbagwu JSC (2008) Biodegradation of waste-lubricating petroleum oil in a tropical alfisol as mediated by animal droppings. Bioresour Technol 99(13):5659–5665

Afzal M, Yousaf S, Reichenauer TG, Kuffner M, Sessitsch A (2011) Soil type affects plant colonization, activity and catabolic gene expression of inoculated bacterial strains during phytoremediation of diesel. J Haz Mater 186(2–3):1568–1575

Agamuthu P, Tan YS, Fauziah SH (2013) Bioremediation of hydrocarbon contaminated soil using selected organic wastes. Procedia Environ Sci 18:694–702

Albergaria JT, Alvim-Ferraz MC, Delerue-Matos CM (2010) Remediation of soils contaminated with trichloroethylene. Environ Biotechnol 6:37–41

Al-Hawash AB, Zhang X, Ma F (2019) Removal and biodegradation of different petroleum hydrocarbons using the filamentous fungus *Aspergillus sp.* RFC-1. Microbiolopen 8(1):e00619

Aliku CB, Madu CN, Aliku O (2021) Organic stimulants for enhancing phytoremediation of crude oil polluted soil: a study on cowpea. Environ Pollut 287:117674

Alotaibi HS, Usman AR, Abduljabbar AS, Ok YS, Al-Faraj AI, Sallam AS, Al-Wabel MI (2018) Carbon mineralization and biochemical effects of short-term wheat straw in crude oil contaminated sandy soil. Appl Geochem 88:276–287

Aloud SS, Alotaibi KD, Almutairi KF, Albarakah FN (2023) Phytoremediation potential of native plants growing in industrially polluted soils of Al-Qassim, Saudi Arabia. Sustainability 15(3):2668

Al-Thukair AA, Malik K (2016) Pyrene metabolism by the novel bacterial strains *Burkholderiafungorum* (T3A13001) and *Caulobactersp* (T2A12002) isolated from an oil-polluted site in the Arabian Gulf. Int Biodeterior Biodegradation 110:32–37

Alvim GM, Pontes PP (2018) Aeration and sawdust application effects as structural material in the bioremediation of clayey acid soils contaminated with diesel oil. Int Soil Water Conserv Res 6(3):253–260

Ambaye TG, Chebbi A, Formicola F, Prasad S, Gomez FH, Franzetti A, Vaccari M (2022) Remediation of soil polluted with petroleum hydrocarbons and its reuse for agriculture: recent progress, challenges, and perspectives. Chemosphere 293:133572

Asemoloye MD, Ahmad R, Jonathan SG (2017) Synergistic action of rhizospheric fungi with *Megathyrsus maximus* root speeds up hydrocarbon degradation kinetics in oil polluted soil. Chemosphere 187:1–10

Aziz S, Ali MI, Farooq U, Jamal A, Liu FJ, He H, Huang Z (2020) Enhanced bioremediation of diesel range hydrocarbons in soil using biochar made from organic wastes. Environ Monit Assess 192:1–14

Babaei AA, Safdari F, Alavi N, Bakhshoodeh R, Motamedi H, Paydary P (2020) Co-composting of oil-based drilling cuttings by bagasse. Bioprocess Biosyst Eng 43:1–12

Balaji V, Arulazhagan P, Ebenezer P (2014) Enzymatic bioremediation of polyaromatic hydrocarbons by fungal consortia enriched from petroleum contaminated soil and oil seeds. J Environ Biol 35(3):521–529

Balseiro-Romero M, Monterroso C, Casares JJ (2018) Environmental fate of petroleum hydrocarbons in soil: review of multiphase transport, mass transfer, and natural attenuation processes. Pedosphere 28(6):833–847

Barathi S, Vasudevan N (2003) Bioremediation of crude oil contaminated soil by bioaugmentation of *Pseudomonas fluorescens* NS1. J Environ Sci Health, Part A 38(9):1857–1866

Barati M, Bakhtiari F, Mowla D, Safarzadeh S (2017) Total petroleum hydrocarbon degradation in contaminated soil as affected by plants growth and biochar. Environ Earth Sci 76:1–11

Barati M, Bakhtiari F, Mowla D, Safarzadeh S (2018) Comparison of the effects of poultry manure and its biochar on barley growth in petroleum-contaminated soils. Int J Phytoremediation 20(2):98–103

Bastida F, Jehmlich N, Lima K, Morris BEL, Richnow HH, Hernández T, García C (2016) The ecological and physiological responses of the microbial community from a semiarid soil to hydrocarbon contamination and its bioremediation using compost amendment. J Proteome 135:162–169

Boopathy S, Appavoo MS, Radhakrishnan I (2021) Sunflower seed husk combined with poultry droppings to degrade petroleum hydrocarbons in crude oil-contaminated soil. Environ Eng Res 26(5):200361

Brassard P, Godbout S, Pelletier F, Raghavan V, Palacios JH (2018) Pyrolysis of switch grass in an auger reactor for biochar production: a greenhouse gas and energy impacts assessment. Biomass Bioenergy 116:99–105

Brimo K, Ouvrard S, Houot S, Lafolie F, Garnier P (2018) Modelling the fate of PAH added with composts in amended soil according to the origin of the exogenous organic matter. Sci Total Environ 616:658–668

Bushnaf KM, Mangse G, Meynet P, Davenport RJ, Cirpka OA, Werner D (2017) Mechanisms of distinct activated carbon and biochar amendment effects on petroleum vapour biofiltration in soil. Environ Sci: Processes Impacts 19:1260–1269

Cai D, Yang X, Wang S, Chao Y, Morel JL, Qiu R (2017) Effects of dissolved organic matter derived from forest leaf litter on biodegradation of phenanthrene in aqueous phase. J Hazard Mater 324:516–525

Cao Y, Yang B, Song Z, Wang H, He F, Han X (2016) Wheat straw biochar amendments on the removal of polycyclic aromatic hydrocarbons (PAHs) in contaminated soil. Ecotoxicol Environ Saf 130:248–255

Carvalho MM, Vila MC, Matos CD, Teles MTO, Fiuza A (2015) Assisted bioremediation tests on three natural soils contaminated with benzene. Eurasian J Soil Sci 4(3):153–160

Chakraborty S, Weindorf DC, Zhu Y, Li B, Morgan CL, Ge Y, Galbraith J (2012) Spectral reflectance variability from soil physicochemical properties in oil contaminated soils. Geoderma 177:80–89

Che R, Qin J, Tahmasbian I, Wang F, Zhou S, Xu Z, Cui X (2018) Litter amendment rather than phosphorus can dramatically change inorganic nitrogen pools in a degraded grassland soil by affecting nitrogen-cycling microbes. Soil Biol Biochem 120:145–152

Chen B, Yuan M (2012) Enhanced dissipation of polycyclic aromatic hydrocarbons in the presence of fresh plant residues and their extracts. Environ Pollut 161:199–205

Chen B, Wang Y, Hu D (2010) Biosorption and biodegradation of polycyclic aromatic hydrocarbons in aqueous solutions by a consortium of white-rot fungi. J Hazard Mater 179(1–3):845–851

Chen B, Liu E, Tian Q, Yan C, Zhang Y (2014) Soil nitrogen dynamics and crop residues. A review. Agron Sustain Dev 34:429–442

Chen X, Yang L, Myneni SC, Deng Y (2019) Leaching of polycyclic aromatic hydrocarbons (PAHs) from sewage sludge-derived biochar. Chem Eng J 373:840–845

Cheng PC, Lin YC, Lin MS, Lin SL, Hsiao YH, Huang CY, Cheng SF (2023) Phytoremediation efficiency of weathered petroleum-contaminated soils by *Vetiveria zizanioides* and *Cymbopogon nardus*. Eng Proc 38(1):63

Chunyan X, Qaria MA, Qi X, Daochen Z (2023) The role of microorganisms in petroleum degradation: current development and prospects. Sci Total Environ 865:161112

Cipullo S, Prpich G, Campo P, Coulon F (2018) Assessing bioavailability of complex chemical mixtures in contaminated soils: Progress made and research needs. Sci Total Environ 615:708–723

Cornelissen G, Hassell KA, Van Noort PCM, Kraaij R, Van Ekeren PJ (2000) Slow desorption of PCBs and chlorobenzenes from soils and sediments: relations with sorbent and sorbate characteristics. Environ Pollut 108(1):69–80

Daassi D, Qabil Almaghribi F (2022) Petroleum-contaminated soil: environmental occurrence and remediation strategies. 3 Biotech 12(6):139

Darma UZ, Aziz NAA, Zulkefli SZ, Mustafa M (2016) Identification of phenanthrene and pyrene degrading bacteria from used engine oil contaminated soil. Int J Sci Eng Res 7:680–686

Deng MC, Li J, Liang FR, Yi M, Xu XM, Yuan JP, Wang JH (2014) Isolation and characterization of a novel hydrocarbon-degrading bacterium *Achromobacter sp.* HZ01 from the crude oil-contaminated seawater at the Daya Bay, southern China. Mar Pollut Bull 83(1):79–86

Di Gregorio S, Siracusa G, Becarelli S, Mariotti L, Gentini A, Lorenzi R (2016) Isolation and characterization of a hydrocarbonoclastic bacterial enrichment from total petroleum hydrocarbon contaminated sediments: potential candidates for bioaugmentation in bio-based processes. Environmental Science and Pollution Research, 23(11):10587–10594

Dindar E, Şağban F, Başkaya HS (2017) Biodegradation of crude oil-contaminated soil using canned-food-industry wastewater sludge for soil application. KSCE J Civil Eng 21:1623

Dores-Silva PR, Landgraf MD, Rezende MO (2018) Humification process in different kinds of organic residue by composting and vermicomposting: have microbioreactors really accelerated the process? Environ Sci Pollut Res 25:17490–17498

Dubrovskaya E, Pozdnyakova N, Golubev S, Muratova A, Grinev V, Bondarenkova A, Turkovskaya O (2017) Peroxidases from root exudates of *Medicago sativa* and *Sorghum bicolor*: catalytic properties and involvement in PAH degradation. Chemosphere 169:224–232

Etim OE, Bassey UE, Nnamudi AC (2019) Impact of seed powder of *Dacryodes edulis* on the levels of some polycyclic aromatic hydrocarbons and heavy metals in crude oil polluted *Telfairiaoccidentalis* model. Int J Sci Res Publ 9(6):545–549

Ezenne GI, Nwoke OA, Ezikpe DE, Obalum SE, Ugwuishiwu BO (2014) Use of poultry droppings for remediation of crude-oil-polluted soils: effects of application rate on total and polyaromatic hydrocarbon concentrations. Int Biodeterior Biodegradation 92:57–65

Falciglia PP, Giustra MG, Vagliasindi FGA (2011) Soil texture affects adsorption capacity and removal efficiency of contaminants in ex situ remediation by thermal desorption of diesel-contaminated soils. Chem Ecol 27(S1):119–130

Faostat (2016) Agriculture Organization of the United Nations Statistics Division 2014. Production Available in: ⟨http://faostat3faoorg/browse/Q/QC/S⟩

Farzadkia M, Esrafili A, Gholami M, Koolivand A (2019) Effect of immature and mature compost addition on petroleum contaminated soils composting: kinetics. J Environ Health Sci Eng 17:839–846

Fatima K, Imran A, Amin I, Khan QM, Afzal M (2018) Successful phytoremediation of crude-oil contaminated soil at an oil exploration and production company by plants-bacterial synergism. Int J Phytoremediation 20(7):675–681

Figueiredo MA, Leite MGP, Kozovits AR (2016) Influence of soil texture on nutrients and potentially hazardous elements in *Eremanthuserythropappus*. Int J Phytoremediation 18(5):487–493

Fine P, Graber ER, Yaron B (1997) Soil interactions with petroleum hydrocarbons: abiotic processes. Soil Technol 10(2):133–153

Garcés Mejía AC, Pino NJ, Penuela GA (2018) Effect of secondary metabolites present in Brassica nigra root exudates on anthracene and phenanthrene degradation by rhizosphere microorganism. Environ Eng Sci 35(3):203–209

Gielnik A, Pechaud Y, Huguenot D, Cébron A, Riom JM, Guibaud G (2019) Effect of digestate application on microbial respiration and bacterial communities' diversity during bioremediation of weathered petroleum hydrocarbons contaminated soils. Sci Total Environ 670:271–281

Gómez-Sagasti MT, Hernández A, Artetxe U, Garbisu C, Becerril JM (2018) How valuable are organic amendments as tools for the phytomanagement of degraded soils? The knowns, known unknowns, and unknowns. Front Sustain Food Syst 2:68

Goswami M, Chakraborty P, Mukherjee K, Mitra G, Bhattacharyya P, Dey S, Tribedi P (2018) Bioaugmentation and biostimulation: a potential strategy for environmental remediation. J Microbiol Exp 6(5):223–231

Gul S, Whalen JK, Thomas BW, Sachdeva V, Deng H (2015) Physico-chemical properties and microbial responses in biochar-amended soils: mechanisms and future directions. Agric Ecosyst Environ 206:46–59

Hajabbasi MA (2016) Importance of soil physical characteristics for petroleum hydrocarbons phytoremediation: a review. Afr J Environ Sci Technol 10(11):394–405

Hamid Y, Tang L, Hussain B, Usman M, Lin Q, Rashid MS, Yang X (2020) Organic soil additives for the remediation of cadmium contaminated soils and their impact on the soil-plant system: a review. Sci Total Environ 707:136121

Harmsen J, Rietra RP (2018) 25 years monitoring of PAHs and petroleum hydrocarbons biodegradation in soil. Chemosphere 207:229–238

Hasan A, Kamal RS, Farag RK, Abdel-Raouf ME (2024) Petroleum sludge formation and its treatment methodologies: a review. Environ Sci Pollut Res 4:1–8

Hegazy AK, Hussein ZS, Mohamed NH, Safwat G, El-Dessouky MA, Imbrea I, Imbrea F (2023) Assessment of Vinca rosea (Apocynaceae) potentiality for remediation of crude petroleum oil pollution of soil. Sustainability 15(14):11046

Hoang SA, Lamb D, Seshadri B, Sarkar B, Choppala G, Kirkham MB, Bolan NS (2021) Rhizoremediation as a green technology for the remediation of petroleum hydrocarbon-contaminated soils. J Hazard Mater 401:123282

Huang Y, Pan H, Wang Q, Ge Y, Liu W, Christie P (2019) Enrichment of the soil microbial community in the bioremediation of a petroleum-contaminated soil amended with rice straw or sawdust. Chemosphere 224:265–271

Huguenot D, Mousset E, van Hullebusch ED, Oturan MA (2015) Combination of surfactant enhanced soil washing and electro-Fenton process for the treatment of soils contaminated by petroleum hydrocarbons. J Environ Manag 153:40–47

Hussain I, Puschenreiter M, Gerhard S, Schöftner P, Yousaf S, Wang A (2017) Rhizoremediation of petroleum hydrocarbon-contaminated soils: improvement opportunities and field applications. Environ Exp Bot 147:202–219

Ikeura H, Kawasaki Y, Kaimi E, Nishiwaki J, Noborio K, Tamaki M (2016) Screening of plants for phytoremediation of oil-contaminated soil. Int J Phytoremediation 18(5):460–466

Jia H, Li J, Li Y, Lu H, Liu J, Yan C (2020) The remediation of PAH contaminated sediment with mangrove plant and its derived biochars. J Environ Manag 268:110410

Khan S, Waqas M, Ding F, Shamshad I, Arp HPH, Li G (2015) The influence of various biochars on the bioaccessibility and bioaccumulation of PAHs and potentially toxic elements to turnips (*Brassica rapa* L.). J Hazard Mater 300:243–253

Khan MA, Biswas B, Smith E, Naidu R, Megharaj M (2018) Toxicity assessment of fresh and weathered petroleum hydrocarbons in contaminated soil-a review. Chemosphere 212:755–767

Kiamarsi Z, Kafi M, Soleimani M, Nezami A, Lutts S (2020) Conjunction of *Vetiveria zizanioides* L. and oil-degrading bacteria as a promising technique for remediation of crude oil-contaminated soils. J Clean Prod 253:119719

Kim SU, Owens VN, Kim SY, Hong CO (2017) Effect of different way of bottom ash and compost application on phytoextractability of cadmium in contaminated arable soil. Appl Biol Chem 60:353–362

Kong L, Gao Y, Zhou Q, Zhao X, Sun Z (2018) Biochar accelerates PAHs biodegradation in petroleum-polluted soil by biostimulation strategy. J Hazard Mater 343:276–284

Kopinke FD, Ramus K, Poerschmann J, Georgi A (2011) Kinetics of desorption of organic compounds from dissolved organic matter. Environ Sci Technol 45(23):10013–10019

Koshlaf E, Shahsavari E, Haleyur N, Osborn AM, Ball AS (2019) Effect of biostimulation on the distribution and composition of the microbial community of a polycyclic aromatic hydrocarbon-contaminated landfill soil during bioremediation. Geoderma 338:216–225

Koshlaf E, Shahsavari E, Haleyur N, Osborn AM, Ball AS (2020) Impact of necrophytoremediation on petroleum hydrocarbon degradation, ecotoxicity and soil bacterial community composition in diesel-contaminated soil. Environ Sci Pollut Res 27:31171–31183

Kuppusamy S, Maddela NR, Megharaj M, Venkateswarlu K, Kuppusamy S, Maddela NR, Megharaj M, Venkateswarlu K (2020) An overview of total petroleum hydrocarbons. In: Total petroleum hydrocarbons: environmental fate, toxicity, and remediation. Springer International Publishing, Cham, pp 1–27 https://doi.org/10.1007/978-3-030-24035-6_1

Lal R (2005) World crop residues production and implications of its use as a biofuel. Environ Int 31(4):575–584

Lamichhane S, Krishna KB, Sarukkalige R (2016) Polycyclic aromatic hydrocarbons (PAHs) removal by sorption: a review. Chemosphere 148:336–353

Li X, Song Y, Yao S, Bian Y, Gu C, Yang X, Jiang X (2019) Can biochar and oxalic acid alleviate the toxicity stress caused by polycyclic aromatic hydrocarbons in soil microbial communities? Sci Total Environ 695:133879

Lily MK, Bahuguna A, DangwalK GV (2009) Degradation of benzo [a] pyrene by a novel strain *Bacillus subtilis* BMT4i (MTCC 9447). Braz J Microbiol 40:884–892

Liu PWG, Chang TC, Chen CH, Wang MZ, Hsu HW (2013) Effects of soil organic matter and bacterial community shift on bioremediation of diesel-contaminated soil. Int Biodeterior Biodegradation 85:661–670

Liu X, Selonen V, Steffen K, Surakka M, Rantalaine AL, Romantschuk M, Sinkkonen A (2019) Meat and bone meal as a novel biostimulation agent in hydrocarbon contaminated soils. Chemosphere 225:574–578

Marchal G, Smith KE, Mayer P, de Jonge LW, Karlson UG (2014) Impact of soil amendments and the plant rhizosphere on PAH behaviour in soil. Environ Pollut 188:124–131

Markway HN, Wolf DC, Davis KJ, Gbur EE (2008) Using biosolids to enhance phytoremediation of oil-contaminated soil. Discovery, The Student Journal of Dale Bumpers College of Agricultural, Food and Life Sciences 9(1):50–56. Retrieved from https://scholarworks.uark.edu/discoverymag/vol9/iss1/10

Martin BC, George SJ, Price CA, Ryan MH, Tibbett M (2014) The role of root exuded low molecular weight organic anions in facilitating petroleum hydrocarbon degradation: current knowledge and future directions. Sci Total Environ 472:642–653

Medina J, Monreal C, Barea JM, Arriagada C, Borie F, Cornejo P (2015) Crop residue stabilization and application to agricultural and degraded soils: a review. Waste Manag 42:41–54

Miller JJ, Beasley BW, Yanke LJ, Larney FJ, McAllister TA, Olson BM (2003) Bedding and seasonal effects on chemical and bacterial properties of feedlot cattle manure. J Environ Qual 32(5):1887–1894

Mohan D, Kumar H, Sarswat A, Alexandre-Franco M, Pittman CU Jr (2014) Cadmium and lead remediation using magnetic oak wood and oak bark fast pyrolysis bio-chars. Chem Eng J 236:513–528

Montreemuk J, Stewart TN, Prapagdee B (2023) Bacterial-assisted phytoremediation of heavy metals: concepts, current knowledge, and future directions. Environ Technol Innov 18:103488

Moubasher HA, Hegazy AK, Mohamed NH, Moustafa YM, Kabiel HF, Hamad AA (2015) Phytoremediation of soils polluted with crude petroleum oil using *Bassia scoparia* and its associated rhizosphere microorganisms. Int Biodeterior Biodegradation 98:113–120
Myazin VA, Korneykova MV, Chaporgina AA, Fokina NV, Vasilyeva GK (2021) The effectiveness of biostimulation, bioaugmentation and sorption-biological treatment of soil contaminated with petroleum products in the Russian subarctic. Microorganisms 9(8):1722
Nedjimi B (2021) Phytoremediation: a sustainable environmental technology for heavy metals decontamination. SN Appl Sci 3(3):286
Nero BF (2021) Phytoremediation of petroleum hydrocarbon-contaminated soils with two plant species: *Jatropha curcas* and *Vetiveria zizanioides* at Ghana Manganese Company Ltd. Int J Phytoremediation 23(2):171–180
Nunes DAD, Salgado AM, Gama-Rodrigues EFD, Taketani RG, Cunha CDD, Sérvulo EFC (2020) Use of plant materials for the bioremediation of soil from an industrial site. J Environ Sci Health A 55(6):650–660
Nwaichi EO, Frac M, Nwoha PA, Eragbor P (2015) Enhanced phytoremediation of crude oil-polluted soil by four plant species: effect of inorganic and organic bioaugumentation. Int J Phytoremediation 17(12):1253–1261
Nwankwegu AS, Orji MU, Onwosi CO (2016) Studies on organic and in-organic biostimulants in bioremediation of diesel-contaminated arable soil. Chemosphere 162:148–156
Olawale O, Obayomi KS, Dahunsi SO, Folarin O (2020) Bioremediation of artificially contaminated soil with petroleum using animal waste: cow and poultry dung. Cogent Eng 7(1):1721409
Oleszczuk P, Kuśmierz M, Godlewska P, Kraska P, Pałys E (2016) The concentration and changes in freely dissolved polycyclic aromatic hydrocarbons in biochar-amended soil. Environ Pollut 214:748–755
Ouvrard S, Chenot ED, Masfaraud JF, Schwartz C (2013) Long-term assessment of natural attenuation: statistical approach on soils with aged PAH contamination. Biodegradation 24:539–548
Panchenko L, Muratova A, Dubrovskaya E, Golubev S, Turkovskaya O (2023) Natural and technical phytoremediation of oil-contaminated soil. Life 13(1):177
Park JH, Lamb D, Paneerselvam P, Choppala G, Bolan N, Chung JW (2011) Role of organic amendments on enhanced bioremediation of heavy metal (loid) contaminated soils. J Hazard Mater 185(2–3):549–574
Patel AB, Shaikh S, Jain KR, Desai C, Madamwar D (2020) Polycyclic aromatic hydrocarbons: sources, toxicity, and remediation approaches. Front Microbiol 11:562813
Phillips LA, Greer CW, Farrell RE, Germida JJ (2012) Plant root exudates impact the hydrocarbon degradation potential of a weathered-hydrocarbon contaminated soil. Appl Soil Ecol 52:56–64
Pierzynski GM, Vance GF, Sims JT (2005) Soils and environmental quality. CRC press
Providenti MA, Lee H, Trevors JT (1993) Selected factors limiting the microbial degradation of recalcitrant compounds. J Indus Microbiol 12:379–395
Ptaszek N, Pacwa-Płociniczak M, Noszczyńska M, Płociniczak T (2020) Comparative study on multiway enhanced bio-and phytoremediation of aged petroleum-contaminated soil. Agronomy 10(7):947
Puglisi E, Cappa F, Fragoulis G, Trevisan M, Del Re AA (2007) Bioavailability and degradation of phenanthrene in compost amended soils. Chemosphere 67(3):548–556
Qin G, Gong D, Fan MY (2013) Bioremediation of petroleum-contaminated soil by biostimulation amended with biochar. Int Biodeterior Biodegradation 85:150–155
Quilty JR, Cattle SR (2011) Use and understanding of organic amendments in Australian agriculture: a review. Soil Res 49(1):1–26
Rahbeh ME, Mohtar RH (2007) Application of multiphase transport models to field remediation by air sparging and soil vapor extraction. J Hazard Mater 143(1–2):156–170
Razafimbelo T, Chevallier T, Albrecht A, Chapuis-Lardy L, Rakotondrasolo FN, Michellon R, Bernoux M (2013) Texture and organic carbon contents do not impact amount of carbon protected in Malagasy soils. Sci Agric 70:204–208

Ren X, Zeng G, Tang L, Wang J, Wan J, Liu Y, Deng R (2018a) Sorption, transport and biodegradation–an insight into bioavailability of persistent organic pollutants in soil. Sci Total Environ 610:1154–1163

Ren X, Zeng G, Tang L, Wang J, Wan J, Wang J, Deng Y, Liu Y, Peng B (2018b) The potential impact on the biodegradation of organic pollutants from composting technology for soil remediation. Waste Manag 72:138–149

Rivera-Espinoza Y, Dendooven L (2004) Dynamics of carbon, nitrogen and hydrocarbons in diesel-contaminated soil amended with biosolids and maize. Chemosphere 54(3):379–386

Rivett MO, Wealthall GP, Dearden RA, McAlary TA (2011) Review of unsaturated-zone transport and attenuation of volatile organic compound (VOC) plumes leached from shallow source zones. J Contam Hydrol 123(3–4):130–156

Robichaud K, Lebeau M, Martineau S, Amyot M (2019) Bioremediation of engine-oil contaminated soil using local residual organic matter. Peer J 7:e7389

Rodriguez-Campos J, Dendooven L, Alvarez-Bernal D, Contreras-Ramos SM (2014) Potential of earthworms to accelerate removal of organic contaminants from soil: a review. Appl Soil Ecol 79:10–25

Ruley JA, Amoding A, Tumuhairwe JB, Basamba TA, Opolot E, Oryem-Origa H (2020) Enhancing the phytoremediation of hydrocarbon-contaminated soils in the Sudd wetlands, South Sudan, using organic manure. Appl Environ Soil Sci 2020:1–8

Sabreena HS, Bhat SA, Kumar V, Ganai BA, Ameen F (2022) Phytoremediation of heavy metals: an indispensable contrivance in green remediation technology. Plants 11(9):1255

Sarkar D, Ferguson M, Datta R, Birnbaum S (2005) Bioremediation of petroleum hydrocarbons in contaminated soils: comparison of biosolids addition, carbon supplementation, and monitored natural attenuation. Environ Pollut 136(1):187–195

Sarma H, Nava AR, Prasad MNV (2019) Mechanistic understanding and future prospect of microbe-enhanced phytoremediation of polycyclic aromatic hydrocarbons in soil. Environ Technol Int 13:318–330

Scotti R, Bonanomi G, Scelza R, Zoina A, Rao MA (2015) Organic amendments as sustainable tool to recovery fertility in intensive agricultural systems. J Soil Sci Plant Nutr 15(2):333–352

Serrano A, Gallego M, González JL, Tejada M (2008) Natural attenuation of diesel aliphatic hydrocarbons in contaminated agricultural soil. Environ Pollut 151(3):494–502

Shahsavari E, Adetutu EM, Anderson PA, Ball AS (2013) Plant residues—a low cost, effective bioremediation treatment for petrogenic hydrocarbon-contaminated soil. Sci Total Environ 443:766–774

Sharma I (2020) Bioremediation techniques for polluted environment: concept, advantages, limitations, and prospects. In: Trace metals in the environment-new approaches and recent advances. IntechOpen

Sharma JK, Kumar N, Singh NP, Santal AR (2023) Phytoremediation technologies and their mechanism for removal of heavy metal from contaminated soil: an approach for a sustainable environment. Front Plant Sci 14:1076876

Sigmund G, Poyntner C, Piñar G, Kah M, Hofmann T (2018) Influence of compost and biochar on microbial communities and the sorption/degradation of PAHs and NSO-substituted PAHs in contaminated soils. J Hazard Mater 345:107–113

Simantiraki F, Kollias CG, Maratos D, Hahladakis J, Gidarakos E (2013) Qualitative determination and application of sewage sludge and municipal solid waste compost for BTEX removal from groundwater. J Environ Chem Eng 1(1–2):9–17

Steliga T, Kluk D (2021) Assessment of the suitability of *Melilotus officinalis* for phytoremediation of soil contaminated with petroleum hydrocarbons (TPH and PAH), Zn, Pb and Cd based on toxicological tests. Toxics 9(7):148

Sui X, Wang X, Li Y, Ji H (2021) Remediation of petroleum-contaminated soils with microbial and microbial combined methods: advances, mechanisms, and challenges. Sustainability 13(16):9267

Tang KHD, Angela J (2019) Phytoremediation of crude oil-contaminated soil with local plant species. In: IOP conference series: materials science and engineering, vol 495. IOP Publishing, p 012054

Tao H, Zhao Z, Bartlam M, Wang Y (2016) Combination of biochar amendment and phytoremediation for hydrocarbon removal in petroleum-contaminated soil. Environ Sci Pollut Res 23:21219–21228

Tejeda-Agredano MC, Gallego S, Vila J, Grifoll M, Ortega-Calvo JJ, Cantos M (2013) Influence of the sunflower rhizosphere on the biodegradation of PAHs in soil. Soil Biol Biochem 57:830–840

Thangarajan R, Bolan NS, Tian G, Naidu R, Kunhikrishnan A (2013) Role of organic amendment application on greenhouse gas emission from soil. Sci Total Environ 465:72–96

Thavamani P, Smith E, Kavitha R, Mathieson G, Megharaj M, Srivastava P, Naidu R (2015) Risk based land management requires focus beyond the target contaminants—a case study involving weathered hydrocarbon contaminated soils. Environ Technol Innov 4:98–109

Tiwari J, Kumar A, Kumar N (2017) Phytoremediation potential of industrially important and biofuel plants: *Azadirachta indica* and *Acacia nilotica*. In: Phytoremediation potential of bioenergy plants. Springer Singapore, pp 211–254

Trinh HT, Duong HT, Ta TT, Van Cao H, Strobel BW, Le GT (2017) Simultaneous effect of dissolved organic carbon, surfactant, and organic acid on the desorption of pesticides investigated by response surface methodology. Environ Sci Pollut Res 24:19338–19346

Truskewycz A, Gundry TD, Khudur LS, Kolobaric A, Taha M, Shahsavari E (2019) Petroleum hydrocarbon contamination in terrestrial ecosystems—fate and microbial responses. Molecules 24(18):3400

Udume OA, Abu GO, Stanley HO, Vincent-Akpu IF, Momoh Y, Eze MO (2023) Biostimulation of petroleum-contaminated soil using organic and inorganic amendments. Plants 12(3):431

Wang Z, Xu Y, Zhao J, Li F, Gao D, Xing B (2011) Remediation of petroleum contaminated soils through composting and rhizosphere degradation. J Hazard Mater 190(1–3):677–685

Wang MC, Chen YT, Chen SH, Chien SC, Sunkara SV (2012) Phytoremediation of pyrene contaminated soils amended with compost and planted with ryegrass and alfalfa. Chemosphere 87(3):217–225

Wang Y, Feng J, Lin Q, Lyu X, Wang X, Wang G (2013) Effects of crude oil contamination on soil physical and chemical properties in Momoge wetland of China. Chin Geogr Sci 23:708–715

Wang J, Liu R, Zhang P, Yu W, Shen Z, Feng C (2014) Spatial variation, environmental assessment and source identification of heavy metals in sediments of the Yangtze River Estuary. Mar Pollut Bull 87(1–2):364–373

Wang S, Wang X, Zhang C, Li F, Guo G (2016) Bioremediation of oil sludge contaminated soil by land farming with added cotton stalks. Int Biodeterior Biodegradation 106:150–156

Waqas M, Khan S, Qing H, Reid BJ, Chao C (2014) The effects of sewage sludge and sewage sludge biochar on PAHs and potentially toxic element bioaccumulation in *Cucumis sativa* L. Chemosphere 105:53–61

Wei Z, Zhao X, Zhu C, Xi B, Zhao Y, Yu X (2014) Assessment of humification degree of dissolved organic matter from different composts using fluorescence spectroscopy technology. Chemosphere 95:261–267

Wei Z, Wang JJ, Gaston LA, Li J, Fultz LM, DeLaune RD, Dodla SK (2020) Remediation of crude oil-contaminated coastal marsh soil: integrated effect of biochar, rhamnolipid biosurfactant and nitrogen application. J Hazard Mater 396:122595

Wen J, Yi Y, Zeng G (2016) Effects of modified zeolite on the removal and stabilization of heavy metals in contaminated lake sediment using BCR sequential extraction. J Environ Manag 178:63–69

Wojtera-Kwiczor J, Żukowska W, Graj W, Małecka A, Piechalak A, Ciszewska L, Tomaszewska B (2014) Rhizoremediation of diesel-contaminated soil with two rapeseed varieties and petroleum degraders reveals different responses of the plant defense mechanisms. Int J Phytoremediation 16(7–8):770–789

Wojtowicz K, Steliga T, Kapusta P, Brzeszcz J (2023) Oil-contaminated soil remediation with biodegradation by autochthonous microorganisms and phytoremediation by maize (*Zea mays*). Molecules 28(16):6104

Wu G, Kechavarzi C, Li X, Sui H, Pollard SJ, Coulon F (2013) Influence of mature compost amendment on total and bioavailable polycyclic aromatic hydrocarbons in contaminated soils. Chemosphere 90(8):2240–2246

Wu T, Xu J, Liu J, Guo WH, Li XB, Xia JB, Wang RQ (2019) Characterization and initial application of endophytic *Bacillus safensis* strain ZY16 for improving phytoremediation of oil-contaminated saline soils. Front Microbiol 10:991

Xiao R, Bai J, Huang L, Zhang H, Cui B, Liu X (2013) Distribution and pollution, toxicity and risk assessment of heavy metals in sediments from urban and rural rivers of the Pearl River delta in southern China. Ecotoxicology 22:1564–1575

Xiao R, Bai J, Wang J, Lu Q, Zhao Q, Cui B, Liu X (2014) Polycyclic aromatic hydrocarbons (PAHs) in wetland soils under different land uses in a coastal estuary: toxic levels, sources and relationships with soil organic matter and water-stable aggregates. Chemosphere 110:8–16

Xu Y, Lu M (2010) Bioremediation of crude oil-contaminated soil: comparison of different biostimulation and bioaugmentation treatments. J Haz Mater 183(1–3):395–401

Xu N, Bao M, Sun P, Li Y (2013) Study on bioadsorption and biodegradation of petroleum hydrocarbons by a microbial consortium. Bioresour Technol 149:22–30

Zhang J, He M (2013) Effect of dissolved organic matter on sorption and desorption of phenanthrene onto black carbon. J Environ Sci 25(12):2378–2383

Zhu X, Chen B, Zhu L, Xing B (2017) Effects and mechanisms of biochar-microbe interactions in soil improvement and pollution remediation: a review. Environ Pollut 227:98–115

Zhu Q, Wu Y, Zeng J, Zhang T, Lin X (2020) Influence of organic amendments used for benz [a] anthracene remediation in a farmland soil: pollutant distribution and bacterial changes. J Soils Sediments 20:32–41

Chapter 10
Electrokinetic-Enhanced Bioremediation of Petroleum Contaminated Soil and Sediment

Hossein Beyrami, Mahdieh Khorshidi, and Flávio Henrique Silveira Rabêlo

Abstract Petroleum hydrocarbons contaminate soil through anthropogenic activities. These substances pose risks to the environment and human health at very low concentrations. Recent research has explored a novel hybrid strategy that enhances petroleum hydrocarbon pollution bioremediation using electrokinetics (EK). Bioelectrokinetic remediation offers effective treatment of petroleum-contaminated soil. Although it has several drawbacks, such as extreme pH levels and rising temperatures, bioelectrokinetic remediation has significant potential in low-permeability soils and sediments, where traditional bioremediation methods are difficult to apply. Soils with high hydraulic conductivity and permeability have low negative impact on microorganisms, soil or sediment characteristics, and are well-suited for bioelectrochemical remediation. Due to limitations of the microbial electron transfer pathway, the treatment period for bioelectrokinetic remediation is long. Comprehensive investigations of specific petroleum-contaminated soils should be undertaken to enhance the equipment, process parameters, and scale up technologies. Further optimization of bioelectrokinetics and remediation strategies necessitates research in environmental engineering, electrochemistry, and biology. Before remediation, a site-specific approach based on the characteristics of petroleum-contaminated soils and sediments is crucial; effective petroleum-degrading bacteria must be isolated; and various technologies employed to meet different objectives of practical remediation. Additionally, remediation must consider mass balance, end product, and overall pollutant removal efficiencies. Ultimately, the primary criteria for restoring petroleum-con-

H. Beyrami (✉)
National Salinity Research Center, Agricultural Research, Education and Extension Organization (AREEO), Yazd, Iran

M. Khorshidi
Department of Plant Protection, Faculty of Agriculture, University of Tabriz, Tabriz, Iran

F. H. S. Rabêlo
Luiz de Queiroz College of Agriculture, University of São Paulo, Piracicaba, Brazil

A. Parnian et al. (eds.), *Bioremediation of Petroleum Contaminated Soil and Sediment*, Environmental Contamination Remediation and Management,
https://doi.org/10.1007/978-3-031-95524-2_10

taminated soils should be adaptability, sustainability, environmental friendliness, and economic viability.

Keywords Bio-electrokinetic remediation · Elctrokinetic · Hydrocarbon · Contaminated

10.1 Introduction

Petroleum hydrocarbons are one of the main energy sources used worldwide. Ensuing hydrocarbon pollution of the environment has grown to be a significant issue, endangering human health and causing severe and persistent environmental damage. Bioremediation is the most economic way to clean up contaminated soils (Chen et al. 2015). In nature, environmental microorganisms are commonplace, especially bacteria. Indigenous bacteria have the innate ability to digest a wide range of substances, including contaminants, as food sources. This offers a fantastic chance to use these microorganisms to clean up contaminated soils and sediments. Environmental factors facilitate specific bio-chemical process and the interactions among microorganisms, pollutants, electron acceptors/donors, and nutrients are necessary for bioremediation (Sturman et al. 1995).

In situ bio-degradation may be inhibited by contaminant bio-availability, the immediate interaction between chemicals and microorganisms required for contaminant bio-degradation, and bio-accessibility; affecting the proportion of components accessible to microorganisms in the environment (Semple et al. 2004). Resultantly, bio-degradation processes take place in the subsurface environment, but do not mitigate risks on site. The following factors typically restrict the effectiveness of in situ bioremediation: (1) limited bio-availability of contaminants (such as in partitioning to aquifer material) for bio-degradation (Lohner and Tiehm 2009); (2) mass transfer of nutrients and electron acceptors to the microorganisms responsible for bio-degradation (Simoni et al. 2001); and (3) adaptation of indigenous microorganisms for specific bioremediation of contaminants (Mrozik and Piotrowska-Seget 2010). Most of the latter factors are highly site-specific (Boopathy 2000). The purpose of coupling electrokinetic to bio-remediation is to curtail restrictions and enhance reclamation measures.

Remediation restrictions are removed by combining electrokinetic methods, techniques that apply low-direct current to subsurface porous materials to effect particular transport phenomena, with bioremediation. The application of EK to enhance the results of bioremediation has been the subject of numerous investigations. Using the concepts of electrokinetic remediation, indigenous soil bacteria may be provided with nutrients supporting their growth and improve their capacity to degrade pollutants (Alshawabkeh 2009; Reddy and Cameselle 2009). Commonly used electrokinetic remediation is known to be effective when used for soils with low hydraulic conductivity (Baek et al. 2009; Darmawan and Wada 2002; Cameselle and Gouveia 2018; Beyrami 2021). It is important to note that the goal of cutting-edge electrokinetic techniques is to extract and separate organic pollutants, radiolucides, and heavy metals from soils, sludge, sediments, and groundwater (Kim et al. 2009; Yang et al. 2020; Cameselle et al. 2020). Under the influence of an electric current applied through an anode (positive electrode) and a cathode (negative electrode) placed in the soil, contaminating ions migrate to the electrode with the opposite charge in the contaminated environment, forming the basis of electrokinetic remediation (Chowdhury et al. 2017; Fu et al. 2017; Yu et al. 2019; Shahmohammadi-Kalalagh et al. 2017). This process happens when a low-voltage electric current charges the soil (Ravera et al. 2006; Al-Hamdan and Reddy 2008; Guedes et al. 2019; Beyrami 2021).

The three processes of ion electromigration (movement in the soil solution), electro-osmosis (movement of fluid across the pores), and electrophoresis (the movement of particles in pore fluid) are collectively referred to as the electrokinetic approach (Fig. 10.1). This technique can be used in both saturated and unsaturated environments to remediate one or more pollutants at the same time (Turer and Genc 2005; Zhu et al. 2015). Meanwhile, the electrokinetic remediation technique is an encouraging approach to reclamation, highly effective in the remediation of pollutants from low-permeability soils and fine textured sediments (Guedes et al. 2019; Beyrami et al. 2021). Water electrolysis near electrodes is a prominent occurrence of soil improvement. Hydrogen ions and oxygen gas are produced when water is oxidized close to the anode, creating an acidic front. Simultaneously, hydroxyl ions and hydrogen gas may be produced in the cathode's vicinity due to water reduction, creating a basic front (Popov et al. 2008; Vázquez et al. 2009). The following denote the reaction:

$$2H_2O \rightarrow 4H^+ + O_2 \uparrow + 4e^- , \quad E^\circ = -1.229 \text{ mV} \quad (\text{anode}) \qquad (10.1)$$

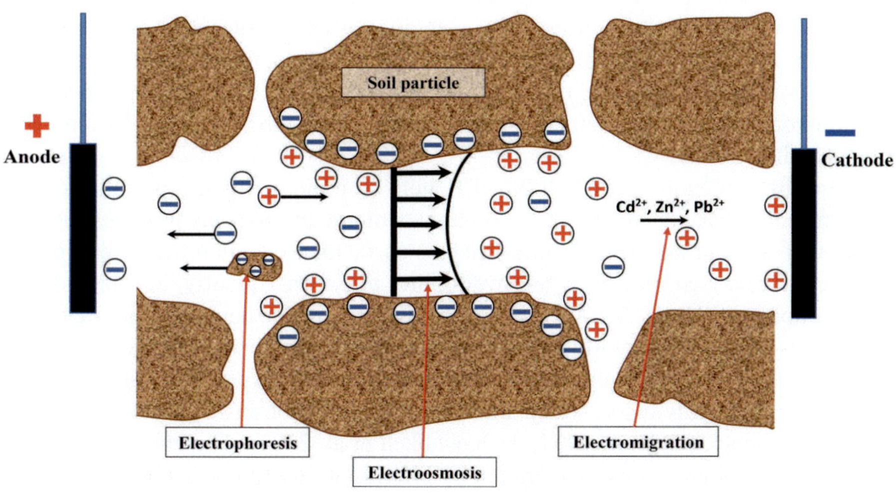

Fig. 10.1 Conceptual model of electrokinetic processes (Beyrami 2021)

$$2H_2O + 2e^- \rightarrow 2OH^- + H_2 \uparrow, \quad E^\circ = -0.828\ \text{mV} \quad (\text{cathode}) \quad (10.2)$$

Conditionally other reactions occur depending on existing chemical species concentration and the reactions' electro-chemical potential. It should be emphasized that the half-water reduction reaction (H_2O/H_2) is prominent during the electrokinetic process, even though secondary reactions may be preferred to the cathode due to their low electro-chemical potential (Lee and Yang 2000; Buchireddy et al. 2009; Yu et al. 2019). Current density, electric field or voltage, and other soil physiochemical parameters, including pH and moisture content, are important determinants of EK efficiency (Haidar et al. 2013; Beyrami 2021). Another is that the joule heating phenomenon causes heat to be created in soil when current is applied affecting the effectiveness of the remediation process.

10.2 Bioremediation Enhanced with EK

By applying a weak electric current (1 V cm^{-1}), bioremediation enhanced with EK, also known as bioelectrokinetic remediation or electro-bioremediation, aims to stimulate the movement and migration of microorganisms, organic pollutants, and nutrients in the subsurface matrix (Hassan et al. 2019a; Kim et al. 2010). Through abiotic effects, the electrokinetic processes accelerate the bio-degradation of organic pollutants in dissolved-phase in the soils. The bio-electrokinetic remediation (BIO-EK) model explains the unique transport phenomena induced by EK, including electromigration, electro-osmosis, and electrophoresis (Saini et al. 2020). However, the electro-osmosis and electromigration rates are strictly related to the transfer medium permeability. Electro-osmosis is important, particularly in low-permeability soils and sediments (Li and Jiang 2021). Electro-osmosis and electromigration are the primary causes of the velocities at which microorganisms migrate (Luo et al. 2006). Other factors will also have an impact on the electrokinetic movement. Microbes, for instance, typically electromigrate quickly in high permeability environments, such as coarse textured soils (0.5 cm^2 V^{-1} h^{-1}) (Suni and Romantschuk 2004).

According to Maini et al. (2000), EK help to remove organic pollutants from soils and sediments that are bound to organic components and clay particles. Studies reveal that without assisting factors (such as surfactants), contaminants will not be completely removed; however, bio-availability and, consequently, bio-degradation can be improved (Niqui-Arroyo et al. 2006). Because of the heterogeneity that naturally arises in electro-osmotic flow from spatial changes in soil pH, electrical conductivity and voltage gradient, whether electro-osmosis assisted desorption without facilitating chemicals successfully improves pollutant biotransformation on a broad scale is unknown. (Eykholt 1997; Alshawabkeh et al. 2004). The efficacy of bioremediation of polluted soils is restricted by bioavailability of contaminants, particularly in clay soils that have a large specific surface area and electronegativity to absorb organic pollutants. This impediment can be circumvented by electrokinetic processes (Kuppusamy et al. 2017). Weak electric field application raises soil temperature and encourages the microbial synthesis of natural biosurfactants (Phulpoto et al. 2023).

Biosurfactants may be efficiently transported into low-permeability soils by electrokinetic injection, making interactions between microorganisms and pollutants more viable. By directing the injection of particular bacteria into contaminated areas, bioaugmentation is enhanced. Wick et al. (2004) investigated the movement and activity of bacteria that degrade Polycyclic aromatic hydrocarbons (PAHs) in electric fields. Contaminant-degrading bacteria might be transported to heavily contaminated locations by combining electro-osmosis and electrophoresis with in situ biological therapy. Nevertheless, the problem with bioelectrokinetic remediation is the accumulation of salinity and harsh pH levels. Problems arise from the fact that water electrolysis will cause base to accumulate near the cathode and acids to form close to the anode (Beyrami et al. 2021). In simple terms, during bioelectrokinetic remediation, devices employed promote interactions between petroleum-degrading bacteria and organic pollutants by speeding up the movement of bacteria, organic components, nutrients, and pore fluids in the soil and sediment.

10.3 Efficiency of Bioelectrokinetic Remediation

Various petroleum hydrocarbons and their derivatives, including diesel, phenol, Total Petroleum Hydrocarbons (TPH), PAHs, and 2,4-dichlorophenol, have been removed from naturally or artificially polluted soil/sediment using bioelectrokinetic remediation in a variety of soil types and electrode materials. Most investigations use TPH and PAHs to track contamination of soil caused by petroleum hydrocarbons. In artificially polluted soils, TPH removal efficiency ranged from 73 to 89% with an initial TPH from 7 to 50 g kg^{-1}. The 1.0 V cm^{-1} was used by Crognale et al. (2020) for the bioelectrokinetic remediation of soils polluted with diesel. The starting concentrations of TPH and PAHs were 0.48 and 0.11 g kg^{-1}, respectively. After 50 days, the removal efficiencies of TPH and PAHs were 76% and 79%, respectively. In sandy loam contaminated with TPH, a maximum removal effectiveness of 89% was attained after 700 hours at an applied electric current of 0.48 V cm^{-1} (Martínez-Prado et al. 2014).

Kim et al. (2010) examined the link between anode distance and the removal efficiency of TPH in diesel-contaminated soils. The greatest removal of TPH was 73% close to the anode after 25 days applying 1 V cm^{-1}, according to the results, which also showed that the removal rate declined with increasing distance between electrodes. For phenanthrene, a common PAH, bioelectrokinetic remediation demonstrated satisfactory removal efficiency. Using electrokinetic injection, electrode polarity reversal, and electrolyte circulation to control pH and achieve uniformity in the inorganic ion distribution, about 80 percent of the initial phenanthrene concentration in soils was removed in 20 days (Xu et al. 2010). In naturally polluted soils containing 78.5 mg kg^{-1} of phenanthrene, removal of phenanthrene improved tenfold when a voltage of 0.6 V cm^{-1} was applied, after 14 days, the maximum removal efficiency was almost 98% (Niqui-Arroyo et al. 2006). Total organic carbon (TOC) can be used to monitor the level of contamination in petroleum-impacted soils. In 25 days, a study with a 1.0 V cm^{-1} voltage gradient in petroleum-contaminated soils demonstrated 82% remediation, as opposed to 36% in the absence of an electric field (Agarry 2017). Additionally, bioelectrokinetic remediation can treat heavy metal and petroleum contaminated soils at the same time. With starting TPH of 12.5 g kg^{-1}, the TPH removal efficiencies under 2.0 V cm^{-1} reached 89% after 30 days (Dong et al. 2013).

One of the main sources of pollution is diesel fuel; bio-electrokinetic remediation of soils contaminated with diesel is research priority. Hassan et al. (2019b) demonstrated that bioelectrokinetics may effectively restore silty clay, in the study diesel removal efficiency varied between 20 and 67%. According to Song et al. (2018), the voltage gradient and diesel elimination has a positive correlation. Diesel elimination effectiveness rose from 44% with an applied voltage of 1 V cm^{-1} to 67% with 2 V cm^{-1} after 15 days of treatment. After 336 hours, a homogeneous removal of diesel roughly 39% was achieved with an applied voltage of 1.0 V cm^{-1} (Ramírez et al. 2015). The effectiveness of bioelectrokinetic remediation in removing particular contaminants, such as chlorinated organics and phenol, from contaminated soils has also been investigated. When the polarity of the electrical field (1.0 V cm^{-1}) was reversed every 3 hours throughout the bio-electrokinetic removal of phenol-polluted soils with 180.2 mg kg^{-1} initial concentration, the remediation effectiveness of phenol reached 58% in 10 days (Luo et al. 2006). In a different investigation, 10 days in a bio-electrokinetic approach with a periodic polarity change every 1.5 hours resulted in a maximum phenol removal efficiency of 67% (Luo et al. 2005). Better performance was demonstrated with a shorter polarity reversal interval, but this requires more energy (Ribeiro and Vara Prasad 2021). Bioelectrokinetic remediation may be used to remove chlorinated organics. By using a bidirectional 1 V cm^{-1} electric current with polarity reversal every 1.5 hours, the bio-degradation of 2,4-dichlorophenol in soils increased to 67% (Fan et al. 2007). By using a combination of bio-remediation and EK using *Dehalococcoides* bacterial strains, perchloroethylene was reduced from 142–464 to 1.7–51 mg kg^{-1} (Mao et al. 2012).

Soil properties, system configuration, electrode material, voltage gradient or electric current intensity, and microbial communities and activity are effective factors and limitations in bioelectrokinetic remediation.

10.3.1 Electrode Material

The initial stage of bioelectrokinetic soil remediation is selection of electrode materials. Three selection criteria should be taken into account in the right electrode materials: (1) high efficiency of the electrode can guarantee a consistent electric energy output to guarantee remediation efficiency; (2) stability, indicating that the electrode resists corrosion well in practical application settings; and (3) economy, electrodes should be inexpensive and appropriate for large-scale uses. Currently metal and inorganic substances, titanium, graphite, stainless steel, and aluminum are frequently used as electrode materials (Mena et al. 2016; Fan et al. 2020; Beyrami 2021). An oxide film develops on the surface of aluminum and titanium electrodes when they were powered, reducing current efficiency and compromising remediation efficency. Stainless steel is easy to install in soils due to its strong mechanical strength, its applicability is limited by its relatively low specific surface area and susceptibility to corrosion in an anoxic medium. Graphite is an excellent electrode material since it is inert and has low cost, consistent power generation, resistance to high temperatures, and resistance to corrosion (Zhang et al. 2021). However, it can be expensive and challenging to install graphite plates in soil, particularly for in situ treatment. Graphite rods have a lower surface area than graphite plates, but are easier to install. For bioelectrokinetic remediation, electrode insertion should be stable, affordable, and simple. It should also be mentioned that certain pollutants, like hexachlorobenzene, polychlorinated biphenyls, and 1-chlorobutane, can be reductively destroyed on the cathode. In the meantime, under alkaline circumstances, the divalent ions of heavy metals in polluted soils will also be deposited close to the cathode region to produce hydroxide precipitates, which will cause bacteria to foul and become poisoned by heavy metals.

10.3.2 System Configurations

In the bioelectrokinetic remediation system, contaminated soils are located in the center of a lab- or bench-scale bioelectrokinetic system, with anodic and cathodic compartments on either side. The system is powered by a direct current power supply, and anode and cathode electrodes are placed inside electrolyte containers. As the systems are often operated in an open-flow configuration, it is feasible to inject processing fluids-such as surfactants, buffers, and fertilizers into contaminated soils (Xu et al. 2010). Single electrode containers are commonly utilized in the treatment of soil contaminated by petroleum because they are straightforward to install and maintain. Multi-electrode container, or bioelectrokinetic remediation, similarly employs a symmetrical configuration of cylindrical electrodes at each of the four corners. Electric field intensity drops from 0.5 V cm^{-1} at the corners to 0 V cm^{-1} in the center (Li et al. 2016). Multi-electrode tanks use many electrode materials, cost is anticipated to be higher than that of a single one. To keep the electrode operating normally, a peristaltic pump has been used to circulate the electrolyte. This device is known as a recirculation tank (Mao et al. 2012).

10.3.3 Voltage Gradient or Electric Current Intensity

Electric field is a crucial component of bioelectrokintic remediation, which includes the voltage gradient and distribution. Microbes, pollutants, and nutrients can all be transported in the soil under the influence of an electric field. Contamination removal is directly impacted by the voltage gradient. According to Agarry (2017), the percentage of total organic carbon (TOC) removed from petroleum-contaminated soils rose as voltages were raised, from 66% in 0.33 V cm^{-1} to 70% in 0.67 V cm^{-1} and 73% in 1 V cm^{-1}. Mena et al. (2016) showed that after 2 weeks, increasing the voltage gradient from 0.5 to 1.5 V cm^{-1}, enables diesel breakdown in polluted soils of 16% to 36%. Li et al. (2020) found a maximum PAH degradation efficiency of 44% at 2 V cm^{-1}. Organic pollutants bio-electrokinetic reclamation from fine-grained soils may only be slightly impacted by applied voltages greater than 1 V cm^{-1} (Yeung and Gu 2011).

Voltage strength and remediation period have an impact on the energy consumption of bioelectrokinetic remediation. The migration of charged organisms, contaminants, and nutrients can be accelerated by a greater voltage gradient, but energy consumed rises quickly. For petroleum hydrocarbon contaminated, bioelectrokinetic remediation, 1 V cm^{-1} voltage gradient is thought to be ideal because a high electric field has a restricted impact, may kill bacteria, and consumes a lot of energy (Ribeiro and Vara Prasad 2021). Electric field operating modes are significant in bioelectrokinetic cleanup. The two primary models used to mimic the real-world distribution of electric fields are uniform and non-uniform electric fields (Hadidi et al. 2020). A two-dimensional electric field with a radial or linear distribution is produced by a matrix of electrodes in a non-uniform electric field. Non-uniform electric fields outperform uniform electric fields in terms of operational stability when it comes to the bioelectrokinetic remediation of polluted soils. Furthermore, fewer electrode materials are needed in non uniform electric field remediation, making it less expensive than uniform electrokinetic treatment (Luo et al. 2005).

10.3.4 Microbial Communities and Activity

Soil microbial populations are an essential component of bioelectrokinetic remediation. Currently, the majority of petroleum-degrading microbes in nature are fungi and bacteria, accounting for over 200 species combined. The primary oxidative genera as TPH degraders were found to be *Streptomyces*, *Nocardioides*, *Arthrobacter*, *Pseudomonas*, and *Bacillus*, with 16S rRNA amplified and sequenced (Ambaye et al. 2022). Certain species, including *Acinetobacter* sp. and *Bacillus* sp., have demonstrated efficacy in bioelectrokinetic remediation of soil contaminated with diesel (Bento et al. 2005). The majority of pure cultures can only digest particular hydrocarbon molecules due to the complexity of the petroleum hydrocarbon component. Mixed microbial populations, therefore, have a better chance of effectively degrading petroleum pollutants. For instance, Chen et al. (2011) remediated petroleum-contaminated soils using the microbial flora *Kineococcus* sp., *Kocuria* sp., and *Acinetobacter* sp., The mixed bacteria had superior removal effects compared to a single bacterium. Exogenous bacteria, which frequently did not even survive in field circumstances, generally perform worse in in situ remediation than local microbes. Nonetheless, some research reveals that in order to remediate soil contaminated by petroleum, exogenous microorganisms including *Bacillus* and *Pseudomonas* may be chosen to consume petroleum hydrocarbons as sources of energy and carbon (Das and Mukherjee 2007).

In the process of remediation, bacteria often create a charged biocolloid that is encased in pollutants and small particles, which they then use electrophoresis to move through soil pores (Ramadan et al. 2018). The transport rate of microbes is typically within the range of 0.06–0.17 cm min^{-1}, depending on biocolloid particle size, charge, and voltage gradient (Mena et al. 2012). A higher voltage gradient will hasten the rate at which bacteria convey information, encourage the use of toxins for their own development and metabolism, and enhance the effectiveness of removing contaminants (Beretta et al. 2019). However, in specific cases it is evident that petroleum-degrading microorganisms may be adversely affected by an extreme voltage gradient (2.0 V cm^{-1}), due to microbial cell membrane penetration (Tiehm et al. 2009). In order to preserve a steady variety and organization of microbial communities, research has determined that a voltage gradient of 1.0–2.0 V cm^{-1} is appropriate (Ramadan et al. 2018). The activities of microorganisms in soil contaminated with petroleum have a significant impact on the removal efficiency of bioelectrokinetic remediation techniques. Bacteria differ in their alkane hydroxylase production and substrate preferences, which results in varying alkane degradation efficiencies for alkanes with varying chain lengths (Liu et al. 2021).

10.3.5 Soil Properties

Moisture content, pH, temperature, and porosity of the soil all impact how well bioelectrokinetic remediation removes contaminants. The concentration of hydrogen ions has a significant impact on microbial biomass, mobility, bio-availability, community structure, and diversity of pollutants and nutrients (Deflaun and Condee 1997). To maintain normal microbial metabolism, pH must be stabalized. Abrupt pH shifts or harsh pH conditions (<3 or >9) have been shown to have detrimental effects on microbial respiration and carbon substrate usage, which in turn inhibit microbial development (Lear et al. 2007; Varia et al. 2013).

After 4 days of operation, Fan et al. (2007) indicated that the pH of soil increased to 9.5 close to the cathode from 7.7 and declined to 0.7 around the anode. Kim et al. (2010) showed that after 25 days of applying a voltage gradient of 1.0 V cm–1, the pH of the soil between the anode and cathode reached 3.5 and 10.8, respectively.

Under a non-uniform electric field, the mobility of phenol and 2,4- Dichlorophenol in soils can be managed by adjusting pH. The mobility of phenol and

2,4- Dichlorophenol significantly increased to 2 and 5 times faster, respectively, when the pH of the soil was raised from 7.7 to 9.3 (Luo et al. 2005). There are a number of ways to prevent or lessen the effects of an extreme pH value, such as adding buffer solutions (such as tris-acetate and bicarbonate buffers), cycling electrolytes, and reversing polarity, but all of these will increase expenses and energy usage (Niqui-Arroyo et al. 2006).

Because of the electric heating brought on by the ohmic drop in low-conductivity soils, EK can raise the temperature of the soil by 1–3 °C (Lear et al. 2007). The soil moisture content and microbial metabolisms may be impacted by temperature changes. The effects of soil heating on bacteria are hotly debated. Some studies assert that increases in soil temperature have no significant effect; however, other studies found that soil heating reduced the efficiency of removing pollutants because adversely impacts microbial metabolism (Ramírez et al. 2015). In bioelectrokinetics remediation, it is also recognized that a rise in temperature accelerates the evaporation of contaminates. Pollutants, particularly volatile organic compounds (VOCs) and semi-VOCs, have the potential to evaporate and create secondary contamination, which calls for bioelectrokinetic remediation (Ward et al. 2003).

In bioelectrokinetic remediation, soil porosity influences the permeability, salinity, water content, temperature, and specific surface area of soils. The method of mass transportation is also impacted by the porosity of the soil (Balseiro-Romero et al. 2018). For instance, the mass transfer rate of oxygen in unsaturated soils is higher due to higher total porosity (Li et al. 2015). Organic substances degrade far more slowly under anaerobic settings than they do under aerobic conditions, despite the fact that nitrates and sulfates can be used as electron acceptors in such scenarios. In general, sandy coarse textured soils, have more permeability than fine-grained clayey material, which has low porosity (Sabari 2021). It is challenging to remove pollutants in low porosity and low permeability soils; nevertheless, biorelectrokinetic removal has emerged as a potentially useful technique for soils of low-permeability. The efficiency of electrokinetic remediation is better in soils with less permeability. The latter is due to that as porosity rises, electro-osmosis influence increases. According to Mena et al. (2016), diesel was removed using a bio-barrier consisting of gravel particles and clay soil, with clay soil demonstrating a greater removal effectiveness of 1.0 V cm^{-1} after 14 days.

Biological degradation and electrokinetic activity require soils that are close to the moisture saturation point, the water content of the soil is a crucial element. According to the literature, the soil moisture decreased under the electric field as the remediation time increased. In the meantime, reports of variations in moisture change over time have been made (Agarry 2017). For instance, the influence of electro-osmosis caused the soil moisture around the anode to decrease by 24% while increasing by 18% around the cathode (Fan et al. 2007). Naturally, adding water and periodically reversing polarity are two popular techniques for keeping soil moisture content at a specific level.

10.3.6 Other Factors

The intrinsic stability of contaminants, which is related to their structural configuration, chemical polarity and molecular weight, has an impact on the remediation of petroleum contaminated soils (Mohan et al. 2006). Since high contamination concentrations in soils would hinder microbial growth and result in less remediation efficiency, contaminant concentration is a further important issue. Specifically, natural soils contaminated with petroleum had more complex contaminant combinations, and TPH remediation efficiency from natural soils on site is often less than that from spiked soils (Barker and Bryson 2002). Additionally, a prior study showed that the effects of electrokinetic remediation of naturally contaminated soil are less severe than in soil that has been intentionally contaminated with petroleum (Mena et al. 2012). Carbon fiber and other conductive particle materials improve electron transfer without interfering with microbial metabolisms. Faster electron transfer pathways are demonstrated by conductive particles and electrodes, which may improve the bioelectrokinetic remediation of PAHs (Mena et al. 2016). Because of their hydrophobic and hydrophilic functional groups, surfactants can dissolve hydrocarbon pollutants in micelle form by lowering surface tension. Surfactants can also increase the bioavailability of contaminants and the effectiveness of their clearance. The type, polarity, dose, and duration of interaction between the surfactants all affect the adsorption capacity of PAHs in practical studies (Yuan et al. 2013). Surfactants that can be employed to make pollutants more soluble include Tween-80, Brij-35, ethylene diamine tetra acetic acid (EDTA), and Triton X-100; nevertheless, they have been shown to have negative impacts on soil microbial activity (Cheng et al. 2017).

10.4 Conclusion

Petroleum hydrocarbons pose a significant threat to human health and the environment, making bioremediation a cost-effective solution for contaminated soils. Bioremediation involves using microorganisms, particularly bacteria, that can metabolize various chemicals as food sources. However, biodegradation can be limited by factors such as contaminant bioavailability and bioaccessibility. Combining bioremediation with EK can overcome these limitations and increase the effectiveness of remediation measures. EK involves applying direct current within subsurface porous media to induce specific transport phenomena, such as electromigration, electro-osmosis, and electrophoresis. Electrokinetic remediation is an advanced technology that separates and extracts organic contaminants, heavy metals, and radiolucides, from soils, sediments, groundwater, and sludge. It can be employed to remediate one or more contaminants simultaneously in both unsaturated and saturated soils/sediments. Bioelectrokinetic remediation, also known as electro-bioremediation, aims to stimulate the movement and transformation of microbes, organic contaminants, and nutrients in the subsurface medium by applying a weak electric current. Bio-electrokinetic remediation method has been used in removal of various petroleum contaminations and their derivatives from natural or artificial contaminated soils, such as diesel, PAHs, TPH, 2,4-dichlorophenol, and phenol. The petroleum-contaminated remediation of soils requires site-specific design, screening for efficient bacteria, and incorporating multiple technologies to meet practical needs. Selection criteria should be economical, flexible, eco-friendly, and sustainable.

10.5 Suggestions

The following suggestions are made to enhance bioremediation efficiency:

I. It is recommended to use electrokinetics combined with bioremediation to increase the removal efficiency of petroleum contaminated soils.
II. Use of graphite electrodes as opposed to other materials in electrokinetic remediation.
III. Optimizing soil properties (such as soil moisture, pH, temperature, and porosity) to increase the efficiency of pollutant removal in bioelectrokinetic remediation.

References

Agarry SE (2017) Enhanced ex-situ bioremediation of soil contaminated with petroleum refinery waste effluents by biostimulation through electrokinetics and inorganic fertilizer. Niger J Technol 36(2):534–542. https://doi.org/10.4314/NJT.V36I2.29

Al-Hamdan AZ, Reddy KR (2008) Transient behavior of heavy metals in soils during electrokinetic remediation. Chemosphere 71(5):860–871. https://doi.org/10.1016/j.chemosphere.2007.11.028

Alshawabkeh AN (2009) Electrokinetic soil remediation: challenges and opportunities. Sep Sci Technol 44(10):2171–2187. https://doi.org/10.1080/01496390902976681

Alshawabkeh AN, Sheahan TC, Wu X (2004) Coupling of electrochemical and mechanical processes in soils under DC fields. Mech Mater 36(5–6):453–465. https://doi.org/10.1016/S0167-6636(03)00071-1

Ambaye TG, Chebbi A, Formicola F, Rosatelli A, Prasad S, Gomez FH, Sbaffoni S, Franzetti A, Vaccari M (2022) Ex-situ bioremediation of petroleum hydrocarbon contaminated soil using mixed stimulants: response and dynamics of bacterial community and phytotoxicity. J Environ Chem Eng 10(6):108814. https://doi.org/10.1016/j.jece.2022.108814

Baek K, Kim DH, Park SW, Ryu BG, Bajargal T, Yang JS (2009) Electrolyte conditioning-enhanced electrokinetic remediation of arsenic-contaminated mine tailing. J Hazard Mater 161. https://doi.org/10.1016/j.jhazmat.2008.03.127

Balseiro-Romero M, Monterroso C, Casares JJ (2018) Environmental fate of petroleum hydrocarbons in soil: review of multiphase transport, mass transfer, and natural attenuation processes. Pedosphere 28(6):833–847. https://doi.org/10.1016/S1002-0160(18)60046-3

Barker AV, Bryson GM (2002) Bioremediation of heavy metals and organic toxicants by composting. Sci World J 2:407–420. https://doi.org/10.1100/tsw.2002.91

Bento FM, Camargo FAO, Okeke BC, Frankenberger WT (2005) Comparative bioremediation of soils contaminated with diesel oil by natural attenuation, biostimulation and bioaugmentation. Bioresour Technol 96(9):1049–1055. https://doi.org/10.1016/j.biortech.2004.09.008

Beretta G, Mastorgio AF, Pedrali L, Saponaro S, Sezenna E (2019) The effects of electric, magnetic and electromagnetic fields on microorganisms in the perspective of bioremediation. Rev Environ Sci Biotechnol 18(1):29–75. https://doi.org/10.1007/s11157-018-09491-9

Beyrami H (2021) Effect of different treatments on electrokinetic remediation of Zn, Pb and Cd from a contaminated calcareous soil. Chin J Chem Eng 38:255–265. https://doi.org/10.1016/j.cjche.2020.09.011

Beyrami H, Shahmohammadi-Kalalagh S, Taran F (2021) Electrokinetic remediation of cadmium contaminated soil in field condition. Environ Eng Manag J 20(12):1883–1891. https://doi.org/10.30638/eemj.2021.177

Boopathy R (2000) Factors limiting bioremediation technologies. Bioresour Technol 74(1):63–67. https://doi.org/10.1016/S0960-8524(99)00144-3

Buchireddy PR, Bricka RM, Gent DB (2009) Electrokinetic remediation of wood preservative contaminated soil containing copper, chromium, and arsenic. J Hazard Mater 162(1):490–497. https://doi.org/10.1016/j.jhazmat.2008.05.092

Cameselle C, Gouveia S (2018) Electrokinetic remediation for the removal of organic contaminants in soils. Curr Opin Electrochem 11. https://doi.org/10.1016/j.coelec.2018.07.005

Cameselle C, Gouveia S, Cabo A (2020) Analysis and optimization of Mn removal from contaminated solid matrixes by Electrokinetic remediation. Int J Environ Res Public Health 17(6):1820. https://doi.org/10.3390/ijerph17061820

Chen J, Yang Q, Huang T, Zhang Y, Ding R (2011) Enhanced bioremediation of soil contaminated with viscous oil through microbial consortium construction and ultraviolet mutation. World J Microbiol Biotechnol 27(6):1381–1389. https://doi.org/10.1007/s11274-010-0589-y

Chen M, Xu P, Zeng G, Yang C, Huang D, Zhang J (2015) Bioremediation of soils contaminated with polycyclic aromatic hydrocarbons, petroleum, pesticides, chlorophenols and heavy metals by composting: applications, microbes and future research needs. Biotechnol Adv 33(6):745–755. https://doi.org/10.1016/J.BIOTECHADV.2015.05.003

Cheng M, Zeng G, Huang D, Yang C, Lai C, Zhang C, Liu Y (2017) Advantages and challenges of Tween 80 surfactant-enhanced technologies for the remediation of soils contaminated with hydrophobic organic compounds. Chem Eng J 314:98–113. https://doi.org/10.1016/j.cej.2016.12.135

Chowdhury AIA, Gerhard JI, Reynolds D, Sleep BE, O'Carroll DM (2017) Electrokinetic-enhanced permanganate delivery and remediation of contaminated low permeability porous media. Water Res 113:215–222. https://doi.org/10.1016/j.watres.2017.02.005

Crognale S, Cocarta DM, Streche C, D'Annibale A (2020) Development of laboratory-scale sequential electrokinetic and biological treatment of chronically hydrocarbon-impacted soils. New Biotechnol 58:38–44. https://doi.org/10.1016/j.nbt.2020.04.002

Darmawan, Wada SI (2002) Effect of clay mineralogy on the feasibility of electrokinetic soil decontamination technology. Appl Clay Sci 20. https://doi.org/10.1016/S0169-1317(01)00080-1

Das K, Mukherjee AK (2007) Crude petroleum-oil biodegradation efficiency of Bacillus subtilis and Pseudomonas aeruginosa strains isolated from a petroleum-oil contaminated soil from North-East India. Bioresour Technol 98(7):1339–1345. https://doi.org/10.1016/j.biortech.2006.05.032

Deflaun MF, Condee CW (1997) Electrokinetic transport of bacteria. J Hazard Mater 55(1–3):263–277. https://doi.org/10.1016/S0304-3894(97)00023-X

Dong ZY, Huang WH, Xing DF, Zhang HF (2013) Remediation of soil co-contaminated with petroleum and heavy metals by the integration of electrokinetics and biostimulation. J Hazard Mater 260:399–408. https://doi.org/10.1016/J.JHAZMAT.2013.05.003

Eykholt GR (1997) Development of pore pressures by nonuniform electroosmosis in clays. J Hazard Mater 55(1–3):171–186. https://doi.org/10.1016/S0304-3894(97)00022-8

Fan X, Wang H, Luo Q, Ma J, Zhang X (2007) The use of 2D non-uniform electric field to enhance in situ bioremediation of 2,4-dichlorophenol-contaminated soil. J Hazard Mater 148(1–2):29–37. https://doi.org/10.1016/j.jhazmat.2007.01.144

Fan R, Ma W, Zhang H (2020) Microbial community responses to soil parameters and their effects on petroleum degradation during bio-electrokinetic remediation. Sci Total Environ 748:142463. https://doi.org/10.1016/j.scitotenv.2020.142463

Fu R, Wen D, Xia X, Zhang W, Gu Y (2017) Electrokinetic remediation of chromium (Cr)-contaminated soil with citric acid (CA) and polyaspartic acid (PASP) as electrolytes. Chem Eng J 316:601–608. https://doi.org/10.1016/j.cej.2017.01.092

Guedes P, Lopes V, Couto N, Mateus EP, Pereira CS, Ribeiro AB (2019) Electrokinetic remediation of contaminants of emergent concern in clay soil: effect of operating parameters. Environ Pollut 253:625–635. https://doi.org/10.1016/j.envpol.2019.07.040

Hadidi H, Kamali R, Manshadi MKD (2020) Numerical simulation of a novel non-uniform electric field design to enhance the electrocoalescence of droplets. Eur J Mech B/Fluids 80:206–215. https://doi.org/10.1016/j.euromechflu.2019.10.010

Haidar M, Dirany A, Sirés I, Oturan N, Oturan MA (2013) Electrochemical degradation of the antibiotic sulfachloropyridazine by hydroxyl radicals generated at a BDD anode. Chemosphere 91(9):1304–1309. https://doi.org/10.1016/J.CHEMOSPHERE.2013.02.058

Hassan IA, Mohamedelhassan EE, Yanful EK, Weselowski B, Yuan Z-C (2019a) Isolation and characterization of novel bacterial strains for integrated solar-bioelectrokinetic of soil contaminated with heavy petroleum hydrocarbons. Chemosphere 237:124514. https://doi.org/10.1016/j.chemosphere.2019.124514

Hassan IA, Mohamedelhassan EE, Yanful EK, Yuan Z (2019b) Mitigation of soil contaminated with diesel fuel using bioelectrokinetics. J Environ Sci Health A 54(5):416–426. https://doi.org/10.1080/10934529.2018.1558903

Kim DH, Jeon CS, Baek K, Ko SH, Yang JS (2009) Electrokinetic remediation of fluorine-contaminated soil: conditioning of anolyte. J Hazard Mater 161. https://doi.org/10.1016/j.jhazmat.2008.03.084

Kim S-H, Han H-Y, Lee Y-J, Kim CW, Yang J-W (2010) Effect of electrokinetic remediation on indigenous microbial activity and community within diesel contaminated soil. Sci Total Environ 408(16):3162–3168. https://doi.org/10.1016/j.scitotenv.2010.03.038

Kuppusamy S, Thavamani P, Venkateswarlu K, Lee YB, Naidu R, Megharaj M (2017) Remediation approaches for polycyclic aromatic hydrocarbons (PAHs) contaminated soils: technological constraints, emerging trends and future directions. Chemosphere 168:944–968. https://doi.org/10.1016/j.chemosphere.2016.10.115

Lear G, Harbottle MJ, Sills G, Knowles CJ, Semple KT, Thompson IP (2007) Impact of electrokinetic remediation on microbial communities within PCP contaminated soil. Environ Pollut 146(1):139–146. https://doi.org/10.1016/j.envpol.2006.06.037

Lee H-H, Yang J-W (2000) A new method to control electrolytes pH by circulation system in electrokinetic soil remediation. J Hazard Mater 77(1–3):227–240. https://doi.org/10.1016/s0304-3894(00)00251-x

Li Y, Jiang L (2021) Comparison of the crude oil removal effects of different surfactants in electrokinetic remediation of low-permeability soil. J Environ Chem Eng 9(4):105190. https://doi.org/10.1016/j.jece.2021.105190

Li X, Wang X, Ren ZJ, Zhang Y, Li N, Zhou Q (2015) Sand amendment enhances bioelectrochemical remediation of petroleum hydrocarbon contaminated soil. Chemosphere 141:62–70. https://doi.org/10.1016/j.chemosphere.2015.06.025

Li T, Wang Y, Guo S, Li X, Xu Y, Wang Y, Li X (2016) Effect of polarity-reversal on electrokinetic enhanced bioremediation of Pyrene contaminated soil. Electrochim Acta 187:567–575. https://doi.org/10.1016/j.electacta.2015.11.097

Li F, Guo S, Wang S, Zhao M (2020) Changes of microbial community and activity under different electric fields during electro-bioremediation of PAH-contaminated soil. Chemosphere 254:126880. https://doi.org/10.1016/j.chemosphere.2020.126880

Liu J, Zhao B, Lan Y, Ma T (2021) Enhanced degradation of different crude oils by defined engineered consortia of Acinetobacter venetianus RAG-1 mutants based on their alkane metabolism. Bioresour Technol 327:124787. https://doi.org/10.1016/J.BIORTECH.2021.124787

Lohner ST, Tiehm A (2009) Application of electrolysis to stimulate microbial reductive PCE Dechlorination and oxidative VC biodegradation. Environ Sci Technol 43(18):7098–7104. https://doi.org/10.1021/es900835d

Luo Q, Zhang X, Wang H, Qian Y (2005) The use of non-uniform electrokinetics to enhance in situ bioremediation of phenol-contaminated soil. J Hazard Mater 121(1–3):187–194. https://doi.org/10.1016/j.jhazmat.2005.02.007

Luo Q, Wang H, Zhang X, Fan X, Qian Y (2006) In situ bioelectrokinetic remediation of phenol-contaminated soil by use of an electrode matrix and a rotational operation mode. Chemosphere 64(3):415–422. https://doi.org/10.1016/j.chemosphere.2005.11.064

Maini G, Sharman AK, Knowles CJ, Sunderland G, Jackman SA (2000) Electrokinetic remediation of metals and organics from historically contaminated soil. J Chem Technol Biotechnol 75(8):657–664. https://doi.org/10.1002/1097-4660(200008)75:8<657::AID-JCTB263>3.0.CO;2-5

Mao X, Wang J, Ciblak A, Cox EE, Riis C, Terkelsen M, Gent DB, Alshawabkeh AN (2012) Electrokinetic-enhanced bioaugmentation for remediation of chlorinated solvents contaminated clay. J Hazard Mater 213–214:311–317. https://doi.org/10.1016/j.jhazmat.2012.02.001

Martínez-Prado MA, Unzueta-Medina J, Pérez-López ME (2014) Electrobioremediation as a hybrid technology to treat soil contaminated with total petroleum hydrocarbons. Revista Mexicana de Ingeniera Quimica 13(1):113–127

Mena E, Rubio P, Cañizares P, Villaseñor J, Rodrigo MA (2012) Electrokinetic transport of diesel-degrading microorganisms through soils of different textures using electric fields. J Environ Sci Health A 47(2):274–279. https://doi.org/10.1080/10934529.2012.640906

Mena E, Villaseñor J, Cañizares P, Rodrigo MA (2016) Effect of electric field on the performance of soil electro-bioremediation with a periodic polarity reversal strategy. Chemosphere 146:300–307. https://doi.org/10.1016/j.chemosphere.2015.12.053

Mohan SV, Kisa T, Ohkuma T, Kanaly RA, Shimizu Y (2006) Bioremediation technologies for treatment of PAH-contaminated soil and strategies to enhance process efficiency. Rev Environ Sci Biotechnol 5(4):347–374. https://doi.org/10.1007/S11157-006-0004-1

Mrozik A, Piotrowska-Seget Z (2010) Bioaugmentation as a strategy for cleaning up of soils contaminated with aromatic compounds. Microbiol Res 165(5):363–375. https://doi.org/10.1016/J.MICRES.2009.08.001

Niqui-Arroyo J-L, Bueno-Montes M, Posada-Baquero R, Ortega-Calvo J-J (2006) Electrokinetic enhancement of phenanthrene biodegradation in creosote-polluted clay soil. Environ Pollut 142(2):326–332. https://doi.org/10.1016/j.envpol.2005.10.007

Phulpoto IA, Yu Z, Qazi MA, Ndayisenga F, Yang J (2023) A comprehensive study on microbial-surfactants from bioproduction scale-up toward electrokinetics remediation of environmental pollutants: challenges and perspectives. Chemosphere 311:136979. https://doi.org/10.1016/j.chemosphere.2022.136979

Popov K, Glazkova I, Yachmenev V, Nikolayev A (2008) Electrokinetic remediation of concrete: effect of chelating agents. Environ Pollut 153(1):22–28. https://doi.org/10.1016/j.envpol.2008.01.014

Ramadan BS, Sari GL, Rosmalina RT, Effendi AJ, Hadrah. (2018) An overview of electrokinetic soil flushing and its effect on bioremediation of hydrocarbon contaminated soil. J Environ Manag 218:309–321. https://doi.org/10.1016/j.jenvman.2018.04.065

Ramírez EM, Camacho JV, Rodrigo MA, Cañizares P (2015) Combination of bioremediation and electrokinetics for the in-situ treatment of diesel polluted soil: a comparison of strategies. Sci Total Environ 533:307–316. https://doi.org/10.1016/j.scitotenv.2015.06.127

Ravera M, Ciccarelli C, Gastaldi D, Rinaudo C, Castelli C, Osella D (2006) An experiment in the electrokinetic removal of copper from soil contaminated by the brass industry. Chemosphere 63(6):950–955. https://doi.org/10.1016/j.chemosphere.2005.08.068

Reddy KR, Cameselle C (eds) (2009) Electrochemical remediation technologies for polluted soils, sediments and groundwater. Wiley. https://doi.org/10.1002/9780470523650

Ribeiro AB, Vara Prasad MN (2021) Electrokinetic remediation for environmental security and sustainability. In: Electrokinetic remediation for environmental security and sustainability. https://doi.org/10.1002/9781119670186

Sabari D (2021) Study on the compaction characteristics of soil and marble dust composite as a potential green building material. Revista Gestão Inovação e Tecnologias 11(4):1252–1263. https://doi.org/10.47059/revistageintec.v11i4.2184

Saini A, Bekele DN, Chadalavada S, Fang C, Naidu R (2020) A review of electrokinetically enhanced bioremediation technologies for PHs. J Environ Sci 88:31–45. https://doi.org/10.1016/j.jes.2019.08.010

Semple KT, Doick KJ, Jones KC, Burauel P, Craven A, Harms H (2004) Peer reviewed: defining bioavailability and bioaccessibility of contaminated soil and sediment is complicated. Environ Sci Technol 38(12):228A–231A. https://doi.org/10.1021/es040548w

Shahmohammadi-Kalalagh S, Beyrami H, Ahmadzadeh F (2017) Electrokinetic remediation of Cd contaminated soil at field condition. In: 15th international conference on environmental science and technology

Simoni SF, Schäfer A, Harms H, Zehnder AJB (2001) Factors affecting mass transfer limited biodegradation in saturated porous media. J Contam Hydrol 50(1–2):99–120. https://doi.org/10.1016/S0169-7722(01)00099-7

Song Y, Cang L, Fang G, Ata-Ul-Karim ST, Xu H, Zhou D (2018) Electrokinetic delivery of anodic in situ generated active chlorine to remediate diesel-contaminated sand. Chem Eng J 337:499–505. https://doi.org/10.1016/j.cej.2017.12.122

Sturman PJ, Stewart PS, Cunningham AB, Bouwer EJ, Wolfram JH (1995) Engineering scale-up of in situ bioremediation processes: a review. J Contam Hydrol 19(3):171–203. https://doi.org/10.1016/0169-7722(95)00017-P

Suni S, Romantschuk M (2004) Mobilisation of bacteria in soils by electro-osmosis. FEMS Microbiol Ecol 49(1):51–57. https://doi.org/10.1016/j.femsec.2004.01.016

Tiehm A, Lohner ST, Augenstein T (2009) Effects of direct electric current and electrode reactions on vinyl chloride degrading microorganisms. Electrochim Acta 54(12):3453–3459. https://doi.org/10.1016/J.ELECTACTA.2009.01.002

Turer D, Genc A (2005) Assessing effect of electrode configuration on the efficiency of electrokinetic remediation by sequential extraction analysis. J Hazard Mater 119(1–3):167–174. https://doi.org/10.1016/j.jhazmat.2004.12.003

Varia J, Martínez SS, Orta SV, Bull S, Roy S (2013) Bioelectrochemical metal remediation and recovery of Au3+, Co2+ and Fe3+ metal ions. Electrochim Acta 95:125–131. https://doi.org/10.1016/j.electacta.2013.02.051

Vázquez MV, Vasco DA, Hernández-Luis F, Grandoso D, Lemus M, Benjumea DM, Arbelo CD (2009) Electrokinetic study of the buffer capacity of some soils from Tenerife.: comparison with a volumetric technique. Geoderma 148(3–4):261–266. https://doi.org/10.1016/j.geoderma.2008.10.010

Ward O, Singh A, Van Hamme J (2003) Accelerated biodegradation of petroleum hydrocarbon waste. J Ind Microbiol Biotechnol 30(5):260–270. https://doi.org/10.1007/s10295-003-0042-4

Wick LY, Mattle PA, Wattiau P, Harms H (2004) Electrokinetic transport of PAH-degrading bacteria in model aquifers and soil. Environ Sci Technol 38(17):4596–4602. https://doi.org/10.1021/es0354420

Xu W, Wang C, Liu H, Zhang Z, Sun H (2010) A laboratory feasibility study on a new electrokinetic nutrient injection pattern and bioremediation of phenanthrene in a clayey soil. J Hazard Mater 184(1–3):798–804. https://doi.org/10.1016/j.jhazmat.2010.08.111

Yang X, Zhou M, Cang L, Ji Q, Xie J (2020) Enhanced electrokinetic remediation of heavy-metals contaminated soil in presence tetrasodium N, N-bis(carboxymethyl) glutamic acid (GLDA) as chelator. Int J Electrochem Sci 15. https://doi.org/10.20964/2020.01.15

Yeung AT, Gu Y-Y (2011) A review on techniques to enhance electrochemical remediation of contaminated soils. J Hazard Mater 195:11–29. https://doi.org/10.1016/j.jhazmat.2011.08.047
Yu X, Muhammad F, Yan Y, Yu L, Li H, Huang X, Jiao B, Lu N, Li D (2019) Effect of chemical additives on electrokinetic remediation of Cr-contaminated soil coupled with a permeable reactive barrier. R Soc Open Sci 6(5):182138. https://doi.org/10.1098/rsos.182138
Yuan Y, Guo S-H, Li F-M, Li T-T (2013) Effect of an electric field on n-hexadecane microbial degradation in contaminated soil. Int Biodeterior Biodegradation 77:78–84. https://doi.org/10.1016/j.ibiod.2012.10.012
Zhang H, Yang Y, Ren D, Wang L, He X (2021) Graphite as anode materials: fundamental mechanism, recent progress and advances. Energy Storage Mater 36:147–170. https://doi.org/10.1016/J.ENSM.2020.12.027
Zhu N, Chen M, Guo X, Hu G (2015) Electrokinetic removal of Cu and Zn in anaerobic digestate: interrelation between metal speciation and electrokinetic treatments. J Hazard Mater 286:118–126. https://doi.org/10.1016/j.jhazmat.2014.12.023

Chapter 11
Bulking Agent Demands for Soil Petroleum Contaminated Co-Composting (Review, Opinions, and a Case Study)

Amir Parnian, Amin Parnian, Gholamreza Zareian, James N. Furze, and Kyanoush Behrahi

Abstract Petroleum hydrocarbon pollution in soils/sediments is a primary concern in ecosystems and occurs as a result of human activity, causing instability in growth and development. Bioremediation is a biotechnological strategy using plants, bacteria, and fungi to stabilize, control, remediate, and restore the environment. Principle treatment targets are soils and sediments polluted by petroleum contaminants. Co-composting technology, an ex-situ bioremediation method, has gained attention for its high-efficiency removal of petroleum-hydrocarbon-based contaminants in soil and sediments. Bulking agents are co-composting amendments and enhancers of bioremediation that are used to create inter-particle voids, providing air space in composting materials, and regulating the water content of waste. In this chapter, different aspects of a suitable bulking agent for co-composting petroleum

---

A. Parnian
National Salinity Research Center (NSRC), Agricultural Research, Education and Extension Organization (AREEO), Yazd, Iran

A. Parnian (✉)
Young Researchers Club, Masjed Soleyman Branch, Islamic Azad University, Masjed Soleyman, Iran

Parnian Sazeh Chehelsotoun (PSC) Company, Isfahan, Iran

G. Zareian
Soil and Water Research Department, Fars Agricultural and Natural Resources Research Center, AREEO, Shiraz, Iran

J. N. Furze
Royal Geographical Society (with the Institute of British Geographers), London, UK

Laboratory of Biotechnology and Valorization of Natural Resources, Faculty of Sciences-Agadir, Department of Biology, Ibn Zohr University, Agadir, Morocco

Control and Systems Engineering Department, University of Technology-Iraq, Baghdad, Iraq

Earth Systems Protection Foundation, Somerset, UK

K. Behrahi
University of Applied Science and Technology, Tehran, Iran

A. Parnian et al. (eds.), *Bioremediation of Petroleum Contaminated Soil and Sediment*, Environmental Contamination Remediation and Management,
https://doi.org/10.1007/978-3-031-95524-2_11

hydrocarbon-contaminated soils/sediments, and a case study regarding the use of sugarcane bagasse in co-composting are reviewed and dissected.

Keywords Bioremediation, Composting, Hydrocarbon, Contaminants, Amendments

11.1 Introduction (Soil/Sediment Petroleum Pollution)

Petroleum hydrocarbon pollution in soil and sediments is a primary concern in ecosystems and happens as a consequence of human activity, causing instability in growth and development. Anthropogenic activities and industrial operations, including oil and gas exploration, production, refinery, storage, and transportation, cause considerable environmental risks through negligence. Operations affect all environments, as well as the surrounding ecosystem. Petroleum hydrocarbons affect DNA, disrupt cellular membrane functions, and threaten the life cycle of organisms. Petroleum hydrocarbon contamination is carcinogenic to humans and animals and jeopardizes plant growth and reproduction. Petroleum contamination threatens ecosystem health and all organisms within. It affects ecosystem vulnerability beyond its capacity to regenerate and has serious negative effects on diversity. The immediate and embedded life-limiting effects of petroleum hydrocarbon contaminants are serious problems. To guarantee ecosystem sustainability and human health, one must avoid environmental contamination and emphasize environmental remediation efforts. Various processes have been developed and applied to manage oil-based pollution in soil and sediments, including bioremediation, fixation, landfill, and thermal processes (Parnian et al. 2022a).

11.2 Soil/Sediment Bioremediation of Petroleum Contamination

Anthropogenic activities are polluting posing a serious threat to the Earth. A portfolio of chemical, physical, biological, and thermal treatments, are available in environmental remediation, to resolve problems. Bioremediation is a biotechnological method employing organisms (plants, bacteria, and fungi), to stabilize, control, remediate, and restore the environment. Soil and sediment matrices contaminated by petroleum pollutants have been successfully remediated using this method. Bioremediation is a green and cost-effective procedure, though has limitations and requires more time than other remediation methods. The efficiency of bioremediation depends on the biological capacity of the active organisms to adapt and proliferate in contaminated environments. Climate and environmental factors determine bioremediation performance. Organisms need sufficient nutrients to survive and

Table 11.1 Soil and sediment bioremediation methods for petroleum contamination removal

Number	Method	Contaminated Medium	Contaminant	Location	References
1	Biosurfactant and commercial mineral NPK fertilizer	Sand and sediment	Crude oil	Guanabara Bay, Rio de Janeiro, Brazil	da Silva et al. (2009)
2	Petroleum-degrading bacteria (*Stenotrophomonas maltophilia*) and soil conditioner	Soil	Crude oil	Guangzhou, China	Liu et al. (2020)
3	Autochthonous bacterial consortium and nutrients	Sediment	Oil spill	Matosinhos, Portugal	Perdigão et al. (2021)
4	Co-composting	Sludge	Total petroleum hydrocarbon (TPH)	China	Xu et al. (2023)
5	Bioslurry	Marine sediments	Crude oil	Augusta Bay, Italy	Avona et al. (2022)
6	Land farming (Cow dung and NPK fertilizer)	Soil	Crude oil spill	Edo State, Nigeria	Okonofua et al. (2021)

degrade pollutants in sediment, soil, water, and air (Parnian et al 2022b, 2023). Bioremediation has different aspects (Table 11.1) with various benefits and weaknesses, distributed across surrounding environmental conditions, the nature and quantity of pollution, and the remediation project goals. Methods offer a comprehensive and sustainable approach to dealing with pollution. They are ecologically friendly alternatives to conventional engineering-based remediation techniques.

11.2.1 Composting (Co-Composting)

Aerobic composting is a controlled, biological process that converts organic materials into nutrient-rich fertilizer, soil amendment, or mulch through natural decomposition. Composting is a natural process that recycles organic matter, such as leaves and food scraps, into valuable fertilizer that enriches soil and plants. Co-composting technology, an ex-situ method, has gained attention for its high efficiency in the removal of various organic contaminants in soil or sediments, such as petroleum hydrocarbon contamination, pesticides, explosives, and chlorophenols. The process relies on the addition of compost, organic co-substrates, or amendments to the contaminated soil or sediment. As the co-substrate matures, the target pollutants are degraded due to the action of various microbial populations within the mixture. Combining petroleum-hydrocarbon-contaminated soil or sediment with composted organic waste is an eco-friendly and sustainable option, enabling disposal and enhancing the biodegradation of petroleum hydrocarbons. The efficiency of the biodegradation process depends on the bioavailability of substrates, environmental conditions, oxygen presence, and nutrient availability. Bioremediation of petroleum hydrocarbon-contaminated soils through composting has proven capability to overcome obstacles and provide a natural and cost-effective solution for remediation efforts. Composting significantly reduces the time and resources needed for remediation and improves soil structure and fertility, promoting overall environmental health and sustainability (Sayara and Sánchez 2020).

11.3 Bulking Agent

Bulking agents are composting amendments used to create inter-particle voids, provide air space in composting materials, and regulate the water content of waste. Numerous bulking agents, including wood chips, straws, sawdust, rice husks, and peanut shells, have been mixed with waste materials to adjust moisture content, C/N ratio, and free spaces between initial particles of mixture piles. Different bulking agents modify the physical properties of composting feedstock, change the biodegradation kinetics, and shorten the composting period (Tabrika et al. 2021; Yang et al. 2013). To balance the procedure in terms of moisture content for aeration, pH for a healthy microbial environment, and carbon and nitrogen for optimum microbial development, it is crucial to characterize food waste and the available bulking agent before composting. Depending on the substrates and composting processes, aeration is aided when the compost mixture provides 30% free air space (Adhikari et al. 2008).

11.3.1 Different Types of Bulking Agents

Several different types of bulking agents can be used in composting, such as straw, wood chips, sawdust, sugarcane bagasse, biochar, and shredded paper. Each of these materials can contribute to the overall structure of the compost pile, helping to create pockets of air for aeration. It is important to choose a bulking agent that is readily available, low-cost, easily transportable, does not add additional pollution to the mixture, does not contain growth inhibitors, has all the technical benefits, and is more likely to be recoverable. By carefully selecting the right bulking agent, site managers can ensure that their pile has the ideal conditions for microbial activity and decomposition (Lu et al. 2020; Damaceno et al. 2021; Zahra et al. 2023; Alves et al. 2023).

11.3.2 Which Bulking Agent Is Suitable for Co-Composting?

When considering which bulking agent is suitable for co-composting, it is important to take into account the specific materials being composted in the mixture pile, e.g., water content, low or high particle size, water permeability, water infiltration and repellency, aeration or compaction, porosity, and carbon-to-nitrogen ratio (Lalthlansanga et al. 2023).

11.3.2.1 The Water Content of the Mixture Pile

The water content of the initial mixture pile should be at an optimal level to promote microbial activity without causing the pile to become too wet and anaerobic. Ideally, the water content of the initial mixture pile should be around 50–60%. This level of moisture allows for the proper decomposition of organic materials by providing a suitable environment for microbial growth. If the pile becomes too wet, it can lead to anaerobic conditions, which slow down the composting process and produce unpleasant odors. On the other hand, if the pile is too dry, microbial activity will be limited, resulting in slower decomposition (Ren et al. 2023). Therefore, maintaining the right water content is important for successful co-composting.

11.3.2.2 Low or High Particle Size of the Mixture

The optimal particle size of the mixture can have a significant impact on the overall quality of the composting process and bioremediation. It is important to carefully consider the specific needs of the materials being composted when choosing a bulking agent for balancing the particle size of the mixture. Finding the right balance of particle sizes mostly leads to providing adequate hydration and aeration, which can create an ideal environment for the decomposition process. Ultimately, the success of the composting operation relies on the careful selection and management of bulking agent size.

11.3.2.3 Water Permeability of the Mixture

The initial water permeability of the mixture and also permeability during the process are important in ensuring that active microorganisms, responsible for breaking down organic matter, have access to essential moisture for their metabolic processes. If the bulking agent is too fine, it may compact and impede water flow, leading to anaerobic conditions that retard decomposition and result in unpleasant odors. Alternatively, if the bulking agent is too coarse water drains quickly and microorganisms are parched, without moisture they need to flourish. Consequently, regular adjustment and balancing of the bulking agent size and water permeability are essential for maintaining optimal conditions in the compost pile.

11.3.2.4 Water Infiltration and Repellency of Piles

The infiltration and water repellency of mixture piles are crucial factors to consider for proper decomposition. Proper water flow allows the microorganisms to thrive and break down organic materials efficiently. By selecting the proper bulking agents and adjusting them in piles, the compost pile could modify the water infiltration and/or cure the water repellency of the mixture initially, which will lead to successful composting by providing adequate water content for the co-composting process.

11.3.2.5 Pile Mixture Aeration or Compaction

Selecting suitable bulking agents and adding an appropriate portion of bulking agent to the initial mixture also plays a key role in the decomposition process by providing the mixture with good airflow and reducing compaction. Adequate aeration ensures that oxygen is available for microorganisms to undergo aerobic decomposition; meanwhile, proper compaction helps maintain the structure of the pile and regulates the flow of water and air. With careful monitoring and adjusting, composters can reach an optimal environment for the decomposition process to occur efficiently and effectively.

11.3.2.6 Pile Porosity

Pile porosity is influenced by the size, shape, and nature of the mixture materials. A well-designed pile with the right amount of bulking agent will provide a good level of porosity, allowing for proper airflow and also water permeability. The good porosity of a mixture pile in co-composting generates a favorable environment for microorganisms to break down organic matter and enhance bioremediation. Achieving proper pile porosity by adding bulking agents is essential for successful co-composting.

11.3.2.7 Carbon-to-Nitrogen Ratio

If a compost pile contains mostly nitrogen-rich materials such as food scraps or grass clippings, a carbon-rich bulking agent like shredded paper or straw is ideal to help balance the carbon-to-nitrogen ratio. Conversely, if a compost pile is predominantly made of carbon-rich materials such as leaves or wood chips, a nitrogen-rich bulking agent like sawdust, sugarcane bagasse, or cow manure will be beneficial.

Importance is given to a bulking agent that complements existing materials in a compost pile and promotes efficient decomposition.

11.4 Use of Sugarcane Bagasse as Bulking to Prepare Petroleum-Contaminated Soil for Co-Composting (a Case Study)

Bioremediation of contaminated soils and sediments is gaining interest as a low-cost and environmentally friendly method to clean up anthropogenic contaminants. However, some organic pollutants, like petroleum hydrocarbons, have low biodegradability or are not bioavailable, hindering their removal. The use of composting is considered promising for the bioremediation of petroleum hydrocarbon-contaminated soils or sediments, with special attention given to compost stability and ex-situ composting. A study investigated (Babaei et al. 2020) the use of sugarcane bagasse as a bulking material in Khuzestan province, Iran, for the bioremediation of petroleum hydrocarbons from oil-based drill cuttings using a co-composting process. Oil-based drill cuttings were collected from an oil drilling rig discharge point, and bagasse was collected from a sugarcane agro-industry near Ahwaz. Experiments were conducted using different bagasse-to-oil-based drill-cutting ratios. The degradation extent and kinetics of total petroleum hydrocarbons (TPHs) and polyaromatic hydrocarbons (PAHs) were studied. The highest PAHs and TPH removals were observed when oil-based drill-cutting was composted with 15% bagasse. After 70 days of incubation, PAHs and TPH were removed up to 24.8% and 67.5%, respectively. The enzyme activities and proportion of heterotrophic bacteria suggest that biodegradation is the main mechanism in the degradation process. The removal performance results of the study were not appropriate due to the bulking agent's improper ratio and unpleasant carbon-to-nitrogen ratio.

11.5 Conclusion

Bioremediation is a high-performance, green, and cost-effective method for treating soil and sediment contaminated by petroleum contaminates. This chapter reviewed and briefly summarized various bioremediation methods for petroleum hydrocarbon bioremediation. Furthermore, we focused on the role of bulking agents in co-composting, these are great agents for enhancing biodegradation of petroleum hydrocarbons from soil and sediments. This chapter explained how different types of bulking agents, such as straw, wood chips, sawdust, sugarcane bagasse, biochar, and shredded paper, can affect the physical, chemical, and biological properties of the composting mixture, and the criteria for selecting a suitable bulking agent for co-composting. This chapter provides useful and practical information and insights for researchers, practitioners, site managers, bioremediation planners, and scientists who are interested in the bioremediation of petroleum-contaminated soils/sediments.

11.6 Suggestion

To ensure the best use of bulking agents the following suggestions are made:

I. Further research is needed to optimize co-composting parameters and evaluate the quality and safety of the final product and outcomes.
II. Potential bulking agents and co-substrates should be tested for co-composting petroleum-contaminated soil/sediment, and the feasibility and scalability of co-composting should be assessed for practical applications.
III. Further research and pilot-scale practice are needed to model the effects of the co-composting parameter and evaluate bioremediation performance.

References

Adhikari BK, Barrington S, Martinez J, King S (2008) Characterization of food waste and bulking agents for composting. Waste Manag 28(5):795–804. https://doi.org/10.1016/j.wasman.2007.08.018

Alves D, Villar I, Mato S (2023) Community composting strategies for biowaste treatment: methodology, bulking agent and compost quality. Environ Sci Pollut Res 31:1–13

Avona A, Capodici M, Di Trapani D, Giustra MG, Lucchina PG, Lumia L, Di Bella G, Rossetti S, Tonanzi B, Viviani G (2022) Hydrocarbons removal from real marine sediments: analysis of degradation pathways and microbial community development during bioslurry treatment. Sci Total Environ 838:156458

Babaei AA, Safdari F, Alavi N, Bakhshoodeh R, Motamedi H, Paydary P (2020) Co-composting of oil-based drilling cuttings by bagasse. Bioprocess Biosyst Eng 43:1–12. https://doi.org/10.1007/s00449-019-02195-6

da Silva AC, de Oliveira FJS, Bernardes DS, de França FP (2009) Bioremediation of marine sediments impacted by petroleum. Appl Biochem Biotechnol 153:58–66. https://doi.org/10.1007/s12010-008-8457-z

Damaceno FM, de Abreu CM, de Souza Batista I, Edwiges T (2021) Centralized treatment of waste through composting: influence of the c/n ratio and bulking agent. Environ Eng Manag J 20(7):1147–1156

Lalthlansanga C, Pottipati S, Meesala NS, Mohanty B, Kalamdhad AS (2023) Evaluating the potential of biodegradation of swine manure through rotary drum composting utilizing different bulking agents. Bioresour Technol 388:129751

Liu H, Tan X, Guo J, Xie Q, Chen S (2020) Bioremediation of oil-contaminated soil by combination of soil conditioner and microorganism. J Soils Sediments 20:2121–2129. https://doi.org/10.1007/s11368-020-02591-6

Lu W, Ye R, Ming Z, Pan C, Abbas Y, Stegmann R, Wang H (2020) Performance evaluation in composting of sewage sludge with different bulking agents. J Environ Eng 146(6):05020002

Okonofua ES, Lasisi KH, Atikpo E (2021) Factorial design study of total petroleum contaminated soil treatment using land farming technique. Sustain Environ Res 31:5. https://doi.org/10.1186/s42834-021-00079-7

Parnian A, Furze JN, Parnian A (2022a) Drilling waste management based on new methods of bioremediation and solar desalination. In: Furze JN, Eslamian S, Raafat S, Swing K (eds) Earth systems protection and sustainability, vol 1. Springer International Publishing. https://doi.org/10.1007/978-3-030-85829-2_4

Parnian A, Parnian A, Pirasteh-Anosheh H, Furze JN, Prasad MNV, Race M, Hulisz P, Ferraro A (2022b) Full-scale bioremediation of petroleum-contaminated soils via integration of co-composting. J Soils Sediments 22(8):2209–2218. https://doi.org/10.1007/s11368-022-03229-5

Parnian A, Mahbod M, Prasad MNV (2023) Microplastics remediation—possible perspectives for mitigating saline environments. In: Microplastics in the ecosphere. Wiley, p 465–476. https://doi.org/10.1002/9781119879534.ch29

Perdigão R, Almeida CMR, Santos F, Carvalho MF, Mucha AP (2021) Optimization of an autochthonous bacterial consortium obtained from beach sediments for bioremediation of petroleum hydrocarbons. Water 13(1):66. https://doi.org/10.3390/w13010066

Ren X, Jiao M, Chen X, Liu T, Zhang Y, Zhang Z (2023) Role of bulking agents and additive on composting. In: Current developments in biotechnology and bioengineering. Elsevier, Amsterdam, Netherlands, pp 127–142

Sayara T, Sánchez A (2020) Bioremediation of PAH-contaminated soils: process enhancement through composting/compost. Appl Sci 10(11):3684. https://doi.org/10.3390/app10113684

Tabrika I, Mayad EH, Furze JN et al (2021) Optimization of tomato waste composting with integration of organic feedstock. Environ Sci Pollut Res 28. https://doi.org/10.1007/s11356-020-12303-9

Xu S, Zhan J, Li L, Zhu Y, Liu J, Guo X (2023) Total petroleum hydrocarbons and influencing factors in co-composting of rural sewage sludge and organic solid wastes. Environ Pollut 319:120911. https://doi.org/10.1016/j.envpol.2022.120911

Yang F, Li GX, Yang QY, Luo WH (2013) Effect of bulking agents on maturity and gaseous emissions during kitchen waste composting. Chemosphere 93(7):1393–1399. https://doi.org/10.1016/j.chemosphere.2013.07.002

Zahra K, Farhan M, Kanwal A, Sharif F, Hayyat MU, Shahzad L, Ghafoor GZ (2023) Investigating the role of bulking agents in compost maturity. Sci Rep 13(1):16003

Chapter 12
Some Tips for Prosperous Soil and Sediment Petroleum Contamination Bioremediation

Amir Parnian, Amin Parnian, and James N. Furze

Abstract Communities today show concern for remediation projects as pollution from rock oil poses considerable environmental problems. Though it comes with difficulties and costs time, bioremediation is economical and environmentally beneficial. This chapter offers a concentrated review of the issue of petroleum hydrocarbon contamination in soil and sediment, its ecosystem effects, and possible bioremediation solutions. Additionally, it provides a brief overview of several bioremediation methods, such as landfarming, phytoremediation, bioaugmentation, and biostimulation, which can be used to treat contaminated soil or sediment. Bioremediation has limitations and restrictions;- long time periods are needed, site-specific requirements must be considered additional to potential hazards. This chapter outlines requirements for successful bioremediation projects, including site evaluation, selection, and management in addition to communication with local communities, authorities, and stakeholders. The goal of this chapter is to inform academics, practitioners, and policymakers with a thorough and up-to-date knowledge framework to design and manage bioremediation initiatives; to clean up soil or sediment contaminated with petroleum hydrocarbons.

A. Parnian
National Salinity Research Center (NSRC), Agricultural Research, Education and Extension Organization (AREEO), Yazd, Iran

A. Parnian (✉)
Young Researchers Club, Masjed Soleyman Branch, Islamic Azad University, Masjed Soleyman, Iran

Parnian Sazeh Chehelsotoun (PSC) Company, Isfahan, Iran

J. N. Furze
Royal Geographical Society (with the Institute of British Geographers), London, UK

Laboratory of Biotechnology and Valorization of Natural Resources, Faculty of Sciences-Agadir, Department of Biology, Ibn Zohr University, Agadir, Morocco

Control and Systems Engineering Department, University of Technology-Iraq, Baghdad, Iraq

Earth Systems Protection Foundation, Somerset, UK

A. Parnian et al. (eds.), *Bioremediation of Petroleum Contaminated Soil and Sediment*, Environmental Contamination Remediation and Management,
https://doi.org/10.1007/978-3-031-95524-2_12

Keywords Environmental pollution · Operation management · Framework

12.1 Introduction

Petroleum hydrocarbon contamination is a serious environmental concern, mostly resulting from human activities in the services, agriculture, and industry sectors. The oil and gas industry, which involves the exploration, production, refinery, storage, and transportation of petroleum products, is a potent source of pollution. Negligence and carelessness in these activities can lead to accidents and spills, posing serious risks to the environment. Contamination affects soil, air, water quality, health, biodiversity, and functioning ecosystems. To manage this problem, strict regulations and guidelines must be implemented, including comprehensive training and education for industry personnel. Periodic control, inspections, and assessments are also necessary to identify potential environmental risks and minimize accidents. In these circumstances of ecosystem contamination, available management strategies should be implemented to mitigate damages and restore the ecosystem and surrounding environments. This may involve specialized equipment and remediation techniques to remediate contaminated soil or sediments with petroleum hydrocarbons (Alexandersen et al. 2014; Cocârţă et al. 2017; Parnian et al. 2023).

12.1.1 Environmental Impact of Petroleum Hydrocarbons

Petroleum hydrocarbons, mostly obtained from crude oil, natural gas, and byproducts, can harm humans and the Earth's ecosystems; damaging DNA structure, and disrupting cellular membranes, which leads to mutations and cancer. Oil contaminants are carcinogenic and toxic hazards for plants, humans, and animals (Fan et al. 2012; Alzahrani and Rajendran 2019). Contamination has serious long-term effects on the environment and public health. Human activities including oil exploration, production, transportation, storage, and consumption lead to petroleum hydrocarbon contamination, posing serious risks to ecosystem health and sustainability, and increasing vulnerability to climate change; reducing environmental diversity and resilience (Parnian et al. 2020, 2021). Petroleum hydrocarbon pollutants, originating from crude oil, can harm ecosystems, human health, and biodiversity. Prevention of petroleum hydrocarbon contamination, reduction of pollution, and enhancement of remediation programs are essential. Stricter regulations, improved monitoring and enforcement, and increased public awareness contribute towards achieving goals (Alzahrani and Rajendran 2019). Various methods have been proposed and adopted to treat petroleum-hydrocarbon-polluted soil or sediment: bioremediation, using living organisms to remove pollutants; fixation, immobilizing contaminants by adding chemical or physical agents; landfill, isolating polluted soil or sediments in a designated area or controlled disposal site; and thermal processing, using thermal energy to evaporate, degrade, decontaminate, or incinerate organic substances (Xuezhi 2020).

12.1.2 Soil Contamination with Hydrocarbons

Oil spills occur during offshore oil drilling, gas well transport, and pipeline transfers, causing significant environmental risks to all life forms, including humans. Petroleum hydrocarbons contaminate soil, causing ecotoxicity and biodiversity loss. Understanding the distribution, environmental fate, transport, and chemical properties of these contaminants is crucial for developing cost-effective, environmentally friendly strategies for remediation, restoration, and reclamation. Bioremediation is a promising method for solving petroleum-hydrocarbon-polluted soil and sediment, using microorganisms to break down and degrade pollutants. Fixation techniques like solidification/stabilization have been employed, immobilizing contaminants and preventing their migration, enabling long-term containment. Landfill and thermal processes have been employed to isolate polluted soil and use heat to evaporate, degrade, decontaminate, or incinerate organic contaminants. Landfill and thermal processes are essential, addressing the complex composition of petroleum materials and their potential environmental disruption (Okoh et al. 2019; Ossai et al. 2020; Xuezhi 2020; Ambaye et al. 2022). These various methods offer potential solutions for remediation efforts and can help protect the ecosystems and communities that rely on them.

12.2 Soil or Sediment Petroleum Contamination Bioremediation

Anthropogenic environmental pollution causes a significant threat to the Earth's ecosystem. Environmental remediation is a promising solution, utilizing chemical, physical, biological, and thermal treatments. Bioremediation is a biotechnological method that uses plants, bacteria, and fungi to stabilize, control, remediate, and restore the environment. It can be applied to soil and sediments polluted by petroleum contaminants. Bioremediation is eco-friendly and cost-effective, although it requires time and faces challenges. The efficiency of bioremediation is dependent on an organism's capacity to survive and expand in contaminated environments. The climate in the surrounding environs is crucial for successful bioremediation operations. The availability of essential nutrients is essential for algae and fungi to survive and degrade pollutants in sediment, soil, water, and air (Parnian et al. 2022a, 2023).

Bioremediation branches regarding active organisms are listed below.

1. Microbial remediation, the use of microorganisms to remediate pollutants.

2. Phytoremediation, the use of plants and algae for environmental remediation.
3. Mycoremediation, the use of fungi and algae for environmental remediation.

Branches may be incorporated together or subjected to other soil and sediment remediation techniques and approaches. Bioremediation consists of various branches with varying advantages and limitations, depending on local conditions, environmental contamination, and remediation goals. Phytoremediation removes organic contaminants through plant degradation, while mycoremediation breaks down complex hydrocarbons and amends heavy metal presence. Approaches provide a comprehensive and sustainable method for addressing pollution. They offer environmentally friendly alternatives to traditional remediation techniques.

12.3 Elements to Perform Bioremediation of Soil/Sediment

In most scientific communities, the elements necessary to perform bioremediation of soil or sediment are limited to selecting appropriate plant species for phytoremediation, utilizing specific fungi for mycoremediation, using bioremediation enhancement techniques or devices, and/or ensuring proper monitoring and maintenance of the remediation site. However, the key is integrating elements to reach an efficient bioremediation plan, which leads to the successful removal of contamination from soil or sediment, whilst considering the ecological impact and sustainability of the consequent ecosystem (Furze et al. 2017; Parnian et al. 2022c). In the following, certain elements that are necessary to have an effective bioremediation project are mentioned.

12.3.1 Petroleum Pollution Quantity and Quality Categorization

This involves determining the level of petroleum pollution in the soil or sediment as well as the specific types of contaminants present. Understanding the quantity and quality of the pollution is crucial in order to develop a targeted and effective bioremediation strategy to achieve a successful project. The success of the bioremediation project ultimately depends on the accuracy of this initial assessment.

On the other hand, if a project manager faces a viable complexity material with different qualities (concentrations of petroleum pollution), they may separate soils/sediments into different piles of contamination concentration levels, and then choose a specific technique for each of them. Moreover, if petroleum pollution quality, crude oil, petroleum-based sludges, diesel oil, and engine oil are different in each pile, different remediation techniques may be required.

12.3.2 Soil and Sediment Quantity and Quality

It is important to conduct thorough testing and analysis of the soil or sediment to determine the level and severity of contamination. This information will help in determining the best approach for bioremediation, be it through ecological mitigation, bioaugmentation, or phytoremediation. By understanding the quantity and quality of the pollution, environmental engineers and scientists can tailor their remediation efforts to effectively restore the affected area to its natural state. By addressing the specific conditions of the site, a successful bioremediation project can not only treat the petroleum pollutants but also prevent future contamination.

Furthermore, soil and sediment quality factors such as texture, permeability, pH, moisture levels, etc., should also be considered when planning a bioremediation project for petroleum pollution. These factors greatly impact the time, cost, performance, and effectiveness of the bioremediation process; a comprehensive and successful remediation plan may be established to minimize or remove contamination.

In the case of a high quantity of pollution, it is better to use a combination of multiple bioremediation techniques; a combination of land treatment, phytoremediation, and bioaugmentation, to build up complex bioremediation systems akin to a factory. Such a complex bioremediation system effectively targets and breaks down various types of pollutants, leading to an overall more successful remediation process in the long term. Conversely, if a bioremediation project planner/designer faces a low amount of contaminated soil, it is better to utilize faster and less labor-intensive remediation methods to manage financial costs.

12.3.3 Climatology and Surrounding Environment

Understanding climatology around and in the surrounding environment is crucial in ensuring the success of a bioremediation project. Temperature, precipitation, soil type, and nearby ecosystems impact how efficiently and smoothly bioremediation techniques work. Environmental experts can optimize conditions for microbial activity, and plant growth, and effect a more efficient and sustainable cleanup process. Additionally, by studying the surrounding environment, scientists, project managers, and field engineers can identify potential risks or threats to the remediation project, allowing them to take proactive measures to mitigate challenges. In conclusion, a comprehensive understanding of the climatology and surrounding environment is essential for the effective and long-lasting restoration of contaminated sites, which might alter the in-situ program to an ex-situ one, or force the project designer to seek alternative bioremediation methods. The choice of remediation technique depends on the type and extent of pollution, the sensitivity of the surrounding environment, and the resources available for remediation. In-situ remediation is often preferred to minimize disturbances to the natural ecosystem.

However, in cases where pollution is severe or widespread, on-situ or ex-situ remediation may be necessary to clean up the contaminated site. Each approach has advantages and challenges, requiring careful consideration to determine the most appropriate method for each unique situation.

12.3.4 Site of Operations

In selecting a site for remediation operations, it is crucial to consider factors such as accessibility, proximity to water sources, and potential impact on local communities. By strategically choosing a site that meets all necessary criteria, environmental experts can ensure that the cleanup process is both effective and safe. Furthermore, proper site selection can help to minimize any potential disruptions to surrounding ecosystems and habitats, ultimately promoting a more sustainable restoration process. Overall, careful consideration of the site of operations is essential for the success of any remediation project.

12.3.5 Authorities, Legal, and Local Communities Desire from Soil and Sediment Bioremediation

Local communities, authorities, and legal entities all play a crucial role in the success of soil/sediment bioremediation projects. Their desire for clean and safe environments drives the need for effective cleanup efforts and provides more and faster funding. By involving stakeholders in the decision-making process and addressing their concerns, environmental experts and scientists can ensure that the remediation project is well-received and supported. Collaboration with local communities can also help build trust and create a sense of shared responsibility for environmental conservation, which helps to establish funding and obtain support from authorities and political figures regarding the bioremediation project demands. Ultimately, the success of any bioremediation project depends on the cooperation and engagement of all relevant parties.

12.3.6 Finance, Labor, Transport Infrastructure, Energy Resources, Market and Supply Chain

The budget of a bioremediation project and also well-supported finances are key to selecting the bioremediation method, techniques, and pathways of project planning. Having a well-supported financial plan is crucial for the success of any bioremediation project, relying on the cooperation and engagement of all parties involved. In the case of low financial support, projects face delays or even failure. The designer should choose a low-cost type of bioremediation method and make plans to provide the project demands within budget constraints. In addition to financial considerations, the availability of qualified labor (experts, highly skilled, and low skilled), transport infrastructure, energy resources, and a reliable market and supply chain are also important considerations when planning a bioremediation project. Without adequate transportation options for materials and equipment, a stable source of energy/fuel to power bioremediation processes, and a good and reliable market for providing/supplying the project demands; devices, apparatuses, materials, food, and fuel for labor, the success of the project may be compromised. A comprehensive analysis of factors is essential in order to ensure the efficient and effective implementation of a bioremediation project.

Effective collaboration with stakeholders, including local communities, is crucial in ensuring the success of any bioremediation project. By addressing concerns and involving key players in the decision-making process, environmental experts gather support and build trust for soil/sediment cleanup efforts. Collaboration can also help create a sense of shared responsibility for environmental conservation, leading to a more sustainable and well-received remediation project. Ultimately, the cooperation and engagement of all relevant parties, including those involved in finance, labor, transport infrastructure, energy resources, market, and supply chain, is essential for project success.

12.3.7 Limiting and Risk Factor Identification, Categorization, and Prioritize

Identifying and categorizing potential risks associated with the soil/sediment cleanup project is crucial for minimizing negative impacts and ensuring its success. By prioritizing bioremediation project risks based on severity and likelihood, bioremediation project managers may allocate resources effectively and develop rescue plans to tackle any potential challenges. This proactive strategy can help mitigate disruptions and setbacks, ultimately leading to more efficient and successful bioremediation.

12.3.8 Bioremediation Types Priorities

The ranking and priority of the bioremediation process is a crucial handle for distinguishing and tending to issues or insufficiencies in a wide extent of areas to improve results and guarantee individuals receive the support they need to succeed. Three distinct types of environmental remediation may be performed; based on management strategies of resources management, environmental disruption, and location of remediation techniques application (Koning et al. 2005; Hiester and Schrenk 2005):

1. The remediation process application is in the situation of pollution occurrence with the lowest environmental disruptions (In-situ).
2. The remediation process application is in the situation of pollution occurrence with high environmental disruptions like soil excavation, deep tillage of soil, etc. (On-situ).
3. Contaminated soil/water transported and remediation process application is in a restricted area and controlled environment (Ex-situ).

The choice of remediation technique depends on the type and extent of pollution, the sensitivity of the surrounding environment, and the resources available for remediation. In-situ remediation is often preferred when possible to minimize disturbances to the natural ecosystem. However, in cases where the pollution is more severe or widespread, on-situ or ex-situ remediation may be necessary to effectively clean up the contaminated site. Each approach has its own advantages and challenges and requires careful consideration to determine the most appropriate method for each unique situation.

12.3.9 Suitable Bioremediation Method Selections for Petroleum Hydrocarbon Contaminated Soil or Sediment

Bioremediation methods have been widely used for the remediation of petroleum hydrocarbon-contaminated soil or sediment. Suitable bioremediation methods include bioaugmentation, biostimulation, phytoremediation, and landfarming. These methods have shown effectiveness in degrading petroleum hydrocarbons in contaminated soil and sediment, providing a sustainable and environmentally friendly solution to pollution remediation. Researchers continue to explore and develop innovative bioremediation techniques to address the challenges posed by petroleum hydrocarbon contamination in soil and sediment.

12.4 Management of a Soil and Sediment Bioremediation Project

Effective management of a soil and sediment bioremediation project involves careful planning, monitoring, and assessment of the remediation process. Key steps include conducting a thorough site assessment to determine the extent of contamination, selecting the most appropriate bioremediation method based on site-specific conditions, and implementing a detailed remediation plan. Regular monitoring of the remediation progress is essential to ensure that the desired outcomes are achieved within the specified timeframe. Additionally, proper documentation and reporting of the project results are crucial for regulatory compliance and stakeholder communication. Overall, successful management of a bioremediation project requires a multidisciplinary approach, strong project coordination, and a commitment to environmental stewardship.

12.4.1 Importance of Project Management in Soil/Sediment Bioremediation Project

Appropriate management of activities of soil and sediment bioremediation is the basic key to successful projects. The reasons for the importance of project management in soil/sediment bioremediation projects are mentioned in Table 12.1.

12.4.2 Principles of Soil/Sediment Bioremediation Project Management

In the management of soil/sediment bioremediation projects, using project management principles can help to improve the management process. Table 12.2 mentions some important principles in the management of bioremediation projects.

Table 12.1 The reasons for the importance of project management in soil/sediment bioremediation project

Number	Reasons for importance	Definition
1	Proper planning	Through project management, it is possible to formulate accurate and coherent plans for carrying out soil and sediment bioremediation activities. This helps to reduce the time and cost required for the project.
2	Coordination/ Harmony	Bioremediation project management makes all team members work well together and work harmoniously. This helps to avoid unnecessary interference and delays in the project.
3	Cost control and management	With project management, costs can be optimized and the budget can be avoided. This is effective in reducing additional costs and increasing project profitability.
4	Time control and management	Soil and sediment bioremediation project management ensures that the project is delivered on time and that the correct schedule is observed. This helps to prevent delays in the delivery of the building and also increases customer satisfaction.

Table 12.2 Some principles in the management of soil/sediment bioremediation projects

Number	Principles in management	Definition
1	Determining the goal and objectives	First, determine the overall goal of the project and then determine the sub-goals. Goals should be measurable, achievable, and time-bound.
2	Effective planning	Accurate and comprehensive planning of project stages and activities, resource allocation, schedule, and budget is essential. Consider all aspects of the project and do logical and balanced planning.
3	Determining the expert team	Forming an expert team with appropriate capabilities is the key to the success of the project. Pay attention to the knowledge, experience, and skills of the team members and select them carefully.
4	Resource management	Manage human, financial, physical, and technical resources efficiently. Ensure that all resources required for project implementation are available and optimally used.
5	Effective communication	Establishing active and effective communication with all members is very important in the management of construction projects. Convey information and messages related to the project in an accurate and timely manner and openly consider the feedback.
6	Risk management	Identifying and evaluating the risks related to the project and predicting their impact on the plan and budget is of great importance. Planning to reduce and manage risks and prepare to face them is essential.
7	Control and monitoring	Continuous monitoring of the progress of the project, quality control, cost control, and scheduling and comparing it with the original plan are of great importance. Progress criteria should be determined and necessary improvements should be made.
8	Compliance with laws and regulations	During the implementation of the project, compliance with relevant laws and regulations; and coordination with regulatory authorities and licensees is essential.

12.5 Soil/Sediment Bioremediation Project Management Steps

Various tools and techniques are used in different parts of soil/sediment bioremediation project management; such as:

1. Identification and estimation of resources
2. Time planning
3. Cost control
4. Risk management
5. Quality control and communication
6. Ability to coordinate between project team members, employer, and other partners

Table 12.3 Soil/sediment bioremediation project management course of actions

Number	Actions	Definition
1	Project definition	In this part, the goals and needs of the bioremediation project are determined and the dimensions, budget, and required resources are determined.
2	Designing/Planning	In this stage, the schedule and breakdown of tasks into smaller activities is done. Furthermore, resources, budget, and activities are also specified.
3	Implementation/ performing	Bioremediation, execution of main activities, and control of project progress are carried out. Coordination with teams and ensuring the quality of the work/project is satisfactory.
4	Control and monitoring	The progress of the project is compared with the plan and if needed, changes are made in the planning. In addition, quality control and error prevention are also taken into consideration.
5	Completion	The project is delivered to the customer and the final evaluation of the quality and goals of the bioremediation is completed.

12.5.1 Soil/Sediment Bioremediation Project Management Processes

Soil/sediment bioremediation project management processes ordinarily include the following many courses of action, defined in Table 12.3.

12.5.2 The Stages of a Soil/Sediment Bioremediation Project

The management of a soil and sediment bioremediation project incorporates specific steps, briefly indicated in Table 12.4.

12.6 How to Succeed in Managing Soil and Sediment Bioremediation Projects?

To succeed in managing soil/sediment bioremediation projects, the following tips and approaches are suggested (Table 12.5).

12.7 Case Study

In 2019, bioremediation methods were introduced to reduce the environmental impact of 1200 cubic meters of crude oil-contaminated saline soils. This involved impregnating soils with bulking agents; sugarcane bagasse and half-composted

Table 12.4 Steps of soil and sediment bioremediation project management

Number	Steps/Stages	Definition
1	Definition and review of requirements	The goals and requirements of the project are reviewed and defined. Including determining project goals, bioremediation site, budget, schedule, and other required items.
2	Project planning	A comprehensive plan is prepared for the implementation of the soil and sediment bioremediation project; creating schedules, allocating resources, budgeting, determining activities and tasks, and specifying guidelines and standards.
3	Analysis	Preliminary studies, including the preparation of maps, tests, measurements, and technical analysis are done to determine project needs.
4	Design, planning, and engineering	Project objectives, schedule, required resources, and budget are determined. In addition, the implementation of the project is reviewed. Site architectural design, demanded structures, providing/construction of facilities, and other apparatuses/components/materials must be considered.
5	Preparation and pre-production	Materials and equipment necessary for the soil/sediment bioremediation are prepared and supported by the bioremediation project management. Raw materials, equipment, provision of temporary site use, and material procurement are considered.
6	Pre-construction	Preparation of the land/site, procurement of materials, making of base sites, and installation of primary equipment/apparatuses.
7	Implementation and soil/sediment bioremediation	Bioremediation procedures start according to the plans and roadmaps to be implemented, respecting schedules of site inspection, soil/sediment bioremediation, and installation of structures, facilities, and equipment, and management of contractors/laborers.
8	Quality control and inspection	The quality of the execution must be continuously controlled. This includes manufacturing quality review, technical inspection, testing, and validation.
9	Delivery and operation	After completion of the soil and sediment bioremediation, technical and structural tests are conducted and inspections are conducted to ensure compliance with standards. Laboratory testing, field tests, periodical tests, training, and handing over the site to the owner are delivered.
10	Warranties/Guaranties/Maintenance	Periodic rechecks and support activities are carried out to maintain and ensure the performance of the bioremediation project and recover possible collateral damage.

agricultural waste to enhance porosity and aeration potential for biological degradation. The treatment process involved three steps: organic semi-composted material production, biological soil water repellency curring, and co-composting of soils. The project was successfully launched, total petroleum hydrocarbon contents of the soil were reduced below 0.109 g kg^{-1}. Recovered soils were released to the neighboring environment to facilitate the growth of the natural ecosystem (Parnian et al. 2022b).

Table 12.5 Some tips to succeed in soil/sediment bioremediation projects

Number	Tips/Approaches	Definition
1	Setting clear goals	Define the goals of the project in a precise and measurable way. Goals should be specific, measurable, achievable, and time-bound.
2	Detailed planning	Detailed planning should be provided based on the project goals, considering timing, resource allocation, and budget. In addition, identifying risks, predictive measurement, and changes are also needed at this stage.
3	Team Coordination	Set project teams with good empathy, teamwork quality, understanding, and cooperation with each other. Effective communication, clear tasks and responsibilities, and continuous interaction with team members improve the quality of cooperation.
4	Progress control	To be successful in soil/sediment bioremediation, continuous control and monitoring of project progress is required. It is imperative to recognize changes in time which occur through unseen/cascaded causes and take necessary measures to solve problems and ensure desired progress.
5	Proactive Communication	Proactive and up-to-date communication with active groups in this job will not only help smooth project management but also schedule and review issues and ensure everyone is on top of changes.
6	Quality control	Define quality standards and criteria well and monitor them throughout the project to ensure that work is done correctly and with acceptable quality.
7	Flexibility	Due to the nature of soil and sediment bioremediation projects, the possibility of changes and the need for quick decision-making and flexibility is high. It is beneficial to be prepared for changes during the project and have options in mind to solve problems and change routes if necessary. To face these situations use brainstorming within the team, encourage high-quality individuals to share new ideas, and utilize their wisdom and work experience.

The project was effectively managed financially with (1) the sugarcane bagasse demands to be delivered to the project site from a local sugarcane agro-industry company at a distance of 40 km; (2) half-composted agricultural waste was provided by outsourcing production to a local subsidiary of a further local sugarcane agro-industry company. These decisions aimed to utilize both resources efficiently in order to meet conditions/limitations (a–f): (a) reduction of costs; (b) minimize environmental impact; (c) compensate for labor challenges due to (d) local community resistance to the bioremediation project; (e) supplement the low quality of locally available labor; (f) provision of food (for workers) in the area.

12.8 Conclusion

This chapter provides a comprehensive overview of the problem of petroleum hydrocarbon contamination in soil and sediment, its environmental impacts, and the potential bioremediation methods to address it. It also discusses bioremediation techniques, which can be employed to contaminate soil or sediment; bioaugmentation, biostimulation, phytoremediation, and landfarming. Challenges and limitations of bioremediation, such as the time required, the site-specific conditions, and potential risks have been mentioned. In this chapter, we conclude by suggesting core elements that are necessary for effective bioremediation projects; proper site assessment, selection, and management, as well as interactions with stakeholders, authorities, and local communities. The chapter aims to provide a comprehensive and updated knowledge framework for researchers, practitioners, and policymakers interested in the matter of planning and management of bioremediation projects of petroleum hydrocarbon-contaminated soil or sediment cleanup.

12.9 Suggestions

To increase the economic and ecologic gain of soil remediation the following suggestions are proposed:

I. Environment and climate are important and bioremediation site managers should take them seriously to lead successful projects.
II. Soil and sediment nature is complex and the bioremediation situation may alter pending minute changes, bioremediation site managers should be open to change their pathways.
III. Bioremediation projects closely depend on employee performance, which managers should consider paramount.

References

Alexandersen DK, Headley TR, Prigent S, Mahmutoglu I (2014) Ex-situ bioremediation of hydrocarbon contaminated soil - an example from Oman. All Days. Retrieved January 2, 2024, from https://doi.org/10.2118/170411-ms

Alzahrani AM, Rajendran P (2019) Petroleum hydrocarbon and living organisms. IntechOpen eBooks. Retrieved January 2, 2024, from https://doi.org/10.5772/intechopen.86948

Ambaye TG, Chebbi A, Formicola F, Prasad S, Gomez FH, Franzetti A, Vaccari M (2022) Remediation of soil polluted with petroleum hydrocarbons and its reuse for agriculture: recent progress, challenges, and perspectives. Chemosphere 293:133572. https://doi.org/10.1016/j.chemosphere.2022.133572

Cocârţă DM, Stoian MA, Karademır A (2017) Crude oil contaminated sites: evaluation by using risk assessment approach. Sustainability. Retrieved January 2, 2024, from https://doi.org/10.3390/su9081365

Fan YY, Wang GC, Fu JH, Zheng XH (2012) Bioremediation of waste drilling fluid: comparison of biostimulation and bioaugmentation. Desalin Water Treat 48(1–3):329–334. https://doi.org/10.1080/19443994.2012.698838

Furze JN, Zhu Q, Hill J et al (2017) Biological modelling for sustainable ecosystems. In: Furze JN, Swing K, Gupta A, McClatchey RH, Reynolds DM (eds) Mathematical advances towards sustainable environmental systems. Springer, Cham. https://doi.org/10.1007/978-3-319-43901-3_2

Hiester U, Schrenk V (2005) In-situ thermal remediation: ecological and economic advantages of the TUBA and THERIS methods. Proceedings of ConSoil 2005

Koning M, Hupe K, Stegmann R (2005) Soil remediation and disposal. In: Environmental biotechnology, Wiley-VCH, Weinheim, Germany, pp 259–274. https://doi.org/10.1002/3527604286.ch9

Okoh E, Yelebe ZR, Oruabena B, Nelson ES, Indiamaowei OP (2019) Clean-up of crude oil-contaminated soils: bioremediation option. Int J Environ Sci Technol. Retrieved January 2, 2024, from. https://doi.org/10.1007/s13762-019-02605-y

Ossai IC, Ahmed A, Hassan A, Hamid FS (2020) Remediation of soil and water contaminated with petroleum hydrocarbon: a review. Environ Technol Innov 17:100526. https://doi.org/10.1016/j.eti.2019.100526

Parnian A, Parnian A, Pirasteh-Anosheh H, Parvizi H (2020) Potential of a pyrolyzed sugarcane industry waste for spillage oil absorption. J Nat Resour Conserv Manag 1(2):163. https://doi.org/10.51396/anrcm.1.2.2020.163-168

Parnian A, Furze JN, Parnian A, Mayad EH (2021) Water purification plantations for oil and gas industries in Iran. Environ Sci Pollut Res 28(45):64193–64198. https://doi.org/10.1007/s11356-021-15819-w

Parnian A, Furze JN, Parnian A (2022a) Drilling waste management based on new methods of bioremediation and solar desalination. In: Furze JN, Eslamian S, Raafat S, Swing K (eds) Earth systems protection and sustainability, vol 1. Springer International Publishing. https://doi.org/10.1007/978-3-030-85829-2_4

Parnian A, Parnian A, Pirasteh-Anosheh H, Furze JN, Prasad MNV, Race M, Hulisz P, Ferraro A (2022b) Full-scale bioremediation of petroleum-contaminated soils via integration of co-composting. J Soils Sediments 22(8):2209–2218. https://doi.org/10.1007/s11368-022-03229-5

Parnian A, Furze JN, Chorom M et al (2022c) Competitive bioaccumulation by Ceratophyllum dimersum L. In: Furze JN, Eslamian S, Raafat S, Swing K (eds) Earth systems protection and sustainability, vol 2. Springer International Publishing. https://doi.org/10.1007/978-3-030-98584-4_1

Parnian A, Mahbod M, Prasad MNV (2023) Microplastics remediation—possible perspectives for mitigating saline environments. In: Microplastics in the ecosphere. Wiley, pp 465–476. https://doi.org/10.1002/9781119879534.ch29

Xuezhi D (2020) Remediation methods of crude oil contaminated soil. World J Agric Soil Sci 4(4). https://doi.org/10.33552/wjass.2020.04.000595

Chapter 13
Soil Petroleum Pollution in South-West of Khuzestan Province, Iran (Case Study)

Amir Parnian, Mohamad Zakipour, James N. Furze, Elaheh Zallaghi, and Nematollah Jaafarazdeh Haghighi Fard

Abstract A huge petroleum reservoir is located in the southwest of Khuzestan province, Iran. The local oil and gas company facilities, pipelines, and oil wells are in the west of the province towards the Iraq border. Facilities are in close proximity to residential areas, agricultural lands, great wastelands, and deserts. In these areas, petroleum leakage and spills accidentally occur and pollute the ecosystem; the company should be informed of potential risks in order to avoid soil pollution. This chapter aims to review cases of oil pollution in the surrounding environment, discusses oil pollution hazards, and introduces various possible soil remediation methods.

Keywords Sustainable environment · Spills · Contamination · Remediation

A. Parnian
National Salinity Research Center (NSRC), Agricultural Research, Education and Extension Organization (AREEO), Yazd, Iran

M. Zakipour (✉)
Arvandan Oil and Gas Company, Khorramshahr, Iran

J. N. Furze
Royal Geographical Society (with the Institute of British Geographers), London, UK

Laboratory of Biotechnology and Valorization of Natural Resources, Faculty of Sciences-Agadir, Department of Biology, Ibn Zohr University, Agadir, Morocco

Control and Systems Engineering Department, University of Technology-Iraq, Baghdad, Iraq

Earth Systems Protection Foundation, Somerset, UK

E. Zallaghi
National Iranian South Oil Company (NISOC), Ahvaz, Iran

N. J. H. Fard
Environmental Technologies Research Center, Medical Basic Sciences Research Institute, Ahvaz Jundishapur University of Medical Sciences, Ahvaz, Iran

A. Parnian et al. (eds.), *Bioremediation of Petroleum Contaminated Soil and Sediment*, Environmental Contamination Remediation and Management,
https://doi.org/10.1007/978-3-031-95524-2_13

13.1 Introduction (Petroleum Pollution)

Human activities in the services, agriculture, and industry sectors result in petroleum hydrocarbon contamination, which affects ecosystem stability and productivity. Stable ecosystems are essential for the well-being and economy of human communities, as they provide services and resources (Parnian et al. 2023). Petroleum hydrocarbon contamination is a major cause of ecosystem degradation and disruption, originating from human activities. One of the main sources of environmental pollution is the oil and gas industry, which involves the exploration, production, refinery, storage, and transportation of petroleum products. These activities are often carried out with the negligence and carelessness of laborers, managers, designers, and/or planners, leading to accidents and spills that pose serious risks to the environment. Contamination affects the soil, air, and water quality and health, as well as the biodiversity and functioning parts of the ecosystems that depend on them (Cocârţă et al. 2017). Petroleum hydrocarbon contamination is a significant environmental problem that needs to be controlled/managed before/after it occurs/happens. To avoid these types of contaminations, it is crucial to implement strict regulations and guidelines to ensure the safe operation of petroleum-related activities. This includes providing comprehensive training and education to all individuals involved in the industry, emphasizing the importance of environmental protection and risk management. Furthermore, regular inspections and assessments should be conducted to identify any potential risks/weaknesses in the system, which allows for timely intervention and the prevention of accidents. Nevertheless, in the unfortunate event of contamination, prompt and effective management strategies should be put in place to mitigate the environmental damage and restore the affected ecosystems and surrounding environment. This may involve the use of specialized equipment/acting groups to control/manage/remediate the spilled petroleum, as well as the implementation of remediation techniques to restore/recover the contaminated soil (Alexandersen et al. 2014).

13.2 Petroleum Hydrocarbons Impact the Environment

Organic substances obtained from crude oil and its byproducts are known as petroleum hydrocarbons. They have the potential to affect living things and the ecosystem. One of the effects is damage to DNA structure and function, which leads to mutations and cancers (Alzahrani and Rajendran 2019). Another negative effect of petroleum hydrocarbon contamination is cellular membrane disruption, which harms critical cell functions and processes and impacts the health, rate of reproduction, and life-cycle sustainability of the organisms. Petroleum hydrocarbon contamination in the environment is considered a carcinogenic and toxic hazard for both humans and animals (Fan et al. 2012).

Petroleum hydrocarbon contamination in the environment (soil, sediment, water, and air) is mainly caused by human activities such as oil exploration, production, transportation, storage, and consumption (Parnian et al. 2020). These activities can result in accidental spills, pipeline/storage tank leaks, and emissions of oil and its derivatives into the soil, sediment, water, and air resources. These pollutants can pose a serious risk to the health and sustainability of ecosystems and the organisms that depend on them. Petroleum hydrocarbon contamination can also increase the vulnerability of ecosystems to other stressors, e.g., climate change, and drought also reduce environmental biodiversity and ecosystem resilience (Parnian et al. 2021).

Petroleum hydrocarbon pollutants are harmful substances that originate from crude oil and its products. They can cause severe damage to the ecosystem and its diversity, as well as to the health and well-being of humans and other living organisms. Petroleum-hydrocarbon pollutants can affect humans directly by exposure through inhalation, ingestion, or skin contact, or indirectly by disrupting environmental balance and stability. Consequently, it is crucial to prevent and reduce environmental pollution by petroleum hydrocarbon pollutants and to enhance environmental remediation by managing the risk of contamination (Alzahrani and Rajendran 2019). This can help preserve ecosystem sustainability and the livelihoods of the human communities that depend on them. Certain methods have been developed and implemented to remediate petroleum-hydrocarbon-polluted soil and sediment. These methods are bioremediation (which uses microorganisms to degrade the pollutants); fixation (which immobilizes the pollutants by adding chemical/physical agents); landfill (which isolates the contaminated soil in a designated area/controlled burial site); and thermal processing (which uses heat to evaporate/degraded/decontaminate/incinerate the organic contaminants) (Xuezhi 2020).

13.3 Soil Contamination with Hydrocarbons

Oil spills happen during offshore oil/gas well drilling, oil/petroleum product transport, and onshore/pipeline transfer. Additionally, the complex composition of petroleum materials is toxic to organisms due to their environmental disruption and polluting nature (Mirjani et al. 2021). Crude oils, once primary energy sources, have raised public concerns about sustainability, price fluctuations, and environmental impact since the 1970s. Petroleum hydrocarbons contaminate soil, causing ecotoxicity to almost all life forms, leading to biodiversity loss (Ambaye et al. 2022).

Petroleum hydrocarbon contaminants in the soil and sediment pose significant environmental risks to humans and other life forms. Persistent and stable organic compounds, such as persistent organic pollutants (POPs) and polycyclic aromatic hydrocarbons (PAHs), are often scattered anthropogenically, consequently leading to ecological/environmental/social disasters/problems. Understanding the distribution, environmental fate, transport, and chemical properties of these contaminants is crucial for developing cost-effective, environmentally friendly strategies/methods for remediation/restoration/reclamation. Comprehending the nature/composition/

properties of these, types of contaminants, in addition to their interactions with microorganisms, can help prevent/mitigate/manage pollution effects (Ossai et al. 2020).

In recent years, bioremediation has emerged as a promising method for solving petroleum-hydrocarbon-polluted soil and sediment. This approach utilizes the ability of microorganisms to break down and degrade pollutants, effectively cleaning up the affected areas. Moreover, fixation techniques such as solidification/stabilization have been employed to immobilize contaminants and prevent their migration, providing long-term containment. These techniques involve the addition of binding agents or materials to polluted soil, forming a solid matrix that reduces the leaching and dispersion into the surrounding environment (Okoh et al. 2019). Techniques have been employed to immobilize contaminants by introducing chemical or physical agents. Landfill and thermal processes have also been utilized, with the former isolating the polluted soil in designated areas and the latter using heat to evaporate, degrade, decontaminate, or incinerate organic contaminants (Xuezhi 2020). These various methods offer potential solutions for remediation efforts and can help protect the ecosystems and communities that rely on them.

13.4 Materials and Methods

Facilities, pipelines, and oil wells owned by Arvandan Oil and Gas Company, Iran, are located near the Iraqi border and in the western region of Khuzestan, Iran. These facilities have a direct impact on the nearby residential areas, vast wastelands, agricultural lands, and deserts. Ecosystems in these locations are contaminated by unintended petroleum spills and leakage. The purpose of this study is to evaluate the sources of petroleum contamination in the surrounding active environment of Arvandan Oil and Gas Company, Iran, during 2023. In this regard, approximately all disasters that led to major petroleum contamination of the soil and surrounding environment were followed and documented by photography. The potential action for possible control, mitigation, and remediation methods are discussed below.

13.5 Results and Discussion

Table 13.1 shows accidental petroleum pollution which occurred in 2023, and how it affected the surrounding environment in the Arvandan Oil and Gas Company's active area.

As is obvious, the main cause of environmental contamination by crude oil is pipeline disruption. Figures 13.1 and 13.2 illustrate the two main oil spills and leakages in 2023.

Table 13.1 Petroleum pollution occurrences in 2023

Number	Cause of pollution	Estimated affected area (square meter)	Date
1	Pipeline corrosion	90	January
2	Pipeline corrosion	25	March
3	Pipeline corrosion	31	April
4	Pipeline corrosion	14	June
5	Pipeline corrosion	14	July
6	Pipeline corrosion	2	August
7	Pipeline corrosion	20	September
8	Pipeline corrosion	2	October
9	Pipeline corrosion	30	November
10	Pipeline corrosion	25	December

Fig. 13.1 Oil spills and leakages, Khuzestan province, Iran, winter 2023

Fig. 13.2 Oil spills and leakages, Khuzestan province, Iran, spring 2023

The first major spill was caused by a rupture in an aboveground pipeline, resulting in the release of huge amounts of crude oil into a nearby desert environment (Fig. 13.1).

The second spill took place following a rupture in an underground pipeline, resulting in a release of crude oil into the neighboring desert environment. To control and mitigate such incidents, it is crucial to implement higher safety regulations and regular inspections of pipelines and transportation machinery. Additionally, the development and use of advanced technologies for early detection and rapid response to any leaks or spills can significantly minimize the impact on the environment (Okoh et al. 2019).

Once a spill occurs, prompt remediation measures should be taken, such as containment and recovery of the spilled oil, as well as the restoration of affected ecosystems. Public awareness campaigns and educational programs can also play a vital role in preventing and reducing petroleum contamination by promoting responsible practices and emphasizing the importance of environmental protection (Ambaye et al. 2022).

In this area, there are several methods for oil spill remediation available. These methods include mechanical remediation (containment, recovery, removal, and transport of oil and contaminated soils), landfills, and biological remediation. Mechanical containment and recovery involve the use of vacuum trucks and heavy machinery, etc., to physically remove the spilled oil and contaminants from the affected lands. Local landfills and temporary reversals, on the other hand, receive the oil-contaminated soils. Biological remediation utilizes living organisms that naturally break down the oil, speeding up the degradation process.

In this region, there was a successful attempt at soil bioremediation in 2019 (Parnian et al. 2022a). Each method has its advantages and limitations, the choice of remediation method depends on various factors such as the type and volume of spilled oil, weather conditions, and the sensitivity of the affected ecosystem (Parnian et al. 2022b).

13.6 Conclusion

This chapter presents a case study of soil petroleum pollution in Khuzestan Province, Iran. It reviewed some of the major oil spills and leakages that occurred in 2023 and discussed the environmental hazards and impacts of petroleum hydrocarbon contamination of soil and sediment and remediation possibilities. The bottom line emphasizes the importance of preventing and reducing petroleum hydrocarbon pollution and enhancing environmental remediation by implementing strict regulations, guidelines, inspections, assessments, technologies, and education is undeniable. Scientists and environmental activists in society should raise awareness and provide information on the sources, risks, and solutions of soil petroleum pollution to contribute to the sustainable development and management of the environment and society.

13.7 Suggestions for Further Investigation and Study

To help inform and prevent petroleum pollution the following suggestions are made:

I. Perform a cost-benefit analysis of various remediation options discussed to compare their effectiveness, efficiency, and feasibility, thereby providing evidence-based recommendations for soil petroleum pollution management.
II. Oil spills and leakages require investigation into their socio-environmental impacts in consideration of local communities' health, well-being, and livelihoods, enabling the design of appropriate interventions and solutions.
III. Implement a participatory approach to involve stakeholders, local authorities, environmental organizations, and affected communities in decision-making and implementation of remediation possibilities, fostering mutual understanding and collaboration.

References

Alexandersen DK, Headley TR, Prigent S, Mahmutoglu I (2014) Ex-situ bioremediation of hydrocarbon contaminated soil - an example from Oman. All Days. Retrieved January 2, 2024, from https://doi.org/10.2118/170411-ms

Alzahrani AM, Rajendran P (2019) Petroleum hydrocarbon and living organisms. IntechOpen eBooks. Retrieved January 2, 2024, from https://doi.org/10.5772/intechopen.86948

Ambaye TG, Chebbi A, Formicola F, Prasad S, Gomez FH, Franzetti A, Vaccari M (2022) Remediation of soil polluted with petroleum hydrocarbons and its reuse for agriculture: recent progress, challenges, and perspectives. Chemosphere 293:133572. https://doi.org/10.1016/j.chemosphere.2022.133572

Cocârţă DM, Stoian MA, Karademır A (2017) Crude oil contaminated sites: evaluation by using risk assessment approach. Sustainability. Retrieved January 2, 2024, from https://doi.org/10.3390/su9081365

Fan YY, Wang GC, Fu JH, Zheng XH (2012) Bioremediation of waste drilling fluid: comparison of biostimulation and bioaugmentation. Desalin Water Treat 48(1–3):329–334. https://doi.org/10.1080/19443994.2012.698838

Mirjani M, Soleimani M, Salari V (2021) Toxicity assessment of total petroleum hydrocarbons in aquatic environments using the bioluminescent bacterium Aliivibrio fischeri. Ecotoxicol Environ Saf 207:111554. https://doi.org/10.1016/j.ecoenv.2020.111554

Okoh E, Yelebe ZR, Oruabena B, Nelson ES, Indiamaowei OP (2019) Clean-up of crude oil-contaminated soils: bioremediation option. Int J Environ Sci Technol. Retrieved January 2, 2024, from https://doi.org/10.1007/s13762-019-02605-y

Ossai IC, Ahmed A, Hassan A, Hamid FS (2020) Remediation of soil and water contaminated with petroleum hydrocarbon: a review. Environ Technol Innov 17:100526. https://doi.org/10.1016/j.eti.2019.100526

Parnian A, Parnian A, Pirasteh-Anosheh H, Parvizi H (2020) Potential of a pyrolyzed sugarcane industry waste for spillage oil absorption. J Nat Resour Conserv Manag 1(2):163. https://doi.org/10.51396/anrcm.1.2.2020.163-168

Parnian A, Furze JN, Parnian A, Mayad EH (2021) Water purification plantations for oil and gas industries in Iran. Environ Sci Pollut Res 28(45):64193–64198. https://doi.org/10.1007/s11356-021-15819-w

Parnian A, Furze JN, Parnian A (2022a) Drilling waste management based on new methods of bioremediation and solar desalination. Earth Syst Protect Sustain 1:79–97. https://doi.org/10.1007/978-3-030-85829-2_4

Parnian A, Parnian A, Pirasteh-Anosheh H, Furze JN, Prasad MNV, Race M, Hulisz P, Ferraro A (2022b) Full-scale bioremediation of petroleum-contaminated soils via integration of co-composting. J Soils Sediments 22(8):2209–2218. https://doi.org/10.1007/s11368-022-03229-5

Parnian A, Mahbod M, Prasad MNV (2023) Microplastics remediation—possible perspectives for mitigating saline environments. In: Microplastics in the ecosphere. Wiley, pp 465–476. https://doi.org/10.1002/9781119879534.ch29

Xuezhi D (2020) Remediation methods of crude oil contaminated soil. World J Agric Soil Sci 4(4). https://doi.org/10.33552/wjass.2020.04.000595

Chapter 14
Environmental Impacts of Petroleum Pollution in South of Iran

Yousef Soleimani, Amir Parnian, Majeti Narasimha Vara Prasad, Flávio Henrique Silveira Rabêlo, and Nematollah Jaafarazdeh Haghighi Fard

Abstract The Persian Gulf coastline and South of Iran is one of country's most important oil producer locations, affecting an area over 5000 km^2. There are rocky valleys and mountains, dam reservoirs, agricultural lands, large permanent rivers and seasonal streams, environmentally sensitive areas (ESAs), deserts, the Persian Gulf coastline, commercial ports, and aquaculture farms in this area. Oil leakage from oil transmission lines caused by corrosion can cause pollution in any of the mentioned ecosystems. The development of an action plan for environmental emergencies suitable for each climate is an important issue in pollution risk management. Pipeline oil spills may be transferred to different ecosystems by draining into permeable soil layers or via flowing into the bedrock or river surface. Although a large amount of the leaked oil can be removed through proper planning and availability of machines and equipment to deal with oil pollution in aqueous and land environments, some of the pollutants will inevitably be absorbed by soil texture. Therefore, local oil and gas company needs to have a soil remediation program. Considering the differences in soil texture and usage, several sites have been selected in various areas to which the removed soil contaminated with petroleum substances

Y. Soleimani (✉)
Gachsaran Oil and Gas Production Company (GOGPC), National Iranian Oil Company (NIOC), Ghachsaran, Iran
e-mail: soleimani.yousef@ut.ac.ir

A. Parnian
National Salinity Research Center (NSRC), Agricultural Research, Education and Extension Organization (AREEO), Yazd, Iran

M. N. V. Prasad
Department of Plant Sciences, School of Life Sciences, University of Hyderabad (An Institution of Eminence), Hyderabad, Telangana, India

F. H. S. Rabêlo
Luiz de Queiroz College of Agriculture, University of São Paulo, Piracicaba, Brazil

N. J. H. Fard
Environmental Technologies Research Center, Medical Basic Sciences Research Institute, Ahvaz Jundishapur University of Medical Sciences, Ahvaz, Iran

A. Parnian et al. (eds.), *Bioremediation of Petroleum Contaminated Soil and Sediment*, Environmental Contamination Remediation and Management,
https://doi.org/10.1007/978-3-031-95524-2_14

will be transferred for remediation. Once the soil remediation process is finished and the soil that was polluted reaches the permissible level, it is returned to its original location. This chapter has reviewed some cases of oil pollution in various environments, discussed oil pollution hazards, and then introduced the various soil remediation sites used by a local oil and gas company.

Keywords Sustainable development · Oil · Remediation · Pollution

14.1 Introduction

Human activities across various economic sectors, particularly services, agriculture, and industry, precipitate petroleum hydrocarbon contamination, and harm ecosystem stability and productivity. Ecosystem stability is important for human prosperity and economic vitality, offering indispensable services and resources. The oil and gas industrial activities can cause environmental pollution e.g. petroleum exploration, production, refining, storage, and transport. Sometimes, operational negligence results in environmental incidents that pollute soil, sediment, air, and water bodies, impairing biodiversity and ecosystem health. Addressing this environmental challenge necessitates preemptive and reactive measures. Proactive strategies include stringent regulations, comprehensive workforce education, and systematic risk assessments to forestall contamination. Conversely, responsive strategies involve utilizing specialized environmental remediation apparatus and ecofriendly techniques to rehabilitate ecosystems post-contamination (Mushtaq et al. 2024; Bakhtiari et al. 2024).

Petroleum is a valuable commodity that humans began extracting in the mid-nineteenth century to meet their needs as an energy source and raw material for many chemical industries. In today's world, the Middle East is known as the global energy hub. Iran became the host of the first oil extractors by drilling well No. 1 in Masjed Soleyman in 1908, earning the title of the Middle East's first oil country. Oil extraction from this well heralded the extensive oil resources in southern Iran. Accordingly, oil exploration in southern Iran expanded, and in 1939, oil extraction began in the Gachsaran region. During those years, oil was used as an unparalleled energy source in an unbridled manner, and oil was extracted from reservoirs at a rapid pace. With technological advancements and increased lifespan of oil reservoirs, the equipment and facilities associated with oil production underwent remarkable progress. As the number of oil wells increased, production units and oil desalination units were constructed one after the other to extract and supply high-quality oil for downstream industries.

Currently, in 2024, Gachsaran Oil and Gas Company (GOGPC), with more than 600 oil wells, 6000 kilometers of pipelines with different sizes and applications, 10 production plants, 6 desalination plants, and 15 associated gas gathering plants, operates in four provinces: Kohgiluyeh and Boyer-Ahmad, Fars, Khuzestan, and Bushehr, covering an area of over 5000 square kilometers, and producing oil.

Over time, humans have come to realize that a better life requires more than unlimited energy resources. The environment has been given special status as the primary biological concern in recent decades. Attention to environmental quality and health (air, water, soil, and sediment) has obligated industries to take measures to operate with minimal environmental impact (Parnian 2024). However, due to the polluting nature of petroleum products, environmental pollution in the oil industry remains unavoidable. This chapter aims to review seven of cases of GOGPC's oil pollution disasters in various environments during 2022–2024. It discusses petroleum pollution hazards and then introduces a related soil remediation technique used for each case.

14.2 Oil and Gas Production Process Description of GSOGPC

14.2.1 Production Process of GSOGPC

Crude oil and a certain amount of saline water, hydrocarbon, and sulfur gases are extracted from oil reservoirs through oil wells. The associated gases are separated from the crude oil in the first stage to meet the desired quality specifications of consumers. For this purpose, oil and gas separators were constructed near oil wells. The role of these separators is to separate the associated gas from the liquids (comprising oil and saline water), as the geographical anomalies in GSOGPC's operating area can cause the "gas lock" phenomenon, halting the transfer of oil to production plants (Brauers et al. 2021; Ranaee et al. 2021).

In the next stage, the liquids from these separators are transferred to production units, where all dissolved gases are removed from the oil. Subsequently, the oil from the production units enters desalting plants to remove water and salt. The salt concentration in the associated water can reach more than three times the seawater's salinity, which can adversely affect downstream industrial processes. Oil enters desalting plants in the next stage, where water, sediments, and heavy metals associated with the oil are completely separated. Due to their detrimental environmental impacts, these materials are then re-injected into the oil reservoir at depths exceeding 2000 meters using high-pressure pumps and corrosion-resistant pipelines. The oil leaving these plants is then supplied to consumers via pipelines (Ranaee et al. 2021).

14.2.2 Environmental Pollution

A notable aspect of this process is the numerous potential points of oil contamination, including oil wells, separators, production and desalting plants, and, more importantly, the extensive pipeline network that transports oil from the wells to the delivery points (Parnian et al. 2022b). All these components can be sources of soil contamination due to sudden leaks at multiple locations.

In GOGPC's operational area, naturally, the widespread distribution of pipelines across an area of over 5000 square kilometers will affect lands with diverse uses. One of the production units is located atop a mountain 900 meters above sea level. At the same time, other pipelines extend to export terminals and commercial ports at sea level, illustrating the variety of affected lands. The pipelines traverse environmentally sensitive areas, posing significant environmental risks to this industry, necessitating a comprehensive emergency response plan.

14.3 Oil Spill Incident and River Pollution

The calcareous and marly soil composition surrounding the company's facilities adds complexity to the transmission of contamination. The experience of an oil spill incident in 2022 showed that after the spill, the oil quickly penetrated the soil from the leak site and followed underground waters, emerging in a valley 100 meters deep and approximately 2 kilometers away from the leak site. Subsequently, as the oil flowed along the riverbed, the rugged terrain and inability to install oil spill response equipment resulted in a 7.5-kilometer stretch of the riverbed becoming contaminated with oil. This river supplies water to approximately 400 hectares of agricultural land, underscoring the urgency of riverbed cleanup. Ultimately, the riverbed cleanup operation was completed within 6 days. Contamination of these areas was fortunately avoided by temporarily halting the water supply to the agricultural lands. However, the surrounding reed beds were not spared, and portions of the riverbed soil and surrounding rocks were contaminated with petroleum products.

At this stage, to prevent the contamination from spreading to clean soil and rocks, the removal of contaminated soil, rocks, and reed beds was carried out manually. The contaminated reed beds were classified as industrial waste according to the Iranian Department of Environment guidelines and were disposed of accordingly. The hydrocarbon-contaminated rocks were stabilized using cement and silicate (Ahmad et al. 2020; Erfani et al. 2024) and rendered harmless. Finally, the oil-contaminated soils were transferred to GOGPC's soil remediation landfarm to be treated and returned to permissible limits. The described case involved the outlet pipeline of one of the company's first-stage separators. Figure 14.1 depicts the riverbed cleanup operations.

14.4 Corrosion of Oil Transfer Pipelines

Various cases of contamination occur at GOGPC, primarily related to leaks from oil transfer pipelines due to corrosion. Due to oil spill incidents vast geographical distribution, the first-stage separators can cause different types of soil contamination.

Fig. 14.1 Oil spill incident caused river pollution, South of Iran, 2022

One specific type of contamination is the scattering of oil onto the ground surface, typically originating from the separators' flares. In this type of contamination, a large area of soil becomes involved in the oil contamination. However, since the contamination's thickness is very low, removing and transferring the soil to remediation sites would cause mixing with clean soil (Fig. 14.2). Therefore, an in-situ remediation method is employed for this type of soil contamination incident.

a b

Fig. 14.2 Massive oil leaks from transfer pipelines due to corrosion, South of Iran, 2022

In this method of environmental restoration, oil-contaminated soil should reach the permissible limits set by the Iranian Department of Environment within a suitable timeframe. In this regard, by spraying a nutrient solution (e.g. commercial NPK 20-20-20 fertilizer provided by local suppliers) (Gadhvi 2019) and enhancing hydrocarbon-degrading microorganisms (e.g. *Trichoderma harzianum*) (Tripathi et al. 2013). Since local wildlife may consume contaminated pastures as food, plowing these areas will aerate the soil and remove it from the reach of animals and nomadic livestock. After a short period, the soil will be fully restored and return to its original condition.

14.5 Oil Spill Incidents Rocky Valleys

Rocky valleys are one of the most inaccessible operational areas for cleanup, where contamination removal is backbreaking. Transporting oil spill response equipment, such as transfer pumps, collapsible tanks, flexible hoses, etc., is not feasible in these areas (Fig. 14.3). On the other hand, leaving the spilled oil in the environment risks contaminating downstream regions that serve as water sources for nomadic livestock and wildlife. Therefore, considering the small volume of spilled oil in the present case and preventing further environmental damage if transferred downstream, the spilled oil will be safely burned on-site. Although this method will result in some air pollution from soot, its adverse environmental impacts will be greatly reduced.

Fig. 14.3 A valley oil spill incidents, Southwest of Iran, 2022

14.6 Torrential Rainfall, Flood, and Oil Spill

Torrential rainfall is a characteristic of the southern regions of Iran. The wet winters in these areas sometimes pose environmental risks to GOGPC. A noteworthy case is the rupture of a pipeline and the consequent spread of oil contamination due to torrential rainfall in 2023 (Fig. 14.4).

Fig. 14.4 Torrential rainfall expands the oil spill-impacted area, South of Iran, 2023

After the pipeline rupture, the oil flowed into the floodwaters and was carried downstream. Raging floodwaters made any oil spill response efforts impossible. Once rain subsided and floodwaters receded, the oil settled in agricultural lands traversed by the flood, contaminating a large portion of these areas with petroleum products. Since the floodwaters spread over a wide area, the contamination was not concentrated; rather, it formed a thin oil film across extensive agricultural lands. Crops were destroyed, and the oil-contaminated lands became unusable.

Regarding environmental remediation necessary, the in-situ ecological remediation method involving commercial NPK 20-20-20 fertilizer provided by local suppliers (Gadhvi 2019) and enhancing hydrocarbon-degrading microorganisms (e.g. *Trichoderma harzianum*) (Tripathi et al. 2013), aeration (Kuppusamy et al. 2020), and controlled moisture addition (Rhykerd et al. 1999) was employed to bring the soil of the agricultural lands within the permissible limits set by the Iranian Department of Environment.

14.7 Oil Spills to Aquatic Environments

The proximity of GOGPC's facilities to dam reservoirs has necessitated the development of diverse emergency response plans for any potential oil contamination in these areas. Trained personnel, cargo and passenger boats, inflatable booms, and oil skimmers (Azizian and Khosravi 2019; Parnian et al. 2022a) are among the necessary equipment for responding to oil spills in aquatic environments (Fig. 14.5). Fortunately, no oil spills into dam reservoirs have been reported so far. However, this issue is so critical that numerous drills are conducted annually to prepare for potential oil spills in these locations.

a b

Fig. 14.5 Oil spills remediation from aquatic environments, South of Iran, 2022

14.8 Corrosion of Main Oil Pipelines

Corrosion of large-diameter pipelines is one of the risks that can lead to extensive soil contamination. In one of the oil spill incidents in 2022, approximately 500 tons of soil were contaminated with petroleum products (Fig. 14.6). In this type of

Fig. 14.6 Oil spill incidents, Southwest of Iran, 2022

contamination, due to the underground location of the pipelines, the spill will continue untapped until soil excavation and installation of a temporary leak clamp. The volume of spilled oil and contaminated soil in this type of contamination is substantial. After repairing the pipeline leak, vacuum trucks collect the spilled oil on-site, remaining contaminated soil is transferred to GOGPC's soil remediation landfarm by heavy machinery for treatment.

14.9 Soil Treatment—Land Farming

Soil contamination in the oil production industry is unavoidable; therefore, the key factor in mitigating adverse environmental impacts is having a proper oil spill response plan (Passow and Lee 2022) for GOGPC. Land farming is a capable soil bioremediation process for petroleum-contaminated soil, sediment, or sludge. This eco-friendly method involves the incorporation of petroleum-polluted materials into the soil matrix and using agricultural devices and methods to stimulate aerobic microbial biodegradation. Soil moisture, oxygen levels, nutrient content, and pH are controlled to optimize the contaminants degradation. The remediation process in this method, mainly uses the soil's natural microbiota for bioremediation. Land farming also supports the recycling of organic carbon and nutrients within the biosphere (Johnsen et al. 2021).

The most crucial aspect of this plan is the establishment of a land farming plantation for remediating oil-contaminated soils. Consequently, the GOGPC created a land farm with an annual capacity of 2000 cubic meters of oil-contaminated soil. At this land farming plantation, the oil-contaminated soil is treated using the biostimulation method and then returned to its original environment. After complete soil remediation, the final environmental impacts are effectively mitigated (Fig. 14.7).

14.10 Conclusion

Many processes in the oil industry result in petroleum contamination, including oil wells, separators, production and desalting plants, and, more importantly, the extensive pipeline network that transports oil from the wells to the delivery points. The local oil and gas producers of Iran has struggled with all of them emphasizing the need for a soil remediation program.

Fig. 14.7 GOGPC's landfarm, South of Iran, 2023

14.11 Suggestion

To emphasize and assist with remediation of petroleum spills the following suggestions are made:

I. The soil/sediment remediation techniques should be utilized regarding the surrounding environment and local requirements.
II. In situ, or ex situ remediation bioremediation methods should be optimized for each petroleum pollution incident.
III. A combination of methods sometimes is more acceptable for environmental remediation.

References

Ahmad S, Al-Amoudi OSB, Mustafa YM, Maslehuddin M, Al-Malack MH (2020) Stabilization and solidification of oil-contaminated sandy soil using Portland cement and supplementary cementitious materials. J Mater Civ Eng 32(8):04020220

Azizian S, Khosravi M (2019) Advanced oil spill decontamination techniques. In: Interface science and technology, vol 30. Elsevier, London, United Kingdom, pp 283–332

Bakhtiari H, Amanipoor H, Battaleb-Looie S (2024) Analysis of heavy metal accumulation and environmental indicators in fluids and drilling cuttings. J Petrol Explor Prod Technol 14(1):41–58

Brauers H, Braunger I, Jewell J (2021) Liquefied natural gas expansion plans in Germany: the risk of gas lock-in under energy transitions. Energy Res Soc Sci 76:102059

Erfani H, Madhu NR, Khodayari S, Qureshi MA, Swetanshu S, P. and Jadoun, S. (2024) Separation and removal of oil from water/wastewater in the oil industry: a review. Environ Technol Rev 13(1):325–343

Gadhvi SN (2019) Remediation of petroleum oil contaminated soil by using organic fertilizer. Acta Sci Agric 3(10):117–180

Johnsen AR, Boe US, Henriksen P, Malmquist LM, Christensen JH (2021) Full-scale bioremediation of diesel-polluted soil in an Arctic landfarm. Environ Pollut 280:116946

Kuppusamy S, Maddela NR, Megharaj M, Venkateswarlu K, Kuppusamy S, Maddela NR et al (2020) Approaches for remediation of sites contaminated with total petroleum hydrocarbons. In: Total petroleum hydrocarbons: environmental fate, toxicity, and remediation. Springer International Publishing, Cham, Switzerland, pp 167–205

Mushtaq Z, Muzammil A, Bellitürk K, Anwar W, Akhter A, Khan HAA et al (2024) Role of Rhizobacteria in phytoremediation of heavy metal. In: Heavy metal remediation: sustainable nexus approach. Springer Nature Switzerland, Cham, pp 183–211

Parnian A (2024) Landfill gas as a source of energy to promote decarbonization: evaluation of Yazd county landfill, Iran. In: Decarbonization strategies and drivers to achieve carbon neutrality for sustainability. Elsevier, Amsterdam, Netherlands, pp 515–526

Parnian A, Furze JN, Parnian A (2022a) Drilling waste management based on new methods of bioremediation and solar desalination. In: Furze JN, Eslamian S, Raafat SM, Swing K (eds) Earth systems protection and sustainability, vol 1. Springer International Publishing. https://doi.org/10.1007/978-3-030-85829-2_4

Parnian A, Parnian A, Pirasteh-Anosheh H, Furze JN, Prasad MNV, Race M, Hulisz P, Ferraro A (2022b) Full-scale bioremediation of petroleum-contaminated soils via integration of co-composting. J Soils Sediments 22(8):2209–2218. https://doi.org/10.1007/s11368-022-03229-5

Passow U, Lee K (2022) Future oil spill response plans require integrated analysis of factors that influence the fate of oil in the ocean. Curr Opin Chem Eng 36:100769

Ranaee E, Ghorbani H, Keshavarzian S, Abarghoei PG, Riva M, Inzoli F, Guadagnini A (2021) Analysis of the performance of a crude-oil desalting system based on historical data. Fuel 291:120046

Rhykerd RL, Crews B, McInnes KJ, Weaver RW (1999) Impact of bulking agents, forced aeration, and tillage on remediation of oil-contaminated soil. Bioresour Technol 67(3):279–285

Tripathi P, Singh PC, Mishra A, Chauhan PS, Dwivedi S, Bais RT, Tripathi RD (2013) Trichoderma: a potential bioremediator for environmental clean up. Clean Techn Environ Policy 15:541–550

Chapter 15
The Role of Earthworms in Petroleum Contaminated Soil Remediation

Shweta Sharma, Nibha Singh, Dilbag Singh, R. K. Gupta, and Rahul Kumar

Abstract Petroleum pollution is a concern due to its extensive use and harm to ecosystems. Among the living organisms found in soils, earthworms are recognized as crucial contributors to soil health and fertility. Various studies have shown that petroleum contamination, including crude oil and its by-products, affects the behavior of earthworms, their population dynamics, and their role in shaping ecosystems. Petroleum contamination can disrupt these essential functions through various mechanisms such as direct toxicity, altered food availability, and changes in soil physicochemical properties. Potential remediation strategies harness earthworms' bioremediation capabilities to mitigate the negative impacts of petroleum contamination. Earthworms' capacity to degrade hydrocarbons and enhance soil microbial communities holds promise for sustainable soil restoration. This chapter delves into the mechanisms behind how earthworms respond to exposure to petroleum substances while emphasizing the significance of understanding both effects and metabolic adaptations. This chapter explores the multifaceted impacts of petroleum pollution at various levels, ranging from cellular responses to biochemical alterations and molecular mechanisms. By dissecting these intricate interactions, This chapter aims to elucidate the far-reaching consequences of petroleum contamina-

S. Sharma
Department of Physiology, AIIMS, New Delhi, India

N. Singh
Department of Biochemistry, Jamia Hamdard University, Delhi, India

D. Singh
School of Agriculture Sciences, Galgotias University, Greater Noida, India

R. K. Gupta
Department of Zoology, Chaudhary Charan Singh Haryana Agricultural University, Hisar, Haryana, India

R. Kumar (✉)
Department of Zoology, Chaudhary Charan Singh Haryana Agricultural University, College of Agriculture, Bawal, Rewari, Haryana, India
e-mail: drrkr321@gmail.com; rahulrohila01@hau.ac.in

A. Parnian et al. (eds.), *Bioremediation of Petroleum Contaminated Soil and Sediment*, Environmental Contamination Remediation and Management,
https://doi.org/10.1007/978-3-031-95524-2_15

tion on vital soil-dwelling organisms, and in turn, on soil ecosystems. There is a need for a comprehensive understanding of the interactions between petroleum pollutants and earthworms to develop effective soil management and remediation strategies that can safeguard soil health and ecosystem sustainability in the face of petroleum contamination challenges.

Keywords Ecosystem · Fertility · Metabolic · Physicochemical · Bioremediation

15.1 Introduction

Earthworms play a crucial role in the remediation of petroleum-contaminated soil, offering a natural and effective solution to this environmental challenge. Research has shown that earthworms, such as *Eisenia fetida* and *Apporectodea caliginosa*, are capable of tolerating and degrading crude oil, making them suitable candidates for bioremediation efforts (Curry and Schmidt 2007; Dada et al. 2021; Kumar et al. 2023). Studies have demonstrated that earthworms can accelerate the degradation of crude oil in the soil, with significant reductions in petroleum concentrations observed in the presence of earthworm species like *E. fetida* and *A. caliginosa* (Dada et al. 2021; Yadav et al. 2022a; Yadve et al. 2023; Yadav et al. 2023). Additionally, vermiremediation which involves the use of earthworms to remove contaminants from soil has shown promising results in various studies with hydrocarbon concentrations decreasing by up to 95% in the presence of earthworms like *Dendrobenaveneta* (Dada et al. 2021; Singh et al. 2022; Tagliabue et al. 2023). The effectiveness of earthworms in bioremediation lies in their ability to enhance soil microbial activity exposing contaminants to intense physical and chemical processes that facilitate degradation (Curry and Schmidt 2007). Earthworms not only ingest soil but also create favorable conditions for microbial degradation of pollutants, making them valuable allies in the restoration of contaminated environments (Tagliabue et al. 2023). Their role in accelerating the breakdown of complex hydrocarbons and promoting the removal of contaminants from soil highlights their potential as a sustainable and eco-friendly remediation strategy (Curry and Schmidt 2007; Sharma et al. 2020; Dada et al. 2021; Yadav et al. 2022b).

The multiple ways in which different contaminants of petroleum contamination in soil arise from various sources, including leaks, spills, and improper disposal of petroleum-based products such as gasoline, diesel, and crude oil (Fig. 15.1). The sources are the root causes of soil contamination, making it difficult to achieve a remediation process. Nonetheless, these challenges are allayed by the contributions of earthworms to bioremediation, which is a process of mitigating the adverse effects of petroleum pollution on soil. These contaminants consist of complex mixtures of hydrocarbons, including volatile organic compounds (VOCs), polycyclic aromatic hydrocarbons (PAHs), and heavy metals. Petroleum contamination poses significant risks to environmental and human health as it can persist in soil for

Fig. 15.1 Illustrates different sources of petroleum pollution in soil

extended periods, leach into groundwater, and disrupt soil ecosystems. The extent of petroleum contamination in the soil varies depending on factors such as the volume of spilled or leaked petroleum, soil type and composition, climate, and proximity to water bodies. In addition to direct contamination from spills, petroleum can also enter soil indirectly through atmospheric deposition and runoff from paved surfaces (Yadav et al. 2022a; Yadav et al. 2023). Once in the soil, petroleum contaminants can have detrimental effects on soil fertility, microbial communities, plant growth, and the health of soil-dwelling organisms. They can also pose risks to human health through exposure pathways such as inhalation of volatile compounds, ingestion of contaminated soil or groundwater, and dermal contact. Petroleum contamination in soil refers to the presence of petroleum hydrocarbons and their derivatives, which are persistent and harmful to the environment. Common petroleum products found in contaminated soil include gasoline, diesel fuel, kerosene, jet fuel, motor oil, hydraulic fluid, and used oil (Curry and Schmidt 2007; Zeb et al. 2020).

When these substances are released at levels exceeding regulatory standards, the soil becomes contaminated, posing risks to human health, animals, and the environment (Rodriguez-Campos et al. 2014). The process involves the activity of microbes, which reside in the gut of earthworms. The complex hydrocarbons then get broken down into their simpler forms when more active. The activity of these microbes enhances the bioremediation of petroleum-contaminated wetlands, facilitating the decomposition of toxic substances for safe living and environmental restoration (Fig. 15.2). Petroleum contamination can harm soil microorganisms, reduce their number and activity, and affect plant health (Dada et al. 2021; Yadav et al. 2022a; Yadav et al. 2023). It can also cause discoloration in the soil and poor vegetative growth (Dada et al. 2021). Petroleum-contaminated soil remediation is necessary, which may involve various techniques such as dredging/excavation, soil vapor extraction, biostimulation, or bioaugmentation (Rodriguez-Campos et al. 2014). Petroleum contamination of soil is a pervasive environmental issue with far-reaching consequences for ecosystems and human health. Hydrocarbons from petroleum products can leach into the soil through various pathways, including accidental spills, leaks from underground storage tanks, and industrial activities. These contaminants persist in the soil matrix, posing challenges for remediation efforts due to their complex chemical composition and longevity in the environment. Traditional methods of soil remediation, such as excavation, soil washing, and chemical treatment, are often costly, energy-intensive, and disruptive to natural habitats. Moreover, these methods may not always achieve complete remediation, leaving behind residual contaminants and disrupting soil structure and microbial communities (Renu et al. 2020; Kumar et al. 2023). As a result, there is a growing interest in exploring alternative remediation approaches that are more sustainable, cost-effective, and environmentally friendly. One such approach involves harnessing the natural bioremediation capabilities of earthworms, which have long been recognized for their role in improving soil health and fertility. Earthworms, through their feeding and burrowing activities, can enhance microbial activity, improve soil aeration and porosity, and accelerate the decomposition of organic matter. These processes contribute to the breakdown and detoxification of petroleum hydrocarbons in

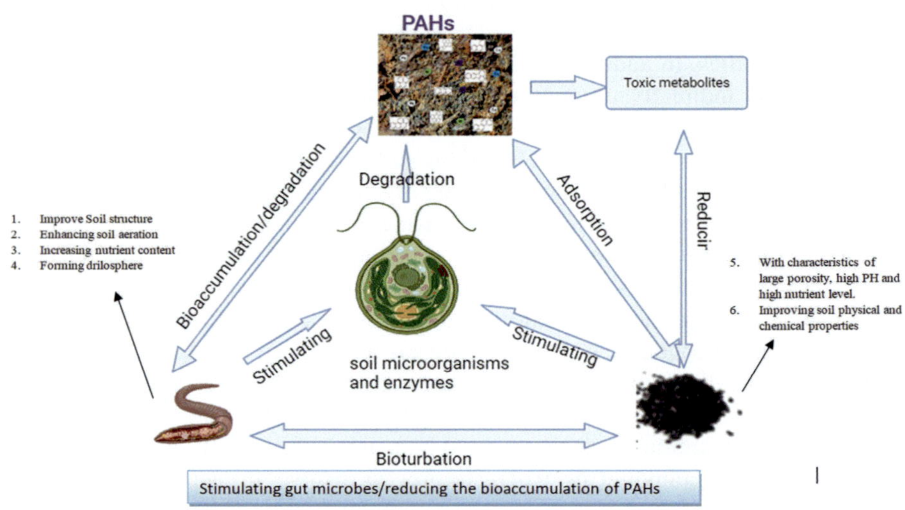

Fig. 15.2 Stimulating gut microbes of PAHs

contaminated soil ultimately leading to the restoration of soil quality and ecosystem function (Chachina et al. 2016; Yadav et al. 2022a; Yadav et al. 2023; Yadav et al. 2022b; Kumar et al. 2023).

Earthworms are soil-dwelling organisms belonging to the class Oligochaeta, comprising thousands of species found in diverse habitats worldwide. By ingesting soil, enhancing microbial activity, and promoting decomposition, earthworms serve as catalysts for the breakdown and detoxification of petroleum hydrocarbons in contaminated soil. Their activities contribute to the restoration of soil quality and ecosystem function, offering a natural and sustainable solution to the pervasive issue of petroleum contamination (Fig. 15.3). They play a crucial role in soil health and ecosystem function through their feeding, burrowing, and casting activities. In the context of petroleum-contaminated soil remediation, earthworms have garnered attention for their ability to enhance biodegradation processes and improve soil quality. Earthworms ingest soil and organic matter, breaking down complex compounds into simpler forms that are more readily accessible to microbial decomposers. As they burrow through the soil they create channels that facilitate air and water movement, enhancing soil aeration and drainage (Kumar and Singh 2022; Kumar et al. 2023). Moreover, earthworms promote the growth and activity of soil microorganisms involved in the degradation of petroleum hydrocarbons. Their gut microbiota can also metabolize certain hydrocarbons, contributing to the detoxification of contaminants. Additionally, earthworms incorporate organic matter into their casts, enriching the soil with nutrients and promoting microbial activity. Earthworms represent a promising natural bioremediation tool for petroleum-contaminated soil, offering potential advantages such as cost-effectiveness, minimal environmental impact, and compatibility with sustainable soil management practices. The

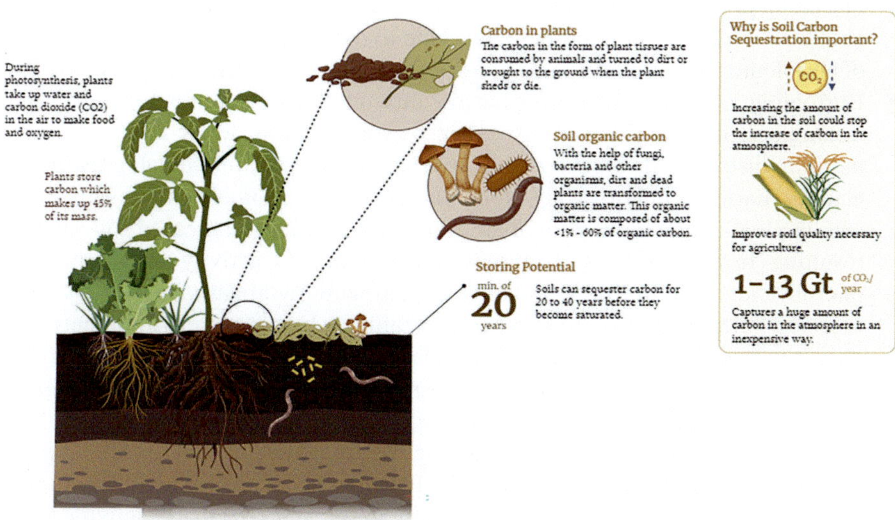

Fig. 15.3 Earthworms as biochemical reactors affect the soil

effectiveness of earthworm-mediated bioremediation depends on various factors, including soil properties, earthworm species, environmental conditions, and contaminant characteristics (Kumar et al. 2022).

15.2 Mechanisms of Bioremediation by Earthworms

The mechanism by which earthworms facilitate petroleum degradation involves a combination of physical, chemical, and biological processes.

- Physical fragmentation: earthworms ingest soil and organic matter, breaking them down into smaller particles. This process increases the surface area available for microbial activity and enhances the accessibility of contaminants to soil microorganisms. Earthworms ingest soil and organic matter, including petroleum-contaminated soil, as they burrow through the soil profile. Inside the earthworm gut, mechanical grinding and mixing occur, breaking down large soil aggregates and contaminants into smaller particles (Kumar et al. 2020; Tang et al. 2011). This physical fragmentation increases the surface area available for microbial colonization and enzymatic degradation of petroleum hydrocarbons.
- Gut microbiota: earthworms host diverse microbial communities in their digestive tracts, which can metabolize a wide range of organic compounds, including petroleum hydrocarbons. These gut microbes contribute to the degradation and detoxification of contaminants as they pass through the earthworm's digestive system. Earthworms harbour diverse microbial communities in their digestive tracts, including bacteria, fungi, and protozoa (Grosser et al. 1991). These gut microbes have been shown to metabolize petroleum hydrocarbons through various metabolic pathways, including aerobic and anaerobic degradation. The metabolic activities of gut microbes contribute to the detoxification and degradation of petroleum contaminants as they pass through the earthworm's digestive system.
- Burrowing activity: earthworms burrow through the soil, creating channels and mixing soil layers. This activity enhances soil aeration, water infiltration, and nutrient distribution, creating favorable conditions for microbial growth and activity. Earthworm burrowing creates a network of channels and tunnels in the soil, facilitating the movement of air, water, and nutrients. Increased soil aeration resulting from burrowing promotes the growth and activity of aerobic soil microbes involved in the degradation of petroleum hydrocarbons. Burrows also enhance water infiltration and drainage, creating dynamic soil environments conducive to microbial degradation processes (Zechmeister-Boltenstern et al. 2015).
- Cast deposition: earthworms excrete nutrient-rich casts, which contain organic matter and microbial biomass. These casts enrich the soil with organic carbon, nitrogen, and other nutrients, stimulating microbial activity and promoting the degradation of petroleum contaminants. Earthworms excrete nutrient-rich casts, which contain organic matter, microbial biomass, and enzymatic activities (Al-Maliki et al. 2020). These casts contribute to the decomposition of organic mat-

ter, including petroleum hydrocarbons, by providing a substrate and inoculum for microbial growth and activity. Nutrients released during cast decomposition stimulate microbial metabolism and enhance the degradation of petroleum contaminants in the soil.

- Co-metabolism and enzymatic activity: earthworms produce enzymes, such as cytochrome P450 monooxygenases, dehydrogenases, and peroxidases, which can metabolize petroleum hydrocarbons (Evans et al. 1965; Kumar et al. 2022). These enzymes are involved in the oxidation, hydroxylation, and cleavage of hydrocarbon molecules, leading to their degradation and detoxification in the soil environment (Tagliabue et al. 2023). Co-metabolic reactions between earthworm-derived enzymes and microbial metabolites further enhance the degradation of petroleum contaminants (Table 15.1).

Earthworms play a crucial role in enhancing microbial activity in soil contaminated with petroleum hydrocarbons. Their activities promote the growth and activity of indigenous soil microorganisms capable of degrading hydrocarbons through various metabolic pathways, including aerobic and anaerobic degradation, as well as co-metabolism. Increasing soil aeration and oxygen diffusion through burrowing activity promotes aerobic microbial degradation of petroleum hydrocarbons. Mixing organic matter and contaminants through ingestion and casting facilitates microbial access to nutrients and hydrocarbons (Bhadauria and Saxena 2010; Kumar and Singh 2022). By incorporating contaminated materials into their digestive systems and subsequently excreting enriched casts, earthworms contribute to the remediation of petroleum-contaminated soil while also enhancing soil fertility and nutrient cycling (Fig. 15.4). Stimulating microbial diversity and metabolic activity in the earthworm gut, where microbial communities can metabolize hydrocarbons and

Table 15.1 The actions of earthworms, their bioremediation potential, and their efficacy in vermiremediation

Number	Earthworm Actions	Bioremediation Potential	Vermiremediation Efficacy	References
1	Enhance soil fertility and nutrition through aeration and bioturbation.	Essential for bioremediation of petroleum-contaminated soils.	Enhance microbial activity and enzyme actions to remove contaminants.	Ekperusi and Aigbodion (2015)
2	Inhibit organic contaminant binding, releasing soil-bound pollutants for degradation.	Impact soil-associated organic contaminant fate, enhancing bioremediation potential.	Facilitate water movement, aeration, and soil particle breakdown for contaminant degradation.	Hickman and Reid (2008)
3	Enhance plant growth through soil ingestion, facilitating particle breakdown.	Remediate diverse organic contaminants: agrochemicals, petroleum, PAHs, and PCBs	Speeds up crude oil degradation, boosting bioremediation efficacy.	Datta et al. (2023)

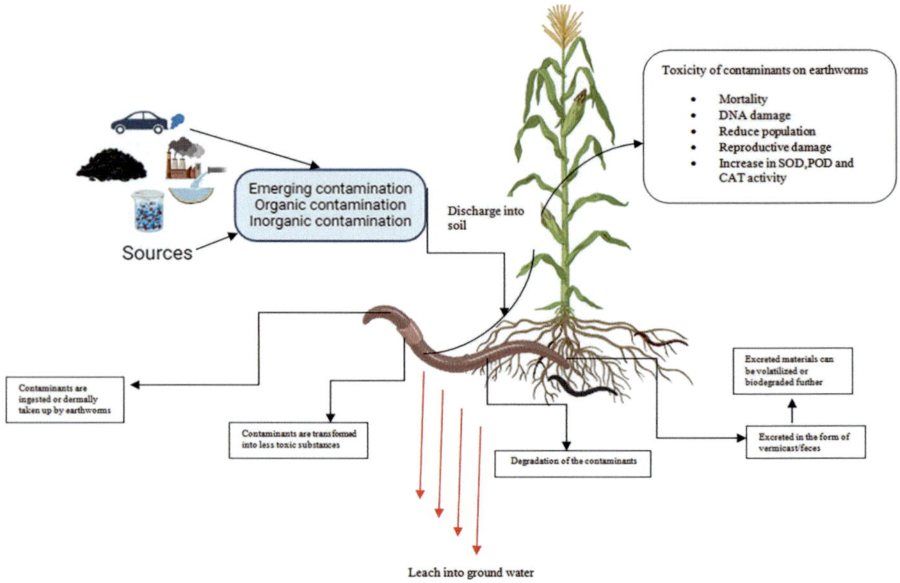

Fig. 15.4 Contaminated ingested or taken by earthworm

contribute to their detoxification. The combined effects of earthworm-mediated physical, chemical, and biological processes result in accelerated degradation and detoxification of petroleum contaminants in soil.

Earthworms enhance soil aeration and porosity through their burrowing activities, which create channels and tunnels in the soil profile (Lavelle and Spain 2002). These channels facilitate the movement of air, water, and nutrients through the soil, creating a more hospitable environment for microbial activity and plant root growth.

- Increase oxygen diffusion into the soil, promoting aerobic microbial degradation of petroleum contaminants.
- Enhance water infiltration and drainage, reducing the risk of waterlogging and creating conditions conducive to microbial growth and activity.
- Facilitate root penetration and nutrient uptake by plants, promoting vegetation establishment and ecosystem recovery in contaminated sites.

Soil aeration and porosity improvement by earthworms contribute to the remediation of petroleum-contaminated soil by creating conditions favorable for microbial degradation and plant growth.

Earthworms play a vital role in organic matter decomposition and nutrient cycling in soil ecosystems. Through their feeding and casting activities, earthworms incorporate organic matter into the soil, where it is decomposed by soil microorganisms. This decomposition process releases nutrients such as nitrogen, phosphorus, and potassium, making them available for plant uptake and supporting plant growth. In petroleum-contaminated soil, earthworms accelerate the decomposition of organic matter, including petroleum hydrocarbons, through their gut microbiota and enzymatic activities. As earthworms ingest contaminated soil and organic matter, they break down complex hydrocarbons into simpler compounds, which are then further degraded by soil microorganisms (Hendrix and Bohlen 2002; Kumar et al. 2022). The casts excreted by earthworms contain nutrients derived from organic matter decomposition, enriching the soil with organic carbon and essential nutrients. This nutrient-rich environment promotes microbial activity and enhances the degradation of petroleum contaminants, contributing to the restoration of soil health and fertility (Sakshi et al. 2020; Kumar et al. 2022).

15.3 Factors Influencing Earthworm-Mediated Remediation

Soil properties such as texture, pH, organic matter content, and nutrient levels influence earthworm activity and the degradation of petroleum contaminants. Sandy soils with low organic matter content may have lower earthworm populations and microbial activity, potentially limiting bioremediation rates. Conversely, soils rich in organic matter provide a favorable habitat for earthworms and microbial communities, enhancing biodegradation processes (Kumar et al. 2020; Geissen et al. 2008; Pan et al. 2016; Sharma et al. 2024). Soil pH also plays a crucial role, as earthworms and soil microbes have optimal pH ranges for activity and metabolism. Different species of earthworms exhibit varying tolerances to environmental conditions and preferences for soil types. Some earthworm species are better adapted to contaminated soils and may exhibit higher tolerance to petroleum hydrocarbons (Bottinelli et al. 2015). The population density of earthworms in contaminated soil can influence the rate and extent of bioremediation, with higher densities potentially accelerating degradation processes. Temperature and moisture levels in soil directly affect earthworm activity, microbial metabolism, and the rate of petroleum degradation (Edwards and Bohlen 1996). Optimal temperatures for earthworm activity and microbial growth typically range from 15 °C to 25 °C, with higher temperatures accelerating biodegradation processes. Adequate soil moisture is essential for earthworm survival and activity, as well as for microbial activity and the diffusion of contaminants (Delgado-Balbuena et al. 2016). The type and concentration of petroleum contaminants in soil influence earthworm behavior, microbial activity, and the feasibility of bioremediation (Table 15.2). Earthworms may exhibit preferences or avoidance behaviors towards certain contaminants, impacting their distribution and activity in contaminated soil. High concentrations of petroleum hydrocarbons can inhibit earthworm activity and microbial degradation processes, limiting the effectiveness of bioremediation.

Table 15.2 The contrasting characteristics of contaminants and pollutants, as well as contamination and pollution, emphasize their differences in terms of their state, nature, concentration, effects, applicability to vermiremediation

Characteristic	Contaminants	Pollutants	References
Substance's State	Variable states: solid, liquid, or gas	Tri-state potential: solid, liquid, or gas	Edwards and Arancon (2022)
Nature of Substance	Inherently/ Potentially Harmful	Potential Harm	Mishra et al. (2022)
Measure of quantity	Below Permissible Levels	Exceeding Tolerable Levels: Potentially Harmful	Meng et al. (2012)
Present Impact on Environment/Life	Non-Harmful: No Adverse Effects	Adverse Effects Causing Harm	Manahan (2017), Smith et al. (2020)
Suitability for Vermiremediation	Applicable	Not applicable	Gupta et al. (2016)
Example	Low-Level Soil Contaminants	Elevated Soil Contaminants	Mani and Kumar (2014)
State of the Environment	Non-harmful Environmental Presence	Harmful Environmental Presence	Briggs and World Health Organization (1999)
Applicability to Vermiremediation	Applicable	Not applicable	Chowdhary et al. (2018)

15.4 Challenges and Limitations

Challenges and limitations associated with earthworm-mediated remediation of petroleum-contaminated soil encompass various factors that may impact the effectiveness, sustainability, and acceptance of this approach.

15.4.1 Soil Conditions

Earthworm activity and effectiveness in bioremediation can be influenced by soil properties such as texture, pH, moisture content, and organic matter content. Soils with unfavorable conditions, such as high clay content or low organic matter (Keilty 1987), may limit earthworm populations and microbial activity, thereby reducing remediation rates.

15.4.2 Contaminant Characteristics

The type, concentration, and persistence of petroleum contaminants in soil can affect earthworm performance. High concentrations of toxic or recalcitrant hydrocarbons may inhibit earthworm activity and microbial degradation processes, limiting the effectiveness of bioremediation.

15.4.3 Earthworm Species

Different earthworm species exhibit varying tolerances to environmental conditions and preferences for soil types. The selection of appropriate earthworm species for specific remediation applications is crucial to ensure optimal bioremediation outcomes.

15.4.4 Soil Recovery

While earthworm-mediated bioremediation can effectively reduce petroleum contamination in soil, the long-term sustainability of soil recovery and ecosystem resilience requires continuous monitoring and management (Vignesh and Manikandan 2023). Monitoring soil properties, microbial communities, and vegetation dynamics over extended periods is essential to assess the stability and health of remediated soil ecosystems.

15.4.5 Residual Contaminants

Despite remediation efforts, residual contaminants may persist in soil at concentrations above regulatory thresholds. Long-term monitoring is necessary to track changes in contaminant levels and ensure compliance with remediation goals and environmental regulations.

15.4.6 Ecosystem Effects

Earthworm-mediated remediation may alter soil food webs, nutrient cycling processes, and plant communities, potentially leading to unintended ecological consequences. Monitoring ecosystem responses to remediation activities is crucial for detecting and mitigating any adverse impacts on soil biodiversity and ecosystem functioning.

15.4.7 Regulatory Compliance

Earthworm-mediated bioremediation may face regulatory challenges related to the approval, monitoring, and reporting of remediation activities. Regulatory agencies may require extensive data on remediation efficacy, environmental impacts, and risk assessment to approve earthworm-based remediation projects.

15.4.8 Public Perception

Public perception of earthworm-mediated bioremediation may vary depending on factors such as awareness; trust in technology, and perceived risks and benefits. Stakeholder engagement and communication efforts are essential for building public trust, addressing concerns, and garnering support for remediation projects (Bhandari et al. 1994).

15.4.9 Liability and Insurance

Liability issues related to the effectiveness and long-term outcomes of earthworm-mediated remediation may pose financial risks for project stakeholders. Insurance coverage for potential environmental liabilities and unforeseen impacts is critical for ensuring project feasibility and sustainability.

15.5 Sustainable Frameworks of Earthworm-Based Bioremediation

Sustainable remediation frameworks aim to address contamination challenges while minimizing environmental impacts and promoting ecosystem resilience. Integrating earthworm-based bioremediation into these frameworks requires considering the ecological, social, and economic dimensions of remediation projects. Earthworm-based remediation should be integrated into adaptive management frameworks that allow for flexibility and iterative decision-making based on monitoring data and stakeholder feedback. Adaptive management enables adjustments to remediation strategies in response to changing site conditions (Zeb et al. 2020; Kumar et al. 2022), emerging scientific knowledge, and stakeholder preferences. Successful integration of earthworm-based approaches relies on effective stakeholder engagement and collaboration. Engaging with local communities, regulatory agencies, industry stakeholders, and scientific experts fosters transparency, trust, and support for remediation projects. Earthworm-based bioremediation can be integrated with other bioremediation techniques, such as phytoremediation, microbial bioremediation, and bioaugmentation, to enhance remediation outcomes. Combining multiple remediation strategies can synergistically accelerate contaminant degradation, improve soil quality, and reduce remediation timeframes. Plant-microbe interactions in phytoremediation can enhance soil microbial activity and nutrient availability, supporting earthworm populations and biodegradation processes. Earthworms, in turn, improve soil aeration and porosity, facilitating plant growth and root penetration in phytoremediation systems (Glick 2003). Supplementing earthworm-mediated bioremediation with specialized microbial consortia or enzymes can enhance the degradation of specific petroleum contaminants. Earthworms create microenvironments within their guts that promote microbial growth and activity, facilitating the bioaugmentation of hydrocarbon-degrading microbes. Earthworm-based bioremediation offers potential cost savings compared to traditional remediation methods, particularly for large-scale projects. Earthworms are inexpensive to procure, require minimal maintenance, and can be applied in situ without the need for costly infrastructure or equipment (Asubiaro et al. 2018). Earthworm-based remediation techniques are scalable and adaptable to diverse soil types, contamination levels, and site conditions. Whether applied at small-scale community gardens or large industrial sites, earthworm-based bioremediation can be tailored to meet specific remediation goals and project requirements. Evaluating the environmental and economic impacts of earthworm-based remediation using life cycle assessment (LCA) can provide insights into the sustainability and cost-effectiveness of these approaches. LCA considers the entire lifecycle of remediation activities, from earthworm production and application to soil recovery and ecosystem restoration.

15.6 The Effect of Earthworm Activity on Petroleum-Contaminated Soil Properties

Earthworm activity plays a pivotal role in altering soil structure and enhancing porosity in petroleum-contaminated soil. Through burrowing and feeding activities, earthworms create a network of channels and tunnels, effectively aerating the soil and improving its hydraulic conductivity. This increased soil porosity facilitates better water infiltration, drainage, and gas exchange, creating favorable conditions for microbial activity and hydrocarbon degradation (Francis and Fraser 1998). Earthworms are known to ingest soil and organic matter, including petroleum-contaminated soil, as they feed and burrow through the soil profile. In the digestive tract of earthworms, organic matter undergoes fragmentation and microbial processing, resulting in the release of nutrients such as nitrogen, phosphorus, and potassium back into the soil. Additionally, earthworm casts, rich in organic carbon and microbial biomass, contribute to the decomposition of organic contaminants and the enrichment of soil with essential nutrients. This stimulation of nutrient cycling enhances microbial activity and promotes the degradation of petroleum hydrocarbons (Nobili et al. 2022). Earthworms influence soil microbial communities and activity through their burrowing, feeding, and casting activities. The gut microbiota of earthworms harbors diverse microbial species capable of metabolizing petroleum hydrocarbons through various metabolic pathways. Moreover, earthworm activities promote the dispersion and mixing of microorganisms throughout the soil profile, increasing microbial diversity and activity. This enhanced microbial activity accelerates the degradation of petroleum contaminants and contributes to soil remediation (Lavelle et al. 2006; Kumar et al. 2022). Earthworms can also influence soil chemistry in petroleum-contaminated soil environments. By altering soil pH through their secretions and burrowing activities, earthworms may create conditions conducive to microbial activity and contaminant degradation. Furthermore, earthworms can affect soil redox potential, shifting conditions towards more favorable aerobic environments for microbial metabolism. These changes in soil chemistry play a crucial role in modulating microbial activity and hydrocarbon degradation processes (Edwards and Bohlen 1996; Yadav et al. 2022a; Yadav et al. 2023; Kumar et al. 2024).

15.7 Conclusion

Earthworms play a crucial role in enhancing soil structure, promoting nutrient cycling, stimulating microbial activity, and modulating soil chemistry, thereby facilitating the degradation of petroleum hydrocarbons. Additionally, earthworm-based bioremediation offers a sustainable and cost-effective approach to soil remediation, with potential applications across various contaminated environments. The implications of earthworm-mediated remediation for sustainable soil management and environmental conservation are significant. By harnessing the natural bioremediation capabilities of earthworms, remediation efforts can minimize environmental impacts, promote ecosystem resilience, and restore soil quality in contaminated sites. Furthermore, integrating earthworm-based approaches into remediation frameworks offers opportunities for synergistic interactions with other bioremediation techniques; enhancing overall remediation efficacy and sustainability. Further research is needed to deepen our understanding of the mechanisms underlying earthworm-mediated bioremediation and its effectiveness across different soil types, contaminant profiles, and environmental conditions. Long-term monitoring studies can provide valuable insights into the sustainability and long-term outcomes of earthworm-based remediation strategies. Efforts should focus on optimizing earthworm-based remediation techniques, including earthworm species selection, application methods, and environmental management practices. By refining these techniques, remediation practitioners can maximize remediation efficiency and minimize potential adverse effects on soil ecosystems. Collaboration and engagement with stakeholders, including local communities, regulatory agencies, industry partners, and scientific experts, are essential for the successful implementation of earthworm-mediated remediation projects. Transparent communication, education, and outreach efforts can foster public acceptance and support for remediation initiatives. Earthworm-mediated soil remediation offers promising opportunities for addressing petroleum contamination and advancing sustainable soil management and environmental conservation efforts. By harnessing the power of earthworms, we can create healthier, more resilient soil ecosystems and contribute to a cleaner and more sustainable environment for future generations.

15.8 Suggestions

The following suggestions are made to assist integration of earthworms into remediation programmes:

I. Detailed understanding of the intricate mechanisms through which earthworms facilitate the breakdown of petroleum hydrocarbons in contaminated soils. Discuss how earthworm activities, such as burrowing and feeding, enhance soil structure, nutrient cycling, and microbial activity, ultimately promoting the degradation of petroleum contaminants. Highlight the importance of understanding these mechanisms for optimizing earthworm-based remediation strategies and improving overall remediation efficacy in petroleum-contaminated environments.

II. Explore strategies for selecting appropriate earthworm species, optimizing application methods, and managing environmental conditions to enhance remediation outcomes. Discuss the potential synergies between earthworm-mediated bioremediation and other remediation approaches, such as microbial degradation and phytoremediation, to achieve comprehensive and sustainable soil remediation solutions.

III. There is a need for further research to deepen our understanding of earthworm-mediated bioremediation and its applicability across diverse soil types, contaminant profiles, and environmental conditions. Advocate for long-term monitoring studies to assess the sustainability and long-term outcomes of earthworm-based remediation strategies. Emphasize the importance of ongoing efforts to optimize earthworm-based remediation techniques through collaboration with stakeholders and engagement with local communities, regulatory agencies, and scientific experts. Suggest avenues for transparent communication, education, and outreach to foster public acceptance and support for earthworm-mediated soil remediation initiatives.

References

Al-Maliki SALWAN, Al-Taey DK, Al-Mammori HZ (2020) Soil microbes, organic carbon protection, and plant production in consideration with earthworms: a review. Plant Cell Biotechnol Mol Biol 21:99–125

Asubiaro DO, Njoku LK, Amechina NS, Musa SI (2018) Enumeration and identification of microbial load in oil-spilled soil remediated using Eudriluseugeniae. J Nat Prod Resour 4:171

Bhadauria T, Saxena KG (2010) Role of earthworms in soil fertility maintenance through the production of biogenic structures. Appl Environ Soil Sci 2010:7

Bhandari A, Dove DC, Novak JT (1994) Soil washing and biotreatment of petroleum-contaminated soils. J Environ Eng 120(5):1151–1169

Bottinelli N, Jouquet P, Capowiez Y, Podwojewski P, Grimaldi M, Peng X (2015) Why is the influence of soil macrofauna on soil structure only considered by soil ecologists? Soil Tillage Res 146:118–124

Briggs DJ, World Health Organization (1999) Environmental health indicators: framework and methodologies (No. WHO/SDE/OEH/99.10). World Health Organization
Chachina SB, Voronkova NA, Baklanova ON (2016) Biological remediation of the petroleum and diesel contaminated soil with earthworms Eisenia fetida. Proc Eng 152:122–133
Chowdhary P, Raj A, Bharagava RN (2018) Environmental pollution and health hazards from distillery wastewater and treatment approaches to combat the environmental threats: a review. Chemosphere 194:229–246
Curry JP, Schmidt O (2007) The feeding ecology of earthworms–a review. Pedobiologia 50(6):463–477
Dada EO, Akinola MO, Owa SO, Dedeke GA, Aladesida AA, Owagboriaye FO, Oludipe EO (2021) Efficacy of vermiremediation to remove contaminants from soil. J Health Pollut 11(29):210302
Datta S, Singh S, Ramamurthy PC, Kapoor D, Dhaka V, Bhatia D et al (2023) Vermiremediation of agrochemicals, PAHs, and crude oil polluted land. In: Bio-inspired land remediation. Springer International Publishing, Cham, pp 287–315
Delgado-Balbuena L, Bello-López JM, Navarro-Noya YE, Rodríguez-Valentín A, Luna-Guido ML, Dendooven L (2016) Changes in the bacterial community structure of remediated anthracene-contaminated soils. PLoS One 11(10):e0160991
Edwards CA, Arancon NQ (2022) The use of earthworms in organic waste management and vermiculture. In: Biology and ecology of earthworms. Springer US, New York, pp 467–527
Edwards CA, Bohlen PJ (1996) Biology and ecology of earthworms, vol 3. Chapman & Hall, London, UK
Ekperusi OA, Aigbodion FI (2015) Bioremediation of petroleum hydrocarbons from crude oil-contaminated soil with the earthworm: Hyperiodrilus africanus. 3 Biotech 5:957–965
Evans WC, Fernley HN, Griffiths E (1965) Oxidative metabolism of Phenanthrene and Anthracene by soil pseudomonads. The ring-fission mechanism. Biochem J 95(3):819–831. https://doi.org/10.1042/Bj0950819
Francis GS, Fraser PM (1998) The effects of three earthworm species on soil macroporosity and hydraulic conductivity. Appl Soil Ecol 10(1–2):11–19
Geissen V, Gomez-Rivera P, Lwanga EH, Mendoza RB, Narcías AT, Marcías EB (2008) Using earthworms to test the efficiency of remediation of oil-polluted soil in tropical Mexico. Ecotoxicol Environ Saf 71(3):638–642
Glick BR (2003) Phytoremediation: synergistic use of plants and bacteria to clean up the environment. Biotechnol Adv 21(5):383–393
Grosser RJ, Warshawsky D, Vestal JR (1991) Indigenous and enhanced mineralization of pyrene, benzo [a] pyrene, and carbazole in soils. Appl Environ Microbiol 57(12):3462–3469
Gupta G, Kumar V, Pal AK (2016) Biodegradation of polycyclic aromatic hydrocarbons by microbial consortium: a distinctive approach for decontamination of soil. Soil Sediment Contam Int J 25(6):597–623
Hendrix PF, Bohlen PJ (2002) Exotic earthworm invasions in North America: ecological and policy implications: expanding global commerce may be increasing the likelihood of exotic earthworm invasions, which could have negative implications for soil processes, other animal and plant species, and importation of certain pathogens. Bioscience 52(9):801–811
Hickman ZA, Reid BJ (2008) Earthworm assisted bioremediation of organic contaminants. Environ Int 34(7):1072–1081
Javed F, Hashmi I (2021) Vermiremediation–remediation of soil contaminated with oil using earthworm (Eisenia fetida). Soil Sediment Contam Int J 30(6):639–662
Keilty TJ (1987) Lethal and sublethal responses to a sediment bound toxicant by two oligochaetes from Lake Michigan: Stylodrilusheringianus and Limnodrilushoffmeisteri. University of Michigan, United States
Kumar R, Singh D (2022) Insecticide-tolerant bacterial population in Eisenia Fetida's gut and Vermicast exposed to Chlorantraniliprole and Fipronil. Appl Biol Res 24(3):273–279

Kumar R, Sharma P, Gupta RK, Kumar S, Sharma MMM, Singh S, Pradhan G (2020) Earthworms for eco-friendly resource efficient agriculture. In: Resources use efficiency in agriculture. Springer Singapore, pp 47–84

Kumar R, Yadav R, Kumar Gupta R, Pooja (2022) A sustainable way for fish health management by replacement of chemical and drugs by earthworm. In: Environmental degradation in Asia: land degradation, environmental contamination, and human activities. Springer International Publishing, Cham, pp 329–352

Kumar R, Yadav R, Gupta RK, Yodha K, Kataria SK, Kadyan P, Sharma P, Kaur S (2023) The earthworms: Charles Darwin's ecosystem engineer. In: Organic fertilizers-new advances and applications. Intech Open, London, UK, pp 1–17

Kumar R, Yadav R, Gupta RK, Pal A, Yodha K, Kumar A (2024) Antioxidant enzyme activities and markers of oxidative stress in the life cycle of different Earthworm species. Indian J Exp Biol 62(July):578–583

Lavelle P, Spain A (2002) Soil ecology. Springer dordrecht, Netherlands

Lavelle P, Decaëns T, Aubert M, Barot S, Blouin M, Bureau F et al (2006) Soil invertebrates and ecosystem services. Eur J Soil Biol 42:S3–S15

Manahan SE (2017) Industrial ecology: environmental chemistry and hazardous waste. Routledger, Taylor & francis group, UK

Mani D, Kumar C (2014) Biotechnological advances in bioremediation of heavy metals contaminated ecosystems: an overview with special reference to phytoremediation. Int J Environ Sci Technol 11:843–872

Meng X, Zhang Y, Zhao Y, Lou IC, Gao J (2012) Review of chinese environmental risk assessment regulations and case studies. Dose-Response 10(2):dose-response

Mishra CSK, Samal S, Samal RR (2022) Evaluating earthworms as candidates for remediating pesticide contaminated agricultural soil: a review. Front Environ Sci 10:924480

Nobili S, Masin CE, Zalazar CS, Lescano MR (2022) Bioremediation of hydrocarbon contaminated soil using local organic materials and earthworms. Environ Pollut 314:120169

Pan LB, Ma J, Wang XL, Hou H (2016) Heavy metals in soils from a typical county in Shanxi Province, China: levels, sources and spatial distribution. Chemosphere 148:248–254

Renu R, Gupta RK, Kumar R, Kaur T (2020) Eco-toxic effects of lead and nickel on survivability, reproduction and growth of earthworm (Eudrilus eugeniae). J Entomol Zool Stud 8(5):1351–1355

Rodriguez-Campos J, Dendooven L, Alvarez-Bernal D, Contreras-Ramos SM (2014) Potential of earthworms to accelerate removal of organic contaminants from soil: a review. Appl Soil Ecol 79:10–25

Sakshi SD, Kumar R, Kumar A (2020) Comparative assessment of bacterial population of gut and cast of *Pheretima posthuma* exposed to Imidacloprid and Pendimethalin. Int J Curr Microbiol App Sci 10:1218–1223. https://doi.org/10.20546/ijcmas.2020.910.146

Sharma S, Gupta R, Singh D (2020) Super oxide dismutase, catalase and glutathione peroxidase activity in heavy metals contaminated earthworm, Eisenia fetida. J Entomol Zool Stud 8:1909–1913

Sharma MMM, Kapoor D, Rohilla R, Loyal A, Sharma P, Husen A (2024) Role of essential minerals and rhizosphere in plant under changing environmental conditions. In: Essential Minerals in Plant-Soil Systems. Elsevier, Netherlands, pp 235–252

Singh D, Sharma S, Kumar R, Dahiya T, Gupta R (2022) Vermitechnology: a review. Pharma Innov 11:142–145

Smith RB, Beevers SD, Gulliver J, Dajnak D, Fecht D, Blangiardo M et al (2020) Impacts of air pollution and noise on risk of preterm birth and stillbirth in London. Environ Int 134:105290

Tagliabue F, Marini E, De Bernardi A, Vischetti C, Casucci C (2023) A systematic review on earthworms in soil bioremediation. Appl Sci 13(18):10239

Tang J, Wang M, Wang F, Sun Q, Zhou Q (2011) Eco-toxicity of petroleum hydrocarbon contaminated soil. J Environ Sci 23(5):845–851

Vignesh K, Manikandan K (2023) Emerging trends in crop protection technology

Yadav R, Gupta RK, Kumar R, Kaur T, Moun R, (2022a) "Biochemical response of earthworm, Eisenia fetida to heavy metals toxicity. J of Applied and Natural Sci 14(3):990–998

Yadav R, Gupta RK, Kumar R, Kaur T (2022b) Assessment of toxicity of lead and nickel on the biochemical and immunological parameters of earthworm, Eudrilus eugeniae.D-5582, 1–8

Yadav R, Kumar R, Gupta RK, Kaur T, Kour A, Kaur S, Rajput A (2023) Heavy metal toxicity in earthworms and its environmental implications: a review. Environ Adv 12:100374

Zeb A, Li S, Wu J, Lian J, Liu W, Sun Y (2020) Insights into the mechanisms underlying the remediation potential of earthworms in contaminated soil: a critical review of research progress and prospects. Sci Total Environ 740:140145

Zechmeister-Boltenstern S, Keiblinger KM, Mooshammer M, Peñuelas J, Richter A, Sardans J, Wanek W (2015) The application of ecological stoichiometry to plant–microbial–soil organic matter transformations. Ecol Monogr 85(2):133–155

Chapter 16
Soil Pollution Risks of Onshore Petroleum Exploration Drilling and Possible Bioremediation Methods

Amir Parnian, Farhad Keykhosravi, James N. Furze, Majeti Narasimha Vara Prasad, Kyanoush Behrahi, and Amin Parnian

Abstract Exploration drilling is the process of drilling wells to explore the presence and extent of oil and gas resources in the subsurface. The process poses risks for soil pollution and environmental degradation. Potential sources of soil contamination from exploration drilling are (1) spills and leaks of drilling fluids, lubricants, fuels, and chemicals used in the drilling operations; (2) disposal of drill cuttings and waste, which may contain traces of hydrocarbons, metals, salts, and other pollutants that can alter the soil properties and pose health risks to humans and animals; (3) flaring and venting of associated gas, burning or releasing natural gas that is produced along with oil affecting the soil microclimate and vegetation. This chapter

A. Parnian
National Salinity Research Center (NSRC), Agricultural Research, Education and Extension Organization (AREEO), Yazd, Iran

F. Keykhosravi (✉)
National Iranian Oil Company (NIOC), Tehran, Iran

J. N. Furze
Royal Geographical Society (with the Institute of British Geographers), London, UK

Laboratory of Biotechnology and Valorization of Natural Resources, Faculty of Sciences-Agadir, Department of Biology, Ibn Zohr University, Agadir, Morocco

Control and Systems Engineering Department, University of Technology-Iraq, Baghdad, Iraq

Earth Systems Protection Foundation, Somerset, UK

M. N. V. Prasad
Department of Plant Sciences, School of Life Sciences, University of Hyderabad (An Institution of Eminence), Hyderabad, Telangana, India

K. Behrahi
University of Applied Science and Technology, Tehran, Iran

A. Parnian
Young Researchers Club, Masjed Soleyman Branch, Islamic Azad University, Masjed Soleyman, Iran

Parnian Sazeh Chehelsotoun (PSC) Company, Isfahan, Iran

A. Parnian et al. (eds.), *Bioremediation of Petroleum Contaminated Soil and Sediment*, Environmental Contamination Remediation and Management,
https://doi.org/10.1007/978-3-031-95524-2_16

explores potential sources of soil pollution and proposes bioremediation, soil washing, and thermal methods as potential solutions for remediation caused by exploration drilling.

Keywords Oil spills · Remediation · Environmental contamination

16.1 Introduction

Petroleum hydrocarbon contamination has significant impacts on ecosystem health and productivity (Sarma and Prasad 2024). A sustainable ecosystem is crucial for human well-being and the economy. The petroleum industry, which involves exploration, drilling, production, refinery, storage, and transportation of oil, gas, and petroleum products, is a major source of environmental pollution. Addressing petroleum hydrocarbon contamination is essential for a sustainable ecosystem (Parnian et al. 2022a). Petroleum contamination originates from oil and gas exploration activities as well as oil spills, which lead to environmental pollution.

Each stage of oil exploration has a potential impact on the environment. The first stage involves reviewing geological maps to identify major sedimentary basins and using aerial photography to identify potential hydrocarbon reserves. More detailed information is collected using magnetic, gravimetric, and seismic survey methods. Geophysical surveying is the most common and is often the first field activity undertaken by oil companies. The information gathered can be used to map the underlying strata and calculate the size and shape of structures. Once a promising geological structure is identified, drilling an exploratory borehole is the only way to confirm the presence of hydrocarbons and the thickness and internal pressure of the reservoir (Services 1991). Exploration drilling is the process of drilling wells to explore the presence and extent of oil and gas resources in the subsurface. Exploration wells, also known as 'wildcats', are drilled to confirm hydrocarbon presence and reservoir thickness. The location depends on geological formations, and a pad is constructed for land-based operations to accommodate drilling equipment and support services (Forum 1997). Exploration drilling plays a crucial role in the oil and gas industry, but it can also lead to environmental pollution in the form of oil spills. These spills can contaminate the soil and pose a significant threat to the ecosystem (Zhichkina et al. 2020).

Poor environmental management practices and unsustainable petroleum exploration and production have negatively impacted the atmosphere, controlled waters, soils, sediments, biological diversity, and sustainability of natural ecosystems (Ite et al. 2013). It is therefore important for companies to prioritize environmental sustainability and implement proper waste management systems to prevent and mitigate the effects of petroleum contamination (Prasad and Katiyar 2010). On top of that, the petroleum industry should invest in research and development of cleaner and more sustainable technologies to reduce its disruptive impact on the Earth's environment. By adopting responsible practices and embracing alternative energy sources, the oil and gas industry can work towards a more eco-friendly future.

Soil remediation techniques, such as bioremediation, soil washing, and thermal techniques, can be employed to effectively remove pollutants to restore, recover, and rehabilitate contaminated soil and sediments. However, it is important to consider the energy-intensive and expensive methods when choosing an appropriate remediation strategy (Ossai et al. 2020).

16.2 Potential Sources of Soil Contamination from Exploration Drilling

Activities including oil and petroleum product spills, leaks of drilling fluids and chemicals, disposal methods of drill cuttings and wasted muds, and drilling-related consequences like flaring and venting of associated gases are the main sources of contaminants related to oil and gas exploration drilling (Xue et al. 2024). Contaminants produced by these processes pose risks to both human health and the environment (Njuguna et al. 2022). Consequently, the oil and gas industry must prioritize responsible exploration drilling procedures and invest in technologies that minimize the risks of soil contamination. Three main sources of soil contamination during exploration drilling are (Fig. 16.1) (1) spills and leaks; (2) disposal of drill waste and cuttings; and (3) flaring and venting of associated gases (Cherepovitsyn and Lebedev 2023).

Spills and leaks can occur due to the release of petroleum or petroleum-contaminated materials into the soil, sediments, and water, which leads to serious environmental pollution. Incidents can occur during various stages of exploring drilling, from the beginning of the drilling process to the end, or related activities during drilling operations (Rana 2008).

The improper disposal of drill cuttings and wasted mud also contribute to soil contamination, as these materials often contain harmful petroleum hydrocarbons,

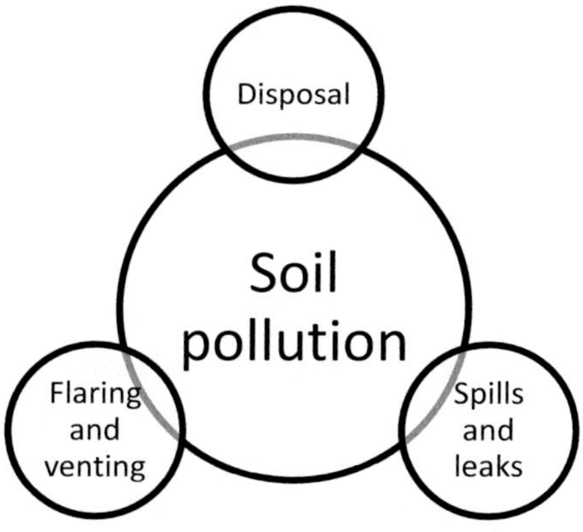

Fig. 16.1 Three main sources of soil contamination during exploration drilling

chemicals, and heavy metals. Additionally, flaring and venting of associated gases can release pollutants such as methane, a potent greenhouse gas, further contributing to climate change. It is crucial for the oil and gas industry to adopt effective measures and technologies to prevent and mitigate soil contamination caused by drilling activities (Cherepovitsyn and Lebedev 2023). Disposal of drill cuttings and wasted mud should be done in a responsible and regulated manner, such as through proper containment and treatment facilities. Implementing advanced drilling techniques, such as closed-loop systems and mud recycling, also helps to reduce the amount of waste generated during drilling operations. Furthermore, implementing regular monitoring and testing of soil and groundwater near drilling sites helps to identify and address any potential contamination issues promptly. By taking these proactive measures, the oil and gas industry can minimize the environmental impact of drilling activities and ensure the long-term sustainability of the land (Paulsen et al. 2003). Drilling mud and cuttings disposal has serious environmental risks and consequences if not properly managed. Improper disposal of waste contaminates nearby water sources, leading to detrimental effects on aquatic ecosystems and local communities that rely on those water sources for their daily needs. Additionally, chemicals present in drilling mud also leach into the soil, damaging the fertility and health of the land. It is crucial for the oil and gas industry to adopt responsible and sustainable practices for drilling mud disposal to mitigate environmental risks and protect surrounding environments. By prioritizing responsible disposal methods, the oil and gas industry may lower its ecological footprint and contribute to the preservation of natural resources for future generations (Alba et al. 2007).

Sometimes it is operationally necessary to incorporate flaring and venting of associated gases during exploration drilling of a well. Lack of oil and gas field special conditions, proper instruments and special devices, disasters and incidents, and also engineering and executive mistakes lead to the demand for running the flaring and venting procedure. However, it is crucial for the industry to ensure that this procedure is only used when absolutely necessary, as flaring and venting contribute to greenhouse gas emissions and air pollution (Asad et al. 2019). To mitigate negative impacts, the industry should invest in advanced technologies and equipment that can capture and utilize these gases instead. Additionally, implementing strict regulations and monitoring systems can help prevent any accidents or mistakes that require the use of flaring and venting procedures. Ultimately, by continuously striving for improved operational practices and investing in sustainable solutions, the oil and gas industry can minimize its environmental impact and work towards a more sustainable future.

16.3 Remediation of Soil Pollution Caused by Exploration Drilling

The oil and gas industry has been criticized for releasing drilling waste into the environment without proper treatment, causing ecological and environmental damage that consequently has negative effects on the earth's ecosystems, animal habitats, and humans in all aspects. Waste disposal is important for both exploration and production drilling operations of oil and gas fields, considering that one well drill produces about 1000–5000 m^3 of waste. Waste management methods can be biological or non-biological, such as landfill disposal, cutting reinjection, stabilization and encapsulation, bioremediation, soil washing, and thermal treatment (Cherepovitsyn and Lebedev 2023). However, effective waste management systems require more new approaches to achieve lower environmental impacts, such as soil pollution. Soil remediation includes ex-situ, on-situ, and in-situ methods. Bioremediation, soil washing, and thermal treatment are common methods for soil remediation resulting from exploration drilling (Ala and Sorkhabi 2024).

16.3.1 Bioremediation

The Earth's ecosystem is in danger from environmental pollution, which is mostly anthropogenic. Environmental remediation is a promising answer to the growing demands of society for the control and management of issues. The approach involves utilizing various technologies like chemical, physical, bio, and thermal bases treatments to combat pollutants from the air, aquatic mediums, and porous environments (Parnian et al. 2023). Bioremediation is a biotechnological method that uses organisms (plants, bacteria, and fungi) to stabilize, control, remediate, and restore the environment. It can be applied to soil and sediments that are subjected to petroleum contamination. Bioremediation is an eco-friendly, advanced environmental management method that can help solve pollution problems at a low cost, but it requires time and faces challenges (Parnian et al. b). Only when the applied organism is able to survive and grow in a contaminated environment can bioremediation be effective. Climatic and elements of the surrounding environment are important for a successful bioremediation operation. Bioremediation's effectiveness relies on the availability of essential nutrients for organisms like algae and fungi to survive and degrade pollutants in sediment, soil, water, and air. Without these nutrients, bioremediation will not be effective. Bioremediation is a process that involves the use of specific techniques and approaches to clean and disinfect various materials, including:

1. Bioventing, an in situ bioremediation method that uses bacteria and archaea to degrade organic contaminants. It intensifies the activities of bacteria and/or archaea, promoting hydrocarbon molecules in situ biodegradation with adequate air, oxygen flow, and nutrient addition. This method helps degrade adsorbed fuels and volatile organic compounds residuals, making pollutants harmless (Van Eyk 2022).
2. Environmental bioattenuation, involving natural pollutant biodegradation through the addition of nutrients, microbes, or other organisms, promoting ecosystem self-remediation. Indigenous microbes determine metabolic activity and act as natural attenuation (Anekwe and Isa 2022).

3. Biosparging, a remediation technique that uses indigenous microorganisms to biodegrade organic constituents in saturated zones. It involves injecting air/oxygen and nutrients to increase their biological activity, reducing petroleum hydrocarbon concentrations in groundwater, soil below the water table, and within the capillary fringe (Kao et al. 2008).
4. Bioaugmentation, the addition of cultured microorganisms to degrade contaminants in solid or aquatic environments. It involves studying native microorganisms to determine if biostimulation is possible. If native microorganisms affect contaminants, more are added to boost degradation, while if not, exogenous microorganisms are introduced. Bioaugmentation can improve efficiency and reduce site mitigation costs and time, ultimately reducing the time and cost of remediation (Poi et al. 2017).
5. Biostimulation, a method that enhances the bioremediation of contaminants by modifying the environment to boost the activity of existing bacteria. It involves adding substances to enrich bioavailable elements like phosphorus, nitrogen, oxygen, or carbon, which act as nutrients or electron acceptors, allowing bacteria to degrade contaminants more efficiently (Qin et al. 2013).
6. Co-composting (composting), a process where the contaminated medium is mixed with bulking agents like wood chips, straw, rice hulls, or husks, increases porosity and aeration potential for the biological degradation of contaminants like petroleum hydrocarbons and organic pollutants (Wang et al. 2011).
7. Landfarming, a bioremediation technology that uses liners and other methods to control the leaching of contaminants, requires excavation and placement of contaminated soils, sediments, and sludges in lined beds and periodic tilling to aerate the waste (Effendi et al. 2022).

As well as microbial remediation (use of microorganisms to remediate pollutants), there are other branch methods of bioremediation, phytoremediation (use of plants and algae for environmental remediation), and mycoremediation (use of fungi and algae for environmental remediation). All branches are somehow incorporated into or subjected to the above-mentioned techniques and approaches. Branches of bioremediation offer different advantages and limitations, depending on the specific local conditions, environmental contamination, and goals of remediation. For example, phytoremediation is effective in removing organic contaminants through the uptake and degradation of pollutants by plants. On the other hand, mycoremediation has shown promising results in breaking down complex hydrocarbons and heavy metals. By incorporating diverse approaches, bioremediation allows for a more comprehensive and sustainable method of addressing different types of pollution.

In 2018, the bioremediation (composting) method was employed to reduce the environmental impact of drilling waste. The operation for solid waste management involves mixing drilling waste with bulking agents like sugarcane bagasse and half-composted agricultural waste to enhance porosity and aeration potential for biological degradation. Treatment of oil and gas well waste involves three steps: organic semi-composted material production, biological dehydration of cuttings and solids, and co-composting of solids. Organic semi-composted material is prepared from agricultural waste with the addition of microorganisms, which is similar to normal composting. Biological dehydration involves microorganisms mixing organic material with slurry or muddy solids and allowing them to dry. Co-composting of solids involves microorganisms, nutrients, and sugarcane molasses, reducing the bioavailability of trace elements and pollutants (Parnian et al. 2022a). These three steps in waste management play a crucial role in reducing the environmental impact of organic waste.

By producing organic semi-composted material, we can effectively recycle agricultural waste and minimize its contribution to landfills. The process of biological dehydration further aids in the efficient decomposition of organic material, while co-composting of solids enhances the breakdown of pollutants. Overall, the holistic approach promotes sustainability and helps create a circular economy where waste is converted into valuable resources.

16.3.2 Soil Washing

Soil washing involves the use of water/solvent and chemicals to remove contaminants from soil particles (Liu et al. 2020). Soil-washing processes are site-specific. Since soil and contaminants are rather different from site to site, soil washing performance and time consumption are related to many of the subjected soil properties: texture, permeability, conductivity, structure, mineralogy, and cation/anion exchange capacity. Clay minerals are crucial in soil washing due to their strong sorption properties, large surface areas, and high cation exchange capacities. The efficiency of removing hydrocarbons from clay-based soil, composed of clay minerals and quartz, may be determined (Li et al. 2016). There are two main systems available for soil washing: agitated soil washing and slow soil washing.

Agitated Soil Washing in agitated soil washing the contaminated soil is placed in a washing chamber, where it is agitated and mixed with water. Water, along with dissolved contaminants, is then separated from the soil particles through a series of filters and settling tanks. This process effectively removes a wide range of contaminants, including heavy metals, pesticides, and petroleum hydrocarbons, from the soil. Soil washing is particularly useful for treating soils contaminated with organic pollutants, as the chemicals used in the process can break down and remove contaminants effectively. Additionally, the treated soil can often be reused for various purposes, reducing the need for landfill disposal and minimizing environmental impact.

Agitated soil washing employs special devices to dewater the slurry of soil and water: filter presses, centrifuges, and screw presses. These devices are essential in separating the solid soil particles from the water, allowing for the removal of contaminants. Filter presses use a series of plates with filter cloths to trap solids and expel water, while centrifuges use centrifugal force to separate the solid and liquid phases. Screw presses, on the other hand, use a rotating screw to squeeze out the water from the slurry. All of these devices play a crucial role in effectively dewatering the soil and ensuring the successful removal of contaminants in the soil-washing process (Son et al. 2011).

To agitate the system, sometimes the use of coagulants and flocculants is inevitable. These chemicals help in the separation of solid particles from the water, making it easier to remove and dispose of the contaminants. Coagulants help in clumping particles together, while flocculants aid in the formation of larger, easily removable flocs. By employing these techniques, soil washing can efficiently remove organic pollutants and ensure the soil is safe for reuse or further treatment. However, it is crucial to carefully manage and dispose of the waste generated during the soil-washing process to prevent any potential harm to the environment (Zoghi and Mafigholami 2023).

Slow Soil Washing sometimes soil washing is incorporated with the drainage system, mostly in-situ or on-site. In this method, a solvent or water is used to wash the contaminants. This method of soil washing demands more area and time; on the other hand, it is more energy efficient and requires less money to operate. Overall, the incorporation of soil washing in the drainage system proves to be a cost-effective and environmentally friendly solution for remediating contaminated soils. The use of solvents or water in the washing process efficiently removes pollutants, allowing the treated soil to be reused rather than discarded in landfills. Despite the need for more space and time, this method offers long-term benefits by reducing energy consumption and operating costs (dos Santos et al. 2017). These types of soil-washing systems, depending on the environment and demands of the operators, use various kinds of drainage systems, open channels, ditches, or underground pipes. Open channels are often used in areas with high rainfall as they provide an efficient way to transport the washed soil to a designated area. Ditches are suitable for use in areas where space is limited, as they can be easily integrated into existing infrastructure. Underground pipes are preferred in urban areas as they minimize visual impact and prevent potential contamination of groundwater sources. Each drainage system has its advantages and disadvantages; the choice depends on site-specific conditions and the goals of the remediation project.

There are many methods to improve soil washing such as the use of surfactants, adjusting water temperature, adjusting chemicals, vibrators, and ultrasonic sources. Incorporating the use of hot water, heating the soil or sediment bed, solvents, and surfactants helps to improve the efficiency and time performance of the soil washing system. These methods aid in breaking down contaminants and/or facilitate their removal from the soil. Hot water can effectively dissolve and remove organic compounds while heating the soil or sediment bed increases the solubility of contaminants and enhances their extraction. Solvents and surfactants, on the other hand, help to separate contaminants from the soil particles, allowing for their removal through the washing process. By incorporating these techniques, the soil washing system can achieve higher levels of remediation and reduce the overall time required for the cleanup process (He et al. 2022; Liu et al. 2021; Heo and Lee 2015; Madadian et al. 2013).

Generally, during soil washing operations, huge amounts of water are utilized; at the end of the operation, polluted effluents and rejected effluents remain, which should be managed and disposed of safely in the environment. To address this issue, proper management and disposal of polluted effluents and rejected effluents are crucial. Effluents often contain high concentrations of contaminants and can pose a risk to the environment if not handled properly. Treatment techniques such as filtration, sedimentation, and chemical precipitation can be employed to remove or neutralize the contaminants in the effluents before their safe disposal. Additionally, implementing recycling and reusing strategies for the treated water can help minimize the overall water consumption and reduce the environmental impact of soil washing operations (Gusiatin et al. 2020).

16.3.3 Thermal Treatment

Thermal treatment of contamination in soil and sediments involves the use of a heat source to remove, recover, fix, encapsulate, or decompose the contaminant (Merino et al. 2003). Thermal treatment includes in situ thermal treatment, ex-situ thermal treatment, microwave heating, smoldering, and vitrification (Table 16.1) (Ding et al. 2019).

Microwave thermal remediation has been used to recover petroleum-contaminated soil (Li et al. 2009). Moreover, smoldering is employed for petroleum contamination remediation practically, as in pyrolytic remediation (Kang et al. 2020) and low-temperature thermal treatment (Alcocer and Chowdhury 1993). It seems that smoldering for soil petroleum contamination remediation is a more practical approach due to the simple and cheaper procedure. On the other hand, thermal treatments, which utilize steam or hot water, may be more convenient and proper approaches in certain situations. Thermal treatments can be easily scaled up to treat large volumes of contaminated soil, making them suitable for industrial sites with high levels of contamination. Overall, the choice of a thermal-based remediation approach depends on the specific circumstances and characteristics of the petroleum-contaminated site (Tyagi and Kaur 2024), soil texture, climate, legal and governmental authorities laws and instructions.

16.4 Conclusion

With the use of soil remediation techniques, the oil and gas industry can contribute to a sustainable future, but they must first be carefully considered for their environmental impact and cost-effectiveness. The industry should invest in advanced technologies, strict regulations, and continuous improvement to mitigate emissions and air pollution. Proper disposal and treatment of polluted and rejected effluents are crucial for protecting the environment. Effective treatment techniques and recycling

Table 16.1 Thermal treatment for soil and sediment remediaton categories (Ding et al. 2019)

In situ thermal treatment		In situ, thermal treatment uses fluid injection or heating elements to heat the soil. Most contaminants are separated from the soil and collected by the above-ground treatment units for disposal.
	Steam-enhanced extraction	Steam-enhanced extraction is a method that combines steam injection with fluid extraction, forming three distinct zones: the steam zone, the variable temperature zone, and the ambient temperature zone. The recovery mechanisms vary, with the steam zone dominated by stripping and co-boiling and the variable temperature zone dominated by viscosity reduction and hydraulic displacement.
	Electrical resistance heating	Electrical resistance heating dissipates heat through subsurface soil, converting moisture to steam, and removing contaminants. Heating enhances polishing mechanisms like hydrolysis, iron reductive dechlorination, and bioremediation.
	Thermal conduction heating	Thermal conduction heating is a method that uses heaters (operated at 400–800 °C) and extraction wells to remediate contaminated sites. It involves heating target areas through thermal conduction and convection, creating temperature gradients around each well. Thermal conduction heating can remove contaminants through steam stripping, thermal desorption, co-boiling, thermal oxidation, pyrolysis, and hydrolysis.
	Large-diameter auger	The large-diameter auger mixing with steam injection uses a Kelly bar with a bladed auger to mix soil in situ, increasing permeability. Steam or compressed hot air is injected into the subsurface, heating the soil. Large-diameter auger operates in batch mode, remediated after each block.
	Radiofrequency heating	Radiofrequency heating is a method of heating soil through dielectric heating, similar to microwave heating. It operates at frequencies between 1 and 27 MHz, reducing heating rate but increasing radiation range, enabling full-scale application.
		Ex situ thermal treatment (ESTT) excavates the contaminated soil and feeds it into a processing unit. The remediated soil can then be backfilled to the site or sent to a landfill if recalcitrant contaminants remain. ESTT can give rise to "secondary pollution" during the stockpiling and transportation of contaminants.
Microwave heating		Microwave heating is a method where materials are heated internally in a microwave field, causing electromagnetic energy to dissipate as heat due to intermolecular friction and the Maxwell-Wagner effect, unlike conventional heating.
Smoldering		Smoldering is a flameless combustion method with peak temperatures between 1500–1800 °C, while those of smoldering are 500–700 °C. It can achieve higher temperatures and heat release through extra oxygen. Smoldering mechanisms involve pyrolysis and oxidation reactions, with forward and opposed propagation modes. The pyrolysis front absorbs energy from the oxidation front, driving smoldering into virgin fuel.
Vitrification		Vitrification is a high-temperature (1000–2000 °C) process that destroys organic contaminants, encapsulates heavy metals and radioactive wastes, and forms a glass-like, chemically inert, and leaching-resistant solid, significantly reducing leaching.

strategies can minimize water consumption and soil washing operations' negative impact. Industries and communities must take responsibility for effluent management and work towards sustainable practices to ensure a cleaner, healthier environment for future generations. Thermal treatments, while more expensive, can efficiently treat large volumes of contaminated soil, though soil texture, climate, and legal requirements must be considered.

16.5 Suggestions

To help alleviate soil pollution risks from petroleum and associated contaminants the following suggestions are made:

I. Before any decision-making to site-remediation, to evaluate the environmental impact and cost-effectiveness of soil remediation techniques regarding exploration drilling risks, a life cycle assessment (LCA) approach is required, which can quantify the environmental benefits and trade-offs of different options, considering the whole process.
II. To enhance the performance and sustainability of soil remediation processes, novel materials, and methods should be considered and explored, such as the application of nanomaterials, biochar, enzymes, and electrokinetics, which can improve the removal, economical, energy efficiency, and also reduce side effects, collateral damages, and/or secondary pollution.
III. To ensure the long-term restoration and reclamation of contaminated soil, the ecological and social aspects of soil remediation should be considered. Soil quality, biodiversity, ecosystem health, human/animal health, and stakeholder participation, support the decision-making and performance of soil remediation projects.

References

Ala M, Sorkhabi R (2024) The modern oil industry in Iran: a historical perspective and review. J Pet Geol 47(1):101–119. https://doi.org/10.1111/jpg.12852

Alba A, Fragachan F, Ovalle A, Shokanov T (2007) Environmentally safe waste disposal: the integration of cuttings collection, transport, and reinjection. All Days. https://doi.org/10.2118/108912-ms

Alcocer CF, Chowdhury HR (1993) Experimental study of an environmental remediation of Gulf Coast crude-oil-contaminated soil using low-temperature thermal treatment. All Days. https://doi.org/10.2118/26107-ms

Anekwe IMS, Isa YM (2022) Bioremediation of acid mine drainage contaminated soils using bio-attenuation, wastewater and air-injection system. Biorem J 27(4):363–381. https://doi.org/10.1080/10889868.2022.2130873

Asad MM, Bin Hassan R, Sherwani F, Soomro QM, Sohu S, Lakhiar MT (2019) Oil and gas disasters and industrial hazards associated with drilling operation: an extensive literature review. In:

2019 2nd international conference on computing, mathematics and engineering technologies (ICoMET). https://doi.org/10.1109/icomet.2019.8673516
Cherepovitsyn A, Lebedev A (2023) Drill cuttings disposal efficiency in offshore oil drilling. J Mar Sci Eng 11(2):317. https://doi.org/10.3390/jmse11020317
Ding D, Song X, Wei C, LaChance J (2019) A review on the sustainability of thermal treatment for contaminated soils. Environ Pollut 253:449–463. https://doi.org/10.1016/j.envpol.2019.06.118
dos Santos EV, Sáez C, Cañizares P, da Silva DR, Martínez-Huitle CA, Rodrigo MA (2017) Treatment of ex-situ soil-washing fluids polluted with petroleum by anodic oxidation, photolysis, sonolysis and combined approaches. Chem Eng J 310:581–588. https://doi.org/10.1016/j.cej.2016.05.015
Effendi AJ, Ramadan BS, Helmy Q (2022) Enhanced remediation of hydrocarbons contaminated soil using electrokinetic soil flushing—Landfarming processes. Bioresour Technol Rep 17:100959. https://doi.org/10.1016/j.biteb.2022.100959
Forum OIIEAP (1997) Environmental management in oil and gas exploration and production. UNEP/Earthprint. http://books.google.ie/books?id=TcvCC-hta9QC&printsec=frontcover&dq=Environmental+management+in+oil+and+gas+exploration+and&hl=&cd=1&source=gbs_api
Gusiatin ZM, Kulikowska D, Klik B (2020) New-generation washing agents in remediation of metal-polluted soils and methods for washing effluent treatment: a review. Int J Environ Res Public Health 17(17):6220. https://doi.org/10.3390/ijerph17176220
He L, Zhou X, Cao J, Shen L (2022) Ultrasound-assisted soil washing for metals-contaminated soil using various washing solutions. Clean (Weinh) 50(12). https://doi.org/10.1002/clen.202100419
Heo H, Lee M (2015) Surfactant-enhanced soil washing using tween and Tergitol series surfactants for Kuwait soil heavily contaminated with crude oil. J Soil Groundwater Environ 20(5):26–33. https://doi.org/10.7857/jsge.2015.20.5.026
Ite EA, Ibok UJ, Ite MU, Petters WS (2013) Petroleum exploration and production: past and present environmental issues in the Nigeria's Niger Delta. Am J Environ Protect 1(4):78–90. https://doi.org/10.12691/env-1-4-2
Kang CU, Kim DH, Khan MA, Kumar R, Ji SE, Choi KW, Paeng KJ, Park S, Jeon BH (2020) Pyrolytic remediation of crude oil-contaminated soil. Sci Total Environ 713:136498. https://doi.org/10.1016/j.scitotenv.2020.136498
Kao C, Chen C, Chen S, Chien H, Chen Y (2008) Application of in situ biosparging to remediate a petroleum-hydrocarbon spill site: field and microbial evaluation. Chemosphere 70(8):1492–1499. https://doi.org/10.1016/j.chemosphere.2007.08.029
Li D, Zhang Y, Quan X, Zhao Y (2009) Microwave thermal remediation of crude oil contaminated soil enhanced by carbon fiber. J Environ Sci 21(9):1290–1295. https://doi.org/10.1016/s1001-0742(08)62417-1
Li G, Guo S, Hu J (2016) The influence of clay minerals and surfactants on hydrocarbon removal during the washing of petroleum-contaminated soil. Chem Eng J 286:191–197. https://doi.org/10.1016/j.cej.2015.10.006
Liu J, Xu L, Zhu F, Jia S (2020) Effects of surfactants on the remediation of petroleum contaminated soil and surface hydrophobicity of petroleum hydrocarbon degrading flora. Environ Eng Res 26(5):200384–0. https://doi.org/10.4491/eer.2020.384
Liu J, Zhao L, Liu Q, Li J, Qiao Z, Sun P, Yang Y (2021) A critical review on soil washing during soil remediation for heavy metals and organic pollutants. Int J Environ Sci Technol 19(1):601–624. https://doi.org/10.1007/s13762-021-03144-1
Madadian E, Gitipour S, Amiri L, Alimohammadi M, Saatloo J (2013) The application of soil washing for treatment of polycyclic aromatic hydrocarbons contaminated soil: a case study in a petrochemical complex. Environ Prog Sustain Energy 33(1):107–113. https://doi.org/10.1002/ep.11759
Merino J, Piña J, Errazu AF, Bucalá V (2003) Fundamental study of thermal treatment of soil. Soil Sediment Contam Int J 12(3):417–441. https://doi.org/10.1080/713610981

Njuguna J, Siddique S, Bakah Kwroffie L, Piromrat S, Addae-Afoakwa K, Ekeh-Adegbotolu U, Oluyemi G, Yates K, Kumar Mishra A, Moller L (2022) The fate of waste drilling fluids from oil & gas industry activities in the exploration and production operations. Waste Manag 139:362–380. https://doi.org/10.1016/j.wasman.2021.12.025

Ossai IC, Ahmed A, Hassan A, Hamid FS (2020) Remediation of soil and water contaminated with petroleum hydrocarbon: a review. Environ Technol Innov 17:100526. https://doi.org/10.1016/j.eti.2019.100526

Parnian A, Furze JN, Parnian A (2022a) Drilling waste management based on new methods of bioremediation and solar desalination. In: Earth systems protection and sustainability, vol 1, pp 79–97. https://doi.org/10.1007/978-3-030-85829-2_4

Parnian A, Parnian A, Pirasteh-Anosheh H, Furze JN, Prasad MNV, Race M, Hulisz P, Ferraro A (2022b) Full-scale bioremediation of petroleum-contaminated soils via integration of co-composting. J Soils Sediments 22(8):2209–2218. https://doi.org/10.1007/s11368-022-03229-5

Parnian A, Mahbod M, Prasad MNV (2023) Microplastics remediation—possible perspectives for mitigating saline environments. In: Microplastics in the ecosphere. Wiley, pp 465–476. https://doi.org/10.1002/9781119879534.ch29

Paulsen JE, Omland TH, Igeltjørn H, Aas N, Solvang SA (2003) Drill cuttings disposal, balancing zero discharge and use of best available technique. All Days. https://doi.org/10.2118/85296-ms

Poi G, Aburto-Medina A, Mok PC, Ball AS, Shahsavari E (2017) Large scale bioaugmentation of soil contaminated with petroleum hydrocarbons using a mixed microbial consortium. Ecol Eng 102:64–71. https://doi.org/10.1016/j.ecoleng.2017.01.048

Prasad MNV, Katiyar SC (2010) Drill cuttings and fluids of fossil fuel exploration in north-eastern India: environmental concern and mitigation options. Curr Sci 98(12):1566–1569. https://doi.org/10.1007/s11356-021-16120-6

Qin G, Gong D, Fan MY (2013) Bioremediation of petroleum-contaminated soil by biostimulation amended with biochar. Int Biodeterior Biodegradation 85:150–155. https://doi.org/10.1016/j.ibiod.2013.07.004

Rana S (2008) Facts and data on environmental risks—oil and gas drilling operations. All Days. https://doi.org/10.2118/114993-ms

Sarma H, Prasad MNV (2024) Phytomanagement of polycyclic aromatic hydrocarbon and heavy metal contaminated sites in Assam, North Eastern India, for boosting the bioeconomy. In: Bioremediation and bioeconomy. Elsevier, pp 705–729. https://doi.org/10.1016/B978-0-443-16120-9.00018-2

Services, I. U. F. C. O. N. A. N. R. E. I. A (1991) Oil exploration in the tropics. World Conservation Union. http://books.google.ie/books?id=BzgG7RLCHDkC&q=OIL+EXPLORATION+IN+THE+TROPICS+Guidelines+for+Environmental+Protection&dq=OIL+EXPLORATION+IN+THE+TROPICS+Guidelines+for+Environmental+Protection&hl=&cd=1&source=gbs_api

Son Y, Cha J, Lim M, Ashokkumar M, Khim J (2011) Comparison of ultrasonic and conventional mechanical soil-washing processes for diesel-contaminated sand. Ind Eng Chem Res 50(4):2400–2407. https://doi.org/10.1021/ie1016688

Tyagi AK, Kaur G (2024) Shale Gas development in India: challenges, legal regulations and way forward. In: Boruah A, Verma S, Ganguli SS (eds) Unconventional Shale Gas exploration and exploitation. Advances in oil and gas exploration & production. Springer, Cham. https://doi.org/10.1007/978-3-031-48727-9_8

Van Eyk J (2022) Petroleum bioventing. Taylor & Francis. http://books.google.ie/books?id=Xr9rEAAAQBAJ&pg=PR17&dq=Van+Eyk,+J.+(2022).+Petroleum+bioventing.+Taylor+%26+Francis.&hl=&cd=1&source=gbs_api

Wang Z, Xu Y, Zhao J, Li F, Gao D, Xing B (2011) Remediation of petroleum contaminated soils through composting and rhizosphere degradation. J Hazard Mater 190(1–3):677–685. https://doi.org/10.1016/j.jhazmat.2011.03.103

Xue T, Yang C, Li Y, Shi X, Ma H, Wei X et al (2024) Disposal of drilling waste in salt mines in China. Sci Total Environ 912:168746. https://doi.org/10.1016/j.scitotenv.2023.168746

Zhichkina L, Nosov V, Zhichkin K, Mirgorodskaya M, Avdotin V (2020) Impact of out-of-service wells on soil condition. IOP Conf Ser Earth Environ Sci 421(6):062021. https://doi.org/10.1088/1755-1315/421/6/062021

Zoghi P, Mafigholami R (2023) Optimisation of soil washing method for removal of petroleum hydrocarbons from contaminated soil around oil storage tanks using response surface methodology. Sci Rep 13(1). https://doi.org/10.1038/s41598-023-42777-9

Chapter 17
Eco-Savvy Solutions: A Deep Dive into Modern Bioremediation Techniques for Soil and Sediment Cleanup

Suranajana Sarkar, Tina Basu, Subhamoy Banerjee, and Ruchira Mukherjee

Abstract Remediating contaminated sediment and soil is a pressing environmental concern that requires effective strategies. This comprehensive review delves into various bioremediation techniques, highlighting six pivotal approaches: biostimulation, nanoremediation, phytoremediation, rhizoremediation, soil microorganisms, and vermiremediation. Biostimulation involves augmenting the growth of indigenous microorganisms to metabolize contaminants effectively, whereas phytoremediation is a predominantly green strategy that harnesses the unique capabilities of plants to carry out three operations: to absorb, accumulate, and detoxify pollutants, thus offering economically viable and sustainable solutions. Rhizoremediation capitalizes on the symbiotic relationship between soil microbes and plant roots to increase contaminant degradation. Microorganisms in the soil play a vital role in pollutant degradation through enzymatic processes, rendering them indispensable in bioremediation endeavors. Vermiremediation employs earthworms to enhance soil structure and microbial activity, thereby facilitating contaminant attenuation. Each technique is meticulously examined in terms of its underlying principles, practical applications, and ecological ramifications. These diverse bioremediation strategies provide invaluable insights for addressing soil and sediment contamination challenges sustainably and efficiently. By leveraging natural processes and organisms, bioremediation offers promising avenues for remediating polluted sites and rejuvenating environmental health.

Keywords Contaminated · Nanoremediation · Soil · Phytoremediation · Microbial

---

S. Sarkar
Department of Microbiology. School of Life Sciences, Swami Vivekananda University, Bara Kanthalia, West Bengal, India

T. Basu · S. Banerjee · R. Mukherjee (✉)
Department of Basic Science and Humanities, Institute of Engineering & Management, University of Engineering and Management, Kolkata, West Bengal, India
e-mail: ruchira.mukherjee@iem.edu.in

A. Parnian et al. (eds.), *Bioremediation of Petroleum Contaminated Soil and Sediment*, Environmental Contamination Remediation and Management,
https://doi.org/10.1007/978-3-031-95524-2_17

17.1 Introduction

Contaminated sediment and soil present formidable environmental challenges, posing risks to ecosystems, human health, and water quality on a global scale. Addressing these challenges requires effective and sustainable strategies to mitigate pollutant impacts and restore ecosystem health. In this comprehensive review, we explore a range of bioremediation techniques, each offering unique advantages in addressing soil and sediment contamination. Bioremediation, a process harnessing the innate abilities of microorganisms, plants, and other biological agents to treat harmful pollutants, is a promising environmental cleanup approach. By harnessing natural biological processes, bioremediation offers sustainable, cost-effective, and environmentally friendly solutions, minimizing the reliance on costly and disruptive conventional remediation methods. This review examines five prominent bioremediation techniques. The techniques exploit different biological mechanisms to effectively remediate contaminated soil and sediment environments (Bala et al. 2022).

Biostimulation involves enhancing indigenous microbial populations via electron acceptors, growth-promoting factors, or the addition of nutrients, stimulating the degradation of contaminants. This approach accelerates microbial activity and metabolism, resulting in increased contaminant degradation rates and reduced pollutant concentrations (Sánchez Mata et al. 2023). Phytoremediation offers both a sustainable and environmentally friendly remediation strategy by harnessing the ability of plants not only to take up and translocate but also to metabolize contaminants (Ma et al. 2023). Plants can accumulate contaminants in their tissues or facilitate their breakdown through various physiological and biochemical mechanisms, effectively removing pollutants from the environment (Peng et al. 2009). Rhizoremediation exploits the symbiotic relationship between plant roots and soil microorganisms, particularly rhizospheric bacteria and fungi, to increase contaminant degradation in the rhizosphere. The carbon source for microbial growth is provided by root exudates, while microorganisms, in turn, facilitate nutrient uptake by plants and degrade contaminants in the vicinity of the roots (Praveen et al. 2019).

Soil microorganisms, including archaea, fungi, and bacteria, play pivotal roles in pollutant degradation through enzymatic processes, making them indispensable components of bioremediation strategies. These microorganisms possess diverse metabolic pathways capable of degrading and converting a wide range of contaminants into less harmful substances (Zhang et al. 2019). Vermiremediation employs earthworms to improve soil structure, increase microbial activity, and promote contaminant reduction through a combination of various processes, including biological, chemical, and physical processes. Earthworms ingest soil and organic matter, increasing mixing and aeration, while their digestive tracts harbour diverse microbial communities capable of degrading organic contaminants (Rajadurai et al. 2022).

Each bioremediation technique is thoroughly discussed in terms of its underlying principles, applications, challenges, and potential ecological impacts. By critically evaluating these diverse approaches, we aim to provide valuable insights for designing and implementing effective bioremediation strategies to address soil and sediment contamination issues sustainably and efficiently. Through the strategic deployment of bioremediation techniques, we can mitigate the environmental impacts of contamination and restore the health and functionality of affected ecosystems, ensuring a clean and healthy environment for generations to come.

17.1.1 Petroleum Waste Composition

A significant volume of petroleum waste is generated and its composition and characteristics are similar to those of the rocks from which it originated (Ozdemir et al. 2020). Although crude oil constitutes the primary component of petroleum waste, trace quantities of water, emulsion, soil, and clay are also present. Improper disposal practices, such as dumping waste into open pits, can pose significant health hazards to humans. The presence of noxious chemicals such as anthracene, benzene, pyrene, chromium (Cr^{+6}), and arsenic (As) in petroleum waste may lead to carcinogenic effects (Mansur et al. 2015). The primary constituents of petroleum waste include water (55.2%), light hydrocarbons (23.2%), asphaltenes (1.9%), clay (9%), and waxes (10.5%) (Islam 2015). The global petroleum reservoirs are dominated by alkanes, cycloalkanes, aromatics, and heteroatomic hydrocarbons, with minor concentrations of nitrogen, oxygen, and sulfur (De Junet et al. 2013). Thus paraffins, naphthalene, cycloalkanes, and NSOs significantly contribute to environmental pollution and ecosystem disruption.

17.2 Effects of Polyaromatic Hydrocarbon Contamination on Soil Features and Attributes

Severe alterations in soil characteristics occur in response to petroleum contamination in the soil. They involve changes in texture, total organic carbon content, conductivity, and moisture levels. Additionally, halogenated hydrocarbons containing nonorganic elements such as bromide, fluoride, chloride, or iodide are commonly identified as toxic, hazardous, and persistent soil pollutants (Klamerus-Iwan et al. 2015). The properties and functions of the soil matrix are intimately linked to a variety of activities within the soil and to the presence of xenobiotic compounds such as polyaromatic hydrocarbons (PAHs). (PAHs) possess distinctive attributes, such as hydrophobic characteristics, which render them resistant to microbial degradation and make them chemically stable. These properties imply that oil spills could have detrimental effects on both the biological and physicochemical properties of the affected soil.

Petroleum is one of the primary sources of PAHs, which are highly hydrophobic, enabling them to be absorbed by soil particles and displacing water molecules, thus reducing water infiltration and the oxygen content in petroleum-contaminated soil.

Furthermore, hydrocarbon contamination can alter soil geotechnical properties such as hydraulic conductivity, permeability, compaction, and biological characteristics, including biomass and enzyme activity, within the soil matrix. Areas affected by petroleum pollution often exhibit a diminished self-purification capacity, leading to a decline in indigenous microbial populations or diversity related to soil purification procedures (Zahermand et al. 2020).

The soil microbial communities are adversely affected by the inherent toxicity and persistence of PAH compounds. Furthermore, toxicity can result in the loss of essential soil functions, creating anaerobic conditions and asphyxia in soil pores, impacting microbial activities. It becomes more severe with increasing molecular weight. Klamerus-Iwan et al. 2015) demonstrated that soil polluted with chainsaw oil experienced remarkable losses in both its enzymatic activities and microbial biomass. The affected enzymes are dehydrogenase and urease (Daâssi and Almaghribi 2022).

17.2.1 Petroleum Contamination Complications

Petroleum contamination in soil and sediment poses a significant environmental threat, impacting ecosystems, human health, and economic stability. Originating from various sources, such as accidental spills, leaks, and improper disposal, petroleum compounds derived from crude oil infiltrate the environment, leading to widespread contamination. The consequences of petroleum pollution are diverse (Ozdemir et al. 2020). First, these hydrocarbons can adhere to soil particles or settle within sediment layers, persisting and disrupting the delicate balance of soil and sediment ecosystems. This persistence not only affects microbial communities but also inhibits plant growth and alters soil fertility (Truskewycz et al. 2019). Additionally, the leaching of petroleum compounds into groundwater further exacerbates environmental degradation, posing risks to both animal and aquatic life. Moreover, petroleum contamination changes the physicochemical properties of both soil and sediment, thus impairing their ability to support life (Islam 2015). This alteration can result in soil compaction, reduced water infiltration, and changes in pH levels, further compromising environmental integrity (Liu et al. 2012).

The remediation of petroleum-contaminated soil and sediment presents significant challenges, given the complexity of petroleum compounds and their interactions with the environment. While conventional techniques such as excavation and soil washing are available, they often prove to be costly, resource-intensive, and environmentally disruptive (Patowary et al. 2023). In contrast, bioremediation techniques harness the inherent capabilities of microorganisms, plants, and other biological agents either to degrade or to remove contaminants, thus offering a sustainable and cost-effective solution (Zhang et al. 2019).

17.2.1.1 Disruption of Soil and Sediment Ecosystems

Petroleum pollutants disrupt the intricate balance of soil and sediment ecosystems by altering their physical, chemical, and biological properties (Klamerus-Iwan et al. 2015). For example, PAHs, known for their strong sorption to soil particles, impede nutrient uptake by plants and hinder microbial activity, which is essential for soil health. Additionally, the presence of VOCs and other hydrocarbons can cause soil compaction and reduce water infiltration rates, augmenting soil erosion and degradation (Tang et al. 2010).

17.2.1.2 Groundwater Contamination

Petroleum contaminants have the propensity to migrate from soil and sediments directly into groundwater, thus posing a significant peril to both human health and aquatic ecosystems (Gao et al. 2011). Contaminant mobility is regulated by hydrophobic interactions, sorption–desorption phenomena, and soil permeability. Once in groundwater, these pollutants can persist for prolonged periods due to their recalcitrant nature, perpetuating contamination and necessitating rigorous remediation efforts.

17.2.1.3 Ecotoxicity and Biological Impacts

Petroleum hydrocarbons have toxic effects on soil and sediment organisms, ranging from microbial communities to higher trophic levels (Ma et al. 2023). PAHs, in particular, are carcinogenic and mutagenic, disrupting cellular functions and metabolic processes in organisms and leading to adverse health effects and population declines (Ma et al. 2015). Chronic exposure to petroleum pollutants can impair reproductive success, alter community structure, and disrupt ecological interactions within affected ecosystems.

17.2.1.4 Air Quality Degradation

Volatile petroleum compounds released from contaminated soil and sediment can volatilize into the atmosphere, contributing to air quality depletion and the risk of exposure (Ekperusi and Aigbodion 2015). Compounds such as benzene, xylene (BTEX), ethylbenzene, and toluene, which are prevalent in petroleum contamination, pose significant health risks, including respiratory issues and neurological disorders. The release of these compounds into the air poses risks to nearby communities, especially those residing near contaminated sites or industrial facilities.

17.2.1.5 Economic Implications

Petroleum soil and sediment pollution incur substantial economic costs, including clean-up expenses, loss of agricultural productivity, and damage to natural resources. The remediation of contaminated sites requires significant financial investments, often spanning years or decades, and may involve complex engineering and biotechnological interventions. Additionally, economic repercussions extend to sectors such as tourism and recreation, where environmental degradation perceptions can deter visitors and impede economic growth.

17.2.1.6 Long-Term Persistence and Remediation Challenges

Petroleum contaminants exhibit varying degrees of persistence in soil and sediment environments and are influenced by factors such as compound volatility, solubility, and microbial degradation rates. High-molecular-weight PAHs, for example, can remain in soil and sediment matrices for a long duration, posing long-term challenges for remediation efforts. Effective remediation strategies necessitate a nuanced understanding of transport mechanisms and contaminant fates, considering site-specific factors and environmental conditions (Truskewycz et al. 2019; Ambust et al. 2021).

17.2.2 Aftermath of an Oil Spill on Land and Alterations in Site Characteristics

Following a terrestrial oil spill event, a cascade of processes unfold, shaping the fate of oil and inducing changes in site characteristics. First, spillage can lead to plant mortality, disrupting the local ecosystem and diminishing the habitat for wildlife. The oil can subsequently obstruct oxygen diffusion into the soil, creating anaerobic zones that alter microbial activity and nutrient cycling processes. Concurrently, the spilled oil modifies the soil structure by affecting properties such as texture, porosity, and permeability, thereby impacting water retention and nutrient availability. The volatile components of the oil may also volatilize, contributing to air pollution and posing health risks. As the oil infiltrates the soil, it may percolate through the profile, potentially reaching groundwater aquifers and contaminating drinking water sources. Despite these challenges, pockets of aerobic zones may persist amidst anaerobic conditions, supporting microbial activity capable of biodegrading hydrocarbons. However, the initial presence of oil can suppress microbial populations and reduce microbial diversity due to its toxicity and unfavorable environmental conditions. Ultimately, hydrocarbons from spilled oil can migrate through the soil, contaminating groundwater and posing risks to human health and aquatic ecosystems. Overall, the aftermath of a terrestrial oil spill involves a complex interplay of processes with profound environmental, ecological, and socioeconomic implications, necessitating effective mitigation and remediation efforts to restore affected ecosystems (Klamerus-Iwan et al. 2015).

17.3 Approaches for Remediating Soils Tainted with PAHs

The remediation of soil petroleum contamination can be achieved through two primary approaches: in situ and ex-situ methods.

17.3.1 In-Situ Remediation

It involves treating contamination at its original location without the need for excavation or removal of the affected soil. This approach often utilizes techniques such as bioremediation, where microorganisms break down petroleum hydrocarbons into less harmful substances. Additionally, in situ methods may include chemical oxidation, where oxidizing agents are injected into the soil to degrade contaminants, or phytoremediation, which employs plants to absorb and metabolize pollutants (Babalola et al. 2017).

17.3.2 Ex-Situ Remediation

Involves the physical removal of contaminated soil from its original location for treatment. Techniques such as soil washing, where the soil is mixed with water or solvents to remove contaminants, or thermal desorption, which vaporizes and removes pollutants via heat, are commonly employed (Patowary et al. 2023). *Ex situ* methods are typically more expensive and disruptive than in situ methods but may be necessary for highly contaminated sites or when in situ treatment is not feasible.

Both ex-situ and in-situ remediation methods are essential for addressing and eradicating soil petroleum contamination. The choice of method varies according to the extent of contamination, site conditions, and regulatory requirements.

17.3.3 In-Situ Bioremediation Strategies

Contaminated sediment and soil pose substantial environmental challenges, necessitating the implementation of effective remediation strategies (Fig. 17.1).

Bioremediation techniques utilize organisms such as plants, bacteria, fungi, or their enzymes to treat environmental contaminants. Notably, bioremediation takes

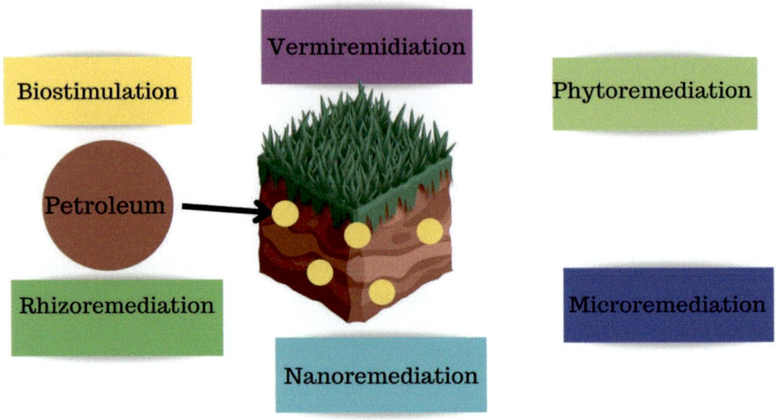

Fig. 17.1 Strategies to treat petroleum-contaminated soil

place in open non-sterile environments that host a variety of organisms. The main players in bioremediation are bacteria, capable of degrading pollutants, whereas other organisms, such as protozoa and fungi, also contribute. These microorganisms can degrade pollutants through oxidation, reduction, or hydrolysis processes (Sánchez Mata et al. 2023). Microorganisms can either be introduced into contaminated sites or stimulated in situ to increase their activity. Techniques such as adding nutrients to promote microbial growth (biostimulation), introducing specific strains of microorganisms known to degrade contaminants (bioaugmentation), or optimizing environmental conditions to increase microbial activity (bioventing or biosparging) are employed. Overall, bioremediation provides a cost-effective, eco-friendly approach to cleaning contaminated sites by harnessing natural processes. It can address contaminants, including heavy metals, petroleum hydrocarbons, industrial chemicals, and pesticides, making it a versatile tool in environmental restoration endeavors (Azubuike et al. 2016; Ojuederie and Babalola 2017).

17.4 Biostimulation: Boosting Indigenous Microorganisms

Biostimulation relies on principles rooted in microbial ecology and biodegradation pathways to amplify the natural capacity of microorganisms to break down petroleum hydrocarbons within contaminated soil. This approach entails augmenting the soil with specific nutrients and amendments to invigorate the development and function of hydrocarbon-degrading microorganisms (Fig. 17.2). The efficacy of biostimulation depends on numerous variables, such as the concentration of petroleum pollutants; soil characteristics, such as texture, pH, organic matter content, and moisture levels; and environmental parameters, such as oxygen availability and temperature. Nutritional requirements such as nitrogen, phosphorus, and potassium

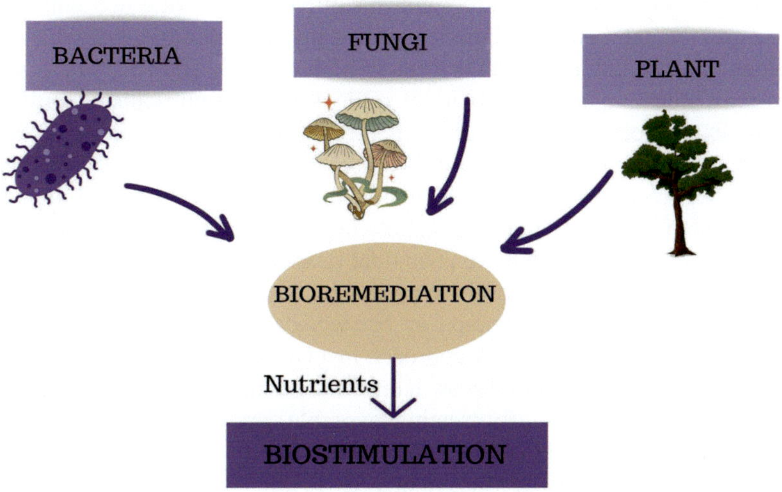

Fig. 17.2 Biological treatments

are typically supplemented during biostimulation to promote microbial proliferation and metabolic activities. These nutrients are integral constituents of cellular structures, enzymes, and metabolic pathways crucial for hydrocarbon degradation (Sánchez Mata et al. 2023). Additionally, trace elements such as iron, magnesium, and sulfur may also be essential for sustaining microbial activity and enzyme functionality. The incorporation of organic amendments such as compost, manure, or biochar can furnish supplementary carbon sources and microbial habitats, further stimulating microbial growth and hydrocarbon decomposition. These organic materials not only provide energy and carbon for microbial metabolism but also enhance soil structure, moisture retention, and nutrient accessibility. Oxygenation techniques, such as aeration or the introduction of oxygen-releasing compounds, can establish aerobic conditions conducive to the activity of hydrocarbon-degrading bacteria. Compared with anaerobic pathways, aerobic degradation pathways typically exhibit superior efficiency and rapidity expediting the remediation process. Biostimulation strategies may also encompass the introduction of specialized microbial consortia or genetically modified microorganisms with enhanced hydrocarbon-degrading capabilities. These microbes exhibit specificity for petroleum hydrocarbons and possess unique enzymatic systems tailored to dismantle complex hydrocarbon structures (Saeed et al. 2021; Macaya et al. 2019).

Monitoring methodologies, including microbial biomass analysis, microbial community profiling, and the quantification of hydrocarbon degradation byproducts, serve to evaluate the efficacy of biostimulation interventions over time. By optimizing nutrient supplementation, environmental parameters, and microbial functions, biostimulation presents a scientifically grounded and sustainable approach to mitigate petroleum-contaminated soil (Simpanen et al. 2016).

In natural settings, native microorganisms often lack the vital nutrients required for their growth and metabolic functions. The provision of nutrients enables microorganisms to break down pollutants through both anabolic and catabolic processes. However, in areas impacted by toxic oil spills, supplying nutrients or fertilizers to foster the expansion of oil-degrading microbial populations can pose challenges. The toxic properties of the spilled oil initially harm or kill numerous indigenous species in the affected vicinity. Consequently, the toxic nature of the oil impedes the effectiveness of nutrient supplementation in stimulating the survival of native microbes. In instances where there is no tidal flushing and where the toxicity of the spilled oil has decreased to a level where indigenous bacteria can survive, containment measures like floating booms or barriers can be utilized to establish oil-containing bioreactor basins. This approach involves supplying additional nutrients or other substances that promote growth to stimulate microbial growth and facilitate degradation of the spilled oil. By creating favorable conditions for microbial growth and metabolism, biostimulation is vital for expediting the bioremediation of oil contamination (Saeed et al. 2021; Macaya et al. 2019).

According to Udume et al. (2023), composting represents an environmentally friendly technique for converting organic matter into biofertilizers, which holds promise for remediating areas contaminated with petroleum. Studies have revealed that water hyacinth is rich in microbial growth stimulating nutrients and, when combined with spent mushroom compost, has shown enhanced degradation of petroleum hydrocarbons. Spent mushroom compost and water hyacinth outperformed commercial fertilizer, indicating the potential of organic compost mixtures for pollutant remediation. Furthermore, this organic blend demonstrated significant degradation of 16 polycyclic aromatic hydrocarbons classified as priority pollutants by the US EPA, suggesting its efficacy in remediating soils contaminated with persistent organic pollutants. This underscores the value of using diverse composted organic materials for effective pollutant cleanup strategies. Work carried out by Simpanen et al. (2016) demonstrated that in biostimulation treatments of a recent oil spill, microbes efficiently degraded the easily accessible oil, reducing contaminant leaching from the soil. Thus, biostimulation not only cleansed the soil but also minimized oil leaching and prevented groundwater contamination.

17.5 Phytoremediation: Harnessing the Power of Plants

Phytoremediation refers to the judicious utilization of plants and related soil microbes to decrease the concentrations and toxicity of pollutants present in the environment (Tripathi et al. 2020). Since phytoremediation is a cost-effective environmental restoration technology, it is widely accepted (Panchenko et al. 2023). Although genetic manipulations are effective, they are harsh and harmful at times to the soil hence, phytoremediation is a solution.

Numerous scientific investigations have demonstrated the effectiveness of phytoremediation across diverse contaminated sites globally. The remarkable capacity of hyperaccumulator plants such as *Thlaspicaerulescens* and *Alyssum bertolonii* to purge pollutants from the soil and accumulate significant amounts of heavy metals in their tissues is a classic example of phytoremediation (Peng et al. 2009). Moreover, in-depth research has delved into the underlying mechanisms of phytoremediation processes, shedding light on the pivotal role of plant physiology, biochemistry, and genetics in the uptake, translocation, and detoxification of contaminants. Molecular studies have pinpointed specific genes and enzymes responsible for metabolizing and neutralizing pollutants within plant tissues. Field trials and pilot-scale investigations have provided invaluable insights into the practical application of phytoremediation techniques. These endeavors have demonstrated the viability of employing phytoremediation in real-world scenarios, encompassing abandoned industrial sites, mine tailings, and wastewater treatment facilities (Ma et al. 2023).

Phytoremediation stands as a well-documented and scientifically backed approach for addressing petroleum-contaminated soil, with extensive research illustrating its efficacy in rehabilitating polluted environments (Ma et al. 2023).

Specific investigations have revealed the presence of enzymes within certain plant species capable of breaking down complex hydrocarbons such as PAHs. These enzymes help transform pollutants that are nontoxic and less damaging, effectively detoxifying the soil. Moreover, research has revealed intricate pathways within plants that contribute to the breakdown and assimilation of petroleum contaminants, shedding light on the underlying mechanisms of phytoremediation processes. Comparative studies have evaluated the remediation efficiency of different grasses, shrubs, and trees in petroleum-contaminated soils, examining factors such as pollutant uptake rates, biomass production, and overall remediation efficacy. These investigations revealed that plant species exhibit high tolerance to petroleum contaminants and demonstrate their capacity to flourish in polluted environments while actively purging pollutants from the soil (Panchenko et al. 2023).

Based on the analysis by Liu et al. (2012) of total petroleum hydrocarbons or TPHs and their composition and pH changes, certain plant species, including *Fawn (Festucaarundinacea Schreb), Gaillardia aristata, Echinacea purpure, a Medicago sativa Linn and Fire Phoenix (a combined F. arundinacea),* have demonstrated significant potential for removing TPH and altering their composition in soil contaminated with petroleum. This research highlights the efficacy of phytoremediation employing specific ornamental species as an effective method for degrading TPHs in petroleum-contaminated soil.

In a recent study by Ma et al. (2023), four varieties of sweet sorghum (KT24, ZK438, KT6, and KT21) from a pool of 28 commercial and research varieties were resistant to petroleum contamination in saline soil. The emergence rate of these varieties was unaffected by high levels of external petroleum contamination (2.0×10^4 mg/kg) in saline soil. Planting these sorghum varieties effectively facilitated the removal of petroleum contaminants from saline soil, with KT21 showing the most promising remediation effects and potential for application in remediating petroleum-contaminated saline soil.

Ongoing research endeavors aim to further refine phytoremediation techniques, identify novel plant species with enhanced remediation capabilities, and devise strategies for upscaling phytoremediation processes to address larger contaminated sites.

17.6 Rhizoremediation: Plant Roots and Microbial Synergy

Rhizoremediation is an advanced strategy aimed at remediating petroleum-contaminated soil by utilizing the interplay among plants, their root systems, and associated soil microorganisms (Praveen et al. 2019). Specifically, certain plant species are carefully chosen based on their ability to withstand and trap contaminants, particularly petroleum hydrocarbons, through their roots. These plants, often termed "hyperaccumulators" or "phytoremediators," possess mechanisms for taking up and transporting pollutants from the soil into their tissues. Once absorbed, these contaminants can trigger the release of root exudates into the rhizosphere—the regional soil adjacent to the roots. Root exudates are by nature a complex blend of organic compounds, enzymes, and other substances, that act both as nutrients and energy sources for various soil microorganisms, including bacteria, fungi, and actinomycetes. Notably, certain microbial species have the metabolic capacity to degrade or transform petroleum compounds into less harmful or inert forms. The rhizosphere environment, enriched by root exudates, creates ideal conditions for microbial activity and the degradation of pollutants. Enhanced soil aeration, increased microbial biomass, and improved moisture levels contribute to the accelerated breakdown of petroleum contaminants (Praveen et al. 2019).

Additionally, plant roots physically entrap and restrict their mobility, preventing their migration and leaching into groundwater. Scientific research has demonstrated the efficacy of rhizoremediation in treating petroleum-contaminated soil. Studies have identified suitable plant species with increased phytoremediation potential and revealed the underlying mechanisms governing contaminant uptake, transport, and degradation (Praveen et al. 2019).

Tang et al. (2010) demonstrated that bioaugmentation with bacterial communities, especially plant growth-promoting rhizobacteria (PGPR), can significantly increase rhizoremediation efficiency, with the degradation rate of contaminants varying among different plant species. *Festuca Arundinacea* presented higher degradation rates than did *Gossypium* sp. and *Lolium perenne* (Tang et al. 2010). Moreover, the composition of microbial communities is strongly influenced by root exudates, which favor the proliferation of specific bacteria such as *Pseudomonas* and *Microbacterium*. These bacteria have been shown to degrade contaminants such as phenanthrene, especially in association with root exudates from certain plants such as *Lolium perenne* (Ce′bron et al. 2011).

Plant roots also release certain enzymes, such as laccases and phenol oxidases, which contribute to hydrocarbon degradation (Martin et al. 2014; Gao et al. 2011). However, microbial-driven degradation remains the primary mechanism for hydrocarbon removal (Martin et al. 2014). Bacterial genera such as *Sphingomonas*,

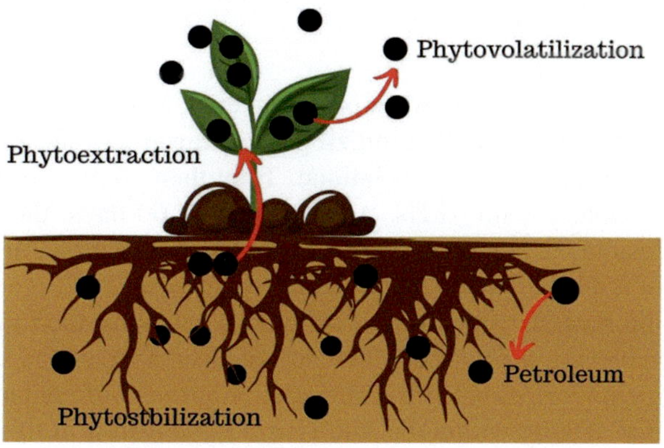

Fig. 17.3 Mechanism of phytoremediation of petroleum contaminants

Mycobacterium, and *Pseudomonas* are capable of effectively degrading an array of organic compounds (Vidali 2001).

While plants alone may play a limited role in rhizoremediation, their symbiotic interaction with microbes significantly enhances this process (Gao et al. 2011). Overall, phytoremediation and rhizoremediation represent promising approaches for addressing petroleum contamination, but further research is needed to fully understand their underlying mechanisms (Fig. 17.3).

17.7 Soil Microorganisms: Enzymatic Clean-up Heroes

Microremediation is a bioremediation approach that harnesses microorganisms such as fungi, bacteria, and algae to degrade pollutants present in petroleum-contaminated soil. Bacteria such as *Pseudomonas, Bacillus*, and *Alcanivorax* are notable for their capacity to produce enzymes such as hydroxylases and oxygenases, which facilitate the breakdown of complex hydrocarbons into simpler, less harmful compounds. Studies indicate that microbial consortia, comprising a variety of bacterial and fungal species, often exhibit superior efficacy in degrading petroleum pollutants compared with individual microbial strains. These consortia offer a broader range of enzymatic activities and metabolic pathways, enabling them to target a wider array of hydrocarbon contaminants. Moreover, microorganisms can adapt to diverse environmental conditions, including fluctuations in temperature, pH, and nutrient availability. This adaptability is crucial for successful remediation across different soil types and climates, making microremediation a versatile solution. In addition to directly degrading hydrocarbons, microorganisms contribute to improving soil quality by facilitating nutrient cycling and decomposing organic matter. This aids in the restoration of soil health and fertility in areas affected by petroleum contamination (Xu et al. 2018).

17.7.1 Bacteria

In the remediation work of Zhang et al. (2019), petroleum-degrading bacteria were captured by scientists utilizing spent mushroom substrate biochar through a straightforward adsorption technique. They then employed this mixture to treat soil contaminated with TPHs. After a span of 60 days, they reported that this approach resulted in the most significant reduction in petroleum levels, especially in the range of n-alkanes with carbon chain lengths between C12 and C18. The treatment involving bacteria immobilized on biochar demonstrated the most promising outcomes, achieving a removal efficiency of over 58% of the contaminants. Moreover, this method accelerated the degradation process of contaminants and enhanced microbial activity within the soil. Overall, the results of this study suggest that the use of bacteria immobilized on biochar has considerable potential for effectively remediating petroleum-polluted soil, although further research is needed to assess its long-term effectiveness and impact.

In another study performed by Hossain et al. (2022), bacterial samples were isolated from soil contaminated with petroleum, and their capacity to degrade two significant petroleum-derived pollutants diesel and burned engine oils, was assessed. Three bacterial *strains—Enterobacter* sp., *Pseudomonas* sp. and *Acinetobacter* sp. were identified through thorough genetic analysis, specifically sequencing the bacterial 16S rRNA gene. Morphological examinations indicated that these isolates were gram-negative, motile cocci, indicative of adaptability to diverse environmental conditions. Biochemical assays further highlighted their metabolic capabilities, including hydrogen peroxide detoxification, citrate utilization, and sugar fermentation. Notably, robust growth in media containing diesel and burned engine oil was observed, with optimal conditions at pH 7.0 and 37 °C. Through systematic experimentation, the ability of the isolates to degrade diesel and burned engine oil was demonstrated, with the highest degradation rates observed at specific concentrations—4% for diesel oil and 8% for burned engine oil. Interestingly, intriguing trends in growth dynamics were revealed: while increased concentrations of diesel oil led to decreased growth rates, increasing concentrations of burned engine oil corresponded to increased bacterial proliferation. These findings highlight the significant potential of *Pseudomonas* sp., *Acinetobacter* sp., and *Enterobacter* sp. in bioremediation efforts targeting oil-contaminated environments. By harnessing the metabolic capabilities of these bacteria, a promising avenue exists for developing sustainable solutions to mitigate the ecological impact of petroleum pollution.

17.7.2 Fungi

Fungal bioremediation of petroleum-contaminated soil involves the use of specific fungi to break down and detoxify petroleum hydrocarbons in the soil. Fungi possess enzymes capable of degrading complex hydrocarbon molecules into simpler compounds, aiding in their degradation process. These fungi can be introduced directly into contaminated soil or through fungal inoculants or mycorrhizal fungi. They then establish a symbiotic relationship with plant roots and increase nutrient uptake and pollutant degradation. During fungal bioremediation, fungi make use of petroleum hydrocarbons for both carbon and energy sources, converting them into harmless byproducts such as carbon dioxide and water through aerobic and anaerobic metabolic processes. Fungi also develop mycelial structures that can penetrate the soil, thus increasing the surface area for pollutant degradation and promoting nutrient cycling. The efficacy of fungal bioremediation depends on multiple parameters, such as moisture content, pH, temperature, and oxygen availability alongside the concentration of contaminants and their types. Additionally, selecting appropriate fungal strains with robust hydrocarbon-degrading capabilities is crucial for successful remediation (Fallahi et al. 2023).

According to a study carried out by Daâssi and Almaghribi (2023), certain fungi, sourced both from contaminated soil and introduced externally, can effectively break down petroleum compounds. They identified specific fungi such as *Aspergillus niger* and *Fusarium chlamydosporum* as particularly efficient in this process. However, further research is necessary to ensure the scalability and safety of this method, particularly with respect to its impact on plant life. In essence, the use of these fungi holds promise for remedying soil contaminated with petroleum, but additional investigations and refinements are essential for their successful application.

Another study by Hamad et al. (2021) successfully identified numerous fungi capable of degrading petroleum hydrocarbons. According to laboratory tests, *Cladosporium sphaerospermum* and *Penicillium chrysogenum* resulted in the greatest degradation of certain hydrocarbons, whereas *Alternari aalternata* and *Penicillium chrysogenum* were the most efficient at breaking down PAHs. They recommend the use of a combination of *Penicillium chrysogenum* and *Curvularia brachyspora* for improved degradation of both types of pollutants. This study supports the idea that the biodegradation of PAHs is a more environmentally friendly alternative to chemical methods. Once these fungi are produced on a large scale, they can be used in real-world applications to remediate petroleum pollution.

However, another study carried out by Fallahi et al. (2023) identified various fungal species capable of effectively degrading petroleum hydrocarbons (PHs), including representatives from the *Aspergillus, Trichoderma, Fusarium,* and *Pleurotus* genera. These fungi demonstrated promising abilities in breaking down different types of PHs, with some displaying high degradation rates. Among the identified fungi, *Aspergillus tenuissima* and *Epicoccum nigrum* have emerged as notable candidates for efficiently degrading pHs both in laboratory tests and in soil environments. The effectiveness of these methods in terms of degradation is independent of the contamination percentage in the soil. The use of fungal inoculum, particularly from *Aspergillus tenuissima* and *Epicoccum nigrum*, could prove to be a potent method for bioremediating crude oil-contaminated soil. However, further research is necessary to refine the formulation of fungal inoculum and assess the efficacy of other fungal isolates in soil settings.

An analysis by Benguenab and Chibani (2021) demonstrated the potential of filamentous fungi for bioremediation, particularly for contaminants that are challenging for other microorganisms to degrade. Two fungi, *Aspergillusustus* and *Purpureocillium lilaacinum* effectively degrade crude oil, diesel, and engine oil in liquid culture. *Purpureocillium lilacinum* exhibited superior hydrocarbon degradation, with removal rates of 44.55% for crude oil and 27.66% for diesel, along with the highest biodegradation constant (0.02 day^{-1}) and shortest half-life (34.66 days) for crude oil degradation. While these findings are promising in liquid cultures, further research is necessary to evaluate their application in field-level bioremediation. Additionally, future studies should explore alternative fungi for soil bioremediation and conduct ecotoxicity tests to assess the efficiency of the treatment in terms of soil protection and healing.

Fungal bioremediation represents a promising and environmentally friendly approach for restoring petroleum-contaminated soil, complementing traditional remediation methods and contributing to sustainable environmental management (Table 17.1).

Table 17.1 Petroleum hydrocarbon-degrading bacteria

Petroleum Contaminant	Organism Type	Bacterial Species	References
Aliphatic Hydrocarbons	Bacteria	*Oleispiraantarctica, Rhodococcusruber, Pseudomonas* sp., *Gordoniasihwensis*	Yakimov et al. (2003), Zhukov et al. (2007), Sugiura et al. (1997), Brown et al. (2016)
Aromatic Hydrocarbons		*Bacillus Licheniformis, Bacillus Mojavensis, Pseudomonas aeruginosa, Achromobacterxylosoxidans, Aeribacilluspallidus*	Eskandari et al. (2017), Mukherjee et al. (2010), Ma et al. (2015), Mnif et al. (2014)
Aromatic Hydrocarbons	Fungi	*Aspergillusniger, Fusariumchlamydosporum, Cladosporiumsphaerospermum, Penicilliumchrysogenum, Alternariaalternata, Curvulariabrachyspora, Trichoderma, Aspergillustenuissima, Epicoccumnigrum, Aspergillusustus and Purpureocilliumlilacinum*	Daâssi and Almaghribi (2023), Hamad et al. (2021), Fallahi et al. (2023), Benguenab and Chibani (2021)
	Earthworms	*Drawidamodesta, Lampitomauritii*	Rajadurai et al. (2022)

17.8 Vermiremediation: Earthworms as Soil Superheroes

Vermiremediation is a process that utilizes earthworms to aid in the degradation and detoxification of contaminants found in soil. Specifically, in the context of petroleum-contaminated soil, vermicomposting, a specialized form of vermiremediation, involves the employment of earthworms to break down complex petroleum hydrocarbons and other organic pollutants into simpler and less harmful compounds (Javed and Hashemi 2021). Throughout the process of vermiremediation, earthworms consume soil particles alongside organic matter, including petroleum hydrocarbons, as they navigate through the soil. Upon ingestion, contaminants undergo enzymatic breakdown within the earthworm digestive system. Key digestive enzymes produced by earthworms, such as cellulases, lipases, and proteases, are critical in facilitating the removal and treatment of these pollutants. Moreover, earthworms contribute to increasing microbial activity in the soil through their excretions, known as castings, which harbor beneficial microorganisms capable of further breaking down contaminants.

Vermiremediation encompasses multifaceted interactions involving earthworms, microorganisms, and soil particles. As earthworms burrow through the soil, they create channels and aggregates, thereby improving soil aeration and fostering microbial activity. This conducive environment promotes the biodegradation and transformation of petroleum hydrocarbons into less complex and less toxic forms (Rajadurai et al. 2022).

Numerous scientific studies have indicated the ability of vermiremediation to reduce the concentrations of petroleum hydrocarbons in contaminated soil. However, the efficiency of this process is contingent upon various factors, including the species and density of earthworms, environmental factors (such as moisture, pH, and temperature), and characteristics of the soil, including texture and organic matter content, as well as the content and type of pollutant in the soil (Sánchez Mata et al. 2023).

In a study by Rajadurai et al. (2022) in India, two earthworm species, *Lampitomauritii* (anecic) and *Drawidamodesta* (epigeic), were utilized to remediate soil contaminated with polycyclic aromatic hydrocarbons (PAHs) and total petroleum hydrocarbons (TPHs) from engine oil. In addition to positive and negative controls, four treatments were established. The treatments involving soil (1 kg), cow dung (50 g), used engine oil (7.5 mL), and earthworms presented the maximum remediation rates, with 68.6% of the PAHs and 34.3% of the TPHs removed after 60 days.

Research conducted by Ekperusi et al. (2015) in Nigeria's Niger Delta region revealed that the earthworm species Hyperiodrilus africanus has the ability to assist in the bioremediation of petroleum-polluted soil by both accumulating and degrading pollutants. The addition of nutrients can further enhance this remediation process. The use of earthworms in later stages of remediation, particularly when soil toxicity decreases, can be beneficial. In addition to pollutant removal, earthworm activity also improves soil quality through the production of vermicast. Research has investigated the efficacy of *Hyperiodrilus Africanus* in treating crude oil-contaminated soil over 12 weeks. Various soil parameters, including calcium, potassium, magnesium, sodium, organic carbon, nitrate, sulfate, and phosphate contents, were assessed. Total petroleum hydrocarbon (TPH) concentrations were measured via atomic absorption spectrophotometry, whereas BTEX compounds and earthworm tissue were analyzed via gas chromatography with flame ionization detection. The results indicated significant improvements in soil conditions with reduced pollutant levels, suggesting the potential of *Hyperiodrilus Africanus* for remediating moderately polluted soils.

Vermiremediation represents a sustainable and environmentally sound approach for remediating petroleum-contaminated soil. By harnessing the natural activities of earthworms and beneficial microorganisms, vermiremediation can effectively degrade pollutants and restore soil quality over time.

17.9 Nanoremediation: Nanosized Solutions for Large Pollution Problems

Nanoremediation involves exploiting the ability of nanoparticles to address environmental contamination issues, such as petroleum-contaminated soil. This method utilizes the unique characteristics of nanoparticles to adsorb, degrade, or immobilize contaminants, thus facilitating their removal from the soil or reducing their harmful effects. NPs possess a high surface area and reactivity, enabling them to efficiently adsorb hydrophobic petroleum compounds from the soil. Some nanoparticles are even engineered for precise targeting and degradation of petroleum contaminants, enhancing the efficiency of the remediation process. However, while nanoremediation holds promise, it remains an evolving field, with ongoing research aimed at optimizing its effectiveness, safety, and environmental impact. Regulatory considerations and risk assessments were crucial factors to address before the widespread adoption of nanoremediation technologies (Vu et al. 2023).

Crude oil pollution is a significant environmental concern, but traditional cleanup methods such as bioremediation have limitations. Nanotechnology, which involves tiny particles called nanoparticles (NPs), offers promising solutions for environmental cleanup. NPs have a large surface area and can efficiently absorb pollutants from soil or water, rendering them more easily available for microbial degradation.

The blend of nanotechnology with bioremediation, known as the NBR process, shows potential for enhancing cleanup efforts. However, further research is needed to understand how NPs and microbes interact in this process to optimize their effectiveness. While some progress has been made in using NBR for oil spill cleanup, more studies are necessary to ensure its applicability in the real world.

Additionally, it is crucial to examine the persisting environmental impacts of NPs and prioritize the use of eco-friendly, bio-based NPs over chemical or metallic ones. Identifying cost-effective methods for NP production and developing nanomembranes for treating oil-contaminated water can also contribute to addressing water scarcity issues (Patowary et al. 2023).

Phyto-nanotechnology is an environmentally friendly approach to harness plants to produce nanoparticles for the treatment of toxic compounds from the surrounding environment. Algae, a type of microorganism, are particularly adept at this process because of their ability to synthesize various metal nanoparticles such as silver, gold, and palladium. They possess characteristics such as high metal uptake capacity, ease of handling, and low toxicity, making them well-suited for nanoparticle production. Algal species from different groups have been utilized in nanoparticle synthesis, which can then be employed to remediate toxic compounds and heavy metals from wastewater. Microalgae have demonstrated efficacy in eliminating heavy metals such as lead, copper, and selenium from wastewater, thereby averting environmental pollution. Techniques such as immobilizing microalgae biomass and employing consortia of microalgae have proven effective for wastewater treatment and heavy metal removal (Singh and Saxena 2022).

17.10 Comparative Analysis: Principles, Applications, and Ecological Impacts

Bioremediation techniques offer multifaceted solutions for addressing soil and sediment contamination, each underpinned by distinct principles, applications, and ecological repercussions. Biostimulation involves the enrichment of indigenous microbial communities through the addition of nutrients or organic compounds, fostering enhanced degradation of contaminants by stimulating microbial metabolism, often targeting hydrocarbon pollutants (Udume et al. 2023). Conversely, phytoremediation employs engineered nanoparticles such as zero-valent iron or carbon nanotubes, exploiting their extended surface area and reactivity to adsorb, degrade, or immobilize contaminants, particularly heavy metals and organic pollutants (Vu et al. 2023). Phytoremediation harnesses the inherent capabilities of plants to uptake, accumulate, and detoxify pollutants (Panchenko et al. 2023). In contrast, rhizoremediation leverages the mutualistic interaction between plant roots and soil microbes, where root exudates stimulate microbial activity, facilitating the degradation of organic contaminants, notably petroleum hydrocarbons (Martin et al. 2014). Through enzymatic processes, soil microorganisms perform critical functions in bioremediation by metabolizing contaminants into less harmful forms, exemplified by microbial consortia that catalyze the removal of polycyclic aromatic hydrocarbons (PAHs) in contaminated soil (Xu et al. 2018). Vermiremediation employs earthworms to enhance soil structure and microbial activity, accelerating the treatment of organic pollutants such as petroleum hydrocarbons (Ekperusi and Aigbodion 2015). Despite their efficacy, these techniques entail ecological implications such as alterations in microbial community dynamics, potential nanoparticle toxicity, or the risk of contaminant entry into the food chain.

17.11 Conclusion and Future Prospects

In summary, bioremediation encompasses a diverse array of techniques aimed at addressing environmental pollution and restoring contaminated sites. This chapter provides a brief comparison of various remediation techniques and their effectiveness in removing contaminants. The impact of environmental pollution is so drastic that it has far-reaching effects, especially in the living world. The complexity of the composition of the pollutant poses an increasing challenge to remediation, both chemically and biologically. The various applied remediation techniques, chemical, physical, and biological, have merits and drawbacks in terms of energy requirements, viability, efficiency, and flexibility. A matter of concern that arises from these observations is that none of the individual techniques can be highly efficient in such a short duration hence an amalgamation of multiple remedial methods occurring either synchronously or consecutively can be applied. Therefore, a hybrid method that takes into account the advantages of two or more biological, chemical, or physical treatment methods is highly recommended for the removal of these complex pollutants.

Another important combination for the efficient removal of contaminants is phytoremediation with bioremediation. Bioremediation reduces the bioavailability of both inorganic and organic pollutants whereas phytoremediation can be employed to reduce the remaining contaminants. Commercialization of this hybrid technique can be achieved via genetic manipulation. Genetic engineering involves the customized production of genetically altered microorganisms and plants according to the necessity of the polluted medium and can be a very potent instrument for more systemic bioremediation.

Although gene manipulation aids immensely in the process of bioremediation, the ultimate fate of these genetically altered organisms is unknown. Studies reveal that these organisms also affect human health and the ecosystem as a whole. Again, nanoremediation involves the application of reactive nanoparticles to mitigate harmful contaminants, and coupling this method with any existing remediation technique will lead to effective sustainable remediation.

Green sustainable remediation involves the use of bionanomaterials, which is an environmentally friendly and economical approach.

The future of bioremediation is promising, with continuous advancements in technology and research driving innovations in environmental clean-up efforts. One exciting prospect relies on integrative collaboration between cutting-edge new technologies such as artificial intelligence (AI) and machine learning algorithms to optimize bioremediation processes. AI algorithms can analyze vast datasets to envisage the best possible parameters for pollutant depletion, leading to more efficient and cost-effective remediation strategies.

By harnessing the power of biology, nanotechnology, and innovative engineering approaches, bioremediation has the potential to attenuate environmental contamination while safeguarding the health of ecosystems for future generations.

17.12 Suggestions

The following suggestions are made to enhance bioremediation implementation:

I. Enhancement of Genetic Engineering Techniques: Research efforts should be directed towards refining genetic engineering methodologies for the creation of genetically modified microorganisms and plants tailored to specific pollutants and environmental conditions. This involves improving gene transfer efficiency, minimizing unintended genetic alterations, and ensuring the stability and containment of engineered organisms in the environment.

II. Advancements in nanoparticle design and characterization: Further studies are needed to advance the design, synthesis, and characterization of nanoparticles intended for nanoremediation purposes. The focus should be on developing nanoparticles with high reactivity, selectivity, and stability while simultaneously minimizing potential toxicity and unintended environmental repercussions. Additionally, innovative approaches for nanoparticle delivery, dispersion, and immobilization should be explored to increase their efficacy in remediation applications.

III. Integration of artificial intelligence (AI) and machine learning: Investigating the integration of AI and machine learning algorithms to optimize bioremediation processes. Predictive models capable of analyzing intricate datasets encompassing environmental parameters, contaminant properties, and microbial dynamics have been developed. These models can identify optimal remediation strategies and forecast long-term outcomes. Additionally, AI-driven decision support systems can be leveraged to facilitate real-time monitoring, adaptive management, and risk assessment throughout remediation endeavors.

References

Ambust S, Das AJ, Kumar R (2021) Bioremediation of petroleum contaminated soil through biosurfactant and Pseudomonas sp. SA3 amended design treatments. Curr Res Microbial Sci 2:100031

Azubuike CC, Chikere CB, Okpokwasili GC (2016) Bioremediation techniques-classification based on site of application: principles, advantages, limitations and prospects. World J Microbiol Biotechnol 32(11):180

Babalola OO, Ayangbenro AS (2017) Microorganisms as tools for bioremediation in polluted environments. In M. A. R. Chukwudi (Ed.), Microbial tools in environmental and remediation biotechnology (pp. 1–20). Elsevier, Amsterdam, Netherlands. https://doi.org/10.1016/B978-0-12-809633-8.09064-2

Bala S, Garg D, Thirumalesh BV, Sharma M, Sridhar K, Inbaraj BS, Tripathi M (2022) Recent strategies for bioremediation of emerging pollutants: a review for a green and sustainable environment. In Toxics 10(8):1–24

Benguenab A, Chibani A (2021) Biodegradation of petroleum hydrocarbons by filamentous fungi (*Aspergillusustus* and *Purpureocilliumlilacinum*) isolated from used engine oil-contaminated soil. Acta Ecol Sin 41(5):416–423

Brown LM, Gunasekera TS, Striebich RC, Ruiz ON (2016) Draft genome sequence of Gordoniasihwensis strain 9, a branched alkane-degrading bacterium. Genome Announc 4:e00622–e00616

Cebron A, Louvel B, Faure P, France-Lanord C, Chen Y, Murrell JC et al (2011) Root exudates affect the bacterial diversity but not the phenanthrene degradation rate of phenanthrene degraders in PAH-polluted soil. Appl Environ Microbiol 13(3):722736

Daâssi D, Almaghrabi FQ (2023) Petroleum-degrading fungal isolates for the treatment of soil microcosms. Microorganisms 11(5):1351

Daâssi D, Almaghribi FQ (2022) Petroleum-contaminated soil: environmental occurrence and remediation strategies. 3 Biotech 12(6):139

De Junet A, Basile-Doelsch I, Borschneck D, Masion A, Legros S, Marol C, Balesdent J, Templier J, Derenne S (2013) Characterisation of organic matter from organo-mineral complexes in an Andosol from Reunion Island. J Anal Appl Pyrolysis 99:92–100

Ekperusi OA, Aigbodion FI (2015) Bioremediation of petroleum hydrocarbons from crude oil-contaminated soil with the earthworm: *Hyperiodrilusafricanus*. 3 Biotech 5(6):957–965. https://doi.org/10.1007/s13205-015-0298-1

Eskandari S, Hoodaji M, Tahmourespour A, Abdollahi A, Mohammadian-Baghi T, Eslamian S et al (2017) Bioremediation of polycyclic aromatic hydrocarbons by *Bacillus Licheniformis* ATHE9 and *Bacillus Mojavensis* ATHE13 as new strains isolated from oil-contaminated soil. J Geograph Environ Earth Sci Int 11:1–11

Fallahi M, Sarempour M, Mirzadi Gohari A (2023) Potential biodegradation of polycyclic aromatic hydrocarbons (PAHs) and petroleum hydrocarbons by indigenous fungi recovered from crude oil-contaminated soil in Iran. Sci Rep 13:22153

Gao Y, Yang Y, Ling W, Kong H, Zhu X (2011) Gradient distribution of root exudates and polycyclic aromatic hydrocarbons in rhizosphere soil. Soil Sci Soc Am J 75(5):16941703–16941703

Hamad AA, Moubasher HA, Moustafa YM, Mohamed NH, Abd-El Rhim EH (2021) Petroleum hydrocarbon bioremediation using native fungal isolates and Consortia. Sci World J 2021:6641533–6641513

Hossain MF, Akter MA, Sohan MSR, Sultana DN, Reza MA, Hoque KMF (2022) Bioremediation potential of hydrocarbon-degrading bacteria: isolation, characterization, and assessment. Saudi J Biol Sci 29(1):211–216

Islam B (2015) Petroleum sludge, its treatment and disposal: a review. Int J Chem Sci 13(4):1584–1602

Javed F, Hashmi I (2021) Vermiremediation–remediation of soil contaminated with oil using earthworm (*Eisenia fetida*). Soil Sediment Contam Int J 30(6):639–662. https://doi.org/10.1080/15320383.2021.1885343

Klamerus-Iwan A, Błońska E, Lasota J, Kalandyk A, Waligórski P (2015) Influence of oil contamination on physical and biological properties of forest soil after chainsaw use. Water Air Soil Pollut 226:1–9

Liu R, Jadeja RN, Zhou Q, Liu Z (2012) Treatment and remediation of petroleum-contaminated soils using selective ornamental plants. Environ Eng Sci 29(6):494–501

Ma YL, Lu W, Wan LL, Luo N (2015) Elucidation of fluoranthene degradative characteristics in a newly isolated *Achromobacterxylosoxidans* DN002. Biotechnol Appl Biochem 175:1294–1305

Ma D, Xu J, Zhou J, Ren L, Li J, Zhang Z, Xia J, Xie H, Wu T (2023) Using Sweet Sorghum varieties for the phytoremediation of petroleum-contaminated salinized soil: a preliminary study based on pot experiments. Toxics 11(3):208

Macaya CC, Durán RE, Hernández L, Rodríguez-Castro L, Barra-Sanhueza B, Dorochesi F, Seeger M (2019) Bioremediation of petroleum. In: Reference module in life sciences. pp 1–6. Elsevier, Amsterdam, Netherlands. https://doi.org/10.1016/B978-0-12-809633-8.90752-2

Mansur AA, Pannirselvam M, Al-Hothaly KA, Adetutu EM, Ball AS (2015) Recovery and characterization of oil from waste crude oil tank bottom sludge from Azzawiya oil refinery in Libya. J Adv Chem Eng 5(1):1000118

Martin BC, George SJ, Price CA, Ryan MH, Tibbett M (2014) The role of root exuded low molecular weight organic anions in facilitating petroleum hydrocarbon degradation: current knowledge and future directions. Sci Total Environ 472:642653

Mnif S, Sayadi S, Chamkha M (2014) Biodegradative potential and characterization of a novel aromatic-degrading bacterium isolated from a geothermal oil field under saline and thermophilic conditions. Int Biodeterior Biodegrad Soc 86:258–264

Mukherjee S, Bardolui NK, Karim S, Patnaik VV, Nandy RK, Bag PK (2010) Isolation and characterization of a monoaromatic hydrocarbon-degrading bacterium, *Pseudomonas aeruginosa* from crude oil. J Environ Sci Health A 45:1048–1053

Ojuederie OB, Babalola OO (2017) Microbial and plant-assisted bioremediation of heavy metal polluted environments: a review. Int J Environ Res Public Health 14(12):1504

Ozdemir A, Karataş A, Palabiyik Y, Yaşar E, Sahinoglu A (2020) Oil and gas exploration in Seferihisar Uplift (Western Turkey) containing an operable-size gold deposit: geochemical evidence for the presence of a working petroleum system. Geomech Geophys Geo-Energy Geo-Resour 6(1):30

Panchenko L, Muratova A, Dubrovskaya E, Golubev S, Turkovskaya O (2023) Natural and technical phytoremediation of oil-contaminated soil. Life (Basel, Switzerland) 13(1):177

Patowary R, Devi A, Mukherjee AK (2023) Advanced bioremediation by an amalgamation of nanotechnology and modern artificial intelligence for efficient restoration of crude petroleum oil-contaminated sites: a prospective study. Environ Sci Pollut Res Int 30(30):74459–74484

Peng S, Zhou Q, Cai Z, Zhang Z (2009) Phytoremediation of petroleum contaminated soils by Mirabilis Jalapa L. in a greenhouse plot experiment. J Hazard Mater 168(2–3):1490–1496

Praveen A, Pandey VC, Marwa N, Singh DP (2019) Rhizoremediation of polluted sites: harnessing plant–microbe interactions. Elsevier, pp 389–407

Rajadurai M, Karmegam N, Kannan S, Yuvaraj A, Thangaraj R (2022) Vermiremediation of engine oil-contaminated soil employing indigenous earthworms, *Drawidamodesta* and *Lampitomauritii*. J Environ Manag 301:113849

Saeed M, Ilyas N, Jayachandran K, Gaffar S, Arshad M, Sheeraz Ahmad M, Bibi F, Jeddi K, Hessini K (2021) Biostimulation potential of biochar for remediating the crude oil contaminated soil and plant growth. Saudi J Biol Sci 28(5):2667–2676

Sánchez Mata O, Aguilera Flores MM, Ureño García BG, Ávila Vázquez V, Cabañas García E, Franco Villegas EA (2023) Bioremediation of automotive residual oil-contaminated soils by biostimulation with enzymes, surfactant, and Vermicompost. Int J Environ Res Public Health 20(16):6600

Simpanen S, Dahl M, Gerlach M, Mikkonen A, Malk V, Mikola J, Romantschuk M (2016) Biostimulation proved to be the most efficient method in the comparison of in situ soil remediation treatments after a simulated oil spill accident. Environ Sci Pollut Res Int 23(24):25024–25038

Singh Y, Saxena MK (2022) Insights into the recent advances in Nano bioremediation of pesticides from the contaminated soil. Front Microbiol 13:982611

Sugiura K, Ishihara M, Shimauchi T, Harayama S (1997) Physiochemical properties and biodegradability of crude oil. Environ Sci Technol 31:45–51

Tang JC, Wang RG, Niu XW, Wang M, Chu HR, Zhou QX (2010) Characterization of the rhizoremediation of petroleum-contaminated soil: effect of different influencing factors. Biogeosciences 7(12):39613969–39613969

Tripathi S, Singh VK, Srivastava P, Singh R, Devi RS, Kumar A, Bhadouria R (2020) Phytoremediation of organic pollutants: current status and future directions. In: Abatement of environmental pollutants. Elsevier, pp 81–105

Truskewycz A, Gundry TD, Khudur LS, Kolobaric A, Taha M, Aburto-Medina A, Ball AS, Shahsavari E (2019) Petroleum hydrocarbon contamination in terrestrial ecosystems-fate and microbial responses. Molecules (Basel, Switzerland) 24(18):3400

Udume OA, Abu GO, Stanley HO, Vincent-Akpu IF, Momoh Y, Eze MO (2023) Biostimulation of petroleum-contaminated soil using organic and inorganic amendments. Plants (Basel, Switz) 12(3):431

Vidali M (2001) Bioremediation: an overview. Pure Appl Chem 73(7):1172

Vu KA, Mulligan CN (2023) An overview on the treatment of oil pollutants in soil using synthetic and biological surfactant foam and nanoparticles. Int J Mol Sci 24(3):1916

Xu X, Liu W, Tian S, Wang W, Qi Q, Jiang P, Gao X, Li F, Li H, Yu H (2018) Petroleum hydrocarbon-degrading bacteria for the remediation of oil pollution under aerobic conditions: a perspective analysis. Front Microbiol 9:2885

Yakimov MM, Giuliano L, Gentile G, Crisafi E, Chernikova TN, Abraham WR et al (2003) *Oleispiraantarctica* gen. nov., sp. nov., a novel hydrocarbonoclastic marine bacterium isolated from Antarctic coastal seawater. Int J Syst Evol Microbiol 53:779–785

Zahermand S, Vafaeian M, Hosein Bazyar M (2020) Analysis of the physical and chemical properties of soil contaminated with oil (petroleum) hydrocarbons. Earth Sci Res J 24(2):163–168. https://doi.org/10.15446/esrj.v24n2.76217

Zhang B, Zhang L, Zhang X (2019) Bioremediation of petroleum hydrocarbon-contaminated soil by petroleum-degrading bacteria immobilized on biochar. RSC Adv 9(60):35304–35311

Zhukov DV, Murygina VP, Kalyuzhnyi SV (2007) Kinetics of the degradation of aliphatic hydrocarbons by the bacteria *Rhodococcus and Rhodococcus erythropolis*. Appl Biochem Microbiol 43:587–592

Chapter 18
Microbioremediation of Contamination of Antarctic Soil by Hydrocarbons Originating from Arctic Diesel Fuel Leaks

Jessé Valentim dos Santos, Aline Oliveira Silva, Davi Santos Tavares, Daniela de Fátima Pedroso, Raquel Milagros Rodríguez-Rodríguez, Juliano de Carvalho Cury, Flávio Henrique Silveira Rabêlo, and Marco Aurélio Carbone Carneiro

Abstract Contamination of Antarctic soil by petroleum hydrocarbons originating from Arctic Diesel Fuel (ADF) leaks is a problem requiring urgent address. Among the technologies that can be employed for decontamination, bioremediation is a viable and low-cost process compared to alternatives. This chapter introduces the potential of microorganisms for bioremediation of areas polluted by petroleum and its derived products, further a case study is presented on the biodegrading capacity of eight bacterial strains previously isolated from Antarctic soil contaminated with hydrocarbons. A microcosm experiment was carried out using sterilized Antarctic soil with and without the addition of 1% ADF (v/w). Microbial respiration was monitored every 72 hours for 30 days, and the carbon content of the microbial biomass was determined after 30 days. The accumulated evolution of C-CO_2 after ADF addition ranged between 92 and 1477 mg C-CO_2 g^{-1} soil, while in the soil without ADF, the variation remained between 90 and 480 mg C-CO_2 g^{-1} soil. The biomass carbon ranged between 12.7 and 90 μg C g^{-1} soil after ADF addition and between 1.6 and 11.7 μg C g^{-1} soil in the treatment without ADF. With emphasis on the use of *Pseudomonas* sp. and *Rhodococcus* sp., findings indicate that the tested strains

J. V. dos Santos · A. O. Silva · D. S. Tavares · D. de Fátima Pedroso · M. A. C. Carneiro (✉)
Federal University of Lavras (UFLA), Lavras, MG, Brazil
e-mail: marcocarbone@ufla.br

R. M. Rodríguez-Rodríguez
Federal University of Lavras (UFLA), Lavras, MG, Brazil

Smithsonian Tropical Research Institute, Panama, Panamá

J. de Carvalho Cury
Federal University of São João del-Rei, São João del-Rei, MG, Brazil

F. H. S. Rabêlo
Luiz de Queiroz College of Agriculture, University of São Paulo, Piracicaba, Brazil

A. Parnian et al. (eds.), *Bioremediation of Petroleum Contaminated Soil and Sediment*, Environmental Contamination Remediation and Management,
https://doi.org/10.1007/978-3-031-95524-2_18

are more efficient in consuming hydrocarbons recently added to the soil and are more viable for use in recent cases of contamination.

Keywords Bacterial · Bioremediation · Petroleum · *Pseudomonas* sp. · *Rhodococcus* sp.

18.1 Introduction

The risk of contamination of natural environments is inherent to anthropic activities and occurs by the most diverse contaminating agents, impacting and altering the quality of the environment. This has serious consequences for ecosystems, especially in very sensitive environments, such as Antarctica, where the situation requires greater care in view of its lower resilience (Beyer et al. 2000).

The degradation of several contaminants by microorganisms (biodegradation) is widely reported as one of the main processes of toxicity removal/attenuation of environmental pollutants (Struthers et al. 1998; Seeger et al. 2001). Petroleum is a complex mixture of hydrocarbons, its derivatives are the major contaminants of soil and water. Spills are due to failures in exploration, pipelines, storage tanks, and transportation (Andrade et al. 2010). Many of its components present high toxicity, mobility, and persistence, posing a great risk to human environmental health (Aislabie et al. 2004; Andrade et al. 2010; Kim et al. 2013).

This chapter addresses contamination and bioremediation strategies for soils contaminated by petroleum and its derivatives, with an emphasis on microorganisms to mitigate possible impact. To facilitate understanding, the text will be divided into the following topics: Environmental contamination by petroleum and its derived products; bioremediation strategies; mechanisms and factors influencing microbial biodegradation. Furthermore, a case study on bioremediation of Arctic Diesel Fuel (ADF) using bacteria will be presented.

18.2 Environmental Contamination by Petroleum and Its Derived Products

Petroleum is complex, predominantly composed of aliphatic, alicyclic, and aromatic hydrocarbons. It also contains additional nitrogen, sulfur, and oxygen, and potentially heavy metals (National Academies of Sciences, Engineering, and Medicine 2022; Zainith et al. 2020).

Soil pollution caused by petroleum and its derivatives (crude oil, leaded gasoline, kerosene, diesel fuel, and motor oil) remains a constant concern due to adverse effects of toxic substances present on terrestrial ecosystems (Frankenberger Jr. and Johanson 1982; Ambaye et al. 2022; Lan et al. 2023). Pollution is typically classified into three types: (i) pollution from prospecting and exploration; (ii) oil pollution during production from petrochemical industries; and (iii) pollution from facilitating and auxiliary processes (Du et al. 2011).

Substances found in petroleum and its derivatives, benzene, toluene, and xylene (BTX), destroy biota and soil quality (Tsao et al. 1998).

The main causes of soil contamination are leaks from oil storage tanks, spills during transport, or inadequate disposal of industrial waste (Dror et al. 2001; Baniasadi and Mousavi 2019; Ambaye et al. 2022). Once introduced into the soil, the low solubility of hydrocarbons results in persistance and environmental devastation (Dror et al. 2001). The impact of soil pollution caused by oil and derivatives depends on chemical and physical properties of the hydrocarbon components, affecting soil through adsorption, biodegradation, and leaching. Toxic components enter the food chain, leading to adverse effects on all organic life (Logeshwaran et al. 2018; Ambaye et al. 2022).

18.3 Bioremediation Strategies

Several strategies have been studied and/or applied to mitigate the negative effects of impacts of oil spills and their derivatives on terrestrial ecosystems (Wu et al. 2019; Mekonnen et al. 2024). Depending on the extent and complexity of pollution on soil due to oil, physical, chemical or biological interventions may be used alone or together (Macaulay and Rees 2014; Sayara and Sánchez 2020; Mekonnen et al. 2024). Bioremediation encompasses biological conversion, disintegration, or transformation of organic contaminants by bacteria and fungi, plants, and their metabolites to detoxify environmental pollutants (Huang et al. 2019). Bioremediation has received greater attention and use because it presents lower operating costs compared to physical and chemical soil detoxification processes (Bala et al. 2022; Mekonnen et al. 2024). The application of bioremediation strategies takes place either *in situ*, involving treatment of oil directly at the contaminated site, or *ex-situ*, which involves the transportation of the contaminated oil away from the site to a treatment facility (Brown et al. 2017; Guarino et al. 2017; Ambaye et al. 2022; Bala et al. 2022). The most common *ex-situ* bioremediation techniques are batteries, bioreactors, composting, and lanADFrming. Conversely, the most common *in situ* techniques are bioremediation, natural attenuation, bioventing, phytoremediation, biostimulation, and bioaugmentation (Chaudhary and Kim 2019; Mekonnen et al. 2024).

18.4 Mechanism and Influential Factors for Microbial Biodegradation

Soil microorganisms play key roles in different techniques involving bioremediation strategies for soils impacted by oil and its derivatives. Bacteria, algae, filamentous fungi and yeast actively participate in soil detoxification, to degrade complex harmful chemical compounds into simpler harmless compounds (Macaulay and Rees 2014; Bala et al. 2022). However, these same organisms can have their metabolisms stimulated or inhibited according to the physical, chemical, and biological conditions of the environment, which consequently affect the results of bioremediation (Bala et al. 2022). Spatial distribution of microorganisms allows greater or lesser availability of nutrients and energy for their metabolism and growth. The main factors that interfere with microbial growth and consequently with bioremediation mechanisms (Bala et al. 2022) are summarized below:

- **Biological factors—**Soil microorganisms compete for carbon sources, affecting the degradation of organic compounds. Enzymatic activities accelerate or delay the degradation of contaminants. The enzymes must have affinity and be available enough to act on the contaminant. Ecological interactions (competition, predation, and succession), population size, and composition are also determining factors for bioremediation success.
- **Oxygen—**Increase in oxygen saturation stimulates microbial metabolism and consequently hydrocarbon degradation.
- **Moisture content—**Adequate moisture levels increase metabolism and microbial growth. Excess moisture compromises biodegradation of target compounds.
- **Nutrients—**Balanced nutrient rates enhance growth and reproduction of microorganisms, enhancing biodegradation. Increasing nutrient availability improves the C:N:P ratio favoring microbial metabolism. At low nutrient concentrations, degradation of hydrocarbons is limited.
- **Temperature—**Temperature physically influences metabolism, microbial survival, and hydrocarbon composition. Degradation is slowed at lower temperatures, accelerated at higher temperatures. Temperature affects enzymes directly involved in degradation. Furthermore, degradation of compounds is temperature specific. Microbial physiological properties accelerate or delay bioremediation.
- **pH—**Microbial growth is influenced by soil acidity. Small variations in pH have significant impact on metabolic processes. A compound's acidity or alkalinity condition affects microbial metabolism and influences removal or degradation rates.
- **Site characterization—**Adoption of bioremediation strategy depends on having knowledge of the site to be treated. Adequately characterizing the extent of contamination in terms of location and the horizontal and vertical extent of contamination are essential.
- **Metal ions—**Although several metals are essential for bacteria and fungi, excessive amounts inhibit cellular metabolism. The presence of metals directly interferes with compound breakdown.
- **Species of microorganisms—**Concentration rates of toxic compounds harm certain species of microorganisms in detriment of others. Growth rates and metabolism vary for different microorganisms. Toxicity varies depending on the compound, concentration, and microorganisms exposed.

18.5 A Case Study: Bioremediation of Arctic Diesel Fuel

18.5.1 Introduction

At the Brazilian Antarctic Scientific Station "Comandante Ferraz" (EACF—Estação Antártica Comandante Ferraz), the source of energy used in installations and land transport is a type of diesel oil, Arctic Diesel Fuel, derived from petroleum. Several studies have reported accidental spillage of ADF in Antarctic soils, representing a major environmental problem (Whyte et al. 2002; Martins et al. 2004; Bargagli 2008).

Studies conducted by Cury et al. (2015) and Vázquez et al. (2017) demonstrated the effect of hydrocarbon contamination on the structure and composition of microbial communities in Antarctic soils and documented the potential biodegradability of their communities. The use of these microorganisms in bioremediation activities would contribute directly to the principles of the Antarctic Treaty by developing a rapid and effective capacity to respond to accidents, especially those that affect the environment (https://www.ats.aq, accessed 02 25).

To evaluate the efficiency of biodegradation of native microorganisms capable of metabolizing hydrocarbons, preliminary tests are required. These studies are the previous stage of selection of bacterial consortia to be tested and/or used in bioremediation processes. In this context, our objective with this case study was to evaluate the potential ADF biodegradation by bacteria isolated from Antarctic-contaminated soils through their metabolic activity. For this, an experiment carried out at the Federal University of Lavras evaluated the ability of different strains of bacteria isolated from contaminated soil in Antarctica to degrade ADF.

18.5.2 Material and Methods

Eight bacterial strains were isolated from ADF-contaminated soil from the EACF adjacency (62 ° 05′ 07.62″ S and 58° 23′ 28.84″ W), in King George Island, at South Shetland Islands archipelago. The strains were obtained from the inoculation of contaminated soil suspension in 0.9% saline solution in Bushnell Haas solid medium (Bushnell and Haas 1941) containing sterilized-by-filtration ADF as the only carbon source. Previously sterilized contaminated Antarctic soil from the adjacency of EACF was used as a substrate in the microcosm experiment. The treatments were: addition of 1% ADF (v:w) and presence of strain (+ADF); no addition of ADF and presence of strain (−ADF); their respective microbiological controls without bacterial inoculation were made with and without ADF.

Fresh suspensions of bacteria in sterile saline solution (0.85%) were added to the soil at the estimated concentration of 1.5 × 106 cfu g^{-1} soil using the McFarland nephelometric scale 0.5 standard (McFarland 1907; Jorgensen et al. 2015). The treatments were applied in 100 g of soil conditioned in 4 L containers with hermetic closure, together with a vial containing 50 mL of water to maintain the humidity of the environment.

The ADF mineralization was evaluated by quantifying the C-CO_2 released from microbial respiration every 72 h (Alef 1995) for 30 days and the amount of C of the microbial biomass at the end of the experiment. At the end of the experiment, the C in the microbial biomass of the conditioned soil was also evaluated by the fumigation-extraction method (Vance et al. 1987).

The extraction of the DNA of the eight isolates and partial amplification of 16S rRNA were performed according to Guimarães et al. (2012). Subsequently, the DNAs were subjected to 16S rRNA sequencing. The electropherogram sequencing files were processed using the Phred program (Ewing and Green 1988) for base calling and trimming low-quality (<20) sequences. The high-quality sequences were used for analyses. The sequences were deposited in GenBank under the accession numbers: MN559053 (S01), MN559054 (S02), MN559055 (S03), MN559056 (S05), MN559057 (S07) and MN559058 (S08).

The phylogram of the sequences and some of their most similar sequences obtained using Blast-NCBI algorithm (Altschu et al. 1990) was inferred using MEGA7 software (Kumar et al. 2016) with the following parameters: Neighbor-Joining method (Saitou and Nei 1987); bootstrap test (Felsenstein 1985) with 1000 replicates; distances computed using the Jukes-Cantor method (Jukes and Cantor 1969). The MALDI-TOF identification of the strains was performed according to Damaceno et al. (2017), in the Maldi Biotyper equipment (Bruker Daltonics, Germany/USA).

18.5.3 Results and Discussions

The total C-CO_2 evolved during the 30 days of the experiment differed among strains in the treatments with and without the addition of ADF (Fig. 18.1a). In the +ADF treatment, the microbial respiration released between 92 and 1477 mg of C-CO_2 kg^{-1} of soil, standing out the strains S01, S02, and S04. In the –ADF treatment, the C-CO_2 released varied between 89.8 and 479.8 mg kg^{-1} of soil.

By analyzing microbial respiration individually with and without ADF addition (Fig. 18.2a), we observed a positive and statistically significant response to the addition of oil for strains S01, S02 and S04, with increases in C-CO_2 evolution of 16.44, 4.68 and 1.99 times in relation to the treatment –ADF, respectively. In turn, strains S05 and S06 evolved less C-CO_2 when ADF was added, while the strains S03, S07 and S08 did not show significant differences between treatments (Figs. 18.1 and 18.2).

The respiratory rates and cumulative evolution of C-CO_2 during the 30 days of incubation were different between strains and treatments. The activity of strains S05, S06 and S07 was evidently depressed by the addition of ADF during the initial tests, showing a slight recovery from the third evaluation, but still negligible. During

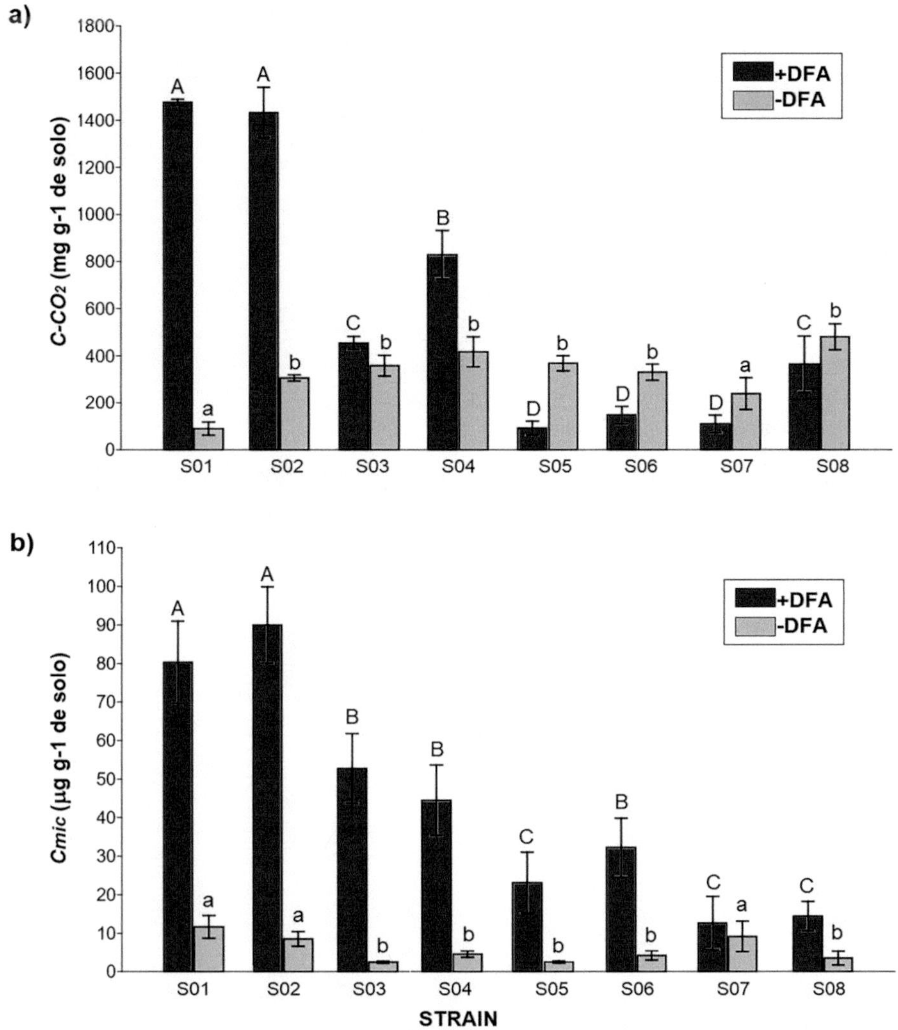

Fig. 18.1 Total evolved C-CO2 (**a**) and Cmic (**b**) of the bacterial strains evaluated in the treatments with and without addition of ADF. Capital letters compare +ADF treatments and lower-case letters compare –ADF treatments. Means with different letters indicate statistical difference by the Scott-Knott test for $p < 0.05$

the experiment, different respiration peaks were detected depending on the strain and treatment (Fig. 18.2a, b).

The cumulative evolution curves (Fig. 18.3c, d) showed that in absence of ADF most strains practically finished the C-CO_2 evolution after 9–12 days of incubation (Fig. 18.3d), probably due to the extinction of the amount of labile C source of the soil. In turn, when ADF was added, the time elapsed until stabilization was greater, with differences between strains (Fig. 18.3c). It should be noted that strain S01 was

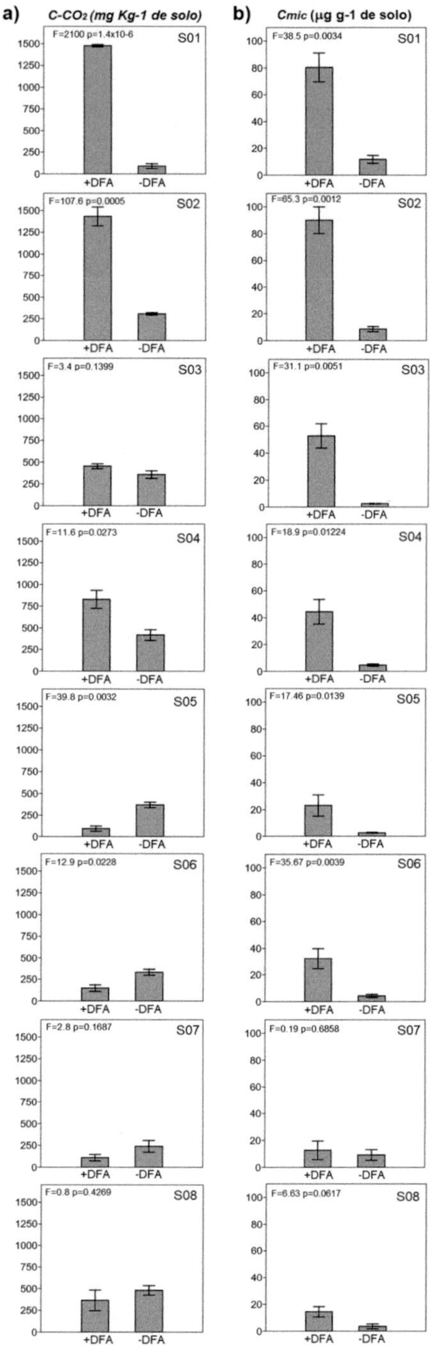

Fig. 18.2 Total evolved C-CO2 (**a**) and Cmic (**b**) of the bacterial strains evaluated with and without addition of ADF

still in the linear growth phase of the C-CO_2 evolution until the end of the experiment (Fig. 18.3c).

Microbial biomass carbon varied between strains in treatments with and without addition of ADF. In the +ADF treatment, the Cmic varied between 5 and 102.6 μg g^{-1} of soil, whereas in the –ADF it varied between 0.9 and 16 μg g^{-1} of soil. Except for strains S07 and S08, the addition of ADF significantly stimulated microbial growth at rates that ranged from 6.9 to 21.6-fold when compared to treatment without ADF (Fig. 18.1b).

Phylogram with the partial 16S rRNA sequences of S01, S02, S03, S05, S07, and S08 strains and some of their most similar sequences obtained using Blastn-NCBI algorithm is presented in Fig. 18.4. There was no success in rRNA amplification 16S of S04 and S06.

Strains S01 and S08 showed high similarities with *Pseudomonas* sp., whereas strains S02, S03, S05, and S07 showed high similarity with *Rhodococcus* sp., *Stenotrophomonas* sp., *Lysinibacillus* sp., and *Microbacterium* sp., respectively (Fig. 18.4). MALDI-TOF confirmed the molecular characterization. Strains S01 and S08 were classified as *Pseudomonas veronii* (score 2.295) and *Pseudomonas putida* (score 2.151), respectively, while strains S02, S03, and S07 were classified as *Rhodococcus erythropolis* (score 2.176), *Stenotrophomonas rhizophila* (score 1.769) and *Microbacterium liquefaciens* (score 2.154), respectively. Strain S06 was classified as *Debaryomyces hansenii* (score 2.058).

The present study evaluated the activity and the potential for biodegradation of hydrocarbons of ADF by eight bacteria isolated from contaminated Antarctic soil allowing the selection of promising strains for future local bioremediation processes.

Considering that any increase in the evolved C-CO_2 and Cmic of the +ADF treatments in relation to the –ADF treatment are due to the use of the added ADF, there

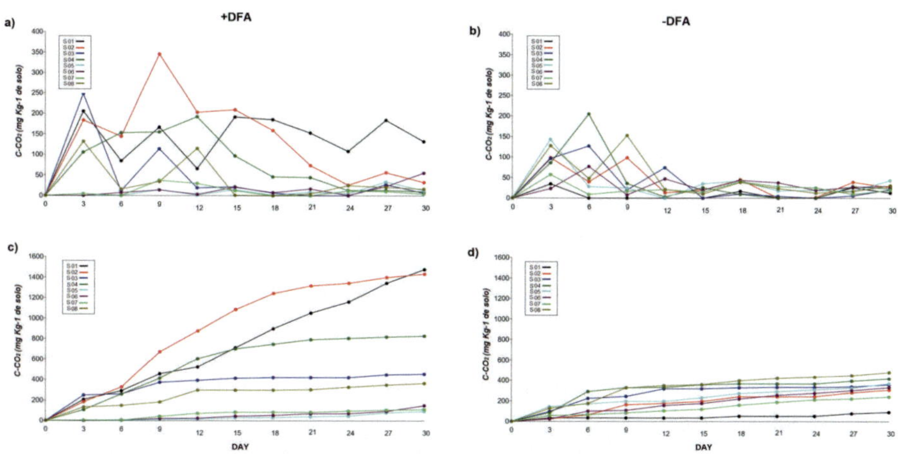

Fig. 18.3 Respiratory rate (**a**, **b**) and accumulated evolution of C-CO2 (**c**, **d**) for the eight strains tested during the 30 days of incubation in treatments with and without addition of ADF (+/– ADF), with interval of analysis of 72 h

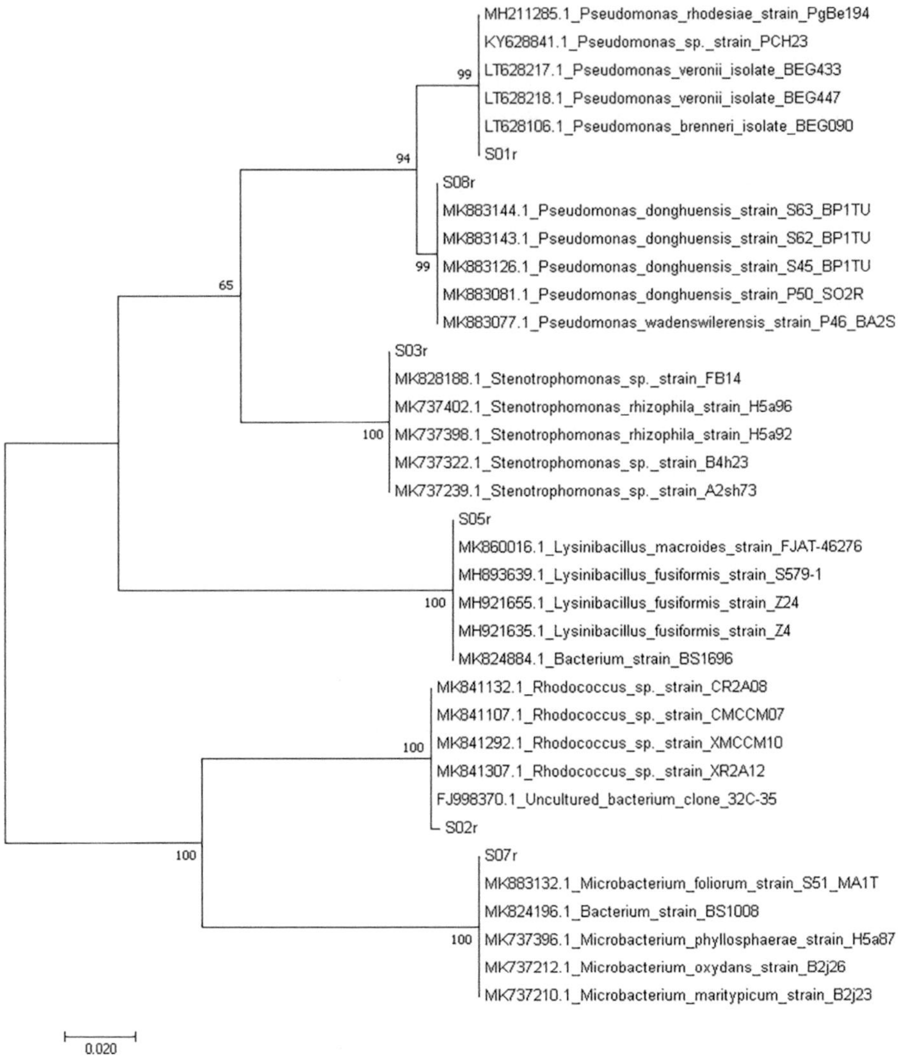

Fig. 18.4 Phylogram of the partial 16S rRNA sequences of S01, S02, S03, S05, S07 and S08 strains and some of their most similar sequences obtained using Blast-NCBI algorithm and inferred using MEGA7 software with the following parameters: Neighbor-Joining method; bootstrap test with 1000 replicates; distances computed using the Jukes-Cantor method

was positive response in the degradation of the oil of most of the strains tested, being the strains S01, S02, and S04 those that showed considerable increases for both variables.

The ability of native microbial communities to adapt to the presence of petroleum hydrocarbons in the environment has been widely reported (Hamamura et al. 2006; Zhu et al. 2015; Ghosal et al. 2016; Liu et al. 2017), which is an important selection factor (Hamamura et al. 2006; Jesus et al. 2015). The use of this capacity to enhance remediation activities against fossil fuel spills is of great interest. Even though the strains S01, S02, and S04 have shown greater activity in the presence of ADF compared to the other strains, it is known that one microorganism does not necessarily present all the genes for the complete mineralization of all the hydrocarbons of such a pollutant as diesel oil (Ghosal et al. 2016).

Therefore, the development of microbial consortia is a good alternative to maximize the biodegradation of hydrocarbons (Richard and Vogel 1999; Ruberto et al. 2005; Deppe et al. 2005; Ghosal et al. 2016). In this sense, strains S01, S02, and S04 showed to be good candidates for consortium for later tests of biodegradation of ADF in Antarctic soils. In addition to future compatibility tests between the strains and the biodegradability of the consortium, evaluations of the degradation products need to be performed.

Another important aspect in the biodegradation of hydrocarbons is related to the imbalance of nutrients, especially of P and N (Ruberto et al. 2005; Wu et al. 2016), being necessary to guarantee higher rates of contaminant degradation. In the present study the soil used contained high levels of P, the addition of N was required. Other experimental studies with strains of Antarctic soils contaminated with hydrocarbons have shown that the combined use of bioaugmentation and biostimulation techniques shorten the response time of native communities in the biodegradation of hydrocarbons, especially in soils without a history of contamination (Ruberto et al. 2005).

The strains identified as more promising in this work belong to the genera *Pseudomonas* and *Rhodococcus*, whose hydrocarbon biodegradation capacities have been widely reported (Stallwood et al. 2005; Calvo et al. 2008; Arulazhagan and Vasudevan 2011; Liu et al. 2017; Kong et al. 2018; Wu et al. 2018). Previous studies in Antarctica also reported the predominance of *Pseudomonas* and *Rhodococcus* in soils contaminated with hydrocarbons (Whyte et al. 2002; Aislabie et al. 2004; Ruberto et al. 2005; Ma et al. 2006).

18.5.4 Conclusion

Bacterial isolates have different capacities to degrade ADF, measured from the evolution of CO_2. The tested proved that bacterial are more efficient in consuming hydrocarbons recently added to the soil and are more viable for use in recent cases of contamination. Isolates of *Pseudomonas* sp. and *Rhodococcus* sp. were those with the greatest capacity to degrade DAF. Therefore, these strains are promising for the possible use of microbioremediation of soil impacted by hydrocarbons in Antarctica.

18.6 Conclusions and Perspectives

Population growth generates great pressure on various mineral resources. Despite global discussions about changes in the main energy matrices in the most diverse countries around the world, the use of oil and its derivatives is likely to last for a few more decades. Attention to the impacts resulting from its extraction, transportation, processing, and uses is a constant global challenge. Considering the imminent risks of soil contamination by oil and its derivatives, in this chapter we present approaches related to bioremediation, highlighting the positive roles that soil microorganisms can play in this process. Furthermore, microorganisms isolated from places that have already suffered some impact (bioprospection) can be a path to the production of inoculants with bioremediation potential in soils impacted by oil product spills.

18.7 Suggestions

To assist microbioremediation of contaminated soils the following suggestions are proferred:

I. Microbial prospecting studies in different environments will present important discoveries of key microorganisms for the degradation of various petroleum-derived compounds.
II. A critical look at the mechanisms played by microorganisms in degradation processes of oil and its derivatives will reveal a range of applications for the bioremediation of areas impacted by these compounds and also for their use in biochemistry for the production of new substances.
III. Currently, several environments are suffering negative impacts resulting from contamination. The unification of knowledge and research areas is essential for partial or total resolution of these impacts.

Acknowledgments We thank the Coordination for the Improvement of Higher Education Personnel (CAPES), the National Council for Scientific and Technological Development (CNPq), and the State of Minas Gerais Research Support Foundation (FAPEMIG) for the financial support and scholarships granted to the authors.

References

Aislabie JM, Balks MR, Foght JM, Waterhouse EJ (2004) Hydrocarbon spills on Antarctic soils: effects and management. Environ Sci Technol 38:1265–1274

Alef K (1995) Soil respiration. In: Alef K, Nannipieri P (eds) Methods in applied soil microbiology and biochemistry. Academic Press, Harcourt Brace & Company Publishers, London, UK, pp 214–219

Altschul SF, Gish W, Miller W, Myers EW, Lipman DJ (1990) Basic local alignment search tool. J Mol Biol 215(3):403–410

Ambaye TG, Chebbi A, Formicola F, Prasad S, Gomez FH, Franzetti A, Vaccari M (2022) Remediation of soil polluted with petroleum hydrocarbons, and their reuse for agriculture: recent progress, challenges, and perspectives. Chemosphere 293:133572

Andrade JA, Augusto F, Jardim ICSF (2010) Biorremediação de solos contaminados por petróleo e seus derivados. Eclet Quim 35:17–43. (In Portuguese)

Arulazhagan P, Vasudevan N (2011) Biodegradation of polycyclic aromatic hydrocarbons by a halotolerant bacterial strain Ochrobactrum sp. VA1. Mar Pollut Bull 62:388–394

Bala S, Garg D, Thirumalesh BV, Sharma M, Sridhar K, Inbaraj BS et al (2022) Recent strategies for bioremediation of emerging pollutants: a review for a green and sustainable environment. Toxics 19;10(8):484

Baniasadi M, Mousavi SM (2019) A comprehensive review on the bioremediation of oil spills. In: Kumar V, Kumar M, Prasad R (eds) Microbial action on hydrocarbons. Springer Nature. Advance online publication, pp 223–254. https://doi.org/10.1007/978-981-13-1840-5_10

Bargagli R (2008) Environmental contamination in Antarctic ecosystems. Sci Total Environ 400:212–226

Beyer L, Bölter M, Seppelt RD (2000) Nutrient and thermal regime, microbial biomass, and vegetation of Antarctic soils in the windmill islands region of East Antarctica (Wilkes Land). Arct Antarct Alp Res 32:30–39

Brown DM, Okoro S, van Gils J, van Spanning R, Bonte M, Hutchings T, Linden O, Egbuche U, Bruun KB, Smith JW (2017) Comparison of land farming amendments to improve bioremediation of petroleum hydrocarbons in Niger Delta soils. Sci Total Environ 596:284–292

Bushnell LD, Haas HF (1941) The utilization of certain hydrocarbons by microorganisms. J Bacteriol 41:653–673

Calvo C, Silva-Castro GA, Uad I, García Fandiño C, Laguna J, González-López J (2008) Efficiency of the EPS emulsifier produced by Ochrobactrum anthropi in different hydrocarbon bioremediation assays. J Ind Microbiol Biot 35:1493–1501

Chaudhary DK, Kim J (2019) New insights into bioremediation strategies for oil-contaminated soil in cold environments. Int Biodeterior Biodegr 142:58–72

Cury JC, Jurelevicius DA, Villela HDM, Jesus HE, Peixoto RS, Schaefer CEGR, Bícego MC, Seldin L, Rosado AS (2015) Microbial diversity and hydrocarbon depletion in low and high diesel-polluted soil samples from Keller Peninsula, South Shetland Islands. Antarct Sci 27:263–273

Damaceno QS, Souza JP, Nicoli JR, Paula RL, Assis GB, Figueiredo HC, Azevedo V, Martins FS (2017) Evaluation of potential probiotics isolated from human milk and colostrum. Probiotics Antimicrob Proteins 9(4):371–379

Deppe U, Richnow HH, Michaelis W, Antranikian G (2005) Degradation of crude oil by an arctic microbial consortium. Extremophiles 9:461–470

Dror I, Gerstl Z, Yaron B (2001) Temporal changes in kerosene content and composition in field soil as a result of leaching. J Contam Hydrol 48:305–323

Du WD, Wan YY, Zhong NN, Fei JJ, Zhang ZH, Chen LJ, Hao JM (2011) Status quo of soil petroleum contamination and evolution of bioremediation. Petrol Sci 8:502–514

Ewing B, Green P (1988) Base-calling of automated sequencer traces using Phred. II error probabilities. Genome Res 8:186–194

Felsenstein J (1985) Confidence limits on phylogenies: an approach using the bootstrap. Evolution 39:783–791

Frankenberger WT Jr, Johanson JB (1982) Influence of crude oil and refined petroleum products on soil dehydrogenase activity. J Environ Qual 11:602–607

Ghosal D, Ghosh S, Dutta TK, Ahn Y (2016) Current state of knowledge in microbial degradation of Polycyclic Aromatic Hydrocarbons (PAHs): a review. Front Microbiol 7:1369

Guarino C, Spada V, Sciarrillo R (2017) Assessment of three approaches of bioremediation (Natural Attenuation, Land farming and Bioagumentation–Assistited Land farming) for a petroleum hydrocarbons contaminated soil. Chemosphere 170:10–16

Guimarães AA, Jaramillo PMD, Nóbrega RSA, Florentino LA, Silva KB, Moreira FMS (2012) Genetic and symbiotic diversity of nitrogen-fixing bacteria isolated from agricultural soils in the Western Amazon by using cowpea as the trap plant. Appl Environ Microbiol 78:6726–6733

Hamamura N, Olson SH, Ward DM, Inskeep WP (2006) Microbial population dynamics associated with crude-oil biodegradation in diverse soils. Appl Environ Microbiol 72:6316–6324

Huang Y, Song Y, Johnson D, Huang J, Dong R, Liu H (2019) Selenium enhanced phytoremediation of diesel contaminated soil by *Alternanthera philoxeroides*. Ecotoxicol Environ Saf 173:347–352. https://doi.org/10.1016/j.ecoenv.2019.02.040

Jesus HE, Peixoto RS, Cury JC, Van Elsas JD, Rosado AS (2015) Evaluation of soil bioremediation techniques in an aged diesel spill at the Antarctic peninsula. Appl Microbiol Biol 99:10815–10827

Jorgensen JH, Pfaller MA, Carroll KC, Funke G, Landry ML, Richter SS, Warnock DW (eds) (2015) Manual of clinical microbiology 1, 11th edn. ASM Press

Jukes TH, Cantor CR (1969) Evolution of protein molecules. In: Munro NH (ed) Mammalian Protein Metabolism. Academic, New York, pp 21–132

Kim KH, Kima KH, Jahan SA, Kabir E, Brown RJC (2013) A review of airborne polycyclic aromatic hydrocarbons (PAHs) and their human health effects. Environ Int 60:71–80

Kong FX, Sun GD, Liu ZP (2018) Degradation of polycyclic aromatic hydrocarbons in soil mesocosms by microbial/plant bioaugmentation: performance and mechanism. Chemosphere 198:83–91

Kumar S, Stecher G, Tamuda K (2016) MEGA7: molecular evolutionary genetics analysis version 7.0 for bigger datasets. Mol Biol Evol 33:1870–1874

Lan J, Wen F, Ren Y, Liu G, Jiang Y, Wang Z, Zhu X (2023) An overview of bioelectrokinetic and bioelectrochemical remediation of petroleum-contaminated soils. Environ Sci Ecotechnol 16:100278

Liu Q, Tang J, Gao K, Gurav R, Giesy JP (2017) Aerobic degradation of crude oil by microorganisms in soils from four geographic regions of. China Sci Rep 7:14856

Logeshwaran P, Megharaj M, Chadalavada S, Bowman M, Naidu R (2018) Petroleum hydrocarbons (PH) in groundwater aquifers: an overview of environmental fate, toxicity, microbial degradation, and risk-based remediation approaches. Environ Technol Innov 10:175–193

Ma Y, Wang L, Shao Z (2006) Pseudomonas, the dominant polycyclic aromatic hydrocarbon-degrading bactéria isolated from Antartic soils and the role of large plasmids in horizontal gene transfer. Environ Microbiol 8:455–465

Macaulay B, Rees D (2014) Bioremediation of oil spills: a review of challenges for research advancement. Ann Environ Sci 8:9–37

Martins CC, Bícego MC, Taniguchi S, Montone RC (2004) Aliphatic and polycyclic aromatic hydrocarbons in surface sediments in Admiralty Bay, King George Island. Antarctica Antarctic Sci 16:122

Mcfarland J (1907) The Nephelometer: an instrument for estimating the number of bacteria in suspensions used for calculating the opsonic index and for vaccines. JAMA J Am Med Assoc 14:1176–1178

Mekonnen BA, Aragaw TA, Genet MB (2024) Bioremediation of petroleum hydrocarbon contaminated soil: a review on principles, degradation mechanisms, and advancements. Front Environ Sci 12:1354422

National Academies Of Sciences, Engineering, and Medicine (2022) Oil in the Sea IV: inputs, fates, and effects. The National Academies Press, Washington, DC. https://doi.org/10.17226/26410

Richard JY, Vogel TM (1999) Characterization of a soil bacterial consortium capable of degrading diesel fuel. Int Biodeterior Biodegradation 44:93–100

Ruberto LAM, Vazquez S, Lobalbo A, Cormack WPM (2005) Psychrotolerant hydrocarbon-degrading *Rhodococcus* strains isolated from polluted Antarctic soils. Antarct Sci 17:47–56

Saitou N, Nei M (1987) The neighbor-joining method: a new method for reconstructing phylogenetic trees. Mol Biol Evol 4:406–425

Sayara T, Sánchez A (2020) Bioremediation of PAH-contaminated soils: process enhancement through composting/compost. Appl Sci 10:3684

Seeger M, Cámara B, Hofer B (2001) Dehalogenation, denitration, dehydroxylation, and angular attack on substituted biphenyls and related compounds by a biphenyl dioxygenase. J Bacteriol 183:3548–3555

Stallwood B, Shears J, Williams PA, Hughes KA (2005) Low temperature bioremediation of oil-contaminated soil using biostimulation and bioaugmentation with a *Pseudomonas* sp. from maritime Antarctica. J Appl Microbiol 99:794–802

Struthers JK, Jayachandran K, Moorman TB (1998) Biodegradation of atrazine by *Agrobacterium radiobacter* J14a and use of this strain in bioremediation of contaminated soil. Appl Environ Microbiol 64:3368–3375

Tsao CW, Song HG, Bartha R (1998) Metabolism of benzene, toluene, and xylene hydrocarbons in soil. Appl Environ Microbiol 64:4924–4929

Vance ED, Brookes PC, Jenkinson DS (1987) An extraction method for measuring soil microbial biomass C. Soil Biol Biochem 19(6):703–707

Vázquez S, Monien P, Minetti RP, Jürgens J, Curtosi A, Primitz JV, Frickenhaus S, Abele D, Cormack WM, Helmke E (2017) Bacterial communities and chemical parameters in soils and coastal sediments in response to diesel spills at Carlini Station, Antarctica. Sci Total Environ 605-606:26–37

Whyte LG, Schultz A, van Beilen JB, Luz AP, Pellizari V, Labbe D, Greer CW (2002) Prevalence of alkane monooxygenase genes in Arctic and Antarctic hydrocarbon-contaminated and pristine soils. FEMS Microbiol Ecol 41:141–150

Wu M, Dick W, Li W, Wang X, Yang Q, Wang T, Xu L, Zhang M, Chen L (2016) Bioaugmentation and biostimulation of hydrocarbon degradation and the microbial community in a petroleum-contaminated soil. Int Biodeterior Biodegradation 107:158–164

Wu T, Xu J, Xie W, Yao Z, Yang H, Sun C, Li X (2018) Pseudomonas aeruginosa L10: a hydrocarbon-degrading, biosurfactant-producing, and plant-growth-promoting Endophytic bacterium isolated from a Reed (Phragmites australis). Front Microbiol 9:1087

Wu M, Wu J, Zhang X, Ye X (2019) Effect of bioaugmentation and biostimulation on hydrocarbon degradation and microbial community composition in petroleum-contaminated loessal soil. Chemosphere 237:124456

Zainith S, Chowdhary P, Mani S, Mishra S (2020) Microbial ligninolytic enzymes and their role in bioremediation. In: Microorganisms for sustainable environment and health. Elsevier, Amsterdam p 179

Zhu X, Yu P, Han N, Jia H, Wang X, Kodirov KG (2015) Monitoring soil respiration dynamics during the biodegradation of crude oil contaminated soil under different environmental conditions. Fresen Environ Bul 24:4087–4094

Chapter 19
Nurturing Nature's Solutions: The Lead of Biofertilizers in Sustainable Agriculture and Soil Remediation

Mudassir Khan, Affan Ahmed, Aleeza Munawar, Bakhtawar Fayyaz, Kaynat Iftekhar, Ghulam Fatima, Muhammad Abu Bakar Saddique, Aftab Ahmed, Ghulam Kubra, and James N. Furze

Abstract Global soil degradation is becoming more prevalent and is a rising concern especially in developing countries. Primary causes of deterioration include farming methods of increasing intensity, inadequate restoration of key nutrients, deforestation, and removal of vegetation from sandy soils. Wind and water erosion are the primary causes of soil loss in terrestrial ecosystems, Inevitably erosion results in the loss of soil nutrients. Poor soil fertility decreases agricultural productivity. The term "bioremediation" refers to the process of removing contaminants from ecosystems utilizing microorganisms. Thus microbial degradation of organic and inorganic impurities improves and boosts the health of the environment. Microbial breakdown aerates soil and adds nutrients encouraging natural biodegradation of hydrocarbons of petroleum by local microbes. Bioremediation techniques are superior to solvent extraction and chemical oxidization. Non-biological treatment costs are exceedingly high in both financial and ecological terms, bioremediation is a viable option for soil remediation. This chapter focuses on the significance of biofertilizers for healthy life and a sustainable environment, as well as how microbial strains improve bioremediation, an intense necessity in modern times and a viable solution in reducing climate change. Microorganism-based biotechnology has been used to solve environmental contamination issues, particularly of petroleum and its byproducts, in recent decades. Fertilizer usage must be expanded to meet the demands of the world's growing population; though should be carefully deployed to avoid environmental concerns, and reliance on the deployment of technological and socioeconomic measures. The ultimate solution is a biofertilizer which remediates pollutants and improves growth and yield of crops.

Keywords Bioremediation · Contamination · Microbial · Fertility · Yield

---

M. Khan · B. Fayyaz · K. Iftekhar · G. Kubra (✉)
Atta Ur Rahman School of Applied Biosciences (ASAB), National University of Sciences and Technology (NUST), Islamabad, Pakistan

A. Parnian et al. (eds.), *Bioremediation of Petroleum Contaminated Soil and Sediment*, Environmental Contamination Remediation and Management,
https://doi.org/10.1007/978-3-031-95524-2_19

19.1 Introduction

The term "bioremediation" refers to a collection of processes that employ biological systems to either clean up or repair polluted areas. Bioremediation usually involves the microbial community playing its role in cleaning the environment. The majority of naturally occurring bacteria are capable of effectively restoring the environment by different approaches like oxidizing, immobilizing, or changing the pollutants (Crawford and Crawford 1996).

The main purpose of bioremediation is to decrease or bring down pollutant levels to levels that are undetectable, harmless, or acceptable (that is, within the limits developed by regulatory agencies). The first time, the concept of bioremediation on a large scale was in Ambler, Pennsylvania in 1972, to clean the Sun Oil pipeline (Kumar et al. 2015). However, George M. Robinson was the first to identify the mechanism while conducting experiments on sewage and oil treatments at a laboratory scale (Sonawdekar 2012). In 1992, based on several cases of bioremediation, the Environmental Protection Agency (EPA) created different bioremediation procedures. The hunt for a new microorganism strain from polluted places is a contemporary bioremediation technique. It is believed that such an isolated microorganism strain has a great potential for pollution remediation. The bioactivity of a bioremediant has been raised either directly or indirectly through the application of microbial consortiums and genetically engineered strains. Involving the elimination of pollutants, a variety of processes have also been studied, comprising bioaccumulation,

A. Ahmed
Environmental and Evolutionary Biology, Department of Biology, University of Louisiana at Lafayette, Lafayette, LA, USA

A. Munawar
Department of Horticulture, Pir Mehr Ali Shah Arid Agriculture University, Rawalpindi, Pakistan

G. Fatima
Institute of Biochemistry and Biotechnology, University of Veterinary & Animal Sciences, Lahore, Pakistan

M. A. B. Saddique
Institute of Urban Agriculture, Chinese Academy of Agricultural Sciences, Chengdu National Agricultural Science and Technology Center, Chengdu, China

A. Ahmed
School of Biological Sciences, University of the Punjab, Lahore, Pakistan

J. N. Furze
Royal Geographical Society (with the Institute of British Geographers), London, UK

Laboratory of Biotechnology and Valorization of Natural Resources, Faculty of Sciences-Agadir, Department of Biology, Ibn Zohr University, Agadir, Morocco

Control and Systems Engineering Department, University of Technology-Iraq, Baghdad, Iraq

Earth Systems Protection Foundation, Somerset, UK

biodegradation pathways, and several modalities for biosorption (Hlihor et al. 2017; Ramírez-García et al. 2019).

As bioremediation is a process involving microorganisms, various techniques and procedures were developed and extensively utilized to minimize pollution. The goal of every technique is to promote the bacterium to produce its maximum level of microbiological enzyme production. The long-evolving enzymatic metabolic pathways are employed in the usage of microbial populations. Therefore, process parameter optimization and control are fundamental to the systematic application of communities of bacteria.

In some circumstances, microorganisms use pollutants as substrates to break down the pollutants and get energy for growth and reproduction. The bonds of chemicals break down to produce free electrons, which are then transported to an electron acceptor like oxygen to develop energy (Ramírez-García et al. 2019).

No hazardous chemicals are used in the process of bioremediation. The process of bioremediation destroys hazardous compounds by converting them into water and innocuous gases (Sharma 2012). Due to their inherent place in the ecosystem, microbes are easy to use, require low labor, and are inexpensive (Dell'Anno et al. 2012). Bioremediation is ecologically benign and sustainable. It has also an advantage in terms of pollutants being eliminated rather than being moved to alternative environmental media.

Plants require essential nutrients for optimal growth and production. The nutrient supply is met by 17 essential elements through fertilizers if the soil is lacking them (Jaswal et al. 2022). Nitrogen, potassium, and phosphorus are necessary for plants to maintain the normal physiology of cells. In the same manner, nitrogen deficiency results in poor growth of plants (Inze and Montagu 1995), however, too much nitrogen causes maturity to occur late and low quality of leaves (Ewulo et al. 2008). Nutrients that are required by plants have specific roles in the physiological functions of plants (Table 19.1) (Jaswal et al. 2022; Uchida 2000). Fertilizers are like food supplements for plants, which are given to improve the fertility of soil as low soil fertility is one of the most significant burdens on the yield (Ayoub 1999). It is important to improve the physical properties such as porosity, friability, and absorption of soil, as well as biological properties that are associated with beneficial microorganisms. Moreover, pH and nutrient levels demonstrate the chemical properties, which are often replenished by applying several organic and inorganic materials. Intensive use of inorganic fertilizers to stimulate plant growth causes ample health issues and environmental pollution. Recently, new practices have been developed in agriculture known as sustainable, organic, or ecological agriculture aiming to eliminate the harmful effects of synthetic fertilizers (Chowdhury 2004). The function of different nutrients in plant growth and development is listed in Table 19.1.

In today's agriculture, different types of fertilizers are used to stimulate plant growth and yield, ranging from traditional inorganic to organic sustainable alternatives and advanced bio-fertilizers.

Table 19.1 Functions of nutrients in plants (Jaswal et al. 2022; Uchida 2000)

Nutrient	Function
Nitrogen	Important constituent in Chlorophyll molecule, making the plant succulent and greenish, promotes vegetative growth and component in Nucleic acids, ATP/ADP, and proteins.
Phosphorus	Root formation, a greater number of tillers, halts the plants ripening and increases grain yield.
Potassium	It promotes plant resistance, strong straw, and decreases lodging. Increases efficiency of starch and sugar production in leaves, needed for stomata functions, and all cell-to-cell gradient creation
Sulfur	Give flavours to vegetables and a constituent in amino acids (Cysteine and methionine)
Calcium	Constituent of the cell wall as calcium pectate, activate enzymes and coordinate cellular activities
Magnesium	Part of the chlorophyll molecule and directly linked with the Phosphorus uptake
Boron	Vital for cell wall formation and pollination
Zinc	Constituent of many enzymes such as Auxin's as a cofactor, and carbohydrate formation
Iron	The buildup of proteins and enhanced metabolic reactions
Molybdenum	Formation of root nodule and Mo is required in the case of legumes, also involved in the synthesis of amino acids
Nickel	Urease enzyme requires Ni to break down urea into useable form, also needed for iron absorption
Chlorine	Role in osmosis and photosynthesis as balancer
Sodium	Balancing ions in cells plays a part in osmosis
Manganese	Part of numerous enzymes in the form of prosthetic group

19.2 Types of Fertilizers

19.2.1 Chemical (Inorganic) Fertilizers

Chemical fertilizers are not natural but synthetically manufactured such as ammonium nitrate, potassium chloride, urea, etc. The nutrients in these fertilizers are readily available because of quick release and absorption by plants (Sharma and Chetani 2017). Much evidence reported that they can significantly enhance crop yield (Ojeniyi 2002). Moreover, they are overpriced and hazardous to the environment if not managed strategically (Morris et al. 2007).

19.2.2 Organic Fertilizers

Organic fertilizers are obtained from natural and biological sources including manure from animals, compost from agricultural waste such as corn stalks, and plant residues also called green manure. They improve the structure of soil and increase microbial activity (Sharma and Chetani 2017). However, they take time to convert into soluble form and release soil nutrients. Organic fertilizers are sustainable, cost-effective, and are available locally as compared to inorganic fertilizers (Solomon et al. 2012).

19.2.3 Biofertilizers

Biofertilizers also called Microbial fertilizers include materials containing microorganisms that are beneficial for soil and plants, such as mycorrhizae and rhizobia. Microorganisms live in symbiosis with plants and help to solubilize phosphorus, fix nitrogen in the atmosphere, and enhance nutrient uptake efficiency (Bashan and Holguin 2002). So, these fertilizers are more environment-friendly and non-bulky as compared to organic and inorganic ones (Mahajan et al. 2008).

19.3 Different Approaches for the Use of Fertilizers in Plant Growth and Yield

Various new approaches to sustainable fertilization have emerged with the goal of enhancing plant growth and yield, thereby affecting nutrient uptake dynamics and responses in plants. Such approaches aim to facilitate large-scale production that may help to improve profitability for farmers, reduce their application footprints on the environment, and achieve food production goals in the future. Some of the approaches used and adopted by farmers for nutrient management are as follows:

19.3.1 Soil Test Crop Response

Another name for this method is "Fertilizer recommendations based on the targeted yield approach". It is one of the most effective and scientific nutrient application approaches for crops as it uses values of soil tests and targeted yield equations. Soil test crop response is necessary to develop a basis for determining the precise quantitative adjustment of doses of fertilizer under varying values of soil test and response. It examines how different soil test results correspond to targeted crop production, considering crop responses to these variable soil conditions.

The application of fertilizer in soil testing for crop response is guided by Liebig's Law of the Minimum. This principle asserts that the growth of crops is determined by the availability of the scarcest resource or nutrient, rather than the total quantity of resources or nutrients present.

19.3.1.1 Targeted Yield

The concept of targeted yield comes under soil test crop response. To achieve a particular yield, a crop must take up a definite amount of nutrients. Once this nutrient demand is established for the desired yield, the necessity for fertilizer can be evaluated by taking into account the efficiency of nutrients present in the soil and the fertilizer requirement.

19.3.2 Site-Specific Nutrient Management

The site-specific nutrient management approach employs field-specific methods tailored to apply nutrients dynamically to rice fields as needed. It emphasizes the optimal utilization of local nutrients from manures, plant residues, soil, and water used for irrigation (Pampolino et al. 2007). It targets nutrient application at optimum times and rates, leading to more yield and efficient use of nutrients by the crop, without essentially increasing or reducing usage of fertilizer usage (Buresh and Witt 2007).

19.3.3 Soil-Testing Based Recommendation

This approach categorizes soils in different classes of fertility concerning their nutrient under consideration as low, medium, and high. Fertilizer applications are recommended and tailored for medium fertility levels and adjustments are later made for low or high fertility classes in this recommended approach. For high or low-fertility soil, it is decreased or increased by 25% respectively. This strategy makes sure that nutrient intakes are optimized as per the demand of different soils.

19.3.4 Soil Test Laboratory

Soil Test Laboratory approach also known as Low-Medium-High method is generally used for fertilizer recommendations in the soil testing centers. However, this approach includes adjusting the recommended dose of fertilizer or general recommended dose for specific crops based on a certain range of test values. This is not considered entirely scientific because it fails to account for the nutrient requirements and its contribution from various sources (Bhavya and Basavaraja 2021).

19.3.5 Ad-hoc Recommendations

Ad-hoc committees formed by various state authorities collectively devised these recommendations based on responses of crops witnessed across extensive regions, targeting medium-fertility soils. However, these commendations are designed only for wider areas and usually overlook the temporal and spatial changes in the nutrient supply capacity of the soil (Majumdar et al. 2014). Consequently, this results in either insufficient or excessive fertilization practices which leads to low yields, and economic loss, and can also decline the efficacy use of nutrients.

19.3.6 Nutrient Management Choices by Farmers

Nutrient management decisions made by farmers are not solely based on their perceptions but can also be influenced by other progressive farmers. Enlightened farmers learn about crop management, particularly nutrient control, from diverse sources like private companies, research institutes, and universities. They share it well with other farmers who trustfully accept and adopt such practices. However, farmer's decision according to their ideas often leads to excessive or inadequate use of fertilizer, affecting balances and timing. Eventually, nutrient efficiency and profit will be reduced, and regions with manifold nutrient problems will be more. These matters can mutilate the environment and food security because of unbalanced fertilizer application (Singh 2021).

19.3.7 Integrated Nutrient Management

The integrated nutrient management approach combines organic and bio-fertilizers with inorganic nutrient sources to enhance soil health, especially in varied climate conditions. A key advantage is its ability to reduce reliance on expensive inorganic fertilizers (Amanullah et al. 2023). Research underscores the importance of environmentally friendly and energy-efficient fertilization practices. A recent study shows that combining conventional fertilizers with nano-urea can improve soil health, boost economic yields, and reduce greenhouse gas emissions (Upadhyay et al. 2023).

19.4 The Chemical Fertilizer and Biofertilizer; Health Adversities

Fertilizers are widely used elements to maximize productivity in conventional agricultural systems. Fertilizers are characterized in two forms i.e. Chemical Fertilizers and Biofertilizers.

19.4.1 Chemical Fertilizer

Higher amounts of chemical fertilizers are being utilized to enhance crop yields. Nevertheless, surpassing the threshold level in fertilizer consumption pollutes water bodies and leads to accumulation in crop plants (Umesha et al. 2018). Chemical fertilizers consist of elements such as nitrogen, phosphorus, and potassium, which, when used excessively, can lead to air, water, and soil pollution. The excessive usage of chemical fertilizers and pesticides contributes to soil acidification, thereby polluting groundwater and the environment as a whole. Additionally, it weakens plant roots, making them susceptible to various pathogens. There is a pressing need for environmentally sustainable alternatives in crop production, which can be achieved by opting for organic products over synthetic chemicals (Singh et al. 2021).

19.4.1.1 Chemical Fertilizers and Their Impacts on the Environment and Human Health

The use of chemical fertilizers can result in enhanced growth and increased yield, but excessive application may lead to soil hardening and a decrease in soil organic matter and pH, ultimately reducing soil productivity. Moreover, a significant portion of chemical fertilizers may be washed away or leached into the environment due to rainfall and intensive irrigation, resulting in environmental pollution and diminished fertilizer efficacy (Wang et al. 2019a). Extensive use of synthetic fertilizers has been employed for centuries to maximize yield per unit area. However, exceeding recommended usage levels can result in numerous issues, including environmental pollution (air, water, and soil), diminished food quality, soil degradation, depletion of soil micronutrients, and toxicity to beneficial soil microbes.

Water Contamination

The chemical fertilizers when applied to the land eventually leach down to the water sources due to excessive irrigation and rainfall which causes water contamination when this contaminated water is used by human beings it may cause several health issues including disturbance in the kidneys, lungs and liver and cause cancer. This is due to the presence of heavy metals such as Mercury, Cadmium, Uranium and Lead that have been found in fertilizers (Savci 2012).

Air Pollution

As Chemical fertilizers are essential for the production of crops, their availability in soil whether low or high can cause problems both to the crops and the environment. When fertilizers are applied in low quantities then the yield will be less but if the fertilizers are available in higher amounts, then they may cause air pollution by the emission of different harmful gases Oxides of nitrogen (NO, N_2O, NO_2), Carbon dioxide, Methane, Hydrogen sulfide with Chlorofluorohydrocarbon. The availability of these gases in the atmosphere causes the greenhouse effect. These gases can also cause health issues including heart and lung diseases in humans. Excessive application of synthetic fertilizers may cause air pollution, inhaling the dust or fumes generated from chemical fertilizers results in respiratory disorders such as coughing, sneezing and difficulty in breathing. Furthermore, the presence of heavy metals can disrupt the functioning of vital organs like the kidneys, lungs, and liver, potentially leading to cancer (Savci 2012).

19.4.1.2 Side Effects of Chemical Fertilizers

Improper application of chemical fertilizers can lead to deterioration in soil and plant health. The primary side effects include:

- The chemicals found in these fertilizers can degrade soil properties, including soil texture, porosity, aeration, and organic matter content.
- The chemical fertilizers when applied to the land eventually causes water contamination and air pollution.
- These fertilizers release significant amounts of harmful gases such as nitrates, carbon dioxide, and methane, all of which are greenhouse gases contributing to global warming by trapping heat in the atmosphere.
- The cost of these fertilizers is steadily increasing, posing challenges for small-scale farmers who struggle to afford them.
- Excessive use of chemical fertilizers results in the destruction of soil-dwelling microorganisms and earthworms (Pahalvi et al. 2021).

19.4.2 Biofertilizer

Biofertilizers are organic preparations of microorganisms that encourage plant growth by improving nutrient acquirement (Maan and Garcha 2021). Organic farming involves cultivating crops using biofertilizers and bio pesticides, which supply essential nutrients to plants while effectively managing pathogens and pests (Umesha et al. 2018; Hadjouti et al. 2022; Basaid and Furze 2024). The following are some examples of biofertilizer:

- Nitrogen fixing bacteria e.g., Azotobacter, Clostridium, *Bacillus polymyxa*, Beijerinckia
- Nitrogen fixing Cyanobacteria: e.g., Nostoc, Anabaena, Aulosira, Totypothrix
- Partial association nitrogen fixing bacteria: e.g., Azospirillum
- Nitrogen-fixing bacteria in symbiosis: e.g., Actinomycete, Rhizobium, Ardisia
- Nitrogen fixing cyanobacteria in symbiosis: e.g., Blue-green algae
- Microphos biofertilizers: e.g., *Pseudomonas striata*, Aspergillus species
- Plant growth promoting rhizobacteria (PGPR): e.g., *Pseudomonas fluorescens*
- Mycorrhiza: Ectomycorrhiza, Endomycorrhiza

19.4.2.1 Biofertilizer Favoring Environment and Human Health

Biofertilizers serve as a sustainable substitute for chemical fertilizers, comprising small microorganisms typically found in the rhizosphere or can be supplied to the plants to provide support to the plants. Plant Growth Promoting Microorganisms play a crucial role in biotic activities inside the rhizosphere for better crop development. By accelerating microbial activity, biofertilizers improve the nutrient availability to the plants. They help in the fixation of atmospheric nitrogen and converting it into a usable form for plants without any significant alteration in natural biological systems of nutrient provision. Biofertilizers significantly enhance soil fertility and crop productivity. Moreover, they have a positive impact on both the environment and human health (Daniel et al. 2022).

Benefits of Biofertilizer

- Biofertilizer tends to increase the productivity of crops by improving the nutrient content and soil fertility.
- A diverse range of microorganisms serve as biofertilizers to enhance yields while minimizing pollution through their biodegradability.
- Biofertilizers contribute to sustainable agricultural development by preserving soil health.
- They help to mitigate the climate impacts and also lower the risk of many serious diseases that result from the pollution caused by synthetic fertilizers.

19.5 Role of Microorganisms in Bioremediation

Microorganisms are the unsung heroes of our ecosystem, which have been used by scientists for the benefit of humanity for ages. They have been used in the medical industry for the treatment of certain ailments, the food industry and detoxification of Earth's atmosphere. It is impossible for humans to live a single day on Earth without them. We are living in the era of a collapsed ecosystem, dealing with climate change, soil erosion, and biodiversity loss so, we need to know how to cater to global food demand. Our land is losing the capacity to produce food at a rate equivalent to global population growth, which has already posed a threat to food security. Microorganism based biofertilizers are game changers in this regard as they act as an excellent alternative to agrochemicals. When we use the term "biofertilizers", we mean the living microbes that fulfil plant's necessities by activating or multiplying available nutrients in the soil (Mitter et al. 2021).

Over the last decade, advancements in technologies such as multi-omics have significantly enhanced our comprehension of the intricate dynamics of microbiomes. These innovations have enabled more refined categorization of both the composition and activities within microbial communities (Kaul et al. 2016). So how do microorganisms remediate the environment? How do they help soil remediate? They promote soil's properties through activating its available nutrients, nitrogen fixation, extracellular secretions and increasing organic content by decomposition of plant and animal residues (Estrada et al. 2012). Bioremediation of soil contaminants such as pesticide residues and other pollutants by microorganisms with the help of redox reactions that convert toxic metallic substances into non-toxic substances shown in the Fig. 19.1.

There are some functional microorganisms with their role in soil remediation specified:

19.5.1 As Nutrient Deficiency Controller in Soil

Soil afflicted with nutrient imbalance can be effectively treated with various microbial species which use existing raw and/or toxic materials present in soil and helps in soil remediation. *Pseudomonas libanensis* EU-LWNA-33, for instance, excels in phosphorous solubilization (Kour et al. 2019). Likewise, *Herbaspirillum* sp. H18 and ZA15, *Burkholderia vietaminensis* AR114 increase Phosphorous solubility and Nitrogen efficiency (Estrada et al. 2012). For Indole Acetic acid production and Nitrogen fixation, *Enterobacter cloacae* RCA25, *Herbaspirillum seropedicae* z67, *Sinorhizobium fredii* NGR234 are used (Defez et al. 2017). *Paenibacillus* spp. catalyze nitrogen fixation, whereas *Pseudomonas* sp. S10-3, *Pantoea agglomerans*, *Rahnella aquatilis*, *Pseudomonas orientalis* and *Burkholderia* spp. increase Potassium solubilization (Xie et al. 2014; Sarikhani et al. 2018). *Providencia rettgeri* TPM23 plays an important role in increasing availability of Nitrogen,

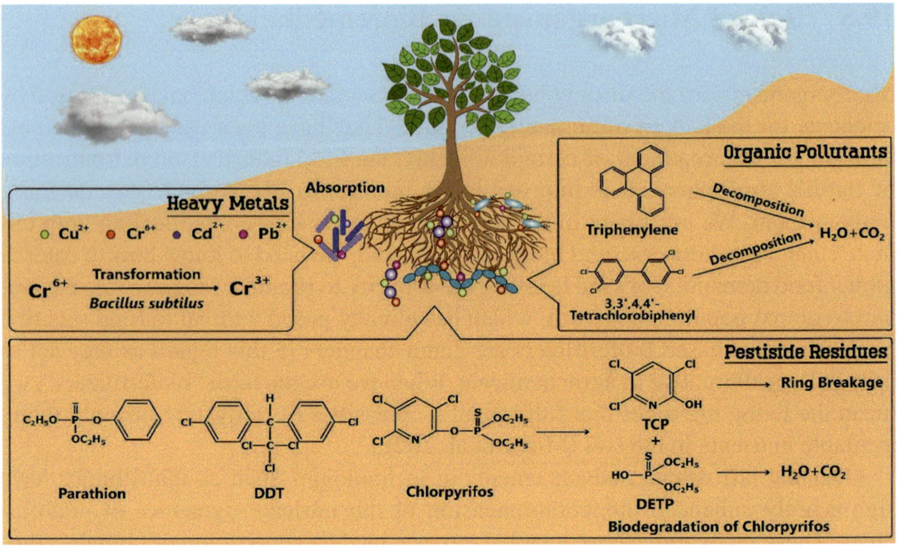

Fig. 19.1 Soil contaminants such as pesticide residues and other pollutants are removed by microorganisms, which are further fixed on their cell surface, redox reaction converts toxic metallic substances into non-toxic substances. (Modified from Wang et al. 2022)

Phosphorous and Potassium contents in the soil (Jiang and Li 2021). *Enterobacter cloacae* HG-1 enhances Nitrogen fixation, produces Iron carriers, ACC (1-aminocy clopropane-1-carboxylate) deaminase and plant hormones (Ji and Liu 2020). *Azospirillium brasilense* improves fertilizer Nitrogen recovery. Rhizobium strains NSFBR-12, and NSFBR-15 enhance Nitrogen fixation (Allito et al. 2020).

19.5.2 As Degraders of Pesticide Residues in Soil

Diverse microbial species play crucial roles in the degradation of various pollutants, helping soil remediation. *Microbacterium* sp. D-2, for instance, contributes to dicol degradation (Lu et al. 2019). Chloroanilines are degraded through *Acinetobacter baylyi* GFJ2 (Hongsawat and Vangnai 2011). DDT (dichloro-diphenyl-trichloroethane) degradation by *Pseudomonas* sp (Kamanavalli and Ninnekar 2004). Moreover, *Pseudomonas putida*, and *Acinetobacter rhizosphaerae* hydrolyze organophosphate and carbamate pesticides (Loh and Wu 2008). *Chryseobacterium* sp. Y16C involved in glyphosate degradation (Zhang et al. 2022), *Arthrobacter* sp. HB-5 in atrazine degradation (Gao et al. 2018), *Trichoderma atroviride* T23 in dichlorvos degradation (Sun et al. 2022), *Alcaligenes faecalis* NBRI OSS2-5 and *Pseudomonas nitroreducens* AR-3 in chlorpyrifos degradation (Yadav et al. 2020; Aswathi et al. 2019) *Bacillus pumilus* W1 degrades organophosphates (Ali et al. 2012), *Pseudomonas putida* ATCC 49451 breakdowns chlorophenols (Loh and Wu

2008), *Bacillus aryabhattai* VITNND15 reduces monocrotophos (Dash and Osborne 2019). *Sphingomonas* sp. AK1 play its role in the degradation of isoproturon (Li et al. 2017), *Bascillus* species in diuron degradation, and *Catellibacterium* sp.CC-5 in cypermethrin degradation (Estrada et al. 2012). Both *Xanthomonas axonopodis* and *Aspergillus niger* are recognized for their capacity to degrade alachlor, and *Bacillus cereus* Y1 is proficient in deltamethrin degradation (Ahmad 2020).

19.5.3 As Heavy Metals Remover in the Soil

Arsenic along with other contaminations from various heavy metals, necessitate the need of soil remediation. *Bacillus thuringiensis* WS3 emerges as a key player in this approach, as evidenced by research endorsing its efficacy in arsenic removal (Altowayti et al. 2019). Moreover, the effects of *Pseudomonas putida* UW4 on polycyclic aromatic hydrocarbons (PAHs) and lead (Pb) in spiked soils have been investigated, shedding light on its potential in remediation efforts (Afegbua and Batty 2019). Additionally, *Kocuria rhizophila* demonstrates promising capabilities in promoting the phytoremediation of Cd, Cr, Cu, and Ni-contaminated soil, offering a potential solution to multi-metal pollution scenarios (Hussain et al. 2019). These findings underscore the importance of microbial interventions in addressing soil contamination issues caused by heavy metals like arsenic and diverse pollutants, paving the way for more effective and sustainable remediation strategies. A recent study showed the activity of *Bacillus cereus* HM5, *Bacillus thuringiensis HM7* in improving Mn phytoremediation (Huang et al. 2020). Copper phytoremediation is improved via *Pseudomonas lurida* EOO26 according to (Kumar et al. 2021). More research works on microorganisms such as *Bacillus subtilus* MAI3's role in reduction of Chromium (VI) and Chromium (III) immobilization has been published in recent periods (Wani et al. 2018). Research highlights the potential of integrating microbial inoculation (*Trichoderma harzianum, Bacillus subtilis*) and biochar application as a holistic and effective strategy for mitigating Cd pollution in agricultural systems, while simultaneously improving soil quality and supporting crop productivity (Haider et al. 2021). Similarly, *Pantoea conspicua* MT5, *Aspergillus niger* CRS3 were seen in the reduction and detoxification of Cr (VI), also these strains exhibit increased production of phytohormones such as indole-3-acetic acid, gibberellic acid, and salicylic acid. These activities position them as excellent phytostimulants, suitable for use as biofertilizers in chromium-polluted soils (Qadir et al. 2022).

19.5.4 As Organic Pollutants Remover in Soil

The research findings indicated that *Hyphomicrobium* sp. GHH, when combined with the cultivation of *Lolium perenne*, displayed significant promise in remediating soil contaminated with 17α-ethynylestradiol, a prevalent environmental endocrine-disrupting compound (He et al. 2018). Similarly, *Arthrobacter* sp. strain HB-5 exhibited exceptional efficacy in atrazine removal, with a degradation half-life in HB-5 inoculated soil three times shorter than that in natural soil (Gao et al. 2018). Additionally, various strains including *Klebsiella pneumonia* PL1, *Pseudomonas mendocina, Brevundimonas olei, Serratia marcescens, Sphingobium yanoikuyae* B1, and *Pseudomonas* sp. USTB-RU have demonstrated effective decomposition of organic pollutants such as pyrene, benzo(a)pyrene, phenanthrene, and naphthalene (Ping et al. 2014; Smułek et al. 2020; Masakorala et al. 2013; Cunliffe and Kertesz 2006). Furthermore, the combined biosorption and biodegradation activities of *Pseudomonas* sp. SDR4 and *Mortierella alpina* JDR7 have been proven to achieve a significant reduction in polycyclic aromatic hydrocarbons (PAHs).

19.6 Healing Environment: The Modern Approaches

Farmers have been using different types of synthetic fertilizers, organic ones, and hybrid fertilizers to meet the nutritional requirements of soils. Biofertilizers are not new to them either. But when we talk about sustainable agricultural practices, we prioritize healing of environment with modern approaches and new goals. Live formulations of bacteria (plant growth promoting rhizobacteria, *Rhizobium, Azotobacter, Azosprillum*, phosphate solubilizing bacteria (species of *Pseudomonas and Bacillus*) and fungi *Mycorrhizae* and blue-green algae are simply termed as bio-fertilizers. To increase their shelf life, a coating technique is applied which helps to release and activate embedded biomaterials efficiently (Thirugnanasambandan 2018). Each type of biofertilizer offers unique benefits and applications, and the efficacy of these methods could fluctuate depending upon variables like soil conditions, climate variations, and the type of crop species utilized. By employing a customized blend of biofertilizers suited to particular requirements, farmers have the opportunity to enhance soil health, boost crop yields, and diminish dependence on synthetic fertilizers.

The world faces a myriad of challenges stemming from global population growth, ecosystem degradation, climate change, soil erosion, and loss of biodiversity. However, innovative solutions are emerging to address these pressing issues. Among them, biofertilizers offer a promising avenue for sustainable agriculture (Mahmud et al. 2021). By reducing dependency on agrochemicals, promoting carbon sequestration (Dębska et al. 2016), enhancing food security, improving soil health, and bolstering abiotic stress tolerance (Rai et al. 2021), biofertilizers contribute to environmentally friendly agricultural practices. Moreover, they facilitate biocontrol,

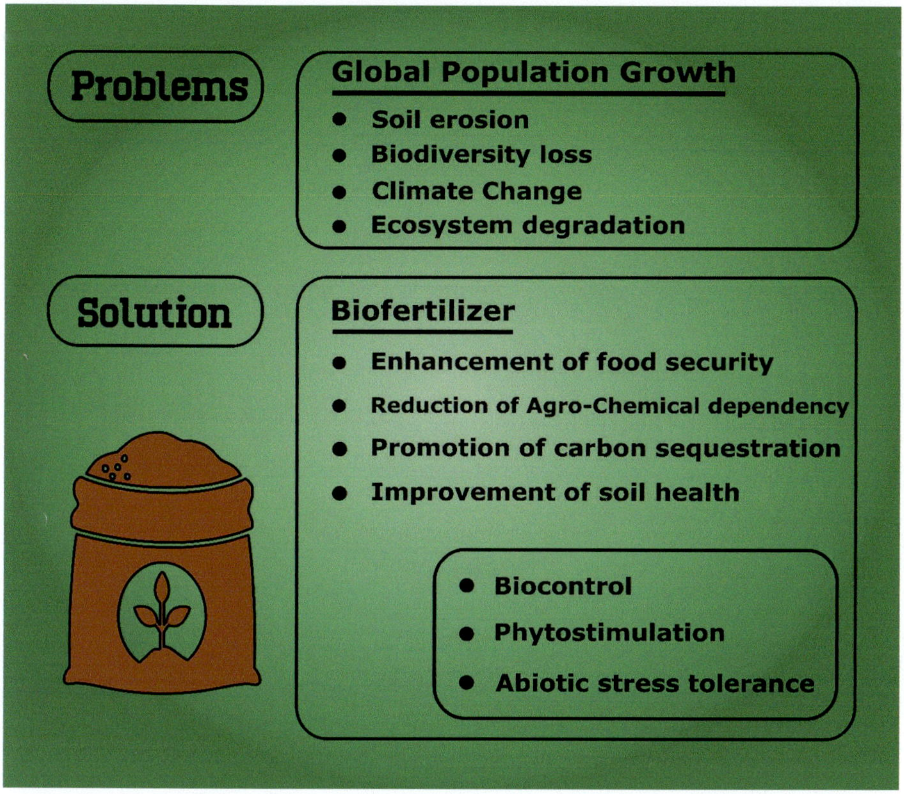

Fig. 19.2 A representation of environment healing with modern age approaches

phytostimulation (Hussain and Hasnain 2011), and overall healing of the environment as shown in Fig. 19.2. Embracing these modern-era goals in agriculture not only addresses current challenges but also fosters a more sustainable future for generations to come.

19.6.1 Classification of Biofertilizers and Their Role in Soil Health Maintenance

The biofertilizers are classified into several types based on function and microorganism type as shown in Fig. 19.3.

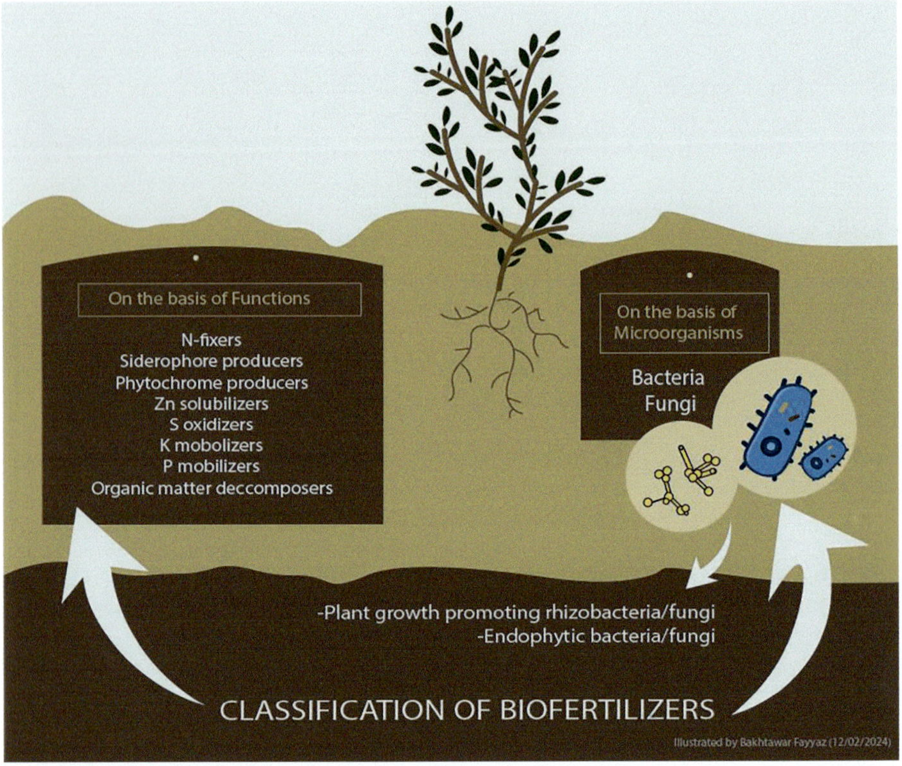

Fig. 19.3 Classification of biofertilizer based on functions and microorganisms. (Modified from Chaudhary et al. 2022)

19.6.1.1 Nitrogen-Fixing Biofertilizers

Nitrogen is required by plants for efficient growth, reproduction, dense content of grains and also for its dark green color (due to the presence of chlorophyll) (Sandhu et al. 2021). Atmospheric dinitrogen which is present in about 78% is converted into non-toxic solvable ammonia by nitrogen fixation (Abbey et al. 2019). Ammonia is then transformed into nitrite and nitrate by bacteria while the rest of the unused nitrates are denitrified into atmospheric nitrogen and become part of the air, it is a classic nitrogen cycle (Roy et al. 2019). For example, N-fixation in forest crops, maize plants *Azotobacter* and *Bacillus* sp. are involved which help in crop productivity and soil fertility (Etesami et al. 2014; Azeem et al. 2022) Nitrogen-fixing microbes can be termed as:

(a) Symbiotic nitrogen-fixing microbes
Examples of this kind of association are: Symbiosis of *Rhizobium* and legume, *Frankia* (actinomycetes) and nonleguminous plants, *Azolla* and *Anabaena azollae,* cyanobacteria and gymnosperms etc. (Ghodhbane-Gtari et al. 2021)

(b) Free living nitrogen-fixing bacteria
Among non-symbiotic free living N-fixer bacteria, *Azotobacter* is mostly studied phototropic bacteria and *Azotobacter chroococcum* can be used as biofertilizers. Azotobacter produces plant hormones, inhibits root pathogens, promotes its growth, helps in mineral absorption and maintains soil health (Mukherjee et al. 2022; Sumbul et al. 2020)

(c) Associative nitrogen-fixing bacteria
Azospirillum is a very well-known biofertilizer that has aerobic-associative nitrogen fixing relation with various crops and produces useful hormones aiding in the uptake of N, P and K (Yasuda et al. 2022).

19.6.1.2 Phosphorus Solubilizing Biofertilizers

After Nitrogen, Phosphorus is the most important macronutrient involved in the synthesis of ATP and amino acid, assisting in the nodulation mechanism and helping proteins in leguminous plants (Wang et al. 2020). Most of it gets transformed into soluble form phosphate anion with the help of phosphorous solubilizing biofertilizers or simply phosphorous solubilizing bacteria and fungi (PSB and PSF) such as, *Bacillus, Aerobacter, Burkholderia, Rhizobium, Aspergillus*, and *Penicillium* (Chaudhary et al. 2022). Arbuscular fungi applied in plant soil can make greater availability of P and shield from stress conditions according to (Nacoon et al. 2020).

19.6.1.3 Phosphorus Mobilizing Biofertilizers

These useful bacteria promptly mobilize ionized phosphorous and do mineralization of organic Phosphorous compounds that are otherwise unavailable to plants. These bacteria are thought to release phosphatases, produce organic acids or may interact with fungal mycorrhiza which helps to mobilize solvable Phosphorous from distant places where the roots of plants are unable to reach (Nassal et al. 2018).

19.6.1.4 Potassium Solubilizing Biofertilizers

Potassium as the third most important nutrient involved in the mechanism of stomatal opening/closing provides a shield against stress situations and helps protein synthesis to be in soluble form in soil. Unavailable Potassium is solubilized with the help of bacteria and fungi either by acidification (release of organic acids), siderophores production, complexation or by exchange reaction. Examples of Potassium-solubilizing bacteria are *Bacillus mucilaginous, B. edaphicus, B. circulans, Acidithiobacillus ferrooxidans, Frateuria aurantia, Herbaspirillum spp.*, and *Clostridium spp.* whereas Potassium-solubilizing fungi contain *Aspergillus spp.*

plus a few arbuscular mycorrhiza fungi (Santosh et al. 2022; Basak et al. 2020; Ali et al. 2020; Sattar et al. 2019; Varga et al. 2020).

19.6.1.5 Potassium Mobilizing Biofertilizers

Potassium-mobilizing microorganisms break down the insoluble potassium in the soil and allow the nutrients to move through the soil freely (Patel et al. 2021). Since potassium is important for the formation of sugar and starches, thus enables high quality and large yield.

19.6.1.6 Sulfur Solubilizing Biofertilizers

Sulfur solubilizers known as Sulfur-oxidizing bacteria increase the uptake of the amino acids to the plants that are important for photosynthesis and make the free soluble sulfur available to plants. These bacteria enhance root development, promote high germination rate and plant growth thus improve plant yield and quality (Wang et al. 2019b).

19.6.1.7 Zinc Solubilizing Biofertilizers

Fertilizers imbued with zinc-solubilizing bacteria facilitate the production of organic acids, which convert insoluble zinc into Zn+2, thereby fostering zinc uptake in plants (Nitu et al. 2020).

19.6.1.8 Phytohormone Producing Biofertilizers

Microorganisms secrete plant hormones or phytohormones i.e. indole-3-acetic acids, gibberellins, cytokinin etc. that have a very promising role in the growth of plants (Eichmann et al. 2021).

19.6.1.9 Siderophores Producing Biofertilizers

Under aerobic conditions, iron typically exists in the Fe+3 form, often as insoluble oxyhydroxides and hydroxide complexes, rendering it inaccessible. However, these bacteria generate low-molecular-weight iron-binding protein molecules, known as siderophores, which facilitate the availability of iron to plants, thereby promoting improved growth and development (Lurthy et al. 2020).

19.6.1.10 Organic Matter Decomposer Biofertilizers

Bacteria such as *Bacillus subtilis* and *Pseudomonas fluorescens* and fungi such as ectomycorrhizal fungi. *Trichoderma* spp. play their part in the deterioration of litter at a faster pace produce antimicrobial compounds which enhance the physicochemical characteristics of soil, and diversify microbes there (Baldi et al. 2021).

19.6.1.11 Endophytic Bacteria As Biofertilizers

Mutualistic microorganisms primarily reside within plant tissues throughout their life cycle, enhancing the nutritional requirements of plants. Endophytes possess the ability to trigger the plant's immune response when confronted with pathogenic threats. Examples of endophytic bacteria encompass *Klebsiella* spp., *Pseudomonas* spp., *Serratia* spp., *Bacillus* spp., *Burkholderia* spp., and *Citrobacter* spp., while endophytic fungi include *Colletotrichum*, *Fusarium*, *Alternaria*, and *Aspergillus*, among others (Fadiji and Babalola 2020; Kang et al. 2014).

19.6.1.12 Plant Growth Promoting Rhizobateria (PGPR)

These biofertilizers represent diverse soil bacteria and in combination with other microorganisms, thus help to enhance the stress tolerance capability of plants. i.e. the research of (Khati et al. 2018) have shown biofertilizers including *Pseudomonas taiwanensis, Bacillus* spp. and *Pantoea agglomerans* enhanced maize growth, production, and soil health (Chaudhary et al. 2022).

19.7 Ethical and Regulatory Considerations In Biofertilizer

19.7.1 Ethical Concerns

Agricultural pollution has become a major concern in many farming regions nowadays. Although agriculture is a critical source of financial development and employment, the pollution it causes can lead to various health and environmental risks. Agricultural contamination refers to the harm and deterioration of the atmosphere and ecosystem, as well as health risks caused by farming practices. Chemical fertilizers are one of the primary contributors to this type of pollution. To manage pollution more economically and ethically, the use of biofertilizers is recommended. Substances containing living microorganisms are known as biofertilizers. These microorganisms, when applied to plant surfaces, seeds, or soil, colonize the interior of the plant or rhizosphere. As a result, they increase the availability or supply of primary nutrients to the host plants, promoting their growth. The application of biofertilizers is an essential approach to enable agriculture management to be free from pollution and guarantee the security of our food supply. Ethics related to agriculture are an integral part of global ethics and are followed across different cultures and religions. Agriculture is a crucial aspect of every culture, making it an important topic for ethical discussions. To fully utilize the innovative potential of sustainable, non-polluting agricultural practices, it would be beneficial to adopt such an approach. Agriculture ethics are related across the other domains of ethics such as political ethics, business ethics, bioethics, environmental ethics, technologies ethics, innovation ethics, and cyber ethics.

19.7.2 Regulatory Considerations in Biofertilizer

Biofertilizers have gained significance in sustainable agriculture as an alternative to chemical fertilizers. Despite facing regulatory hurdles and technical complexities, the potential for growth in this industry is significant. A thorough knowledge of rhizosphere engineering can improve the proficiency of biofertilizers and ensure full crop gains. Genetic engineering of bioinoculants is a potential path to customize biofertilizers according to specialized crop requirements, which can increase their efficiency. The biofertilizer market can be transformed by creating microbial formulations that have multiple strains, traits, and nutrients. These formulations can be tailored to meet various agricultural requirements and provide personalized solutions. The integration of nanotechnology has enabled the enhancement of biofertilizer reach and performance, and this is further supported by market dynamics and innovations. Opportunities like these suggest that a prosperous future awaits the commercialization of biofertilizers. With enhanced formulations and an improved comprehension of soil-plant interactions, sustainable agriculture can be substantially benefitted. Biofertilizers offer a favorable prospect for a more environment-friendly and sustainable approach to sustaining the world's expanding population (Yadav and Yadav 2024).

19.8 Soil/Sediment Petroleum Pollution

Increased industrialization and more demand for petroleum products have adverse effects on ecosystems including the environment and human health. Due to the activities related to petroleum exploration, processing and transport, the main concern is soil pollution which causes severe environmental risks that draw the attention of scientists and the public. In 2005, the daily reported oil pollution incidents were around nine worldwide. These incidents are in addition to the yearly estimated oil spills of one million tonnes in the UK alone (Stroud et al. 2007). 90% of contaminated areas in the USA are of petroleum hydrocarbon contaminated soil (Das and Chandran 2011; Stroud et al. 2007).

The high rate of cancer cases is being observed in the workers associated with petroleum industries including liver, bladder, skin, lungs, and stomach cancers in adjunct to developmental, sexual and neuro effects. All these effects are considered to be associated with certain petroleum hydrocarbons that have severe carcinogenic effects (Das and Chandran 2011; Samanta et al. 2002). So, it is necessary to remediate these hydrocarbons from contaminated areas worldwide. There are different physical and chemical traditional methods to remediate these contaminants including soil vapour extraction, soil washing, oil boom usage, incineration, and solidification. Although, these practices are very laborious, difficult and costly which require a lot of time and resources (Huang et al. 2004). For instance, the estimated cost for the removal of soil contaminants is more than 1 trillion US dollars in the USA (Stroud et al. 2007).

The use of remediation approaches using microorganisms (bioremediations) has been extensively used since 2000. These processes are becoming current research interest specifically in terrestrial environmental remediation. These approaches have significant benefits in comparison to traditional physical and chemical techniques as they are easy to use, eco-friendly, can be applied over large areas, are inexpensive and can degrade several contaminants completely (Guo et al. 2014). Chemically, crude oil consists of a complicated combination of organic compounds that are composed of carbons and hydrogen atoms. Along with that, they also have traces of nitrogen, sulphur, oxygen and minor quantities of metallic particles (Yemashova et al. 2007). For the identification, characterization and separation of oil compounds, the most important used techniques are High-resolution Gas Chromatography (GC). GC techniques are used well equipped with capillary GC-Mass Spectrometry (MS) and flame-ionization detection (FID) (Head et al. 2006). Crude oil is one of the most complicated natural combinations of organic compounds with more than seventeen thousand different chemical components (Yasin and Bhanger 2013).

The petrogenic hydrocarbons are subdivided into four classes (Yasin and Bhanger 2013):

- The Aliphatic Hydrocarbons (Saturates): e.g. Hexadecanes etc.
- The Aromatic Hydrocarbons: e.g. Naphthalene, Pyrene etc.
- The Asphaltene Hydrocarbons: e.g. Fatty Acids, Phenol, Porphyrins etc.
- The Resins Hydrocarbons: e.g. Pyridines, carbazoles, sulfoxides etc.

19.8.1 Bioremediations

Microbes use organic compounds as energy sources and then convert them to less toxic products like biomass, water and carbon dioxide. For the biotransformation of aliphatic (saturates) and aromatic (ringed) hydrocarbons, different electron acceptors are required including iron, oxygen, manganese, sulphate and nitrate. Microorganisms are involved in the oxidation and activation of different hydrocarbons. The first step involved in the aerobic catabolism of hydrocarbon compounds is the addition of one or more hydroxyl groups to the skeleton. Different enzymes are involved in the process of remediation of petrogenic hydrocarbons. For instance, oxygenases are involved in the activation of aerobic catabolism (the addition of O_2 to the substrate), dioxygenases are used in the addition of two hydroxyl groups and for the addition of one hydroxyl group monooxygenases are involved (Rojo 2009).

The activation of hydrocarbon degradation is achieved by two different mechanisms (Wentzel et al. 2007):

- Aerobic catabolism: Oxygen is used as an electron acceptor.
- Anaerobic catabolism: In the absence of Oxygen.

There are several strategies for the bioremediation of petrogenic compounds in soil that are being used currently including bioaugmentation, natural attenuation, phytoremediation and biostimulation. The main features, advantages and limitations of different methods are summarized in the Table 19.2 (Koshlaf et al. 2017).

Table 19.2 Different strategies for bioremediation of petrogenic compounds in soil

Bioremediation strategies	Main features	Advantages	Limitations
Bioaugmentation	Addition of effective pollutants of hydrocarbon-degrading microorganisms	Use of a high biomass of microbes	Immense long-term supervising Not always efficient Poor adjustment of microbes to the contaminated site Co-pollutants or native microorganisms can inhibit the activity
Microbial consortium	Speed up the degradation of complicated pollutants blend		
Isolated strain	Speed up the degradation of a single pollutant		
Natural Attenuation	Applying native microbial populations under natural conditions	Economical	Immense long-term supervising Not always efficient
Phytoremediation	Use of plants and their related microorganisms	Help hydrocarbon-degrading microbes within plant root	Pollutants lethal to the plant
Biostimulation	Managing environmental factors	More effective than natural attenuation	Not always efficient Optimal C/N/P ratio and pollutant bioavailability must be established
Fertilizers	Optimization of C/N/P ratio		
Bio-surfactants	Stimulation of contaminant bioavailability		

19.9 Regulatory Hurdles

19.9.1 Bio-Safety Concerns

Proficient microbial strains in biofertilizers have a significant impact on soil health and crop yield, which leads to commercial success. It is important to meet regulatory requirements and market demands. However, the commercialization of biofertilizers realizes multiple challenges, including bio-safety concerns and regulatory hurdles etc. There are concerns regarding the possible impact of biofertilizers on human health and ecosystems when used in agriculture and environmental applications. Ensuring the safety of biofertilizer products and minimizing any potential risks they may pose to the environment and human health is of utmost importance. Due to these concerns, evaluating safety factors, determining efficacy, and performing thorough assessments are indispensable to commercialize the use of biofertilizers in the market.

The process of commercializing products can be time-consuming and challenging due to navigation through regulations. Therefore, to overcome these complications, it is crucial to estimate the safety and effectiveness of biofertilizer products. These biofertilizers provide plants with the essential nutrients required for the growth of plants. For instance, a researcher screened a strain of *Pseudomonas fluorescens* (PSM1), that was found to be effective in promoting the growth of wheat plants by solubilizing phosphate (Yadav et al. 2016). This bacterium is an eco-friendly and efficient strain as it solubilizes tricalcium phosphate and increases the free phosphorous for plant absorption. To guarantee the safe commercialization of biofertilizers, technical challenges must be considered. One of the most important challenges is the optimization of the biofertilizer production process and the preparation of guaranteed stability and effectiveness. To address safety concerns, the researchers have introduced biofertilizer-based methods that use ultraviolet (UV) mutagenesis and immobilization technology which enhances nitrogen fixation and phosphate solubilization process (Walpola et al. 2014).

19.9.2 Quality Control

Quality control is the major challenge regarding the use of biofertilizers for commercial purposes. The main hurdle is the production cost of biofertilizers, which must be competitive with traditional chemical fertilizers to ensure market viability (Malusá et al. 2012). Due to pre-existing microbial communities in agriculture setup, the adoption of plant growth-promoting microbes (PGPMs) is limited and usually outcompete and exclude newly introduced PGPMs (Ke et al. 2021). The existing challenges can be worse if the regulations are not followed consistently. Following international standards like ISO is crucial to ensure the quality and effectiveness of biofertilizers (Soumare et al. 2020). The number of colony-forming units

(CFU) is the key standard for the quality assessment of biofertilizers but in many countries, the comprehensive regulatory framework is noticeably absent (Kassem et al. 2021). The production and preservation of biofertilizers also come under the umbrella of quality control problems. The production of biofertilizers without a strong quality control system results in inconsistent composition and limited knowledge about the best storage and application practices (Lauriano-Barajas and Vega-Frutis 2018).

19.10 Technical Challenges

To use biofertilizers in agriculture, it is important to understand their relation with plants and other microorganisms present in the rhizosphere. Biofertilizers interact metabolically with plants in different ways, for example, by producing amino acids outside of their cells (Smirnova et al. 2022).

19.10.1 Strain Stability

For the commercialization of biofertilizers, there are several technical challenges that need to be addressed before use. One of the most important challenges is microbial strain stability. These can only be successfully used in agriculture practices by ensuring that biofertilizers strains remain effective when stored and during application in the field. The genetic variability present within bacterial population is the main reason for the difficulty in maintaining the stability of biofertilizer strains. The variability is a significant hurdle in achieving the desired level of stability. The biofertilizer's performance can vary depending on the measure of plant growth promotion that influences its efficiency in different soil conditions. For successful agricultural operations, it is important to choose those biofertilizer strains that demonstrate consistent plant growth promoting abilities. The environmental factors like temperature, pH, and nutrient accessibility may affect the stability of these stains (Yadav 2013). Due to different soil types and agricultural operations, these factors may have significant importance because these can cause the inconsistency in performance of biofertilizer strains (Purwaningsih et al. 2022). For the promotion of the development and functioning of biofertilizers, it is critical to have thorough knowledge and understanding of the optimum environmental conditions required for strains. Formulating plants to maintain the favorable condition is important to confirm the stability of stains. Moreover, the interaction between biofertilizers and other soil microorganisms adds up an extra level of complexity. Different strains of biofertilizers are always in competition with other fungi, soil bacteria and actinomycetes for resources and space, which ended up in the decrease in their population and activity (Dandessa and Bacha 2018). The existence of microbial strains in the soil is important for durable benefits to plants but nutrient depletion, predation and alternation in soil conditions can negatively affect their persistence (Mahdi et al. 2020). A comprehensive approach can overcome these challenges. Strict strain selection and screening processes are the parameters for the identification of perfect microbial strains that have constant plant growth promoting capacity. The genetic stability and adaptability of the strains to different soil conditions should be in consideration during these processes (Motamedi et al. 2016). To improve the stability and effectiveness of biofertilizer strains, different genetic editing techniques like genetic engineering and mutagenesis can be used but it is necessary to follow the ethical concerns and potential hazards of using these methods.

19.10.2 Bioformulation

According to the research by Broeckx, it is important to develop appropriate biofertilizer carriers to maintain microbial viability and commercial success (Broeckx et al. 2016). The viability of biofertilizers is the main factor affecting the efficacy of microbial inoculants. Several techniques have been studied to improve the viability and stability of biofertilizers for durable storage (Broeckx et al. 2016; Safari et al. 2020). One of the commonly used methods is to cover the microbial suspension or pellet with a protective shell (Broeckx et al. 2016). Polysaccharides, proteins, fats, and other coating materials can be used to form this protective capsule that improves the stability and survival of biofertilizers. In biological concrete applications, researchers have examined coated and expended clay particles as possible transporters for biofertilizers, to guarantee the endurance of bacteria for long period.

19.11 Economic and Healthcare Implications

Biofertilizers are ecofriendly and economical related to synthetic fertilizers. They differ from organic and chemical fertilizers because they are alive and active microorganisms, like fungi and bacteria that help plants to intake nutrients from soil. At the same time, Biofertilizers are not capable of providing nutrients directly to the crops. They are relatively inexpensive and can enhance efficiency per unit area in a comparatively short time. They have lesser manufacturing costs and cheap usage costs, especially when it comes to nitrogen and phosphorus use. The easy application process of biofertilizers consumes less energy, which translates to lower fertilization costs and greater profitability for farmers. Biofertilizers are a renewable source of plant nutrients and can replace chemical fertilizers in green agriculture. They are most usually available in powder form and mixed with carriers such as lignite, which contains high amounts of organic matter and water. The high water content of lignite boosts the growth of microorganisms in biofertilizers. To apply such biofertilizers, farmers can create a slurry and apply it to the seeds. Though this method was once considered universal, newer, and more efficient methods are now available for applying biofertilizers (Table 19.3).

A new method of applying fertilizer directly to soil has been developed. This method involves using granules made from tank bed clay that are 1–2 mm in size

Table 19.3 Comparative analysis of biofertilizers: mechanisms and illustrative instances

Biofertilizers	Mechanism	Groups	Examples	References
Phosphorus-mobilizing	Phosphorus is transferred from the soil to the root cortex. These are bio-fertilizers with a wide range of applications.	Mycorrhiza	*Glomus* spp., *Acaulospora* spp., Arbuscular mycorrhiza, *Scutellospora* spp., *Gigaspora* spp., and *Sclerocystis* spp.	Nosheen et al. (2021)
Nitrogen-fixing	Increase the amount of N2 in the soil by fixing atmospheric nitrogen and making it available to plants.	Free-living, symbiotic, and associative symbiotic	*Nostoc*, *Aulosira bejerinkia*, *Klebsiella*, *Desulfovibrio*, *Azotobacter*, *Clostridium*, *Rhodospirillum*, *Stigonema*, *Anabaena*, and *Rhodopseudomonas Rhizobium*, *Anabaena azollae*, *Frankia*, and *Trichodesmium Herbaspirillum* spp., *Azospirillum* spp., *Azoarcus* spp., Alcaligenes, Enterobacter, and *Acetobacter diazotrophicus*	Nosheen et al. (2021)
Potassium mobilizing	They transfer potassium from the soil's inaccessible forms.	Bacteria	*Bacillus* spp.	Jha (2017)
Potassium solubilizing	Produce organic acids that degrade silicates and aid in the removal of metals to solubilize potassium (silicate) ions and make them available to plants.	Bacteria	*Arthrobacter* spp., *B. edaphicus*, *Bacillus*, *B. circulanscan*, and *B. mucilaginosus*	Etesami et al. (2017)
Phosphorus solubilizing	To dissolve bound phosphates, they secrete organic acids and lower soil pH by converting insoluble forms of P in the soil into soluble forms.	Bacteria	*Bacillus circulans*, *Pseudomonas striata*, *Penicilium* spp., *Bacillus subtilis*, *B. polymyxa*, *Micrococcus*, *Agrobacterium*, *Aereobacterium*, *Flavobacterium*.	Nosheen et al. (2021)

(continued)

Table 19.3 (continued)

Biofertilizers	Mechanism	Groups	Examples	References
Micronutrient	Protons, chelated ligands, acidification, and oxidoreductive systems can all be used to dissolve zinc.	Zinc-solubilizing	*Pseudomonas* spp., Mycorrhiza, and *Bacillus* spp.	Kamran et al. (2017)
Plant growth-promoting	Produce hormones that encourage root growth, increase nutrient availability, and boost crop yields.	Plant growth-promoting rhizobacteria	*Pseudomonas fluorescens*, Agrobacterium, *Erwinia*, *Arthrobacter*, *Rhizobium*, *Bacillus*, *Enterobacter*, *Pseudomonas* spp. *Streptomyces*, and *Xanthomonas*	Nosheen et al. (2021)

Modified from Nosheen et al. (2021), Jha (2017), Etesami et al. (2017), Kamran et al. (2017)

and baked at 200 °C in a muffle furnace. The material is sterilized and porous granules are produced through a specific process. These granules are then immersed in a suspension of bacteria and left to grow in a suitable medium overnight. Following this, the clay granules are air-dried at room temperature under sterile conditions. The granules consist of approximately 109 bacteria per gram and are appropriate for use in the field when sowing seeds. However, the quantity of biofertilizer required for this method is slightly higher than that for seed application (Ogwu et al. 2023).

19.12 Conclusion and Future Perspectives

The doors have been opened for creative techniques to improve crop yield with growing attention towards sustainable agriculture, lowering dangerous environmental impact caused by conventional farming (Chandran et al. 2021). The combination of biofertilizers with integrated nutrient management (INM) structure is an innovative strategy which provides a complete solution for sustainable farming, covering different aspects such as nutrient accessibility, soil well-being, ecological sustainability, and crop production and quality. The soil health has been found to have a positive impact because of the treatment of biofertilizers in conjunction with INM. The physical and physicochemical properties of the soil are enhanced when a mixture of biofertilizers and INM is added to the soil; better soil structure, improved water retention, and increased nutrient availability are additional benefits. Overall, for the promotion of soil health and improving agricultural yield, the combination of biofertilizers and INM can be a powerful tool (Maleva et al. 2018). Due to their renewability and slight environmental impact, biofertilizers are becoming popular as a green substitution for chemical fertilizers. From an environmental perspective, they offer an appealing choice for sustainable agriculture (Otaiku et al. 2019; Basaid and Furze 2024).

Due to the critical factor of understanding rhizosphere dynamics, the commercialization of biofertilizers is a challenging task. Various factors such as soil type, microbial interactions, and plant species influence the rhizosphere, which is a dynamic and complex environment. Seasonal changes complicate the dynamics of temperature, soil moisture, and microbial activity. These fluctuations affect the effectiveness of biofertilizers and make it difficult to find a solution that works for every condition. For revolutionizing sustainable agriculture, the integration of biofertilizers with INM is a potential approach. However, several challenges need to be addressed before the achievement of the commercialization of biofertilizers. Creating affordable methods to produce biofertilizers, guaranteeing their effectiveness, and encouraging the farmers to adopt them for use are important parameters. Additional research is required to understand the interaction of biofertilizers with other components of INM, such as organic and inorganic fertilizers, for the further improvement of nutrient management strategies.

In addition, the management of nutrients and efficiency of resource utilization is potentially enhanced by the combination of biofertilizers and precision agriculture methods, such as data analysis and remote sensing. Overall, agricultural productivity can be increased without harming the environment by integrating biofertilizers with INM. Moreover, the world is facing soil degradation as a growing concern specifically in developing countries. Bioremediation, i.e. removing contaminants from ecosystems by use of microorganisms, is a viable and cost-effective method for soil remediation. To reduce climate change and for a sustainable environment, biofertilizers are a necessity of the modern era to improve bioremediation. Therefore, to meet the requirements of the fast-growing global population, the use of biofertilizers should increase. Proper deployment of appropriate technological and socio-economic measures is required to address potential environmental concerns of biofertilizers.

19.13 Suggestions

To assist the increased implementation and use of biofertilizers the following suggestions are put forward:

I. Conducting educational sessions and programs to raise awareness of using biofertilizers in the agriculture sector, highlighting their role in enhancing nutrient availability, improving soil structures and promoting sustainable farming practices.
II. The established regulatory framework should be considered thoroughly for the quality, safety and efficacy standards for the production, distribution, and use of biofertilizers.
III. Incorporation of ethical consideration into biofertilizer development and deployment strategies adhering to principles of social responsibility, equality and environmental stewardship.

References

Abbey L, Abbey J, Leke-Aladekoba A, Iheshiulo EM, Ijenyo M (2019) Biopesticides and biofertilizers. In: Byproducts from agriculture and fisheries, Wiley, Hoboken pp 479–500. https://doi.org/10.1002/9781119383956.ch20

Afegbua SL, Batty LC (2019) Effect of plant growth promoting bacterium; Pseudomonas putida UW4 inoculation on phytoremediation efficacy of monoculture and mixed culture of selected plant species for PAH and lead spiked soils. Int J Phytoremediation 21(3):200–208

Ahmad KS (2020) Environmental contaminant 2-chloro-N-(2, 6-diethylphenyl)-N-(methoxymethyl) acetamide remediation via *Xanthomonas axonopodis* and *Aspergillus niger*. Environ Res 182:109117

Ali M, Naqvi TA, Kanwal M, Rasheed F, Hameed A, Ahmed S (2012) Detection of the organophosphate degrading gene opdA in the newly isolated bacterial strain Bacillus pumilus W1. Ann Microbiol 62:233–239

Ali AM, Awad MY, Hegab SA, Gawad AM, Eissa MA (2020) Effect of potassium solubilizing bacteria (*Bacillus cereus*) on growth and yield of potato. J Plant Nutr 44(3):411–20

Allito BB, Ewusi-Mensah N, Logah V (2020) Legume-rhizobium strain specificity enhances nutrition and nitrogen fixation in faba bean (*Vicia faba* L.). Agronomy 10(6):826

Altowayti WA, Algaifi HA, Bakar SA, Shahir S (2019) The adsorptive removal of As (III) using biomass of arsenic resistant Bacillus thuringiensis strain WS3: characteristics and modelling studies. Ecotoxicol Environ Saf 172:176–185

Amanullah OG, Al-Tawaha AR (2023) Editorial: integrated nutrients management: an approach for sustainable crop production and food security in changing climates. Front Plant Sci 14:1288030

Aswathi A, Pandey A, Sukumaran RK (2019) Rapid degradation of the organophosphate pesticide–Chlorpyrifos by a novel strain of *Pseudomonas nitroreducens* AR-3. Bioresour Technol 292:122025

Ayoub AT (1999) Fertilizers and the environment. Nutr Cycl Agroecosyst 55:117–121

Azeem M, Haider MZ, Javed S, Saleem MH, Alatawi A (2022) Drought stress amelioration in maize (*Zea mays* L.) by inoculation of *Bacillus* spp. strains under sterile soil conditions. Agriculture 12(1):50

Baldi E, Gioacchini P, Montecchio D, Mocali S, Antonielli L, Masoero G, Toselli M (2021) Effect of biofertilizers application on soil biodiversity and litter degradation in a commercial apricot orchard. Agronomy 11(6):1116

Basaid K, Furze JN (2024) Botanical-microbial synergy-fundaments of untapped potential of sustainable agriculoture. J Crop Health 76:1263–1280. https://doi.org/10.1007/s10343-024-01008-5

Basak BB, Maity A, Ray P, Biswas DR, Roy S (2020) Potassium supply in agriculture through biological potassium fertilizer: A promising and sustainable option for developing countries. Arch Agron Soil Sci 68(1):101–14

Bashan Y, Holguin G (2002) Plant growth-promoting bacteria: A potential tool for arid mangrove reforestation. Trees 16(2–3):159–166

Bhavya N, Basavaraja PK (2021) Influence of different approaches of nutrient recommendations on growth and yield of aerobic rice. Mysore J Agric Sci 55(4):61

Broeckx G, Vandenheuvel D, Claes IJ, Lebeer S, Kiekens F (2016) Drying techniques of probiotic bacteria as an important step towards the development of novel pharmabiotics. Int J Pharm 505(1–2):303–318

Buresh RJ, Witt C (2007) Site-specific nutrient management. In: Fertilizer best management practices, International Fertilizer Industry Association, Paris, France pp 47–55

Chandran H, Meena M, Swapnil P (2021) Plant growth-promoting rhizobacteria as a green alternative for sustainable agriculture. Sustainability 13(19):10986

Chaudhary P, Singh S, Chaudhary A, Sharma A, Kumar G (2022) Overview of biofertilizers in crop production and stress management for sustainable agriculture. Front Plant Sci 13:930340

Chowdhury R (2004) Effects of chemical fertilizers on the surrounding environment and the alternative to the chemical fertilizers. IES-ENVIS Newsl 7:4–5

Crawford RL, Crawford DL (eds) (1996) Bioremediation: principles and applications, XII, 400P. Cambridge University Press. Journal of the Marine Biological Association of the United Kingdom. 1997;77(2):569–569

Cunliffe M, Kertesz MA (2006) Effect of Sphingobium yanoikuyae B1 inoculation on bacterial community dynamics and polycyclic aromatic hydrocarbon degradation in aged and freshly PAH-contaminated soils. Environ Pollut 144(1):228–237

Dandessa C, Bacha K (2018) Review on role of phosphate solubilizing microorganisms in sustainable agriculture. Int J Curr Res Acad Rev 6(11):48–55

Daniel AI, Fadaka AO, Gokul A, Bakare OO, Aina O, Fisher S, Burt AF, Mavumengwana V, Keyster M, Klein A (2022) Biofertilizer: the future of food security and food safety. Microorganisms 10(6):1220

Das N, Chandran P (2011) Microbial degradation of petroleum hydrocarbon contaminants: an overview. Biotechnol Res Int 2011:1–13. https://doi.org/10.4061/2011/941810

Dash DM, Osborne JW (2019) Biodegradation of monocrotophos by a plant growth promoting *Bacillus aryabhattai* (vitnndj5) strain in artificially contaminated soil. Int J Environ Sci Technol 17(3):1475–1490

Dębska B, Długosz J, Piotrowska-Długosz A, Banach-Szott M (2016) The impact of a bio-fertilizer on the soil organic matter status and carbon sequestration—results from a field-scale study. J Soils Sediments 16:2335–2343

Defez R, Andreozzi A, Bianco C (2017) The overproduction of indole-3-acetic acid (IAA) in endophytes upregulates nitrogen fixation in both bacterial cultures and inoculated rice plants. Microb Ecol 74(2):441–452

Dell'Anno A, Beolchini F, Rocchetti L, Luna GM, Danovaro R (2012) High bacterial biodiversity increases degradation performance of hydrocarbons during bioremediation of contaminated harbor marine sediments. Environ Pollut 167:85–92

Eichmann R, Richards L, Schäfer P (2021) Hormones as Go-Betweens in plant microbiome assembly. Plant J 105(2):518–541

Estrada GA, Baldani VL, de Oliveira DM, Urquiaga S, Baldani JI (2012) Selection of phosphate-solubilizing diazotrophic *Herbaspirillum* and *Burkholderia* strains and their effect on rice crop yield and nutrient uptake. Plant Soil 369(1–2):115–129

Etesami H, Mirseyed Hosseini H, Alikhani HA (2014) Bacterial biosynthesis of 1-aminocyclopro pane-1-caboxylate (ACC) deaminase, a useful trait to elongation and endophytic colonization of the roots of rice under constant flooded conditions. Physiol Mol Biol Plants 20:425–434

Etesami H, Emami S, Alikhani HA (2017) Potassium solubilizing bacteria (KSB):: mechanisms, promotion of plant growth, and future prospects A review. J Soil Sci Plant Nutr 17(4):897–911

Ewulo BS, Ojeniyi SO, Akanni DA (2008) Effect of poultry manure on selected soil physical and chemical properties, growth, yield and nutrient status of tomato. Afr J Agric Res 3(9):612–616

Fadiji AE, Babalola OO (2020) Elucidating mechanisms of endophytes used in plant protection and other bioactivities with multifunctional prospects. Front Bioeng Biotechnol 8:467

Gao J, Song P, Wang G, Wang J, Zhu L, Wang J (2018) Responses of atrazine degradation and native bacterial community in soil to Arthrobacter sp. strain HB-5. Ecotoxicol Environ Saf 159:317–323

Ghodhbane-Gtari F, D'Angelo T, Gueddou A, Ghazouani S, Gtari M, Tisa LS (2021) Alone yet not alone: Frankia lives under the same roof with other bacteria in actinorhizal nodules. Front Microbiol 12:749760

Guo J, Feng R, Ding Y, Wang R (2014) Applying carbon dioxide, plant growth-promoting Rhizobacterium and EDTA can enhance the phytoremediation efficiency of ryegrass in a soil polluted with zinc, arsenic, cadmium and lead. J Environ Manag 141:1–8. https://doi.org/10.1016/j.jenvman.2013.12.039

Hadjouti R, Mohand Kaci H, Benzina F, Furze JN (2022) Enhancing recovery of *Phaseolus vulgaris* L. and *Cucrbita pepo* L. with *Olea europea* L. plant growth promoting rhizobacteria. Soil Res 60(8):850–863. https://doi.org/10.1071/SR21320

Haider FU, Coulter JA, Cheema SA, Farooq M, Wu J, Zhang R, Shuaijie G, Liqun C (2021) Co-application of biochar and microorganisms improves soybean performance and remediate cadmium-contaminated soil. Ecotoxicol Environ Saf 214:112112

He S, Guo H, He Z, Yang C, Yu T, Chai Q et al (2018) Interaction of *Lolium perenne* and *Hyphomicrobium* sp.. GHH enhances the removal of 17α-ethinyestradiol (EE2) from soil. J Soils Sediments 19(3):1297–1305

Head IM, Jones DM, Röling WF (2006) Marine microorganisms make a meal of oil. Nat Rev Microbiol 4(3):173–182. https://doi.org/10.1038/nrmicro1348

Hlihor RM, Apostol LC, Gavrilescu M (2017) Environmental bioremediation by biosorption and bioaccumulation: principles and applications. Enhancing cleanup of environmental pollutants: volume 1: biological approaches, Springer, Cham pp 289–315. https://doi.org/10.1007/978-3-319-55426-6_14

Hongsawat P, Vangnai AS (2011) Biodegradation pathways of chloroanilines by *Acinetobacter baylyi* strain GFJ2. J Hazard Mater 186(2–3):1300–1307

Huang X-D, El-Alawi Y, Penrose DM, Glick BR, Greenberg BM (2004) A multi-process phytoremediation system for removal of polycyclic aromatic hydrocarbons from contaminated soils. Environ Pollut 130(3):465–476. https://doi.org/10.1016/j.envpol.2003.09.031

Huang H, Zhao Y, Fan L, Jin Q, Yang G, Xu Z (2020) Improvement of manganese phytoremediation by *Broussonetia papyrifera* with two plant growth promoting (PGP) Bacillus species. Chemosphere 260:127614

Hussain A, Hasnain S (2011) Phytostimulation and biofertilization in wheat by cyanobacteria. J Ind Microbiol Biotechnol 38(1):85–92

Hussain A, Kamran MA, Javed MT, Hayat K, Farooq MA, Ali N, Ali M, Manghwar H, Jan F, Chaudhary HJ (2019) Individual and combinatorial application of *Kocuria rhizophila* and citric acid on phytoextraction of multi-metal contaminated soils by *Glycine max* L. Environ Exp Bot 159:23–33

Inze D, Van Montagu M (1995) Oxidative stress in plants. Curr Opin Biotechnol 6(2):153–158

Jaswal A, Prasad D, Singh A, Singh M (2022) Fertilizers and their role in plant growth. In: Advanced research and review in agronomy, Bright Sky Publications, Dehli pp 71–88

Jha Y (2017) Potassium mobilizing bacteria: enhance potassium intake in paddy to regulates membrane permeability and accumulate carbohydrates under salinity stress. Braz J Biol Sci 4(8):333–344

Ji C, Liu Z (2020) Effects of *Enterobacter cloacae* HG-1 on the nitrogen-fixing community structure of wheat rhizosphere soil and on salt tolerance. Front Plant Sci 11:510947

Jiang H, Li S (2021) Interaction between halotolerant phosphate-solubilizing bacteria (*Providencia rettgeri* Strain TPM23) and rock phosphate improves soil biochemical properties and peanut growth in saline soil. Front Microbiol 12:777351

Kamanavalli CM, Ninnekar HZ (2004) Biodegradation of DDT by a *Pseudomonas* species. Curr Microbiol 48:10–13

Kamran S, Shahid I, Baig DN, Rizwan M, Malik KA, Mehnaz S (2017) Contribution of zinc solubilizing bacteria in growth promotion and zinc content of wheat. Front Microbiol 8:322772

Kang SM, Radhakrishnan R, You YH, Joo GJ, Lee IJ, Lee KE, Kim JH (2014) Phosphate solubilizing *Bacillus megaterium* mj1212 regulates endogenous plant carbohydrates and amino acids contents to promote mustard plant growth. Indian J Microbiol 54:427–433

Kassem HS, Alotaibi BA, Aldosri FO, Muddassir M (2021) Exploring the relationship between information-seeking behavior and adoption of biofertilizers among onion farmers. Agronomy 11(6):1258

Kaul S, Sharma T, Dhar KM (2016) "Omics" tools for better understanding the plant–endophyte interactions. Front Plant Sci 7:183603

Ke J, Wang B, Yoshikuni Y (2021) Microbiome engineering: synthetic biology of plant-associated microbiomes in sustainable agriculture. Trends Biotechnol 39(3):244–261

Khati P, Parul BP, Nisha KR, Sharma A (2018) Effect of nanozeolite and plant growth promoting rhizobacteria on maize. 3 Biotech 8:1–2

Koshlaf E, Ball S, A. (2017) Soil bioremediation approaches for petroleum hydrocarbon polluted environments. AIMS Microbiol 3(1):25–49. https://doi.org/10.3934/microbiol.2017.1.25

Kour D, Rana KL, Sheikh I, Kumar V, Yadav AN, Dhaliwal HS et al (2019) Alleviation of drought stress and plant growth promotion by *Pseudomonas libanensis* EU-LWNA-33, a drought-adaptive phosphorus-solubilizing bacterium. Proc Natl Acad Sci India Sect B Biol Sci 90(4):785–795

Kumar A, Govil M, Singh S, Sharma KK, Tripathi SK, Tiwari RK, Tripathi AN, Singh S (2015) Role of micro-organisms in bioremediation: A comprehensive model using *Trichoderma* spp. In: Handbook of research on uncovering new methods for ecosystem management through bioremediation. IGI Global, New York pp 29–50

Kumar A, Voropaeva O, Maleva M, Panikovskaya K, Borisova G, Rajkumar M, Bruno LB (2021) Bioaugmentation with copper tolerant endophyte *Pseudomonas lurida* strain EOO26 for improved plant growth and copper phytoremediation by *Helianthus annuus*. Chemosphere 266:128983

Lauriano-Barajas J, Vega-Frutis R (2018) Infectivity and effectivity of commercial and native arbuscular mycorrhizal biofertilizers in seedlings of maize (*Zea mays*). Botan Sci 96(3):395–404

Li R, Dörfler U, Munch JC, Schroll R (2017) Enhanced degradation of isoproturon in an agricultural soil by a *Sphingomonas* sp. strain and a microbial consortium. Chemosphere 168:1169–1176

Loh K, Wu T (2008) Cometabolic transformation of 2-chlorophenol and 4-chlorophenol in the presence of phenol by *Pseudomonas putida*. Can J Chem Eng 84(3):356–367

Lu P, Liu HM, Liu AM (2019) Biodegradation of dicofol by Microbacterium sp. D-2 isolated from pesticide-contaminated agricultural soil. Applied. Biol Chem 62(1):72

Lurthy T, Cantat C, Declerck P, Gallardo K, Leroy F, Ourry A, Salon C, Mazurier S (2020) Impact of bacterial siderophores on iron status and ionome in pea. Front Plant Sci 11:523812

Maan PK, Garcha S (2021) Chapter 3 - Production technology, properties, and quality management. In: Biofertilizers, Woodhead Publishing, Sawston, Cambridge pp 31–43

Mahajan A, Gupta RD, Sharma R (2008) Bio-fertilizers-A way to sustainable agriculture. Agrobios Newslett 6(9):36–37

Mahdi I, Fahsi N, Hafidi M, Allaoui A, Biskri L (2020) Plant growth enhancement using rhizospheric halotolerant phosphate solubilizing bacterium *Bacillus licheniformis* QA1 and *Enterobacter asburiae* QF11 isolated from *Chenopodium quinoa* willd. Microorganisms 8(6):948

Mahmud AA, Upadhyay SK, Srivastava AK, Bhojiya AA (2021) Biofertilizers: A Nexus between soil fertility and crop productivity under abiotic stress. Curr Res Environ Sustain 3:100063

Majumdar K, Dey P, Tewatia RK (2014) Current nutrient management approaches. Indian J Fertil 10:14–27

Maleva M, Borisova G, Koshcheeva O, Sinenko O (2018) Biofertilizer based on silicate solubilizing bacteria improves photosynthetic function of brassica juncea. AGROFOR 2(3):13

Malusá E, Sas-Paszt L, Ciesielska JJ (2012) Technologies for beneficial microorganisms inocula used as biofertilizers. Sci World J 2012:491206

Masakorala K, Yao J, Cai M, Chandankere R, Yuan H, Chen H (2013) Isolation and characterization of a novel phenanthrene (PHE) degrading strain *Psuedomonas* sp. USTB-RU from petroleum contaminated soil. J Hazard Mater 263:493–500

Mitter EK, Tosi M, Obregón D, Dunfield KE, Germida JJ (2021) Rethinking crop nutrition in times of modern microbiology: innovative biofertilizer technologies. Front Sustain Food Syst 5:606815

Morris M, Kelly VA, Kopicki RJ, Byerlee D (2007) Fertilizer use in African agriculture: lessons learned and good practice guidelines. The World Bank

Motamedi H, Aalivand S, Varzi HN, Mohammadi M (2016) Screening cabbage rhizosphere as a habitat for isolation of phosphate-solubilizing bacteria. Environ Exp Biol 14(4):173

Mukherjee A, Gaurav AK, Singh S, Yadav S, Bhowmick S, Abeysinghe S, Verma JP (2022) The bioactive potential of phytohormones: A review. Biotechnol Rep 35:e00748

Nacoon S, Jogloy S, Riddech N, Mongkolthanaruk W, Kuyper TW, Boonlue S (2020) Interaction between phosphate solubilizing bacteria and arbuscular mycorrhizal fungi on growth promotion and tuber inulin content of *Helianthus tuberosus* L. Sci Rep 10(1):4916

Nassal D, Spohn M, Eltlbany N, Jacquiod S, Smalla K, Marhan S, Kandeler E (2018) Effects of phosphorus-mobilizing bacteria on tomato growth and soil microbial activity. Plant Soil 427:17–37

Nitu R, Rajinder K, Sukhminderjit K (2020) Zinc solubilizing bacteria to augment soil fertility—A comprehensive review. Int J Agric Sci Vet Med 8:38–44

Nosheen S, Ajmal I, Song Y (2021) Microbes as biofertilizers, a potential approach for sustainable crop production. Sustainability 13(4):1868

Ogwu MC, Izah SC (eds) (2023) One health implications of agrochemicals and their sustainable alternatives. Springer Nature, Singapore

Ojeniyi SO (2002) Soil management, national resources and environment. Nigeria J Soil Sci 16:131–135

Otaiku AA, Mmom PC, Ano AO (2019) Biofertilizer impacts on cassava (Manihot esculenta Crantz) cultivation: improved soil health and quality, Igbariam. Nigeria Soil Sci 4:1–4

Pahalvi HN, Rafiya L, Rashid S, Nisar B, Kamili AN (2021) Chemical fertilizers and their impact on soil health. In: Microbiota and biofertilizers, Vol 2: ecofriendly tools for reclamation of degraded soil environs, Springer, Cham pp 1–20. https://doi.org/10.1007/978-3-030-61010-4_1

Pampolino MF, Manguiat IJ, Ramanathan S, Gines HC, Tan PS, Chi TT, Rajendran R, Buresh RJ (2007) Environmental impact and economic benefits of site-specific nutrient management (SSNM) in irrigated rice systems. Agric Syst 93(1–3):1–24

Patel SH, Viradiya MB, Prajapati BJ (2021) Effect of potassium and potassium mobilizing bacteria (KMB) with and without FYM on yield of wheat (Triticum aestivum L.). J Pharmacogn Phytochem 10(1):1615–1620

Ping L, Zhang C, Zhang C, Zhu Y, He H, Wu M, Tang T, Li Z, Zhao H (2014) Isolation and characterization of pyrene and benzo [a] pyrene-degrading Klebsiella pneumonia PL1 and its potential use in bioremediation. Appl Microbiol Biotechnol 98:3819–3828

Purwaningsih S, Sutisna E, Nugroho AA (2022) Characterization, diversity, and effectiveness phosphate solubilizing bacteria from the soil and rhizosphere on the growth of Glycine max L. in greenhouse. In: IOP conference series: earth and environmental science, vol 976, No. 1. IOP Publishing, Banjarbaru City, Indonesia pp 1–8

Qadir M, Hussain A, Shah M, Lee IJ, Iqbal A, Irshad M, Sayyed A, Ahmad A, Hamayun M (2022) Comparative assessment of chromate bioremediation potential of *Pantoea conspicua* and *Aspergillus niger*. J Hazard Mater 424:127314

Rai N, Rai SP, Sarma BK (2021) Prospects for abiotic stress tolerance in crops utilizing phyto-and bio-stimulants. Front Sustain Food Syst 5:754853

Ramírez-García R, Gohil N, Singh V (2019) Recent advances, challenges, and opportunities in bioremediation of hazardous materials. Phytomanag Pollut Sites:517–568. https://doi.org/10.1016/b978-0-12-813912-7.00021-1

Rojo F (2009) Degradation of alkanes by bacteria. Environ Microbiol 11(10):2477–2490. https://doi.org/10.1111/j.1462-2920.2009.01948.x

Roy S, Liu W, Nandety RS, Crook A, Mysore KS, Pislariu CI et al (2019) Celebrating 20 years of genetic discoveries in legume nodulation and symbiotic nitrogen fixation. Plant Cell 32(1):15–41

Safari M, Motamedi E, Dolatabad HK, Sanavy SA (2020) Nano-carriers effects on the viability and efficiency of *Pseudomonas* strains as phosphate solubilizing bacteria. Heliyon 6(10):e05076

Samanta S, Singh O, Jain R (2002) Polycyclic aromatic hydrocarbons: environmental pollution and bioremediation. Trends Biotechnol 20(6):243–248. https://doi.org/10.1016/s0167-7799(02)01943-1

Sandhu N, Sethi M, Kumar A (2021) Biochemical and genetic approaches improving nitrogen use efficiency in cereal crops: A review. Front Plant Sci 12:657629

Santosh S, Velmourougane K, Idapuganti RG, Manikandan A, Blaise D (2022) Potassium solubilizing potential of native bacterial isolates from cotton rhizosphere of rainfed vertisols. Natl Acad Sci Lett 45(3):209–212

Sarikhani MR, Oustan S, Ebrahimi M, Aliasgharzad N (2018) Isolation and identification of potassium-releasing bacteria in soil and assessment of their ability to release potassium for plants. Eur J Soil Sci 69(6):1078–1086

Sattar A, Naveed M, Ali M, Zahir ZA, Nadeem SM, Yaseen M, Meena VS, Farooq M, Singh R, Rahman M, Meena HN (2019) Perspectives of potassium solubilizing microbes in sustainable food production system: A review. Appl Soil Ecol 133:146–159

Savci S (2012) Investigation of effect of chemical fertilizers on environment. Apcbee Proc 1:287–292

Sharma S (2012) Bioremediation: features, strategies and applications. Asian J Pharm Life Sci 2231:4423

Sharma A, Chetani R (2017) A review on the effect of organic and chemical fertilizers on plants. Int J Res Appl Sci Eng Technol 5:677–680

Singh D, Thapa S, Geat N, Mehriya ML, Rajawat MV (2021) Biofertilizers: mechanisms and application. In: Biofertilizers. Woodhead Publishing, Sawston, Cambridge pp 151–166

Smirnova I, Sadanov A, Baimakhanova G, Faizulina E, Tatarkina L (2022) Metabolic interaction at the level of extracellular amino acids between plant growth-promoting rhizobacteria and plants of alfalfa (*Medicago sativa* L.). Rhizosphere 21:100477

Smułek W, Sydow M, Zabielska-Matejuk J, Kaczorek E (2020) Bacteria involved in biodegradation of creosote PAH–A case study of long-term contaminated industrial area. Ecotoxicol Environ Saf 187:109843

Solomon GOW, Ndana RW, Abdulrahim Y (2012) The comparative study of the effect of organic manure cow dung and inorganic fertilizer N.P.K on the growth rate of maize (Zea Mays L). Int Res J Agric Scie Soil Sci 2(12):516

Sonawdekar S (2012) Bioremediation: A boon to hydrocarbon degradation. Int J Environ Sci 2(4):2408–2424

Soumare A, Boubekri K, Lyamlouli K, Hafidi M, Ouhdouch Y, Kouisni L (2020) From isolation of phosphate solubilizing microbes to their formulation and use as biofertilizers: status and needs. Front Bioeng Biotechnol 7:425

Stroud JL, Paton GI, Semple KT (2007) Microbe-aliphatic hydrocarbon interactions in soil: implications for biodegradation and bioremediation. J Appl Microbiol 102(5):1239–1253. https://doi.org/10.1111/j.1365-2672.2007.03401.x

Sumbul A, Ansari RA, Rizvi R, Mahmood I (2020) Azotobacter: A potential bio-fertilizer for soil and plant health management. Saudi J Biol Sci 27(12):3634–3640

Sun J, Karuppiah V, Li Y, Pandian S, Kumaran S, Chen J (2022) Role of cytochrome P450 genes of *Trichoderma atroviride* T23 on the resistance and degradation of dichlorvos. Chemosphere 290:133173

Thirugnanasambandan T (2018 Dec 27) Advances and trends in nano-biofertilizers. Available at SSRN 3306998

Uchida R (2000) Essential nutrients for plant growth: nutrient functions and deficiency symptoms. Plant Nutr Manag Hawaii's Soils 4:31–55

Umesha S, Manukumar HM, Chandrasekhar B (2018) Sustainable agriculture and food security. In: Biotechnology for sustainable agriculture. Woodhead Publishing, Sawston, Cambridge pp 67–92

Upadhyay PK, Dey A, Singh VK, Dwivedi BS, Singh T, GA R, Babu S, Rathore SS, Singh RK, Shekhawat K, Rangot M (2023) Conjoint application of nano-urea with conventional fertilizers:

an energy efficient and environmentally robust approach for sustainable crop production. PLoS One 18(7):e0284009

Varga T, Hixson KK, Ahkami AH, Sher AW, Barnes ME, Chu RK et al (2020) Endophyte-promoted phosphorus solubilization in *Populus*. Front Plant Sci 11:567918

Walpola BC, Kim AY, Jeon JH, Yoon M-H (2014) Application of immobilization technology in solubilization of rock phosphate. Korean J Soil Sci Fertiliz 47(4):249–253

Wang Q, Awasthi MK, Zhang Z, Wong JW (2019a) Sustainable composting and its environmental implications. In: Sustainable resource recovery and zero waste approaches. Elsevier, St. Louis, Missouri pp 115–132

Wang R, Lin J-Q, Liu X-M, Pang X, Zhang C-J, Yang C-L et al (2019b) Sulfur oxidation in the acidophilic autotrophic *Acidithiobacillus* spp. Front Microbiol 9:3290

Wang Y, Kong Y, Li X, Li W, Du H (2020) GmPAP12 is required for nodule development and nitrogen fixation under phosphorus starvation in soybean. Front Plant Sci 11:498343

Wang G, Ren Y, Bai X, Su Y, Han J (2022) Contributions of beneficial microorganisms in soil remediation and quality improvement of medicinal plants. Plants 11(23):3200

Wani PA, Wahid S, Singh R, Kehinde AM (2018) Antioxidant and chromium reductase assisted chromium (VI) reduction and Cr (III) immobilization by the rhizospheric Bacillus helps in the remediation of Cr (VI) and growth promotion of soybean crop. Rhizosphere 6:23–30

Wentzel A, Ellingsen TE, Kotlar H-K, Zotchev SB, Throne-Holst M (2007) Bacterial metabolism of long-chain N-alkanes. Appl Microbiol Biotechnol 76(6):1209–1221. https://doi.org/10.1007/s00253-007-1119-1

Xie JB, Du Z, Bai L, Tian C, Zhang Y, Xie JY, Wang T, Liu X, Chen X, Cheng Q, Chen S (2014) Comparative genomic analysis of N2-fixing and non-N2-fixing *Paenibacillus* spp.: organization, evolution and expression of the nitrogen fixation genes. PLoS Genet 10(3):e1004231

Yadav A (2013) Seasonal population dynamics of rhizosphere and non-rhizosphere soil microorganisms of Chir pine seedlings (pinus roxburghii sarg.). Br Microbiol Res J 3(4):664–677

Yadav A, Yadav K (2024) Challenges and opportunities in biofertilizer commercialization. SVOA Microbiol 5(1):01–14

Yadav A, Yadav K, Vashistha A (2016) Phosphate solubilizing activity of *Pseudomonas fluorescens* PSM1 isolated from wheat rhizosphere. J Appl Nat Sci 8(1):93–96

Yadav U, Kushwaha S, Anand V, Kumar S, Prakash O, Singh PC (2020) Chlorpyrifos degradation by plant growth-promoting *Alcaligenes faecalis* bacteria isolated from oil-contaminated soil. Biorem J 25(1):80–88

Yasin G, Bhanger MI, Ansari TM, Naqvi SMSR, Ashraf M, Ahmad K, Talpur FN (2013). Quality and chemistry of crude oils. J. of Pet. Technol. Altern. Fuels 4(3): 53-63

Yasuda M, Dastogeer KM, Sarkodee-Addo E, Tokiwa C, Isawa T, Shinozaki S, Okazaki S (2022) Impact of *Azospirillum* sp. B510 on the rhizosphere microbiome of rice under field conditions. Agronomy 12(6):1367

Yemashova NA, Murygina VP, Zhukov DV, Zakharyantz AA, Gladchenko MA, Appanna V et al (2007) Biodeterioration of crude oil and oil derived products: A review. Rev Environ Sci Biotechnol 6(4):315–337. https://doi.org/10.1007/s11157-006-9118-8

Zhang W, Li J, Zhang Y, Wu X, Zhou Z, Huang Y, Zhao Y, Mishra S, Bhatt P, Chen S (2022) Characterization of a novel glyphosate-degrading bacterial species, *Chryseobacterium* sp. Y16C, and evaluation of its effects on microbial communities in glyphosate-contaminated soil. J Hazard Mater 432:128689

MIX
Papier aus verantwortungsvollen Quellen
Paper from responsible sources
FSC® C105338

If you have any concerns about our products,
you can contact us on
ProductSafety@springernature.com

In case Publisher is established outside the EU,
the EU authorized representative is:
Springer Nature Customer Service Center GmbH
Europaplatz 3, 69115 Heidelberg, Germany

Printed by Libri Plureos GmbH
in Hamburg, Germany